The Human Geography of East Central Europe

The Human Geography of East Central Europe examines the geography of the transition economies that were not formerly part of the Soviet Union: Albania, Bosnia & Hercegovina, Bulgaria, Croatia, The Czech Republic, East Germany, Hungary, Macedonia, Poland, Romania, Slovakia, Slovenia, and Yugoslavia. There is a thematic treatment beginning with the landscape and historical background, which moves on to the social and economic geography (industry, agriculture and infrastructure) and to issues concerning regional development and environmental protection.

The book is all about the transitions that have followed the collapse of the communist system. In political terms this means the development of civil society, based on pluralism, and the changed territorial basis with the collapse of the federations of Czechoslovakia and Yugoslavia. The social geography covers demography and ethnic relations against a background of stress arising from high unemployment and low incomes. Industrial restructuring depends heavily on foreign direct investment, while agricultural modernisation is constrained by the social importance attached to small family farms. The overhaul of the infrastructure is replacing links with the Former Soviet Union by Europe-wide transport and energy systems; reflecting the changing trade relationships and strong desire for EU membership among the candidate countries. Regional variations reflect the discrimination of investors and here the poverty problems of the southeast European countries is potentially destabilising, despite the generally modest environmental problems and the climatic advantages which boost the potentials for agriculture and tourism.

The book highlights the problems of transition which have been most evident through the ethnic tensions of the Balkans. However, the change in government in Yugoslavia points to a consensus in favour of a single Europe and is a step away from the 'third way' syndrome: an existence separate from both Soviet-style communism and a capitalism rooted in the EU. Ethnic tolerance is widely seen as the only way forward and a spate of measures to help the states of southeastern Europe gives some ground for optimism that the gap can be reduced. The stark question of democracy versus authoritarianism has to an extent been resolved, but politics will remain volatile as long as there are large numbers of poor people with no optimism over their future prospects.

David Turnock is a Reader in Geography at the University of Leicester.

For a set of additional references, a list of books published since 1989 (arranged by country) and a list of websites, related to the material covered in this book, please go to the following website: www.reference.routledge.com/research. Then click on the subject line for 'Environment & Society' series, then choose the link to the 'Routledge Studies in Human Geography series' and then click on the links to the extra material, which are beside the book's title.

Routledge Studies in Human Geography

This series provides a forum for innovative, vibrant, and critical debate within Human Geography. Titles reflect the wealth of research which is taking place in this diverse and ever-expanding field. Contributions are drawn from the main sub-disciplines and from innovative areas of work which have no particular sub-disciplinary allegiances.

1 A Geography of Islands
Small island insularity
Stephen A. Royle

2 Citizenships, Contingency and the Countryside
Rights, culture, land and the environment
Gavin Parker

3 The Differentiated Countryside
T. Marsden, J. Murdoch, P. Lowe, N. Ward and A. Taylor

4 The Human Geography of East Central Europe
David Turnock

5 Imagined Regional Communities
Integration and sovereignty in the global South
James D. Sidaway

6 Mapping Modernities
Geographies of Central and Eastern Europe 1920–2000
Alan Dingsdale

7 Rural Poverty
Restricted lives and social exclusion
Paul Milbourne

The Human Geography of East Central Europe

David Turnock

London and New York

First published 2003
by Routledge
11 New Fetter Lane, London EC4P 4EE

Simultaneously published in the USA and Canada
by Routledge
29 West 35th Street, New York, NY 10001

Routledge is an imprint of the Taylor & Francis Group

© 2003 David Turnock

Typeset in Baskerville by Florence Production Ltd,
Stoodleigh, Devon
Printed and bound in Great Britain by MPG Books Ltd, Bodmin

British Library Cataloguing in Publication Data
A catalogue record for this book is available from the British
Library

Library of Congress Cataloging in Publication Data
A catalog record for this book has been requested

ISBN 0–415–12191–4

For Marion

Contents

List of figures   viii
List of tables   x
List of abbreviations   xii
Acknowledgements   xvi
Foreword   xviii

1 Introduction: the political and economic context   1

2 Aspects of social geography   56

3 Production: industry and agriculture   111

4 Tertiary sector geographies: transport, energy and tourism   187

5 Urban and rural settlement   252

6 Regions of East Central Europe   307

7 Conclusion   365

  References   391

  Index   404

Figures

0.1	The states of East Central Europe: pre-1989 and present	xx
1.1	War and separation in Bosnia & Hercegovina	28
2.1	Romania: natural increase by counties 1975–93	80
2.2	Population distribution and inferred migration in the Romanian Carpathians	81
3.1	Slovakia's communications and business innovation centres	141
3.2	Poland's Upper Silesian Industrial Region	142
3.3	Transport axes in Cluj County, Romania	145
3.4	Romanian cereal yields by county 1970–94 (a) 1990–94 (b) 1970–84	159
3.5	Agriculture in Slovakia	160
3.6	Land use in the Polish Sudetes in the 1980s	161
3.7	Romania's counties and planning regions	179
3.8	Romania's Less Favoured Areas	182
4.1	Aspects of employment in the Romanian Carpathians 1966–92	190
4.2	Eurocorridors and some additional Balkan routes	196
4.3	Optical-fibre networks in Hungary 1998	220
4.4	Hydropower in the Romanian Carpathians	233
4.5	The concept of the tourist axis applied to Bulgaria's Central pre-Balkan Zone	250
5.1	Commuting in the Carpathians	253
5.2	Warsaw, showing land values across the city and the development of business in the Central area	268
5.3	Shopping malls in Budapest	276
5.4	The infrastructure of Wrocław, Poland	283
5.5	Cooperatives and key villages in the Nitra region of Slovakia	291
5.6	The village of Pătârlagele, Romania	298
5.7	Dispersed settlement in the Arieş Valley of Romania's Apuseni Mountains	305
6.1	Regional variations in employment at the end of the communist era	309
6.2	The regions according to per capita GDP 1996	312
6.3	Poland's administrative regions in the 1990s	316
6.4	Slovakia's regional reform	317

6.5   Hungary Southern Great Plain Region   326
6.6   Hungary's border regions   336
6.7   Euroregions and 'Ecological Bricks'   337
6.8   The Berlin region   345
6.9   Romania's West Region   352
6.10 Planning the road system of Timişoara   355
6.11 Danube–Criş–Mureş–Tisza Euroregion   363
7.1   The environment of the Sudetes in the 1980s   379
7.2   Danube Basin: pollution and wetland conservation   380
7.3   Settlement in the Pătârlagele area of Romania's Buzau
      subcarpathians showing land use and dispersed settlement on
      unstable terrain prone to landslides and mudflows   384
7.4   Hazardous waste storage in the Tisza Basin   386
7.5   Carpathian Ecoregion: protected areas   389

Tables

0.1	Area, population and transition indicators	xxi
1.1	States of East Central Europe	5
1.2	Membership of international organisations	7
1.3	Political stability 1989–2001	11
1.4	The East German economy: lessons for the transition	14
1.5	Major trading partners 1994–8	43
1.6	Annual percentage change in gross domestic product 1990–2002	44
1.7	Foreign direct investment 1991–3 to 2000–2	45
1.8	International trade 1989–2002	47
1.9	Consumer retail price inflation 1990–2002	48
1.10	Currency exchange against the dollar 1990–2002	49
2.1	Social indicators 1994–8	65
2.2	Unemployment rates 1992–2002	68
2.3	Population 1950–2050	75
2.4	Population growth and migration 1994–5 and 1997–8	76
2.5	Fertility, marriage, divorce and life expectancy 1993–8	77
2.6	Age structure 1993–8	79
2.7	Population by gender 1950–2050	82
2.8	Gender 1998: demographic, social and economic aspects	83
2.9	Ethnic minorities c.1990	93
2.10	Summary of general elections 1990–2001	97
2.11	External debt 1988–2002	107
3.1	Percentage change in industrial production per annum 1989–2002	118
3.2	Monthly production of basic commodities 1982–2000	120
3.3	Percentage change in agricultural production per annum 1990–2002	157
3.4	Livestock numbers 1985–2000	157
3.5	Land use change 1950–99	163
3.6	Cereal production and net trade 1975–99	166
3.7	Trade in agricultural commodities 1965–99	167
3.8	Livestock production indexes 1988–2000	168
3.9	Crop production indexes 1988–2000	168
3.10	Rural population and agricultural population 1990–2000	169

4.1 Major economic sectors: (a) By national income and gross value
added; (b) By employment 1989–99 188
4.2 Romania: the service sector by regions 1998–9 189
4.3 Transport systems and road vehicles 1994–8 195
4.4 Railway traffic, monthly average 1986–2000 198
4.5 Freight transport 1994–8 199
4.6 Long-distance passenger transport 1994–8 200
4.7 Telecommunications, computing and the Internet 1994–9 219
4.8 Energy balance for EU candidate countries 1980–98 225
4.9 Energy intensity 1980–98 230
4.10 Nuclear power 1980–95 236
4.11 Tourism arrivals and overnights 1990–8 246
5.1 Urban population 1950–2030 254
5.2 Urban population size groups 1996 262
5.3 Capital cities 1950–99 263
5.4 Rural population 1950–2030 288
5.5 Rural issues in the Czech Republic 1993 297
6.1 Regional variations on the eve of transition 1988 310
6.2 Regional variations in EU candidate countries 1996 313
6.3 Units of regional development and local government 315
6.4 Hungarian regional development policy 1985–96 323
6.5 Romania's West Region: a profile 351
6.6 Less Favoured Areas in Romania's West Region 357

Abbreviations

bn.	billion
DM	Deutsch Mark
E	Euro
ECU	European Currency Unit
Ft	Forint
g	gram
ha	hectare
km	kilometre
km^2	square kilometre
kwh	kilowatt hour
l	litre
m	metre
m^2	square metre
m^3	cubic metre
m.	million
mg	milligram
min	minute
MW	megawatt
na	not available
oe	oil equivalent
pc	per capita
ptp	per thousand of population
t	tonne
th	thousand
tpa	tonnes per annum
tpd	tonnes per day
$	United States Dollar
Zl	Złoty

ABB	Asea-Brown Boveri
AECL	Atomic Energy of Canada Ltd
AEP	Agri-Environmental Programme
BANU	Bulgarian Agrarian National Union
B&H	Bosnia & Hercegovina

BCE	Business Central Europe
CANDU	Canadian Deuterium Uranium Reactor
CAP	Common Agricultural Policy
CBC	Cross-Border Cooperation
CBD	Central Business District
CCI	Chamber of Commerce and Industry
CDU	Croatian Democratic Union (B&H and Croatia)
CEFTA	Central European Free Trade Association
CEME	Central European Media Enterprises
CER	Carpathian Ecoregion
CIS	Confederation of Independent States
CNG	Compressed natural gas
CNN	Cable News Network
CoE	Council of Europe
Comecon	Council for Mutual Economic Assistance
CPEs	Centrally-Planned Economies
CSCE	Conference for Security and Cooperation in Europe
C*SOP	Czech Union for Nature Conservation
DDBR	Danube Delta Biosphere Reserve
DOS	Democratic Opposition of Serbia
DP	Democratic Party (Albania)
DRPC	Danube River Protection Convention
EAP	Environmental Action Plan
EBRD	European Bank for Reconstruction and Development
ECE	East Central Europe
ECfE	Economic Commission for Europe
ECEC	East Central European Country
EEA	European Environmental Agency
EIA	Environmental Impact Analysis
EIB	European Investment Bank
EIU	Economist Intelligence Unit
EMBO	Employee–Management Buy-Out
ENGO	Environmental Non-Governmental Organisation
EU	European Union
FAO	Food and Agriculture Organisation
FB&H	Federation of Bosnia & Hercegovina (within the confederation of Bosnia & Hercegovina)
FCS	Former Czechoslovakia
FDI	Foreign Direct Investment
FFRG	Former Federal Republic of Germany (West Germany)
FFRY	Former Federative Republic of Yugoslavia
FGD	Flue Gas Desulphurisation
FGDR	Former German Democratic Republic (East Germany)
FSU	Former Soviet Union
GDP	Gross Domestic Product
GEC	General Electric Company

GHG	greenhouse gases
GMO	Genetically-Modified Organism
GNP	Gross National Product
GOP	Gornoslaskie Okreg Pzemyslowy (Upper Silesian Industrial Region)
GSM	Global System for Mobile Communications
HCP	Hungarian Civic Party
HDF	Hungarian Democratic Forum
HUF	Hungarian Forint
IBM	International Business Machines
ICTY	International Criminal Tribunal for the Former Yugoslavia
IMF	International Monetary Fund
IMRO	Internal Macedonian Revolutionary Organisation
ISDN	International System of Digital Networks
ISP	Independent Smallholders Party (Hungary)
ISPA	Instrument for Structural Policies for Pre-Accession Aid
IT	Information Technology
IUCN	World Conservation Union
JNA	Yugoslav National Army
KLA	Kosovo Liberation Army
KFOR	NATO-led Kosovo Peace Enforcement Force
LCY	League of Communists of Yugoslavia
LFA	Less Favoured Areas
LPH	Lithuania–Poland–Hungary
MAV	Magyar Államvasutak (Hungarian State Railways)
NATO	North Atlantic Treaty Organisation
NEM	New Economic Mechanism (Hungary)
NGO	Non-Governmental Organisation
NUTS	Nomenclature of Territorial Units for Statistics
OECD	Organisation for Economic Cooperation and Development
OHR	Office of the High Representative
OSCE	Organisation for Security and Cooperation in Europe
PDA	Party of Democratic Action (B&H)
PDP	Party of Democratic Prosperity (Macedonia)
PET	Polyethylene terephtalate
PHARE	Pologne-Hongrie: Actions pour la Reconversion Economique
PJ	Petajoule
PLA	Protected Landscape Area
PLAb	Party of Labour of Albania
PPP	Polluter Pays Principle
R&D	Research & Development
REC	Regional Environmental Centre for Central and Eastern Europe
RERP	Regional Environmental Reconstruction Programme
RS	Republika Srpska (i.e. Serb Republic within the confederation of Bosnia & Hercegovina)

SAPARD	Special Accession Programme for Agriculture and Rural Development
SDA	Social Democratic Alliance (Macedonia)
SEE	South East Europe
SEECs	South East European Countries
SFOR	Stabilisation Force (B&H)
SME	Small- and Medium-Sized Enterprise
SOE	State-Owned Enterprise
SOF	State Ownership Fund
SP	Socialist Party (Albania and Hungary)
SPSEE	Stability Pact for South Eastern Europe
TEM	Trans-Europe Motorway
TINA	Transport Infrastructure Needs Assessment
TIR	Transport International des Marchandises par la Route
TNC	Trans-National Corporation
TRACECA	Transport Central Europe–Central Asia
UDF	Union of Democratic Forces (Bulgaria)
UN	United Nations
USAID	United States Agency for International Development
VAT	Valued Added Tax
VVER	Vodo Vodyanoy Energeticheskiy Reaktor (type of Soviet reactor)
WB	World Bank
WWF	World-Wide Fund for Nature

Acknowledgements

This book would not have been possible without the surge in research carried on in transition countries over the past decade and hence my thanks to all the authors whose essays and articles inevitably constitute the bedrocks for works of synthesis. I am grateful to many friends in the UK, USA, Western Europe and above all in East Central Europe who have helped with encouragement, along with many ideas and useful pieces of information: none more so than my erstwhile partner in many previous literary ventures – Frank Carter – whose death in 2001 has cut off my ever-dependable 'hot line' of many years' standing. However, there are several other colleagues who have been particularly helpful in providing material and facilitating my field studies, often over a lengthy period: Ioan Abrudan and Florin Ioraş (Braşov), Stefan Buzarovsky (Skopje), Remus Cretan (Timişoara), Branislav Djurdjev (Novi Sad), Serban and Simina Dragomirescu (Bucharest), Vladimir Drgona (Nitra), Derek Hall (Ayr), Martin Hampl (Prague), Wilfried Heller (Potsdam), Nicolae Hillinger and Martin Olaru (Reşiţa), Peter Jordan (Vienna), Włodzimierz Kurek (Kraków), Max Linke (Weissenfels), Krzysztof Mazurski (Wrocław), Roy Mellor (Aberdeen), Cristina and Nicolae Muica (Bucharest), Erika and Gabor Nagy (Békéscsaba), Richard Osborne (Nottingham), Mirko Pak (Ljubljana), Gheorghe Ploaie (Râmnicu Vâlcea), Grigor Pop and Vasile Surd (Cluj-Napoca), Chad Staddon (Bristol), Istvan Suli-Zakar (Debrecen), Wolf Tietze (Helmstedt) and Herman van der Wusten (Amsterdam). I should really include virtually all present and recent members of the Romanian Academy's Geography Institute and I hope they will understand why I mention only the Dragomirescu and Muica families. My warmest thanks to Ruth Pollington who has drawn virtually all the maps included in this book (I am grateful to Erika and Gabor Nagy for Figures 4.3 and 6.5). And an inestimable debt of gratitude to my wife Marion who has generously shared my commitment to the region and has therefore experienced the full range of satisfactions and tribulations over a period of 35 years. But finally my thanks to Routledge for providing me with a principal outlet over the past 15 years and, in particular, for setting up this project in the first place and maintaining support during the many lean spells encountered during six years of preparation.

The author and publishers would like to thank the following for granting permission to reproduce material in this work (both in this book and on-line):

The Afold Institute and Gabor Nagy (Director), for Figure 4.3 'Optical-Fibre Networks in Hungary 1998'.

Arnold Publishers Ltd, for table 20.1 from p. 237 of *East central Europe and the former Soviet Union*, edited by David Turnock, 2001 (Table 0.1 in this book).

Kluwer Academic Publishers, for figures 2, 3, 4 and table 5 from pp. 75–90; for figure 3 from pp. 157–72; for figures 1 and 2 from pp. 235–47; for figures 1, 3 and 5 from pp. 255–71; for figures 2 and 3 from pp. 273–84; for figure 4 from pp. 305–9 and table 1 from pp. 109–25, all by David Turnock and from *Geojournal*, 50 (2–3), 2000. (In the book these are Figures 3.1, 3.3, 3.6, 3.7, 3.8, 5.5, 5.6, 6.1, 6.3, 7.1, and Tables 2.10 and 6.2.

Pion Limited, London, for the map 'Commuting the Romanian Carpathians', from 'The Changing Romanian Countryside', by David Turnock, in *Environment and Planning C*, volume 9 (3), p. 333, 1991 and the table on foreign direct investment on pp. 854–5, figure 1 on p. 856, figures 2–3 on pp. 858–9, figure 4 on p. 864 and figure 5 on p. 868 from 'Location Trends for Foreign Direct Investment in East Central Europe', by David Turnock, in *Environment and Planning C*, 19 (6), pp. 849–80, 2001. (In the book these are Figures 3.2, 5.1, 5.2, 5.3, and Table 1.7.)

Every effort has been made to contact copyright holders for their permission to reprint material in this book. The publishers would be grateful to hear from any copyright holder who is not here acknowledged and will undertake to rectify any errors or omissions in future editions of this book.

David Turnock, Leicester
November 2001

For a set of additional references, a list of books published since 1989 (arranged by country) and a list of websites, related to the material covered in this book, please go to the following website: www.reference.routledge.com/research. Then click on the subject line for 'Environment & Society' series, then choose the link to the 'Routledge Studies in Human Geography series' and then click on the links to the extra material, which are beside the book's title.

Foreword

East Central Europe (ECE) (Figure 0.1) is not substantially better known now than it was in 1939 when a confused British public was told its country was at war with Germany in support of the territorial integrity of Poland. However, the region remains very important to us, especially in view of recent events and those scheduled for the near future. As the former communist states have re-established their traditionally close links with Western Europe, the EU and its member states have supported the restructuring process and have also resolved, along with non-European NATO members, to intervene against unacceptable uses of state power in several areas of former Yugoslavia, especially Bosnia & Hercegovina and Kosovo where UN protectorates have now been set up. More generally, the process of EU enlargement eastwards is gathering momentum rapidly while the acute instability in the Balkans makes it all too evident that there must be an end to the marginalisation of this part of the continent. History demonstrates all too clearly that within ECE as a whole there is no basis for a coherent grouping of southeast European countries (SEECs) separate from a European Union (EU) rooted in the West. Again, the painful transitions which are taking place from centrally planned to market economies – and from totalitarian one-party states to pluralist systems – introduce a new element to development studies as the transition economies navigate uncharted waters. There are many states that can demonstrate the workings of a market economy but none that can chart a crash course in comprehensive system change.

Following the author's broad contextual study prepared during the early transition years (Turnock 1997), this book profiles the essentials of what has proved to be a highly dynamic human geography, very much in keeping with the spirit of 'transformation'. However, anyone approaching the task must cope with several basic questions. The first is how large an area the book should cover, given that the old definition of Eastern Europe as comprising eight communist states on the western border of the Soviet Union is no longer appropriate. Not only has this area been transformed by the reunification of Germany and the breakup of the Czechoslovak and Yugoslav federations but the collapse of the Soviet Union has produced a new tier of independent states in the east of Europe lying adjacent to Russia. And if the Caucasian and Central Asian states are considered, along with Russia, a total of 28 transition economies can be enumerated. This book happens to deal only with those territories that were not part of the Soviet Union, but this has to be regarded as an arbitrary decision, especially in view of the involvement of three former Soviet states (Estonia, Latvia and Lithuania) in the EU enlargement process. On the other hand, the coverage does include the transition states historically most closely associated with Central and Western Europe – and currently most important for its security – while also introducing wide diversity in terms of transition geographies. There are wide variations in fitness for early accession to the EU and NATO (with the Czech Republic, Hungary and Poland already members of the latter). However, at least it can be said that none of the states covered by this book should be excluded on the grounds of Russian sensitivity over Western penetration of her 'near abroad'. For progress in reconstituting civil society and building a market economy may not at the end of the day be sufficient to permit the extension of organisations in a way that Russia would perceive as provocative. It may well be that 'Europe' will not extend far beyond the 'Habsburg' domains and that even 'Byzantine' territory in the Balkans may be debatable. In this context the Eastern Europe of the Cold War still retains some viability.

Another question concerns the balance between global and local. Regional geography can no longer be approached through mosaics of natural regions because not only are the fortunes of states bound up with a global world, but regional development is very much in the hands of foreign investors. The book has therefore been written on a thematic basis, drawing appropriate examples without any pretence of an even distribution of empirical detail between the 13 states involved (including East Germany), however this might be calculated – by reference to area and/or population weight Slovenia is smallest by area (20,300 km² and 1.99 m. persons) while Poland is plainly the largest (312,700 km² and 38.76 m. persons). The danger here is that the local is obscured, especially in view of contemporary geography's concern for the dynamics of change rather than the comprehensive treatment of static patterns. It is immediately apparent from Table 0.1 that the ECECs have experienced widely different fortunes during the transition. So the book seeks a compromise by including national profiles in the first three chapters and providing a review of the local in the conclusion to help demonstrate that there is not so much a homogenous transformation process going on as a countless number of locally-based scenarios which undermine any attempt at generalisation. Of course, there were sharp differences under communism but geographers as well as other social scientists generally preferred to see them as nuances beneath the umbrella of central planning (Karasimeonov 1998). This fitted in with the geopolitical polarisation of the world into two power blocs under respective superpower leadership and the primary interest of a Western leadership in the broad characteristics of what was a fundamentally different system. Now local identity has a more central importance in a global world.

There is also the balance between concepts and empirical detail. It so happens that several conceptual works have already appeared, typically as essay collections, compared with a dearth of literature on the geography of the region as a whole. Few monographs deal with the range of countries covered by this

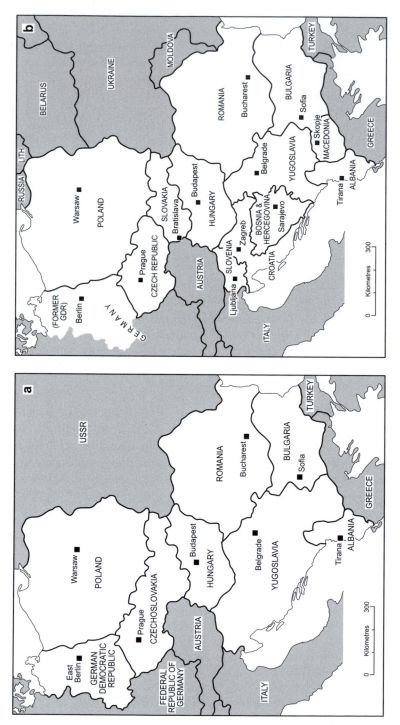

Figure 0.1 The states of East Central Europe: pre-1989 and present

Source: Administrative maps

Table 0.1 Area, population and transition indicators

	Area	Population		Transition indicators													
	A	B	C	D	E	F	G	H	I	J	K	L	M	N	O	P	Q
Albania	27.4	3113	114	2.6	86	1992	1992[1]	1993[1]	1992	60.4	75	1.7	132	na	na	na	cf
Bosnia & Herceg.	51.1	3972	78	2.0	na	na	na	na	na	na	35	1.6	na	na	na	na	cf
Bulgaria	111.0	8225	74	2.7	66	1991	1991[1]	1995[1]	1997	63.2	60	2.7	159	15	1.4	16.1	cn
Croatia	56.5	4473	79	3.0	78	1993	1993	1994	1993	59.5	60	2.5	444	na	1.4	15.2	cf
Czech Republic	78.9	10244	130	3.5	95	1991	1991	1992	1992	84.6	80	2.9	967	[1]	1.6	20.0	vm
Hungary	93.0	10036	108	3.7	95	1990	1990	[2]	1993	81.9	80	3.7	1627	4	2.0	41.8	m
Macedonia	25.7	2024	79	2.7	72	1994	1992	1995	1995	55.1	55	1.9	121	na	na	na	ci
Poland	312.7	38765	124	3.5	117	1990	1990	1993	1991	82.2	65	3.4	389	20	1.7	15.0	cv
Romania	238.4	22327	94	2.8	76	1993	1993[1]	1995[1]	1992	75.9	60	3.1	200	59	1.1	23.3	ci
Slovakia	40.0	5387	135	3.3	100	1991	1991	1992	1993	75.9	75	2.2	326	[1]	1.7	43.4	iv
Slovenia	20.3	1986	98	3.2	104	1992	1991	1993	1992	82.0	55	3.0	596	[1]	2.0	26.5	vm
Yugoslavia	102.2	10640	104	na	na	na	na	na	na	na	na	na	na	na	na	na	na

Source: EBRD 1999 and FAO database <http://www.apps.fao.org>

Notes

A Area ,000km²
B Estimated population 2000 (,000)
C Density 2000: persons/km²
D Transition Indicator (privatisation; markets and financial institutions +/– add/subtract 0.3; 4+ deemed 5, pp.24–5)
E Estimated real GDP 1998 where 1989 = 100
F Year of stabilisation programme launch
G Year inflation peaked ([1] indicates a second peak in 1997 for ECE and 1998 for CIS)
H Year inflation fell below 40 per cent ([1] indicates a second cycle starting in 1998 except for Romania 1999; [2] always below 40 per cent)
I Year of lowest output
J Lowest registered GDP in relation to 1989 per cent
K Private sector share of GDP

L Infrastructure 1999 (combination of indicators for telecommunications, electric power, railways, roads and water/waste water on a five point scale – with modifications as above)
M Cumulative FDI 1989–98 $/per capita
N Poverty 1993–5 based on household budget surveys
O Quality of governance (firms' perception of obstacles to business on a 0 (major obstacles) to 3 (no obstacles) averaging nine factors covering tax, inflation, physical infrastructure, law and order and corruption)
P State intervention in enterprise decisions (based on firms' reports on intervention in investment, employment, sales, mergers, dividends, wages and prices). Averages across the range reported by firms
Q Main export commodity groups: c clothing; f footwear; i iron and steel; m machinery; n non-ferrous metals; v vehicles
[1] less than 0.1

volume and essay collections rarely cover all 13 territories. But while the book concentrates on what is happening and the reasons why there should be both virtuous and vicious circles in the region today, the complexities of the empirical profile are balanced by some reference to the principles underpinning the transition. The significance of the transition as a major hiatus requires a balance to be struck between 'shock therapy' and gradualism, while recognising the crucial importance of central government regulation despite the instinct of decentralisation to erode the infrastructures of communism. Then there is the rebuilding of civil society in the context of 'winners' and 'losers' which forms the basis of the second chapter, while the third is concerned with restructuring in agriculture and industry in the context of sweeping privatisation and the fourth concentrates on the reorientation of transport, power and tourism from the east to the west. The fifth chapter raises the question of a new approach to the planning of settlements and the need for local authority spending to maintain services, including the public housing sector which has been much depressed during the past decade. The following chapter on the regions deals with the contested issue of spatial units for administration and strategies to moderate difference, with particular attention to remote rural areas and trans-frontier regions. The final chapter attempts to balance the rich diversity of the local against the broad flow of the transition, with reference to the special problems of the SEECs and the challenge of sustainability which means continuing relevance for environmental perspectives. Efforts have been made to indicate the extent of supporting materials in the form of books, journals and websites though much is available only on the website, while statistical coverage leans heavily on the EIU, Eurostat and the FAO.

The change taking place today is but an interlude in the shaping of a region which had stood for so long on the edge of Europe. A detailed historical context is resisted in favour of a launchpad comprising the communist era. This has the merit of staying within the span of recent time and providing contrasts with a system of coercive regulation unprecedented in its concern to internalise the problems of peripherality and suppress the real costs of distance in the interest of the former Soviet state. Over the past decade, the revolution in space relationships has been every bit as radical as the change in ideology, as the region seeks to rejoin Europe after the removal of Soviet communism, unfairly imposed on the region after the Second World War to satisfy Stalin's irresistible security requirements. Nevertheless, this is a region over which history hangs heavily and the pathways being negotiated under transition are in various ways dependent on the experiences of the communist period and those of earlier periods as well. While restitution has corrected many of the worst excesses of communism, surviving elite networks have been able to exercise influence and ensure an eminently 'path dependent' restructuring process. At the same time, the national sentiments grounded in a much longer historical timespan are at work in guiding the fortunes of states which enjoyed enlightenment and modernisation in the context of nineteenth-century imperialism as well as others, previously marginalised, which are now struggling to consolidate within their present boundaries. Extending European socio-economic and security structures over this region is a radically new project which doubtless deserves to succeed in the estimation of enlightened opinion. But it requires a commitment which few in the West have thought through, in terms of the financial and business obligations, while the peoples of the region will have to swallow prejudices which many have been brought up to associate with national identity. It can no longer be the 'natural' order of things to expect Transylvania to be administered from Budapest and equally for the Hungarians of Transylvania to have their cultural development constrained by Romanian sensitivities.

1 Introduction

The political and economic context

Historical overview

The foreword to this book has suggested how the eastward expansion of the EU and NATO is a logical sequel to the aberration of communism. Yet the accession states have historically comprised a backward area – part of Western Europe's periphery – which the communist system was, in some respects, intended to reinvigorate through the creation of a world region where the spatial logic of capitalism would be overridden by central planning and artificial pricing in the interest of regional equality. It will be shown how this socialist dream was tarnished by Soviet self-interest and a failure to adjust as the population born after the Second World War inevitably came to form the majority. But if the West is to succeed with its recipe for redistribution moderated by the EU it is worth remembering that the politics of the shatter zone – the region's historic status – are still within living memory. While medieval feudal states emerged on the basis of prehistoric tribal organisation and Dark Age migrations, these did not fairly reflect ethnicity (given the enormous variations in identity and organisation) and they were in any case swamped by imperialism of the Ottoman, Russian, Habsburg and Prussian empires. Independent East Central Europe (ECE) disappeared, apart from a few limited instances where imperial power was exerted indirectly through suzerainty. Instead there was colonialism with much instability and ethnic diversity where imperial frontiers were in a state of flux. As East (1961: 14) explains, the individuality of the shatter zone derives from the existence of many national groups as well as 'the persistence there of politico-territorial organisations essentially multinational and imperialist in character'. However, it is worth adding that German influence was particularly strong in terms of settlement, culture and trade: traditional links that underpinned the notion of Central Europe or 'Mitteleuropa' as an informal German commonwealth.

Modern nation states emerged in the region in the nineteenth century, first in the Balkans through the decay of the Ottoman Empire at a time of particular rivalry among the great powers. Independent Bulgaria, Montenegro, Romania and Serbia were joined by Albania in 1912. The process was extended to the Habsburg Empire – and to parts of the German and Russian empires – after the First World War. However, there was no unity of purpose among the new states because of territorial disputes, often grounded in the fate of ethnic minorities stranded by the demise of imperialism and economic competition among a tier of primary producers. With the shatter zone divided against itself, there was no effective buffer between Germany and USSR, though the region could command some support through its function as a 'cordon sanitaire' against Bolshevism – albeit for a short period dominated by insecurity and depression. The situation was much discussed by geographers at the time including L.W. Lyde (1926: 394) who wrote of a 'belt of continual political instability' comprising the isthmus between the Danube and Volga and between the Black and Baltic Seas ('Phoenicia and Amberland'), with cultural diversity and states too 'incoherent and impotent to win self-government sooner'. H.G. Wanklyn then examined Europe's 'eastern marchlands' in 1941, with somewhat greater sensitivity. But the Second World War saw the shatter zone partitioned between Germany and Russia under the Ribbentrop–Molotov Pact only for Germany to invade the USSR in 1941 and then for the Red Army to advance towards Berlin in 1944–5, buoyed by Western aid and an allied agreement to maintain a united front until Germany's unconditional surrender was assured.

Communism and the transition

Allied victory effectively gave the Soviet Union a free hand over the region, a situation which Stalin naturally exploited to the full. The Western Allies' dependence on Soviet manpower and the informal horse-trading of the Churchill–Stalin 'percentages agreement' are important factors in understanding why Western guarantees (e.g. for Poland) could not be sustained when the war ended. We do not know if Stalin worked to a secret agenda during the war or if his actions post-1945 were opportunistic. It was understandable that he would wish to extend the Soviet sphere beyond 'socialism in one country' to capture resources for a dynamic world region carved out of capitalism's periphery and dedicated to the defence and security of the Russian state. Stalin was even looking to extend his influence in Greece, Italy and West Germany until the NATO defensive shield established the 'Iron Curtain' as the world's most heavily fortified frontier and one which seemed to deny any logic to a 'Middle Tier' between Russia and the West. Austria was reunified in 1955 and the Western position in Berlin was eventually guaranteed, but the Soviets were able to establish their occupation zone in Germany as a separate state and to impose their system of communist party monopoly and central planning in virtually all territories occupied by the Red Army. Sophisticated NATO weaponry was balanced by Warsaw Pact superiority in tanks and infantry while in the 1980s US Pershing II and cruise missiles deployed in Western Europe brought a response through the deployment of nuclear weapons in Czechoslovakia in 1983.

Scarcely a decade has passed since the momentous events of 1989 and it is worth reflecting on the circumstances. We now know that the Soviet Union was not sustainable but Mikhail Gorbachev showed extraordinary courage in working for change and establishing a regime that prepared for the open dissent and political mobilisation that became evident in the late 1980s. He 'had a knack for sizing up problems quickly and was ready to consult to attempt novel solutions' (Keep 1997: 278) and when it was evident that the Soviets could not match Western technological skills (especially electronics), Gorbachev advocated 'glasnost' (frankness) and 'perestroika' (restructuring) to overcome Western sanctions linked with the Helsinki human rights agenda and mobilise the Soviet people for improved economic performance, though he believed 'over-confidently that the party had much stronger public support than was the case' (ibid.: 276). But there had to be a group of people to spearhead reform: Stalin had destroyed civil society and hence Khrushchev's reforms failed to progress. Moves towards political pluralism were previously blocked by the 1968 Warsaw Pact invasion of Czechoslovakia which led to the 'Brezhnev Doctrine' of limited sovereignty which validated collective action against threats to the status quo. However, in the 1980s the Soviet leader had the advantage of a new cohort in the nomenklatura 'better educated, more specialized and identified with particular interests rather than the overarching project of communist construction' (ibid.: 277–8). He was able to change the rules of the game by dismantling the Brezhnev Doctrine in 1985, giving the green light (probably late in 1988) for the accommodation with the Polish 'Solidarity' Movement negotiated the following year and finally in refusing to support a crack-down in East Germany shortly before the Berlin Wall collapsed.

Radical change began in Poland in 1987, in connection with the need for consensus over economic reform which brought widespread strikes in 1988. This led to the toleration of opposition parties after 'roundtable' discussions; an important criterion for progress towards pluralism. This stage was reached in Poland in 1988, followed by Czechoslovakia, East Germany, Hungary and parts of Yugoslavia in 1989; and Albania, Bulgaria and Romania in 1990. Free elections mark the second phase: in 1989 elections in Poland allowed the opposition parties to contest most parlimentary seats while the communists remained in power (in deference to Soviet sensitivities) but in most other cases governments were broadened in advance of comprehensive parlimentary elections. It was hoped that communist regimes could become truly legitimate through reform but they were greeted by rising expectations which precipitated the 1989 revolutions and took the ECECs out of the Soviet camp, ultimately destroying Gorbachev's own credibility and the Soviet Union that he sought to maintain. Comecon failed to survive the trade shocks associated with political change and was dissolved in 1991 when the remaining members agreed to trade in convertible currencies, while the Warsaw Pact was wound up in the same year and Soviet troops left the region (including East Germany) by the end of 1994. Social divisions were much reduced in comparison with the nineteenth century and idealists who were thinking of better ways of building socialism were effectively marginalised by a silent majority that saw a better future with the EU which practically all states now seem keen to join. Social change has arisen through the creation of a more urban, more educated and more 'aware' population prepared to question the legitimacy of Marxist-Leninism as a range of economic, cultural and environmental issues came to the fore. In East Germany, those who wanted to improve the socialist experiment in the context of an ecological–pacifist utopia were dismayed to find that the majority wanted to enjoy the benefits of West Germany's welfare capitalism. And similar preferences came to the fore elsewhere except in parts of the Balkans where the nationalist agenda took precedence for a time.

Changes in the states system

There were significant changes in the political geography of states. While unification in Germany was initially regarded as a long-term process, an immediate solution soon appeared the only option in a situation where the East German population was deserting in droves. The process was also expedited by the postwar occupation powers (including the Soviet Union, on the eve of disintegration) and by the readiness of the German nation as a whole to accept the Oder–Neisse boundary as the centrepiece of the two German–Polish treaties 1990–1. There was also reconciliation with the Czech Republic in 1997 over the Nazi occupation and the subsequent expulsion of the German population from the so-called Sudetenland. But other potential unions have proved more elusive and the momentum that developed in Chişinău to take Moldova into Romania was soon blocked by the implacable opposition of the minorities that still maintain Transnistria as a separate entity through a Cyprus-style partition of a former Soviet Socialist Republic. The possibility of a link between Poland and Lithuania has also been a matter for speculation in view of the commonwealth which existed in medieval times. However, the possibility has not been seriously explored and seems highly unlikely in view of the tension between the two states after the First World War when the Vilnius area was disputed. However, the countries could move closer together in the context of some wider association perhaps involving the Baltic states as a whole.

Meanwhile, unification in Germany has been balanced by the breakup of the two federations – Czechoslovakia and Yugoslavia – which means that there are 12 ECECs apart from East Germany, which is not covered by most of the statistical tables in this book since only somewhat unrepresentative data for the whole country can be quoted in most instances (Table 1.1). Czechoslovakia split in 1993 following the transitional arrangements arising out of the 'Velvet Revolution' of 1989 (the federal system dating back ultimately to 1969). There was little controversy over this separation of the Czech Republic from Slovakia after the 1992 elections endorsed the platform of the nationalists in Slovakia (Pavlinek 1995). However, the division has become more complete than was envisaged with the abandonment of a common currency. In Yugoslavia too relations between a total of five successor states have been strained, although arguably this is more understandable because the changes took place unilaterally and to the dismay of Serbia and Serb minorities, especially in Bosnia & Hercegovina and Croatia. This provided the basis for civil war not simply to halt the breakup of the federation but to contest the process of disaggregation by rejecting the boundaries of the republics as sacrosanct and envisaging a Serb nation state enlarged by ethnically-cleansed territories in eastern and western Bosnia, Krajina and parts of Slavonia. Further fragmentation now seems

Table 1.1 States of East Central Europe

Country and official name	Head of State/Government and next election			
Albania: Republika e Shqipërisë	R. Mejdani	2002[2]	I. Meta	2005
Bosnia & Hercegovina:				
Republika Bosna i Hercegovina	B. Belkić[1]	2002	A. Domazet[3]	2002
Federation of Bosnia & Herceg.	K. Filipović	2002	A. Behmen	2002
Serb Republic/Republika Srpska	M. Sarović	2002	M. Ivanic	2002
Bulgaria: Republika Bulgaria	P. Stoyanov	2001	S. Saxe-Coburg	2005
Croatia: Republika Hrvatska	S. Mešić	2005	I. Račan	2004
Czech Republic: Česká Republika	V. Havel	2003[2]	V. Spidla	2006
Hungary: Magyar Köztársaság	F. Mádl	2005[2]	P. Medgyessy	2006
Macedonia: Republika Makedonija	B. Trajkovski	2004	L. Georgievski	2002
Poland: Rzeczpospolita Polska	A. Kwaśniewski	2005	L. Miller	2005
Romania: Republica România	I. Iliescu	2004	A. Nastase	2004
Slovakia: Slovenská Republika	R. Schuster	2003[2]	M. Dzurinda	2002
Slovenia: Republika Slovenija	M. Kučan	2002	J. Drnovšek	2004
Yugoslavia:				
Savezne Republika Jugoslavija	V. Koštunica	2004	D. Pesić	2004
Montenegro	M. Djukanović	2002	F. Vujanović	2005
Serbia	M. Milutinović	2002	Z. Djindjić	2004

Source: Administrative handbooks

Notes
[1] There is three-man rotating Presidency for Bosnia & Hercegovina which also includes K. Križanović (Bosniak) and Z. Radišić (Serb).
[2] Indicates a parliamentary election for the presidency.
[3] The Prime-Ministerial role is played by a Chairman and the office rotates among the members of the Council of Ministers. The other ministers are A. Hadziahmetović, Z. Lagumidzija, S. Mihajlović, D. Mikerević and K. Zubek.

inevitable in view of the UN protectorate over Kosovo which may well end in independence. However, the Serb concentration round Mitrovica could suggest a Cyprus solution while Kosovo's separation in any form would make it diffi-cult to resist a similar scenario for Bosnia & Hercegovina's 'Republika Srpska'.

At the same time, a prolonged crisis in Kosovo could also drive Montenegro further towards secession and also rouse the Muslims in Sandzak, where the local Party of Democratic Action under R. Lajiić provides a focus for discon-tent. Yugoslavia remained unstable in the aftermath of the Kosovo War with a split in public opinion between censure of the former nationalist president S. Milošević and a Western conspiracy theory garnished by criticism of the townspeople who couldn't cope with the NATO pressure. Milošević's removal from the scene provides a breathing space for moderates who will need Western support to demonstrate that joining Europe offers the best prospect. The scope for consolidation seems remote although after further fragmentation in Yugoslavia, a limited federation to act as a counterweight to Serbia might conceivably interest Bosnia & Hercegovina, Macedonia and Montenegro. And pressure on the Macedonian government by Albanians from Kosovo could be the start of a Greater Albania movement. Such a reordering of political forces is an encouragement to those who would leave the Balkan peoples to get on with their struggles and accept the revised territorial arrangements.

Although the events in Czechoslovakia and Yugoslavia indicate extreme forms of fragmentation they are not untypical of experiences around the world which show, paradoxically, how closer global interrelations seem to generate stronger local identities as each 'member' of the world community tries to create its own niche in the system of global competition. In the region of ECE, which has suddenly found itself free to operate unhindered on the world stage, the local now provides substance for the notion that there are countless transformations occurring under the umbrella of a broad transition from one-party states to plural societies and from state capitalism, fundamentally rooted in the strategic requirements of the FSU, to market capitalism. The former centrally-planned economies are seeking a place in the global economy geared to efficient and profitable production to meet consumer demand. Every area will reflect, in its own way, an ongoing restructuring which will reflect the starting position and the strategies being adopted on the route to a largely unknown destination. System substitution has unleashed a complex range of adjustments, occurring at local, regional and national levels, as the economic and social issues are balanced through the political process. And all of them are in some way embedded in the experiences of the past. Local culture is a significant force and will surely remain so.

The volatility of the local in the early phase of the transition points to the crucial importance of effective national government. While early post-communist regimes sought some attenuation of the state's power to prevent any return to communist totalitarian power (a view which happened to coincide with neo-liberal thinking in Western Europe), effective regulation at the national level is now being emphasised as a corrective to the slide towards unbridled localism and illegality which reached extreme positions in the Yugoslav civil war and the chaos in Albania in 1997. Yet, stable national administration cannot be taken for granted: it depends on resources and institutions that were, in varying degrees, lacking at the onset of the transition and still remain in short supply. At one stage it seemed that transformation would see political life polarised on Latin American lines with economic policy veering to interventionism of the kind familiar to the Asian 'tiger' economies. ECE is a region of great importance for the socio-economic progress and political cohesion of the continent. It is very much in the interest of the West that there should be stability, especially in the present situation, in which independence for the region is reinforced by the Soviet Union's collapse which leaves a Russian state that is incapable of controlling these shatterbelt territories.

International institutions

While international financial institutions like the IMF have great importance for the region, the ECECs are being drawn into a wide range of international organisations (Table 1.2). Global perspectives point first and foremost to membership of the EU and the NATO as a means of enhancing security and as the most promising way to securing some amelioration of peripherality and as a launching pad for wider economic activities. Except for Albania and the Yugoslav successor states (other than Slovenia) all the countries of the region have entered into Europe agreements. They are now negotiating for full EU membership

Table 1.2 Membership of international organisations

	A	B	C	D	E	F	G	H	I	J	K	L	M	N	O	P	Q	R	S	T	U	V	W	X	Y
Albania	*	*		*		*		*	*									*	*	*	*	*	*		*
Bosnia & Herceg.		*				+		*	*					+				*		*	*	*	*		*
Bulgaria	*	*	*	*		*	*	*	*			*		+		*		*		*	*	*	*	a	*
Croatia				*		*	*	*	*	*		*		#		*		*		*	*	*	*	a	*
Czech Republic			*	*		*		*	*	*	*				*	*	*	*				*	*	a	*
Germany					*	*	*	*	*	*	*		*		*	*	*	*				*	*	*	*
Hungary		*	*	*		*	*	*	*	*	*		*	+	*	*	*	*		*	*	*	*	a	*
Macedonia			*	*		*		*	*			*	*			*		*		*	*	*	*		*
Poland	#	*	*	*	*	*	*	*	*	*	*		*	+	*	*	*	*		*	*	*	*	a	*
Romania	*	*	*	*		*	*	*	*	*		*		+		*		*		*	*	*	*	a	*
Slovakia	#	*	*	*		*	*	*	*	*		*				*	=	*		*	*	*	*	a	*
Slovenia			*	*		*	*	*	*	*	*			+		*		*			*	*	*	a	*
Yugoslavia						*	*							*				*				*	*		*

Source: Administrative handbooks

Notes

* Member; a Associate member; = Partner in transition; # Observer; + Guest.

A Agence de la Francophone
B Black Sea Economic Cooperation Group (under the Declaration on Black Sea Economic and Environmental Cooperation 1992)
C Central European Free Trade Association (originally the Visegrád Group)
D Central European Initiative (originally the Pentagonale 1990)
E Council of Baltic Sea States
F Council of Europe
G Danube Commission (1948)
H European Bank for Reconstruction and Development
I European Free Trade Association (Agreement for Trade and Cooperation)
J European Organisation for Nuclear Research
K European Union (Member State)
L European Union (Candidate Country)
M Missile Technology Control Regime
N Non-Aligned Movement (1961)
O North Atlantic Treaty Organisation
P North Atlantic Treaty Organisation Partnership for Peace
Q Organisation for Economic Cooperation and Development
R Organisation for Security and Cooperation in Europe (originally Conference on Security and Cooperation in Europe 1975, renamed 1995)
S Organisation of the Islamic Conference
T Security Pact for South Eastern Europe
U South East Europe Cooperation Initiative
V United Nations
W United Nations Economic Commission for Europe
X West European Union (defence component of the EU 1991)
Y World Trade Organisation

and the 'Luxembourg Group' of 'fast track' countries (including the Czech Republic, Hungary, Poland and Slovenia) may join soon after negotiations finish around the end of 2002. Meanwhile, the 'Helsinki Group', which includes Bulgaria, Romania and Slovakia, could experience some delay although Slovakia might catch up with the front-runners. It is possible, however, that the potential problems of institutional digestion (bearing in mind that the Baltic States, Cyprus, Malta and Turkey are also in the frame) might suggest a series of small waves drawing from both groups. But in any event this enlargement process – involving 80,000 pages of laws and regulations regarding the 30 'chapters' of the acquis communautaire – will reinforce the restoration of traditional links by ECECs with Austria, Germany and Italy and further legitimise the concept of Central Europe as a region of cultural and ethnic diversity in which Germans (and to a lesser extent Italians) have played a key role in the spread of technology and the growth of industry and trade. F.W. Carter (1996) suggests that despite ethnic and environmental tensions there is an underlying tolerance and a willingness to cooperate which can be seen in the opening up of border regions. He thinks that the EU can still learn something from the Austro-Hungarian empire's experiment with flexible federalism and its liberalism in respecting the rights of small language groups. However, tough EU border controls to regulate illegal entry, drugs and car theft will tend to separate the members from their eastern neighbours, even where there are Euroregions to facilitate cross-border links. This will be unfortunate if it interferes with the development of close economic and cultural relations between ECECs and Russia which could avoid mutual hostility and contribute to the region's identity.

NATO is crucially important for securing democracy and for integration in a Europe to which ECECs were denied access in 1945. The Czech Republic, Hungary and Poland became members in 1999, allowing 'normal relationships' between Germany and her eastern neighbours and easing residual fears that transformation could be endangered by internal unrest or an irredentist Russia. There is a desire for wider membership although public opinion has expressed some concern over nuclear weapons. Government elites are inclined to ignore these concerns in order to expedite entry into NATO and provide a residual deterrence against the danger of becoming the target of a Russian nuclear attack, notwithstanding issues of civil defence preparedness and security for nuclear materials. The present situation is unsatisfactory and, given the sensitive issue of Hungarian minorities in neighbouring countries, it is important that NATO does not appear to support any Hungarian move towards greater assertiveness in the Danube Basin. At the same time, it is suggested that Romania should be a pivotal member like Poland. For Romania, with a very well-developed military infrastructure and a transport system offering fast deployment of troops in the region, is a very active partner of NATO programmes with financial contributions pledged to all activities from the beginning. It might be argued that NATO should expand while Russia is relatively weak, for it is difficult to imagine that Russia herself would be interested, given her global situation and great power pretensions (quite apart from anti-Russian sentiments in NATO). However, scant respect for Russian strategic interests could exacerbate tensions and encourage reintegration in the form of possible unions with Belarus and Kazakhstan and it has therefore been suggested that despite American assurances that 'no European democracy will be excluded because of its position on the map', the alliance will not extend into the tier of states from Latvia to Bulgaria. There may however be a sincere belief that after the inter-war experience Europe's new democracies should not be consigned to a buffer zone and left to fend for themselves.

ECECs are keen to join the Council of Europe (CoE), described as a 'gate-keeper' responsible for checking the credentials of states seeking participation in the integration process and reference should also be made to the creation of a 'Visegrad Group' (Czechoslovakia, Hungary and Poland) in 1992 as a free trade organisation anticipating early EU accession. The group was extended to four countries in 1993 with the breakup of Czechoslovakia and it has successfully encouraged the other states in the region to project their reformist credentials. Renamed 'Central European Free Trade Association' (CEFTA), the club was joined by Slovenia in 1996, Romania in 1997 and Bulgaria in 1999. The Organisation for Security and Cooperation in Europe or OSCE (formerly CSCE) is concerned with economic cooperation as well as security, science and technology, environment and human rights. It monitors the 1990 Treaty on Conventional Forces in Europe and monitors minority issues through a High Commissioner on National Minorities first appointed in 1992. Meanwhile, all ECECs except Yugoslavia are members of the European Bank for Reconstruction and Development (EBRD) established in 1991 while appropriate groupings of ECECs combined with other European and Asian states to form the Black Sea Economic Cooperation Group and the Council of Baltic Sea States both of which date to 1992 and have a range of cultural, economic and environmental interests.

STABILITY IN SOUTH EASTERN EUROPE

However, a big question concerns coordination among the SEECs, especially in the light of wars in former Yugoslavia which have left several states out of the EU accession process. Criminalisation of government and dependence on fragile coalitions have increased the perception of instability, compared with the states in the north of the region (Table 1.3). A limited grouping has always been unsatisfactory because of a lack of economic complementarity and an overriding desire by the states involved to be part of a wider European grouping. An American concept for Europe South East builds on the Black Sea Economic Cooperation Group and involves a larger group (including non-ECE states like Greece, Moldova and Turkey) and includes political as well as economic objectives combining rights and representation for minorities with integrity for states. More tangible however is a Stability Pact for South Eastern Europe (SPSEE) set up after a meeting in Cologne in 1999, followed by a summit at Sarajevo later in the year. With membership including all the SEECs, including the two EU candidate countries Bulgaria and Romania, the aim is to seek cooperation and development, working through roundtables. An 'Association and Stabilisation Process' will be expedited by an official international body – a Balkan

Reconstruction Agency – established to address the damage arising from both the Kosovo War and previous neglect. It requires recognition of each member's territorial sovereignty and the return of refugees. A Regional Environmental Reconstruction Programme (RERP) was launched under Stability Fund auspices in January 2000 with E28 m. of funding for environmental rehabilitation following wars and previous neglect. The resources will be used for priority local and national projects to strengthen EIA procedures, improve environmental monitoring and project preparation and enhance regional cooperation of the kind currently evident at Lake Ohrid and Lake Prespa. The reopening of the Danube shipping lane should be achieved during 2002 now that the Pact's Economic Development Working Group has approved the RERP. More generally there will be emphasis on strengthening institutions and policy development and on raising awareness among civil society. With the goal of regional infrastructure overhaul by 2005 and European integration by 2010, the Stability Pact could contribute to a single effective long-term Western policy if it can overcome organisational weaknesses and gain credibility in the region. The 2000 Zagreb summit pointed the way to EU candidate status, with Croatia supported by Italy and Macedonia by Greece; making full membership achievable by 2006 and 2010 respectively. The EU insists first on good relations between pact members with cooperation in the context of confirmed borders and there is some progress in this domain. Yugoslavia initially declined observer status and the stability of the region seemed compromised as long as this state remained aloof from the international community. However, the political changes in Belgrade during 2000–1 have overcome this obstacle.

The SEECs have remained on the fringe of Europe since the Ottomans withdrew. There have been occasional dramatic interventions but no effective integration. Economic and democratic deficits have encouraged irrational nationalism; yet except for the Roma, there is no clear pattern of oppression in the midst of traditional feuding and 'the indisputable reality of the Balkans is that none of its peoples have been altogether innocent victims or vicious neighbours' (Binder 2000: 3). Interventions require allies and implicit support for certain factions against others which generate a new stock of grievances as Serbs demonise the NATO aggressor states. There are huge costs arising from protectorates in Bosnia & Hercegovina and Kosovo – and proto-protectorates in Albania, Macedonia and Montenegro; not to mention the wider implications through the Danube closure. The Lisbon Summit in 2000 declared that peace, prosperity and stability for the SEECs are a now strategic priority of the EU; so special stabilisation and association agreements were drawn up with five countries with prospects for EU membership. Meanwhile, Bulgaria and Romania stand to receive $5.9 bn. in combined EU assistance over six years, in recognition of heavy losses sustained through sanctions against Yugoslavia in the early 1990s as well as the Kosovo War. The American financier and philanthropist George Soros would like to see a pan-Balkan free trade zone to increase intra-regional trade and attract FDI – an idea attractive to the smaller countries like Albania and Macedonia – but it would require budget subsidies to compensate for lost customs revenue.

Table 1.3 Political stability 1989–2001: short (a) medium (b) and long term (c) [†]

Year	Albania			Bulgaria			Croatia			Czech Rep.*			Hungary			Poland			Romania			Slovakia			Slovenia			Yugoslavia		
	a	b	c	a	b	c	a	b	c	a	b	c	a	b	c	a	b	c	a	b	c	a	b	c	a	b	c	a	b	c
1989[2]	2	3	6	3	4 +	5	na			4	5	6	4	6	7	1	3	5	2	3	5	na			na			2	5	7
1990[2]	1	4	6	0 +	3	5	na			3	6	8	2	5	7	1	2	4	2	5 +	6	na			na			1	4	7
1991[2]	0 +	3 +	6	2 +	4	5	na			2	5	8	3 +	5	7	0 +	2	4 +	3	4 +	6	na			na			0 +	2	5
1992[2]	0 +	3 +	6	2	4	5	na			3 +	6	8	3	4 +	7	1 +	3	4 +	2 +	4 +	6	1	2 +	6	na			0 +	2	4
1993[2]	1	3 +	6	1 +	3 +	5	na			5	6 +	8	3	4 +	7	3	5	6	2 +	4 +	6	1 +	2 +	5	na			na		
1994[2]	1 +	2 +	6	2	4	5	na			5 +	7	8	4	5 +	7	4 +	5 +	6 +	2 +	4 +	6	1 +	2 +	5 +	4 +	6	8	na		
1995[2]	0 +	2 +	6	3	4 +	5 +	na			6 +	7	8 +	3 +	5	7	4 +	6	7	3	4 +	6	1 +	3	5 +	6	6 +	8	na		
1996[2]	0 +	1 +	5	1	2 +	5 +	na			4 +	5 +	8 +	5 +	5 +	7	5 +	6 +	7 +	4	5	7	1	3 +	5 +	5 +	6 +	8	na		
1997[2]	1 +	3	4 +	2 +	3 +	5 +	na			4	5 +	8	5 +	6 +	7 +	5	5 +	7 +	4 +	5 +	7	2 +	4	6	5 +	6	8	na		
1998[2]	0 +	1	4 +	3	4	5 +	na			4	5	8	5	5 +	7 +	5	5 +	8	2 +	3 +	6	3 +	5	7 +	5 +	6 +	8	na		
1999[2]	0 +	1	4 +	3	4	5 +	na			4	5	8	5 +	6 +	8	5 +	5 +	8	1 +	2 +	5 +	3 +	5	7 +	5 +	6 +	8	na		
2000[2]	0 +	1	4 +	2 +	3 +	5 +	na			3 +	5	8	5 +	6 +	8	5 +	5 +	8	1	2 +	5	3 +	5	8	5 +	6 +	8	na		
2001[1]	0 +	1	4	2 +	3 +	5 +	na			3 +	5	7 +	5 +	6 +	8	5 +	5 +	8	1	2 +	5	3 +	5	8	5 +	5 +	8	na		

Source: *Eastern Europe: The Fortnightly Political Briefing* (last appropriate issue for each year)

Notes
[†] Using a ten-point scale (10 is high) e.g. 5 indicates 5.0; 5 + indicates a rating between 5.1 and 5.9
[1] First half-year; [2] Second half-year
* Former Czechoslovakia from 1989 to 1991 inclusive
East Germany was rated at 5.6.8 for the second half of 1989

State profiles: the unification of Germany

The outpouring of population from East Germany and the breaching of the Berlin Wall in 1989 made for rapid unification to stop the massive population transfers which would have created enormous problems for both parts of Germany. However, 'Die Wende' (the change) began as a spontaneous reform movement underpinned by the Lutheran Church aiming at democratic involvement in the state. As late as November 1989 it was assumed that the GDR would survive, albeit with a reformist government. However, the collapse of the communist regime made these objectives untenable and public opinion swung sharply in favour of unification. West German policy had favoured unification ('Wiedervereinigung') since the 1950s, within the framework of the Western alliance, although the 'Ostpolitik' was based on links between two viable states (hence the Honecker quasi-state visit to the West in 1987). In November 1989 the West German chancellor, Helmut Köhl, issued a ten-point plan which aimed at rapprochement between the two states. But events in the East led to what was in effect a Western takeover, carried out with great urgency in view of the impending collapse of the USSR. After a freely-elected parliament ('Volkskammer') appointed a new government in East Berlin, the two Germanies signed three treaties between May and August 1990: a 'Staatsvertrag' (economic and currency union), 'Wahlvertrag' (settling the federal framework for all-German elections) and 'Einigungsvertrag' or unification treaty to create a single state on 3 October 1990. Meanwhile, in the international sphere both Germanies renounced claims to territory east of the Oder–Neisse line (thereby accepting the GDR–Polish frontier as permanent) and all Germany's frontiers were then accepted by the 'Two plus Four' settlement between the two Germanies and the four occupying powers. Unification with full sovereignty was thereby achieved after almost half a century of occupation.

Events could have moved more slowly, but quite apart from the Soviet factor, it is debatable whether a 'gradual merging' (provided for under the West German constitution, as was the 'takeover model') would have better served the objective of full unification. It was particularly significant that the Soviets accepted a united Germany within NATO, whereas unification on Soviet terms had been the objective in Stalin's time. However, it was agreed that 360,000 Soviet troops would be retained in East Germany for a short transition period, while Germany promised the Soviets substantial economic aid. The year of 1990 saw full economic and social union plus the first all-German elections. The government remained in Bonn in the interim but Berlin is now, once again, the capital and the seat of government. Meanwhile, the administrative regions of the GDR ('Bezirken') were abolished in favour of a return to the traditional regional system: Brandenburg, Mecklenburg-Vorpommern, Sachsen, Sachsen-Anhalt and Thüringen which, along with the unified Berlin, comprise the 'New Länder' of the enlarged Federal Republic.

Economic issues

There was major economic dislocation as East German industries, with low labour productivity, were suddenly thrust into open competition with the highly capitalised businesses of West Germany at a time when Comecon was collapsing. The 1:1 currency exchange resulted in a massive over-valuation of East German industry, politically necessary to avoid job losses in the West. Whole enterprises were wiped out: thus the traditional textile and footwear industries of Weissenfels were totally eliminated. Rural textile industries in the Harz Mountains, enlarged in the 1960s to cushion the isolation of the 'closed' frontier zone, were also forced into liquidation. The large chemical complexes of Buna and Leuna near Merseburg saw their employees reduced from some 50,000 to 15,000. The newer plant in these complexes has been taken over by private companies but some of the old factories have been difficult to dispose of; underlining the weakness of the central planning era in concentrating on increases in capacity rather than modernisation and retooling across the board. In the case of the motor vehicle combine ITA a good deal of new investment is now under way at the old locations; Opel and BMW have taken over the former Wartburg plant at Eisenach while Volkswagen acquired an interest in the Zwickau Trabant assembly unit and the Chemnitz (formerly Karl Marx Stadt) engine factory which used to turn out two-stroke engines with attendant pollution hazards. Mercedes have acquired the Ludwigsfelde lorry plant but the factory at Zittau, which produced the now-obsolete Robur lorry, is now reduced to the production of spare parts only. Even with takeovers there will be a substantial fall in employment which will not be balanced out entirely by growth in the tertiary sector. However, eastern Germany may benefit in the longer term by having the most modern industrial base in the EU.

The economic reform had to be quite radical on account of the highly centralised 'Kombinate' model in force in East Germany. In 1989 there were 126 'Kombinate', each with 20–40 plants and 20,000 employees on average. Privatisation through the 'Treuhandanstalt' was a creation of the GDR, established three days before the first free elections. Treuhand's headquarters were in East Berlin, with branches in the districts that managed all firms with less than 1,500 employees (for the East German Länder did not then exist in a functional sense). As the former 'Kombinate' were broken up into separate pieces, many isolated units had to be liquidated when they found themselves cut off from central R&D, distribution and administrative functions with little chance of survival. In any case, in view of the over-valuation of the Ostmark, almost two-thirds of 'Kombinate' were insolvent within ten months of monetary union, following an overall decline in production of similar magnitude. Thus from March 1991 (to late 1994) Treuhand went in for 'active decentralised restructuring' through 15 regional restructuring agencies (largely independent of the centre) assisted by some 5,000 experienced managers brought in from the West. There was a balance between takeover by West German and foreign firms on the one hand and EMBO on the other. Where units were successfully privatised there was an interim 'joint corporation' stage involving Treuhand and the investor – with Treuhand covering the main costs of reorganisation, retraining, modernisation and rectification of ecological damage – before the investor took over all the shares and integrated the eastern plant into the corporate network. The operation was very costly and, given the low market value of the former SOEs, it produced a deficit of DM250 bn. instead of the DM600 bn. antici-pated. The new structures were integrated into regional development plans, while federal institutions were now able to provide administrative and welfare back-up (Table 1.4).

Thus the perceived advantages of unification for the 'Easterners' ('Ossies'), through better services, an ongoing overhaul of dilapidated infrastructure and freedom to travel westwards, were tempered by economic restructuring which has resulted in a great increase in unemployment. This came as a shock to a people conditioned by communist propaganda to associate the lack of a job with 'parasitism' and was especially agonising for those found unsuitable for work in the new Germany (in the armed services or education) because of links with the East German establishment (through party membership and/or links with the 'Stasi' security police). The inclusion of some 'bitter fruit' in the unifi-cation package has therefore produced a perception that it has been, in military terms, a triumph for the Bonn government. With the 'Westerners' sensitive to the financial costs of unification and indignant about continued overmanning in the East, the old inner-German boundary will evidently disappear from public perceptions much more slowly than it is doing from the landscape: the 'wall in

Table 1.4 The East German economy: lessons for the transition

The German experience may offer some insights into the general problems of transforming the countries of ECE from command to market economies:

- The 'big bang' strategy was politically determined and fully justified But restruc-turing takes time, which was 'the scarcest factor in the process of political and economic unification'. But other countries can buy time, fixing exchange rates low enough to stimulate exports and restrict imports: thereby they can maintain low real wages as long as necessary.

- The 'big bang' is more costly than a gradualistic strategy in the short run but may be cheaper in the long run. Opening up the economy to foreign trade is essential: there is no transformation without change and without change only slight improve-ments can be expected.

Restructuring former socialist economies is a major challenge requiring:

  - Splitting and privatising state enterprises, with plant closures and unemploy-ment unavoidable.

  - New capital from financially strong investors who can only be found abroad. So foreign ownership should be encouraged in the context of restructuring old enterprises and creating new ones.

  - New management: once again a strong foreign partner can help with produc-tion (installing modern technologies) or with new sales outlets.

  - New patterns of specialisation linked with open and internationally linked markets so as to overcome the distorted trade structures of Comecon that were often unrelated to comparative advantages.

Source: Various sources including Schmidt and Naujoks 1996

the mind' syndrome. By 1994, however, unemployment had peaked at 17.8 per cent and many of those initially displaced had found rewarding situations in the new order, though the communist regime that is stereotyped as backward, alien and dependent still invokes a degree of humour and nostalgia. Of course the level of support from West Germany and the EU as a whole (for Germany's New Länder qualify for regional/structural assistance) has been of a far higher order than for other former communist states. Nevertheless, the economic stress in East Germany has provided an opportunity for the communists to make a comeback as a radical left-wing party advocating eastern interests. It is now a coalition partner in some of the New Länder governments.

Migration and regional trends

Migration certainly accelerated the unification process and it overshadowed the increasing concentration of population that had taken place in East Germany during the 1970s and 1990s, nationally and regionally: cities grew at the expense of surrounding communities, especially in the north and centre, and there was a general decline in the old industrial areas in the south. After 1989, most out-migration originated in (a) regions with fewest opportunities for finding government-subsidised employment and on-the-job training, and also places in higher education; (b) rural/agricultural regions; and (c) regions with poorest housing quality. There were low rates in the southeast, especially Dresden (more attractive after becoming the Land capital for Sachsen) and Zwickau-Plauen (where out-migration is traditionally low): so the old industrial regions lost population before unification but relatively slowly after. Meanwhile, there has been positive change in Berlin, with suburbanisation, extending further into Brandenburg where East Germans chose to commute to employment opportunities in the Berlin region rather than migrate to the West. All this correlates with industrial potential, with unemployment lower in the cities and higher rates in the rural north (also in the rural border areas between Sachsen-Anhalt and Thüringen), though some rural areas have benefited from a 'bridgehead' function where main lines of communication cross into the east from Hamburg, Hannover and Munich.

Development of leisure and retail establishments has taken advantage of the slow transition to strict West German planning systems. But the main feature in the geography of transition is the contrast between the south, with proven industrial potential and an 'image' to attract new investment, and the rural north with a recent history of state-directed industrial development best illustrated by the Rostock shipyards. The 'far east' (exemplified by the town of Stralsund) is gravely disadvantaged at the 'end of world' in contrast to Schwerin which is close to the Berlin–Hamburg axis and the 'Wirtschaftsstandort' of Rostock which has motorway access. A motorway from Hamburg to meet the Berlin–Szczecin autobahn near the Polish border has therefore been proposed as a regional development measure. But a low level of investment in Mecklenburg-Vorpommern in general reflects the prominence of EMBO in privatisation (privatisation is thus no guarantee of investment). Meanwhile, it can be expected that the industrial heartlands of the pre-war period in Sachsen, Sachsen-Anhalt, Thüringen and

South Brandenburg (including the area immediately south of the capital) will have the best prospects for reindustrialisation. Notwithstanding chronic structural problems, they are attractive locations for investment. The most important advantage of these regions is that they are geographically well placed at the crossroads between Europe's West and East. Regional aid programmes have discriminated according to potential and in 1997 aid was reduced in Berlin, Dresden, Leipzig, Schwerin and the Thüringian towns of Erfurt, Jena and Weimar.

State profiles: the breakup of the federal systems

The outstanding territorial changes have involved disintegration rather than union; for both the region's federations have broken up. The Czechoslovak and Yugoslav states were not creations of communism, but the federal systems were. So, it is arguable that national self-confidence grew through limited decentralisation under the umbrella of monopoly parties. Pluralism has certainly contributed to a growth of nationalism but – more significantly – it has also offered the possibility of secession as major stresses have come to the surface. The option has been taken all the more seriously because the security system of the modern world (through the UN, the NATO umbrella and the expectation of eventual European integration through the enlargement of the EU) gives small states an effective guarantee against external military intervention. In both cases the momentum for breakup developed in areas away from the centres of federal government (Belgrade and Prague), although in Czechoslovakia it was the less-developed Slovakia that sought independence whereas in former Yugoslavia it was the more advanced republics of Croatia and Slovenia. Very significantly, however, the Czech government was happy to negotiate a 'Velvet Divorce' whereas – notwithstanding the nationalist turn in Serbia under Milošević – Belgrade was implacably opposed and used all its resources for armed intervention to frustrate disaggregation.

Former Czechoslovakia

Czechoslovakia split in 1993 following the transitional arrangements arising out of the 'Velvet Revolution' of 1989 (and the federal system dating back ultimately to 1969). Historical factors have some bearing because although both the Czech and Slovak Lands lay within the Habsburg Empire before the First World War, the former was closely tied to Austria and the latter with Hungary. Moreover, Slovakia enjoyed some years of independence (with German support) during the Second World War at a time when the Czech Lands were stripped of the border territories (transferred to 'Grossdeutschland' and occupied by the Germans as the Protectorate of Bohemia and Moravia). Under communism the relatively backward Slovakia received considerable economic assistance – and the capital city of Bratislava grew from 173,000 in 1947 to 435,000 in 1989, much faster than Prague or any other city in the Czech Lands – the adequacy of this help was always a matter of dispute because of Slovak dissatisfaction over the extent of their influence over the government in Prague.

Despite some initial police brutality, the transition in Czechoslovakia began peacefully for the entire communist leadership resigned in the face of a massive pro-democracy campaign coordinated by Civic Forum, leading to a 'government of national understanding' under M. Calfa with a majority of candidates nominated by the non-communist parties, while the communist president resigned in favour of the former dissident V. Havel. In deference to Slovak sensitivities the country was renamed 'Czecho-Slovakia' and subsequently the 'Czech and Slovak Federal Republic'. Germany quickly replaced the Soviet Union as the country's most important trading partner but Western trade was not enough to compensate for the decline in former Comecon markets and the president expressed concern about the social problems arising from a rapid transition in all parts of the country. But it became increasingly difficult to reconcile the dash for private enterprise in the Czech Lands (Bohemia and Moravia) with the interventionist policies that were preferred in Slovakia where there was a determination to protect the large defence industries from neo-liberal policies spiced by an ethical foreign policy where arms sales were concerned. When it became impossible to formulate a programme, after the 1992 elections in Slovakia endorsed a nationalist platform, complete separation seemed inevitable and there was a will on both sides to negotiate a 'Velvet Divorce' which became effective at the beginning of 1993.

Slovakia

Post-communist politics took a strongly nationalist line as a reformist party 'Public Against Violence' crystallised into a nationalist Movement for a Democratic Slovakia (MDS) under the charismatic V. Mečiar who gained a firm ally in the Slovak National Party. However, Slovakia found it difficult to stabilise the economy while attracting foreign investment and defending its key interests, for unemployment has been high and the anticipated parity between Czech and Slovak currency has not been maintained. It is also worth mentioning that war in former Yugoslavia complicated the economic situation in Slovakia given the UN trade embargo and the disruption of overland transport which was damaging for all neighbouring countries. So the country became more oriented towards the east: recognising the value of the Caspian region, association with major Russian companies like Gasprom, and utilisation of the best Russian technology for the revamping of the country's engineering and armaments industries, including production of DV-2 and JAK-130 aircraft (partly for the Slovak air force). There was friction over the Hungarian minority, but after Hungary turned down the idea of voluntary repatriation, the two countries signed a Treaty of Friendship and Cooperation in 1995, guaranteeing the rights of ethnic minorities. A subsequent amendment passed in the Slovak legislature rejected the concept of autonomy, but a joint committee was set up in 1997 to monitor minority human rights in both countries. There is now an end to forcible assimilation through the education system.

Mečiar's style of government involved crude nationalist rhetoric and a highly politicised approach to economic management – where 'cronies' were rewarded in the privatisation stakes while 'non-sympathetic' local government officials were removed from office – caused dismay in the West and Slovakia fell behind in the drive for EU accession. A divided opposition allowed for only a brief phase of government for a broad-based coalition under J. Moravčik before Mečiar regained power in 1994 with particularly strong support from the poor and the unemployed. However, extremism helped to draw opposition into a Slovak Democratic Coalition in 1997, while the three main ethnic Hungarian parties merged into a Party of Hungarian Coalition (PHC) with a conservative-populist and civic-liberal platform. Along with the Party of the Democratic Left (PDL) and the Party of Civic Understanding (led by the then Košice mayor – now state president – Rudolf Schuster), they refused to cooperate with the Mečiar government and in the 1998 election gained a total of 93 seats – with a particularly strong urban vote alienated by Mečiar's authoritarian style and overwhelming support from first-time voters. NGOs were also influential in launching the civic campaign 'Občianská Kampan 1998', described as an open, independent, nonpartisan initiative to boost voter awareness through a better supply of information and increase participation in the election process.

With an experienced 'technocratic' cabinet, the new government under M. Dzurinda has restored democratic normality and administrative regions now have elected councils. The coalition of nine parties – including Dzurinda's own newly-formed Slovak Democratic and Christian Union – is inevitably prone to squabbling. The country now has a 'functioning market economy' which is expanding with low inflation thanks to a growth in exports which seems sustainable. Foreign investment is reviving companies, e.g. US Steel at the Košice steelworks (VSŽ) – driven to bankruptcy by Mečiar-friendly managers, the Hungarian oil company MOL has taken over Slovnaft while Slovak Telecom has been taken over by Deutsche Telekom and the car factory in Bratislava belonging to the German company Volkswagen is expanding. President Schuster is looking to EU and NATO membership rather than isolation: there is pressure to join NATO at the Prague summit in 2002 while the EU now has an office in Slovakia and progress is being made over approximation, especially in social security, pensions and health care. However, inherited structural deficiencies and unemployment of nearly 20 per cent maintain Mečiar's MDS as the largest single party. The PDL is anxious to defend jobs, incomes and welfare standards, but it is a part of the governing coalition and the new but popular centre-left party ('Smer'), which is making gains in local and regional government, could veer towards authoritarianism and link with Mečiar.

Czech Republic

Despite some regional problems a Civic Forum government in Prague implemented market reforms and this political stance was confirmed in 1992 with a victory for the Civic Democratic Party (CDP) of the charismatic V. Klaus which emerged out of the Civic Forum. A massive programme of voucher privatisation was launched, giving each citizen a personal stake in the country's assets. However, despite a right-wing image, the CDP proved to be a pragmatic party showing a commitment to the free market unique in ECE, yet veering towards social democracy to maintain low unemployment because there was enough investment coming into the country to delay restructuring in the state-owned enterprises while controlling the money supply to contain inflation. So despite neo-liberal rhetoric Klaus believed in gradualism and encouraged the development of small businesses while avoiding the social conflict of restructuring in order to win tolerance for further policy initiatives. It was fortunate that the boom in tourism and other services, along with sustained German investment, generated growth without the need for rapid restructuring. In 1996 the CDP lost ground slightly but although Klaus remained in power, his government presided over rising inflation – calling for austerity measures – and a series of corruption scandals, while the premier's vacillation over restructuring increased friction with his coalition partners.

Meanwhile, parties on the left failed to build a common platform because the Social Democratic Party (SDP) – seeking a more socially- and ecologically-oriented market economy – would not cooperate with the communists who retained their Stalinist credentials. But M. Zeman's SDP gained ground in the 1996 election in the areas of relatively high unemployment in northern Bohemia and in Moravia where there was resentment over government complacency. However, in 1998 the SDP was returned as the largest party (but without an overall majority) and in a situation where Zeman would still not countenance a coalition with the communists, yet could not attract any of Klaus's erstwhile allies, Klaus himself amazed the establishment by accepting a cohabiting role. Thus, given the personal animosity between Klaus and his former associates, it was possible for the SDP to run a minority government tolerated by the CDP. With Klaus apparently taking a cynical attitude for purely personal reasons, the arrangement was initially deplored by the president as a perversion of the popular mood. But stability was maintained and the CDP had sufficiently recovered its position by the end of 1998, through successes in the Senate elections, for Klaus to lecture the Social Democrats over economic policy.

However, the economic 'miracle' of the early days has faded, particularly in view of the currency crisis of 1997, linked with weak domestic demand, poor export performance, low productivity and the failure to develop functioning capital markets. GDP declined – and unemployment rose sharply – in 1998 and 1999. Growth resumed in 2000, although the rate is low – fuelled by FDI and the exports of foreign-owned companies – and the lack of a functioning bankruptcy law restricts restructuring. The economy still needs firm management to complete the privatisation and restructuring in the case of Tatra trucks, and the gas and telephone monopolies CEZ and České Telecom. But with influence over the SDP, Klaus could see deregulation of rents and energy prices – and privatisation of two-thirds of remaining state property – by the 2002 election. Assimilation of EU legislation may also help to accelerate growth, for the country is pressing ahead with European cooperation and integration, with the prospect of joining the EU in 2004 after NATO membership in 1999. An important milestone was the agreement with Germany in 1997 expressing mutual regrets over the Nazi occupation and the subsequent deportation of Germans from the Sudetenland. Meanwhile, a new alignment has occurred in politics through the formation of a 'Quad Coalition' in 1998 by four small opposition parties: Christian Democratic Czechoslovak People's Party; Freedom Union; Democratic Union; and Civic Democratic Alliance. They have remained untainted by the atmosphere of mistrust and did well in the Senate elections of 2000. It is conceivable that they could win a majority without the need for CDP support in 2002. But the instability of leadership evident in 2001, with the rapid replacement of C. Svoboda by K. Kuhn after two months, may indicate a loose political grouping rather than a truly cohesive coalition.

Former Yugoslavia

Disintegration on a larger scale has occurred in former Yugoslavia, a country that used to consist of six republics held together by a single party – League of Communists of Yugoslavia (LCY) – under the charismatic leadership of the late Marshal Tito. However, there was an 'extraordinary decentralisation' under Yugoslavia's 1974 constitution (in contrast to Czechoslovakia – and even the Soviet Union for that matter). Although Serbia was the dominant republic it had the same institutional endowments as all the others and so the Serbian elite could not control the central political and economic institutions like the Czechs in Czechoslovakia (or the Russians in the Soviet Union). Moreover, the national army (JNA) existed alongside republic-based militias. Culturally, the 1954 Novi Sad agreement provided for a single Serbo-Croat language, but with two equal pronunciations (the 'ijekavian' or western form used in Zagreb, and an eastern 'ekavian' equivalent used in Belgrade). Then the 1974 constitution allowed each republic to establish a 'standard linguistic idiom' in Serbo-Croat, as occurred in Bosnia & Hercegovina, Croatia and Montenegro. As such they served as precursors for new standard and potential successor languages. A separate Montenegrin language has been advocated, though it has not yet emerged officially. 'The universalist goals of a federal Yugoslavia and the control functions of the party state have, in this transition, given way to localisms of violence based on ethnicity and religion' (Pavlinek and Pickles 2000: 4).

However, given the federal government's inability to implement economic reform, autonomy veered towards separation, especially in Croatia and Slovenia which delivered substantial payments to the federal exchequer. After the Cold War Yugoslavia was exposed to the full rigours of IMF conditionality: economic austerity was highly divisive and accentuated nationalistic tendencies. As the communists lost their monopoly the constituent republics exercised their constitutional right and voted for independence. Croatia and Slovenia defected in 1991, driven by the possibility of state affirmation, linked to the Central European sphere of civilisation rather than the problems of the Balkans. For while Western Europe was keen to maintain the federation, this ceased to be an overriding objective after the collapse of the USSR deprived Yugoslavia of its geostrategic function as a Western buffer zone against the East. Bosnia & Hercegovina and Macedonia left the federation in 1992: economic reform was less pressing, but with the defection of Croatia and Slovenia these republics were unwilling to remain in a federation dominated by Serbia. This left Montenegro and Serbia in a greatly reduced Yugoslav Federation: two countries which had collaborated as independent states after the 1913 Balkan War to form the basis of a wider Yugoslav/Southern Slav state.

Disintegration is an indication of the strength of nationalism among federal units that rejected the compromise of federation worked out by Tito. But it is curious – in view of their weak position – that Serb leaders opposed economic and political liberalisation (unlike the Czechs in Czechoslovakia, and the Russians in the Soviet Union). Indeed the Serbs used all available force to stop the breakup, including the JNA which was largely officered by Serbs. However, Serbs comprised significant minorities in other Yugoslav provinces (especially Bosnia & Hercegovina and Croatia) while not enjoying an overwhelming majority in Serbia itself. Despite the looseness of the federal structure, Serbs retained powerful vested interests since a large Yugoslavia brought all Serbs together in a single state. While there was a surge of nationalism in Serbia – which brought S. Milošević to power in 1987 – this did not make for an independent Serbia, but rather for tighter Serb control in Serbia (ending autonomy for Albanians and Hungarians in Kosovo and Vojvodina respectively in 1989) and maximum Serb influence throughout the federation. Serbia would not even entertain the idea of converting Yugoslavia into a confederate state (discussed by the Croats and Slovenes in 1990) since federal institutions would cease to be primary and would act merely as agencies of the republics holding the real power. In line with a deeply ingrained 'Pan-Serb' (Greater Serbian) ideology, Belgrade's response to the breakup of Yugoslavia could only be an armed struggle to achieve a union of Serbia and majority Serb areas detached from Bosnia & Hercegovina and Croatia where local Serbs fully supported Belgrade radicals and made their own contributions to hostilities through local militias. And since Serb nationalism was expansionist and anti-liberal there could be no compromise until their institutional resources – especially the military option – had been fully exhausted.

Doubtless the scale of war was not foreseen while Serbia believed that the right of republics to secede unilaterally was not clear-cut and that there was a justification for military intervention by the JNA. But while the hostilities have been exceptional by general ECE standards, ethnic cleansing can be seen in other areas of tension in the wider Mediterranean theatre. There is also a history of mutual Croat–Serb animosity, for the long association of Croatia and Slovenia with the Habsburg Empire (also Vojvodina and for a short time Bosnia & Hercegovina) and the connections with the Ottoman Empire elsewhere make for particularly sharp cultural/religious and economic divisions. Croatia sought autonomy from Belgrade during the inter-war years while the Second World War experience brought independence (and German support) for a large Croatian state at a time when Serbia was subjected to a harsh occupation regime. Thus the civil war in the 1990s quickly revived memories of the militias of a generation ago: the Croat 'Ustaša' and the Serbian 'Cetniks'. Nevertheless despite these special considerations there was widespread apprehension during 1992–4 that the conflagration in Bosnia & Hercegovina might spread to parts of Serbia and Macedonia; precipitating a Balkan War that might have pitched Albania, Bulgaria and Turkey against Greece and Serbia. But fortunately the situation was contained, not least because the tragedy of Bosnia & Hercegovina made for restraint in other countries with similar ethnic diversity.

Slovenia

Slovenia made an early and relatively peaceful exit, for public opinion swung in favour of independence in the post-Titoist reform climate of the 1980s and the pace accelerated in 1988 when the military tried to suppress nationalist literature containing evidence of planned militia moves against Slovene liberals. Alternative political parties emerged during the 'Slovene Spring' of the following year and tension with Serbia pushed the Slovenes further towards secession. The year of 1989 saw a sharp polarisation of opinion in the former Yugoslavia as pluralism – linked with autonomy and liberal reforms – was embraced by the Democratic Alliance launched in Ljubljana (Slovenia), contemporaneous with the Movement for a Yugoslav Democratic Initiative in the Croatian capital of Zagreb. Lack of consensus across the whole of Yugoslavia caused the Slovenes to leave the LCY Congress in 1990 after rejection of their plan for eight independent parties. They subsequently renounced links with the LCY altogether and set up the Party of Democratic Renewal (later Social Democracy of Slovenia), subsequently joined by other parties during the elections of 1990. It was immediately after these elections that Slovenia made a successful break for independence (July 1990) and the JNA was obliged to withdraw after only a few days of hostilities. There was no significant Serb minority to create dissention and Slovenia had a clear advantage in its low level of dependency on all other parts of the federation (rejecting any idea of being a Balkan nation). The country enjoyed close links with foreign neighbours, especially Austria, Hungary and Italy, and could take full advantage of its central position on the European transport networks.

Slovenia made a success of stabilisation despite a fall in production and a rise in unemployment. The growth of business has led to an International Executive Development Centre in Kranj and the formation of the Slovene Association of Entrepreneurs which includes Chambers of Commerce, private capital groups and individual business owners, involving many from small firms. Slovenia joined the Visegrád Initiative in 1992 and despite claims from Italians expelled after the Second World War a cooperation agreement with the EU was signed in 1993. By the middle of 1993 Slovenia had gained admission to almost all international financial organisations and had negotiated a number of bilateral economic agreements. The pro-European policy is popular and the system of Eurocorridors promises much in terms of infrastructure which should benefit Koper. EU wants to see faster privatisation and more administrative reform to remove barriers in the way of FDI. This reflects a degree of resistance in Slovenia to allowing foreign investment taking advantage of cheap sell-offs and causing turbulence in financial markets. However, given substantial economic growth, Slovene politics have been remarkably stable with a succession of coalitions which since 1992 have all been headed by J. Drnovšek, apart from a short break in 2000. Drnovšek is a former member of the Yugoslav collective presidency and heads the Liberal Democratic Party, which developed out of the communist youth organisation and works closely with the Party of Democratic Renewal (emerging from the League of Communists of Slovenia). By contrast, the coalition that replaced him temporarily is now badly split, but A. Bajuk (who was premier in 2000) wants to do better by combining his New Slovenia–Christian People's Party with the impressively-robust centre-right Social Democrats of J. Jansa. This could provide an attractive alternative at the next election.

Croatia

Croatia is one of the larger of the Yugoslav successor states and it has the benefit of ready access to the Mediterranean by virtue of the Adriatic coastline. However, it has an extremely long boundary in relation to its area, given its distinctive butterfly shape, and has a potentially destabilising ethnic balance by virtue of the large Serb minority. Although the traditional borders were slightly modified under the 1974 Yugoslav constitution – ceding Srijem to Serbia (Vojvodina) and Bokakotorska to Montenegro – Croatia still faced a formidable challenge when the Yugoslav federation started to break up, with territory to defend in Eastern Slavonia (little more than 100 km from Belgrade but up to 250 km from Zagreb, in a straight line) and also at Prevlaka, just over 50 km from Podgorica but over 450 km from Zagreb. Croatia is not well endowed in terms of energy. There are hydropower stations in the mountains which featured in the war through the control exerted over them by the Krajina Serbs and through the damage caused in the case of the first such installation (Peruca) to be taken from them. But there has been heavy dependence on Bosnia (both the Neretva hydropower plant and the thermal stations of Kakanj and Tuzla) and Croatia's chemical, shipbuilding and metallurgical industries were disrupted when these supplies failed. But now Croatia plans a number of environmentally-friendly thermal stations on its own territory. She is also working on motorways to the Hungarian border and to Rijeka while the isolated railways of the Istrian peninsula are being connected to the national network. Nevertheless, 'Croatia saw the possibility of its own state affirmation, based on the Central European sphere of civilisation and divided from the problems of the burdened neighbouring Balkans' (Topalovic and Krleza 1996: 404). But when independence was declared in 1991, the Serbs immediately separated themselves from the rest of the population. 'The poorly-equipped and ill-trained Croatian forces therefore faced the twin threats of an interior uprising by well-armed local Serb irregulars and a major attack by regular JNA forces in the east' (Klemencic and Schofield 1996: 393).

The situation was stabilised during 1992 by UN peacekeepers operating in three discrete zones separated from the Zagreb administration: Krajina, East Slavonia (Daruvar and Pakrac) and West Slavonia (Vukovar). But normalisation was prevented by the Serb refusal to accept that their territories lay within the Croatian state: the option of autonomy for Krajina within Croatia was rejected.

Serb intransigence understandably placed Croatian politics firmly on a nationalist basis. The party of the president (F. Tudjman) – the Croatian Democratic Union (CDU) – won a sweeping victory with its right-wing, nationalistic platform in the first free parliamentary elections in 1990 and the party was again successful in 1992 (despite the loss of territory), though the lead was reduced and there were gains by both the right-wing Croatian Party of Rights (linked with the paramilitary 'Ustaša') and the main opposition Croatian Social Liberal Party. Croatian nationalism was not only geared to defence of the existing Croatia but to possible enlargement through a partition of Bosnia & Hercegovina linked with Serb aggression in the east. However, a new policy of alliance with the Bosniaks in 1993 put an end to the idea of a Croat state (Herceg-Bosna) based on Mostar. Then Croatia's prospects were transformed during 1995 when the build-up of the military establishment enabled Western Slavonia and Krajina to be overrun, whereupon the near-certainty of a further Croat success in Eastern Slavonia brought a negotiated settlement for the restoration of the devastated Vukovar area to Croatian control in January 1998. By this time rail links between Croatia and Serbia had been restored and through trains now operate between Belgrade and Zagreb. Whereas the Serbs initially fled Krajina, the more orderly transfer of power in Eastern Slavonia ensured that the majority of Serbs stayed on and local elections have given the Independent Serb Democratic Party appropriate influence over local government, especially in Vukovar.

Meanwhile, despite a successful stabilisation programme in 1993 and progress in the textile and clothing industries by companies with long-standing Western contacts, the war created stress through the bombing of factories, not to mention the loss of some 200,000 homes and hyperinflation caused through finance of the war effort. Although there was potential investment by the large Croat diaspora in Germany (initially 270,000 but rising to 600,000 as a result of the war), only a few relatively prosperous areas removed from the fighting – like Istria and the northern Adriatic, along with Medjimurje – could make any progress. The ruling CDU became linked with corruption in banking (Dubrovačka Banca but also Vukovarska Banca and Istarska Banca) through siphoning off of funds and party 'mafiosi' gained a major stake in an economy through 'crony privatisation'. As the CDU maintained a stifling control of the media, the educational system and the privatisation process (partly a response to movements in Dalmatia and Istria seeking autonomy), popular distrust of government was heightened by decline in GDP and increases in unemployment and foreign debt, while the peasantry was discouraged by low-cost agricultural imports which provided the substantial customs revenue. Tudjman's CDU lost further ground in the 1995 elections, being defeated decisively in Zagreb and in most towns apart from Osijek. This reflected public concern over economic and social problems, despite successful prosecution of the 'Homeland war'.

THE INTERNATIONAL COMMUNITY

The years since 1993 have seen Croatia's gradual transformation as a member of the international community with UN confirmation of the integrity of its borders. Croatian politics have also adjusted to a conventional Western model, beginning in 1995 when the left wing of the CDU hived off as the Croatian Independent Democrats. But Tudjman's regime – sustained by an unwieldy group of ideological factions, including conservative nationalists, former communists, intellectuals and technocrats – was adjudged to have acted in too draconian a manner in Krajina: failing to minimise the exodus of 120,000–180,000 Serbs and take positive measures to facilitate their early return. Membership of the CoE was held up as a result. Indeed, the Croatian regime became marginalised in Europe as corrupt and unreliable in not carrying out its Dayton obligations regarding the International Criminal Tribunal for the Former Yugoslavia (ICTY) in respect of activities by Croatian militias in Bosnia & Hercegovina and the Croatian army in the 'Homeland War' of 1995. A clean break with nationalist politics seemed to be on the cards during 1998–9 in view of the president's terminal illness, and in the elections at the end of 1999 the CDU was replaced by a centre-left coalition comprising the Social Democrats, the Social Liberal Party and four small parties with I. Racan as premier. The death of Tudjman was followed by presidential elections at the beginning of 2000 which returned S. Mešić (a former president of Yugoslavia), selected by the group of four small parties already referred to.

The impact of the new government could be considerable. Serbs are returning and there are majority Serb towns again in Krajina, including Knin. Also, a retreat from a nationalist policy could help in stabilising Bosnia & Hercegovina – reorientating Bosnian Croats towards Sarajevo and easing hard-line Serb influence through restoring the traditional links between Banja Luka and Zagreb. However, Croats have not abandoned all their nationalist sensitivities and it did not go unnoticed that the change in the Yugoslav presidency in 2000, when V. Koštunica replaced Milošević, brought immediate UN recognition without accommodations of the kind made by the Racan government over war criminals and the return of Serb refugees. Croatia feels she is under greater pressure than Yugoslavia to cooperate with ICTY and is warning that the West should not behave too leniently towards Belgrade. 'Some Croats openly express fear that once Serbia regains favour with the West, Croatia will lose its leadership role in the region' (Lindstrom 2001: 7). Since Croatia's 'Homeland War' of 1995 is widely seen as a defensive action, the government is still constrained by deposed nationalists critical of compromises that pander to the perceived 'whims' of the West. Croatia is also exercised by the territorial adjustments sought by three of her neighbours. Montenegro is unhappy about the southernmost extremity (Prevlaka) which overlooks the Bay of Kotor while Slovenia feels compromised because her territorial waters do not include the navigation channel into the port of Koper given the alignment of the frontier on the Bay of Piran. Finally Bosnia & Hercegovina would like a territorial extension to cement its hold on the port of Ploče. But Croatia was reluctant to yield in any one theatre for fear of establishing a precedent when the priority is to consolidate national unity.

However, Croatia has been pulled fully out of isolation since the elections of 2000 and negotiations have been completed with the EU on a Stabilisation and

Association Agreement – effective in 2001 – which could lead forward to application for full membership. The governing coalition has survived some strain over the extradition of two generals to ICTY (strongly opposed by the Croatian Social Liberal Party). And with peace restored there is a brighter prospect of economic growth to take advantage of the country's position between Europe, the Middle East and Asia. In addition, Croatia could be a springboard for investment in Bosnia & Hercegovina, while relations with Slovenia have improved with the settling of the land border and a Croatian concession over the Bay of Piran which gives Slovenia control over the shipping lane to Koper. There is also agreement over Krsko nuclear power station from which Croatia will draw power in 2002. However, the economy needs great improvement to reduce debt (new money does little more than service existing loans) and revive industrial production. Economic reform (taxes and pensions) has also been slow, though growth is boosted by exports and tourism. Austrian, German and Italian capital is moving into banking and insurance but not industry to any extent, still owned by the state and individuals associated with the late President Tudjman. Greater transparency is needed in government in general and in the privatisation process especially. And the overhaul of the infrastructure – energy and transport – to achieve greater national cohesiveness within the present boundaries remains a priority to a greater extent than in the other successor states of the former Yugoslavia.

Bosnia & Hercegovina

As a republic within Tito's Yugoslavia there was peaceful coexistence among the Bosniak (Muslim), Croat and Serb populations and considerable economic benefit arose from centrality within the federation: accessible by a standard-gauge railway and surfaced highway from the Sava to the Adriatic. But when disaggregation became inevitable with the failure to negotiate a new confederal arrangement Bosnia & Hercegovina faced perhaps the stiffest challenge, given the three leading communities and the traditional high degree of interdependence: there were certainly no urban functional regions where one group was totally dominant. The Bosniaks were determined to advance their interests against a historical background of strong Serb resistance to national status, reflected in trials of Bosnian Muslims in 1983 because of the threat posed by the large Muslim population in the Serbian province of Kosovo where autonomy was subsequently revoked in 1989. The Bosniaks mustered an overall majority in favour of independence in 1992 with the support of the Croatian Democratic Union (CDU), but they faced determined opposition from the extreme Serbian National Renewal Party. There was a rapid descent into armed struggle with the Serbs enjoying an early advantage because the scale of their military resources available through their militias and the JNA. The fact that many of their majority areas lay adjacent to Serbia made the goal of unification all the more appealing. This left the Croats in a dilemma: to ally with the Bosniaks in defence of a unified state or to cooperate with the Serbs in partition by connecting most of Hercegovina (Herceg-Bosna) with Croatia. Mounting chaos in 1992–3 followed from the attempts of all the three main ethnic groups to establish clearly-defined ethnically-cleansed sectors. Although UN protection forces arrived on the scene, as in Croatia, their role was to monitor ceasefires rather than engage directly with the warring factions.

Although the Croats have still not fully resolved their dilemma, the situation was simplified in 1993 when they were encouraged by the United States to support the Bosniaks in a war which then turned dramatically in their favour during 1994–5 through improved military supplies for a coordinated offensive against Serb territory in western Bosnia and Hercegovina. Diplomatic solutions were advanced through the Vance–Owen 'canton' plan (April 1993) for a unified B&H and the 'Contact Group' confederation plan (July 1994) for a Bosniak–Croatian core with areas of Serb control to the east and west – apart from three eastern Bosniak enclaves (Goražde, Srebrenica and Žepa) connected with the core by a corridor under international control. Initially, the Serbs saw no need to settle for territory commensurate with their population share while they enjoyed military superiority and hoped to press more heavily on the industrial heartland of central Bosnia. But matters were brought to a head during 1995 – after the fall of Srebrenica and Žepa (accompanied by genocide) – through the decision to use NATO air power to check Serb pressure on Sarajevo and other vulnerable areas such as Bihać, Goražde and Tuzla. In the aftermath of Croatia's 'Homeland War' in Krajina, Bosniak and Croat forces launched a successful offensive in western Bosnia and Hercegovina in the summer of 1995 for a continuous strip of territory – including Gornji Vakuf, Jajce, Mrkonjić Grad, Ključ and Sanski Most – connecting the Bosniak–Croat heartland with the Bihać 'pocket'. The deteriorating military situation, coupled with the bombing of Serb positions in Sarajevo in response to the deliberate shelling of civilians, created the basis for a settlement which acknowledged the sovereign status of Bosnia & Hercegovina while providing for a confederation between a Muslim–Croat Federation of Bosnia & Hercegovina (FB&H) – comprising 10 cantons, and 'Republika Srpska' (RS) (Figure 1.1).

THE DAYTON AGREEMENT

Agreement was reached between the Bosniak leader (A. Izetbegović) and the presidents of Croatia and Serbia (Tudjman and Milošević) who met at Dayton, Ohio in December 1995. The military situation has largely dictated the peace deal but territory around Mrkonjić Grad was exchanged for the corridor to Goražde and the Serb-held areas of Sarajevo. Bosniak constitutionalists deplore the effective partition of the country. Yet Bosnia & Hercegovina exists within its historic frontiers and the net result has been a disaster for the Serbs since their goal of a Greater Serbia has not been achieved. As was the case with Hungary after 1918, world opinion has accepted the principle of self-determination on a 'meso-regional' scale (favouring all the leading nations of former Yugoslavia) rather than a 'mega-regional' scale relevant to pan-Serb aspirations. Although Milošević initially sided with the nationalists he settled at Dayton in view of the military situation in B&H and the pressure of economic sanctions

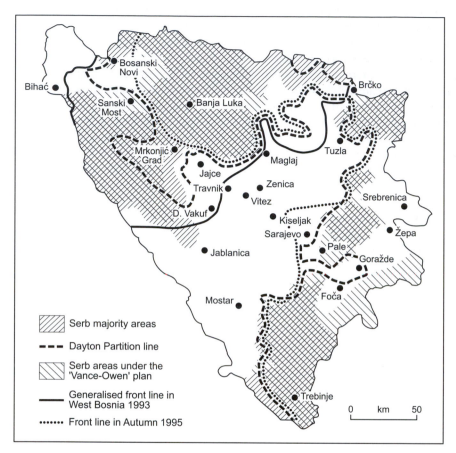

Figure 1.1 War and separation in Bosnia & Hercegovina

Source: Administrative maps

against Serbia: as a skilled 'crisis manager' he was reluctant to see more refugees in Serbia supporting ultra-nationalist politicians who could erode his own power base. But although the fighting has stopped B&H is not yet the multi-ethnic, economically viable society envisaged at Dayton. Political life is dominated by ethnically-based parties and although there are small parties and coalitions operating in both entities they do not yet have a substantial base. Further dislocation occurred when all Serbs left the suburbs of Sarajevo before they were handed over to the FB&H, and because refugees still cannot return in safety to the RS, accommodation in the FB&H cannot always be given back to Serbs who have returned in considerable numbers.

The central government is responsible for foreign policy and adjudication between the entities which comprise a loose confederation with a common citizenship law, a unified passport system and a single currency: the 'Konvertibilna Marka' tied at parity with DM. A unified customs regime is inhibited by the political strength of some local networks in Herceg-Bosna and RS financed through trade in alcohol, cigarettes, oil and arms. On the other hand, the Croat and Serb sides have accepted trade laws that rule out customs unions with Croatia and Serbia respectively and this is an important step in the creation of a single Bosnian economy. Agreement over a common border police should make customs more effective and cut down on smuggling, though illegal trade remains an issue in the Croatian majority area of Herceg-Bosna and also in the western part of RS where new petrol stations are set up for the purpose. The two entities are working together over environmental policy and a joint railways corporation. Infrastructure is being rebuilt, with a bridge over the Sava between Orašje in Bosnia and Županja in Croatia. The crisis over Brčko, which has a Bosniak–Croat majority yet provides a link between the RS territories in east and west Bosnia, has been solved by placing it in both entities. This opens up the possibility of both Brčko as a river port and Ploče as a sea port being generally available to the whole country. Currently there is agreement with Croatia for customs-free transit to Ploče (also a free trade zone) while Croatia has free transit through the country's coastal strip: a territorial anomaly of no economic value which survives from the days of the Dubrovnik city state. Now it is important to restore the railway between Ploče and Sarajevo which was destroyed during the war at Jablanica in central Bosnia. In future it is likely that a motorway from Osijek to Ploče will provide B&H with a modern highway, while also giving Croatia a direct connection between eastern Slavonia and Dalmatia. A Zagreb–Split motorway might also use a route through the Bihać area of western Bosnia.

Within the FB&H, the Bosniaks and Croats work together in a state of 10 cantons: five have strong Bosniak majorities (Goražde, Sarajevo, Tuzla, Una-Sava and Zenica-Doboj), three Croat (Livno-Tomislavgrad, Posavina and Western Hercegovina), leaving Central Bosnia (Travnik) and Hercegovina-Neretva (Mostar) with majorities of 50–66 per cent for Bosniaks and Croats respectively. Political life is dominated by the two ethnically-based parties: the Bosniak Party of Democratic Action (PDA) and the Croatian Democratic Union (CDU – which has always closely integrated with its counterpart in Croatia) in a Coalition for a Single and Democratic Bosnia & Hercegovina. However, there are marked contrasts in the political mood among the cantons. Tuzla shows commendable ethnic tolerance for the Bosniaks here are westernised, unlike the refugees from Srebrenica who are socially conservative and hostile to Serbs. Croats and Serbs make up only a tenth of the population compared with a third before the war, though many would return if there were houses and jobs. With Bosniaks vacating apartments to returning Serb owners, this seems the best place to build an ethnically integrated community. On the other hand, nationalism remains very strong in Mostar where the local council was for long deadlocked between the CDU and the PDA. However, shops have reopened and there is movement between the two sides (Bosniak east of the Neretva and Croatian to the west) over the suspension bridge built by the Spanish army to replace the sixteenth-century bridge destroyed in the civil war: the exclusion of Muslims from the west has ended. However, segregation remains in education because after the ceasefire Bosniaks used the old Yugoslav syllabus and Croats adopted material from Zagreb and Catholic religious instruction became standard. Reintegration has stalled over Croat insistence on using their own material for music, language, literature and history. At the same time, both sides have rejected an EU offer of a joint university and both groups seek their own successor languages to Serbo-Croat, with Arabic and Turkish borrowings in the Bosniak case.

While the CDU has been eclipsed in Croatia it remains strong in FB&H where there is much anxiety over Zagreb's cooperation with the ICTY and many Croats would still like to establish their own entity: indeed the call by one Croat leader (A. Jelavić) for an end to ethnically-mixed cantons is seen as a strategy for Croat independence. The mood has strengthened since the party's setback in the 2000 elections and it is paralleled by Croat penetration of FB&H enterprises like Aluminij Mostar where state ownership has been heavily diluted disproportionately by investment from Croatia (after the book value was cut suspiciously during the war from DM1.2 bn. to 0.2). But the Croatian president will not support the breakup of FB&H and the CDU bid for separation is not going well. The West wants the electoral process to bring to power politicians who can run the country without international involvement and thereby enable the UN military operation to be run down. The international community is doing all it can to accelerate integration. A stabilisation force (SFOR) of 20,000 troops maintains free movement of people and goods. The military have also intervened on behalf of moderate politicians. Under Dayton, state and entity governments pass legislation but the Office of the High Representative (OHR) has taken decisions over a flag, national anthem, passport, vehicle licence plates and a common currency when the three groups could not agree. The OHR has also removed politicians representing all three ethnic groups for obstructing the peace process, notably over the return of refugees. Progressive elements would like to end rotation at the head of the confederation government because it is so difficult to implement legislation but this is not likely to happen soon. Most conspicuous has been OHR support for a pro-Dayton government (under M. Dodik) in RS between the elections of 1998 and 2000, combined with the removal of a hardline president (N. Poplasen) whose Radical Party was also banned from contesting the local elections of 2000. The third post-war general election in 2000 was disappointing for those who expected a decisive swing away from nationalism. In RS, the nationalist Serbian Democratic Party regained power, polling more votes than all its rivals put together with a programme of 'virulent opposition' to the Dayton Agreement.

In the FB&H the PDA took Zenica-Doboj and Una-Sava while the CDU won in the Croat cantons of Hercegovina-Neretva, Livno-Tomislavgrad, Posavina and Western Hercegovina as well as the mixed district of Central Bosnia. However, the multi-ethnic Social Democratic Party of B&H clearly topped the poll in the cantons of both Sarajevo and Tuzla and also won narrowly in Goražde. This has led to an Alliance for Change coalition with another significant non-nationalist party (Party for Bosnia & Hercegovina) and some smaller groups to challenge the dominance of the two nationalist parties. This contrasts with the results in RS (already noted) which returned the Serb nationalists to power. Thus the West is becoming increasingly impatient with the country's political class and has adopted the tactic of changing the electoral system to encourage local politics. But it came too late to affect the 2000 election and indeed there is always a risk that attempts to manipulate politics through economic inducements will backfire. On balance the experience since Dayton 'raises severe doubts about the willingness of local elites and the broader public to develop the political maturity and sophistication needed to establish democratic institutions and seek cooperation schemes' (Altmann 2001: 115). Although nationalist parties are plainly not acceptable for foreign investors there is still a powerful non-progressive faction which not only points to a long stay for the international community but fuels EU scepticism towards enlargement in favour of different forms of participation. On the other hand economic pressure could help dismantle hardline networks such as the Serbian Democratic Party hierarchy in RS descended from the wartime leadership under R. Karadžić, still sought – along with his general R. Mladić – by the ICTY. Economic aid could woo the RS electorate in a situation where the Yugoslav economy cannot be relied upon to take all the production. While components from Cajovec in Banja Luka used to go to Zastava assembly line in Kragujevac, markets are now being sought in Croatia, FB&H and Italy. The bottom line is that the Serbs no longer have a military option: further fighting would immediately mean loss of western Bosnia, which is economically closely bound to Croatia as well as being strategically vulnerable.

THE ECONOMY

Reconstruction is proceeding slowly in FB&H where the metallurgical industry at Zenica has received an injection of funds from Kuwait, while the Srednja Bosna coalmine and the Kakanj coal-fired power station have been repaired, as has the Krivaja timber-processing works. Another significant achievement has been the Volkswagen project at Vogašćá near Sarajevo: assembling 1,000 Felicias from kits delivered by Škoda during 1998 but intending only to double that figure in the future – with a workforce of 60 – after turning out as many as 40,000 a year in the past. At the beginning of 2000 production was still only around 40 per cent of the pre-war level. There is too much dependence on aid which cannot maintain the 30–40 per cent growth of recent years. Sarajevo is a particular problem since the city is dependent on spending – some DM50 m. a month – by a 15,000-strong foreign community, plus money sent home by émigrés. Much industry is inefficient, with coal extracted at DM5/Gigajoule when imported coal would cost DM4 or less. An FDI of $4.7 m. per year since Dayton – the lowest level among the ECECs – is quite insufficient to cut the 30 per cent unemployment rate. This is unfortunate since there is obvious scope for food processing at a time when most foodstuffs are imported. But entrepreneurs in Sarajevo find themselves forced into a grey economy given the administrative complications of the confederation boundary in the eastern suburbs and the web of cantons and municipalities. Restructuring is very necessary and privatisation has started with small firms in designated sectors (driven in RS by vulnerable politicians seeking economic assets as security in the event of electoral defeat). Over the longer term greater integration between the two entities could reduce the level of dependence on foreign aid. The process could be assisted by complementarity since the RS economy is dominated by Bosanski Brod and Modrica refineries while FB&H is strong on heavy industry and textiles. A negative aspect is the departure of young, skilled professionals and intellectuals (the cornerstone of the rebuilding process), including many who lived undisturbed in FB&H-controlled areas throughout the war. Meanwhile, it is the elderly and the sick who move in the other direction from refugee status abroad. Faster return of refugees could trigger more IMF finance while the recent acquisitions by foreign companies: Austrian banks, Heidelberger Zement – acquiring the Kakanj cement plant – and Slovenian brewers and retailers (Pivovarna and Mercator respectively) suggest a faster tempo of change.

Macedonia

Macedonia's decision to opt for independence in 1992 reflected Tito's success in creating national consciousness among the Macedonians as a people distinct from Bulgarians, Slavophone Greeks and Serbs. Macedonia was able to secede without major conflict when the JNA agreed to pull out, although Serbia was ambivalent over accepting Macedonian nationality (some nationalists would claim the whole of Macedonia or advocate partition between Serbia and Greece) and Macedonia is crucial for Serbia's communications with Thessaloniki. The situation was also problematic in view of internal ethnic and religious divisions, given the Albanian-Muslim minority and economic isolation as a result of UN sanctions imposed against Serbia in connection with the war in Bosnia & Hercegovina. There were other external threats since Bulgaria declined to recognise Macedonian nationality because of the implications regarding Macedonians within Bulgaria, although she did recognise the Macedonian state in 1992 and entered into bilateral economic relations. However, Greece showed extreme belligerence towards Macedonia because the name Macedonia is seen as a focus for a larger state that could have brought the frontiers of Greece into question. There is a fear that the Macedonian minority in Greece (the so-called 'Slavophone Greeks') might secede if there was a Macedonian state in existence. Greece is also concerned at the impact of Macedonian nationalists on the Turkish minority of 'Greek Muslims' who might also demand secession. For this reason Greece also objected to the Macedonian flag with its fourth-century star, said to be a symbol of Philip II and hence an appropriation of Greek heritage.

Meanwhile, a further threat came in 1992 from the Serb leader Milošević in the shape of a proposal for a Serbian–Greek confederation. This clearly posed a threat to Macedonia which lay between these two states and a UN package was formulated to overcome the risk of a diplomatic 'black hole' and forestall a Serb invasion which might have followed any move by Macedonian Albanians to enter Kosovo. Good relations were established with a UN Protection Force on the border with Kosovo. Under the 1992 agreement Macedonia formally denied territorial claims and any policy of interference in the internal affairs of neighbouring states. Greece would accept 'Vardar-Macedonia' or 'Republic of Skopje', but the Skopje government insisted on 'Macedonia', backed by the nationalistic opposition of the Internal Macedonian Revolutionary Organisation (IMRO). The interim solution of 'Former Yugoslav Republic of Macedonia' paved the way for UN membership and recognition by Russia, USA and major West European states at the end of 1993. It has also enabled the EU to support Macedonian independence with its PHARE programme and ensured continued Macedonian support for the sanctions against Serbia in force at the time. However, Macedonia declined to remove the Vergina Star from its flag and this eventually prompted a Greek blockade during 1994–5 and activity in the port of Thessaloniki (a major potential business centre for the region as a whole) was much reduced while trade flowed through Bulgaria (via Kumanovo-Kjustendil) to Turkey instead. The crisis was solved when Macedonia agreed to remove the star (substituting an eight-rayed sun) and removed sections from its constitution considered 'irredentist'. Meanwhile, Macedonia was suffering additional burdens through support of the blockade against Serbia and accommodation of 30,000 refugees from Bosnia & Hercegovina, which were controversial to people in Skopje where unemployment has risen and incomes have fallen.

Domestic politics have strongly reflected these pressures. There were tensions over foreign policy, for the nationalist IMRO was particularly sensitive towards the policies of the old federal capital – seeing the leaders of the rival Social Democratic Alliance (SDA) as 'Belgrade agents'. Indeed, the SDA, as successors to the LCY, inherited the former Yugoslav policy of placing Macedonian nation building on an anti-Bulgarian agenda in response to Sofia's traditional stance that Macedonia, Moesia and Thrace constitute the three units of Bulgaria. In 1990 the IMRO combined with the Party for Macedonian National Unity and gained a small lead over the SDA, leaving the Albanian-supported Party of Democratic Prosperity (PDP) holding the balance. However, given the critical situation arising out of the Greek blockade, the three parties came together for the 1994 elections under the leadership of the Social Democrats and B. Crvenkovski headed an Alliance for Macedonia government in 1994. With the easing of tension by 1996, Crvenkovski went on to head a Social Democrat administration. The end of the Greek embargo signalled the take-off of the Macedonian economy, while the easing of sanctions has brought a boost to trade with Serbia. In 1995 Russian natural gas distribution started with the arrival of the pipeline from Bulgaria. However, ethnic tension arose in 1997 over the desire to display the Albanian flag which was legally permitted only on the occasion of national holidays. The mayors of Gostivar and Tetovo ignored this ruling and the removal of the offending flags led to violence in Gostivar and the imprisonment of the mayor R. Osmani. However, amidst the political tensions there was presidential stability through the leadership of the experienced K. Gligorov although he was fortunate to survive an assassination attempt in 1995.

Crvenkovski was defeated in 1999 by IMRO (now the Democratic Party of Macedonian National Unity) allied with the new Democratic Alternative, the

Democratic Party of Albanians and former apparatchiks marginalised by the Social Democrats. Led by L. Georgievski, the new coalition has proved cohesive in the face of IMF pressure to liquidate or privatise 12 failing companies. An amnesty has overcome the sensitive issue of Osmani's imprisonment while on the economic front an Agency for Reconstruction and Development has been set up to encourage investment and much-needed economic support has been obtained from Taiwan in exchange for recognition. Foreign investment is much needed in the tobacco business and other staple industries. Relations with neighbouring states have improved, especially Bulgaria where the basis for a friendship treaty has been laid, reflecting Georgievski's close personal links with his Bulgarian counterpart. Bulgaria's endorsement of the Macedonian language allows only that it is a technical variation of Bulgarian, in return for which Macedonia has agreed (as with Greece in 1995) that its constitution will never be used as a pretext for interference in Bulgaria on behalf of that country's Macedonian population. But Bulgaria has given Macedonia expert advice to speed up WTO membership and has helped strengthen Macedonia's armed forces that were denuded of equipment when the JNA left.

Macedonia has gained much from a change of attitude by Greece in 1999 which produced a 5.1 per cent growth rate the following year. Greek capital has been invested in the Stopanska Bank while El Pet Balkanike is developing Macedonia's oil refinery and intends to build a pipeline to Thessaloniki. A French company has taken over the Feni ferro-nickel combine. Macedonia is well placed to do business with the EU and with all its neighbours since the languages of Albania, Bulgaria and Serbia are well understood. There is a young, well-educated population to attract investors and a South Korean electronics firm has started production of telecommunications equipment at a state-owned plant at Lake Ohrid. Transport links have a high priority especially the east–west routes to link the Adriatic with the Black Sea. A rail link with Bulgaria is under way and there is great potential for tourism in Ohrid, a UNESCO-protected world heritage site. Unemployment hovers around 30 per cent and yet the country seemed to be approaching the end of a tumultuous decade of independence. There are few outstanding external problems other than the need for clarification of the borders with Albania and Serbia where small modifications are sought in Macedonia's favour. As the only Balkan state with successful multi-ethnic democracy – with an Albanian political party in every post-independence government, Macedonia seemed on the verge of a more stable future that is very much tied to Europe through a Stabilisation and Association Agreement and eventual EU/NATO membership.

However, Macedonia is exposed to mafia activity on the part of Albanians, Bulgarians and Russians in connection with the drugs trade and smuggling of all kinds with its implications for violent crime. There was a setback in 1999, affecting the industrial sector especially, when the Kosovo conflict disrupted trade through Yugoslavia and foreign help was necessary in respect of the 25,000 Albanian refugees who arrived in the country. More seriously the Kosovo crisis has been followed up in 2001 through a 'Spring offensive' by Albanians entering the country from Kosovo. This raises the issue of border security and manipulation of Macedonian Albanians by radical Kosovo Liberation Army (KLA) elements – dubbed a new European Taliban incorporating drug traffickers and other criminal elements – with longer-term 'Greater Albania' aspirations. Moderate Albanian leaders are being 'black-listed' for betraying the ideals of Albanians, although the Democratic Party of Albania gained concessions over education and civil service jobs when it became the main coalition partner in 2000. Albanians now seek constituent status in governance – but there is the threat of a Greater Albania and the danger that the country may be destabilised at any time. NATO has not been able to seal the border in Kosovo but has offered to help Macedonia if the government chooses to make political concessions outright and provide greater equality for Albanians, including 'official' status for the Albanian language in some circumstances. This seems to reward violence when – in contrast to the Serbs in Kosovo – there was no significant violation of human rights in Macedonia and the new agreement merely corrects some instances of discrimination. Certainly some elements in the new government of national unity, formed in 2001, felt that concessions were 'extorted'. While the churches had little influence, a partisan media reflected the lack of trust between an 'oppressed' Albanian minority and the 'besieged' Macedonian majority, though most think they can still live together and a token Western presence (probably involving no more than 1,000 troops) may be sufficient to keep the peace.

Yugoslavia

This final section deals with the component of the federation which has, superficially, seen the least change because it has been abandoned by the other nationalities and left as a 'rump' Yugoslavia. Yet Serbia, which is the stronger partner in the new federation, has by no means been a passive bystander. It was against a background of rising nationalism – expressed dramatically through the 1986 'Memorandum' by Serb intellectuals asserting the continued validity of the traditional goal of Greater Serbia – that S. Milošević came to power in 1987 and won great acclaim by removing the autonomous status of the Kosovo and Vojvodina provinces as the culmination of a 'reign of terror' between 1987 and 1990 to subjugate Serb moderates. The new, smaller Yugoslavia – comprising a Serbian–Montenegrin federation – was hastily established in 1992 by Serbian pressure on Montenegro when the old Yugoslavia was falling apart. Serbia was unusual among ECECs in that ultra-nationalist views have tended to take precedence over individual civic liberties and in the face of extreme nationalism the moderate/liberal opposition parties seem unable to form a stable coalition. As war enveloped other parts of the old Yugoslavia no politicians could afford to condemn the activities of Serb minorities in newly-independent states when a Greater Serbia seemed a possible outcome. With ultra-nationalist factions on the right, Milošević's new Serbian Socialist Party (developing out of the LCY) appeared to hold a populist, central position and could thereby eclipse its main rival: V. Drašković's Movement for Serbian Renewal (MSR) which is nationalistic, conservative and monarchist. The only setback arose when the 'Zajedno'

(Together) coalition scored local election victories in Belgrade and 13 other towns in 1997. The disruption of war on her borders and sanctions imposed by the international community in recognition of Belgrade's indirect involvement in much of the hostility undermined rational economic policy by the central bank based on 'ten commandments' including a balanced budget, price and wage freeze, export stimulation and control banks and financial institutions. As the economy collapsed there was a rise of illegal trading activity which helped to create a 'gangster politician' element.

However, the downfall of Milošević in 2000 did not arise through failures in B&H and Croatia. He emerged from Dayton as an astute crisis manager convincing Serbs that their country's problems were the result of an international conspiracy. A situation where RS remained separate from Serbia proper avoided the potential embarrassment of radical Bosnian Serbs voting with his ultra-nationalist opposition. Instead Serbia has been destabilised by events in the southern Kosovo province where the Albanian majority lost their autonomous status. Although Serbs were concerned at the disproportionate level of welfare spending in Kosovo on account of a very high birthrate among Albanians this did not prevent above-average infant mortality and illiteracy and should be seen against a background of misguided investment in manufacturing: notably an unsuccessful aluminium industry based on low-grade bauxite. During the 1990s the Albanians felt obliged to operate an illegal shadow administration, which endeavoured to run services and conduct a referendum seeking independence. Opinion became increasingly radicalised as leaders (like I. Rugova) seeking diplomatic solutions were sidelined in favour of military action by the KLA which developed through 1998–9 and attracted offensives by the JNA that were condemned by the international community advocating peaceful solutions by negotiation, though Serbs would see the 'efficient' clearance of Albanians as a carbon copy of the procedure adopted by Croats in Krajina in 1995. The failure of Belgrade to halt its military campaign attracted NATO bombing in the spring of 1999: the result of a US foreign policy change regarding the KLA, previously seen as a terrorist organisation linked with Albanian heroin trafficking rings. Despite the rejection of a ground offensive in favour of an air war, with bombing from the very high level of 15,000 ft to minimise casualties, NATO's first 'out of area' operation has – controversially – been deemed a success. It represents a substantially higher scale of commitment than in Bosnia & Hercegovina, following the European Security and Defence Initiative as a basis for coercive diplomacy backed by the use of force, with a UN mandate providing clear rules of engagement. However, air power limited NATO casualities at the expense of civilians who could not be protected from expulsion by the Serbs. Hence the realities of the bombing campaign undermined the humanitarian pretensions.

However, under the NATO protectorate there is a greater chance of humane governance. In contrast to the situation in B&H, the UN Interim Administration Mission in Kosovo placed itself in a strong position to rebuild civilian structures by holding executive authority, with a Transitional Administrative Council, to bring together competing power centres, including Serb representatives.

Meanwhile, the KLA has become the civilian-based Kosova Protection Corps. Yet the NATO-led Peace Enforcement Force (KFOR) has not been able to prevent the ethnic cleansing of Serbs and other minorities in eastern Kosovo, or to provide an effective local police force and operative legal system. Fortunately the population is generally law abiding and local elections in 2001 have delivered a comeback for I. Rugova's Democratic League of Kosovo – previously discredited by its disposition to negotiate with Milošević – over the political wing of the KLA under H. Thaci which is divided and tainted by criminal activities though still the most powerful force. Massive reconstruction/humanitarian assistance has been pledged under the Stability Pact, including money for 'quick start' initiatives, but it has been slow in arriving and the economic problems of the province are dire. An Albanian travel agent who initially helped fellow Albanians circumvent the Serb-controlled banking system – evading financial controls by running a lucrative but dangerous smuggling system linked with wealthy Albanian communities across Europe – is now trying to establish an official organisation, though it will take time for confidence to be established. The 'big question' over the final status of the province has not been taken, though it seems unlikely that Albanians – who have been estranged from the Serb language and culture for over a decade – will ever accept a future within Yugoslavia. Nevertheless, despite Rugova's moderation (albeit limited by the KLA which stands in the wings) Belgrade will not recognise Kosovo's de facto independence as the basis for negotiation.

The extensive bomb damage throughout Yugoslavia increased the economic privations of the population only half as rich as it had been a decade ago, with services devastated and unemployment soaring to 50 per cent. This provided a basis for a coalition representing the Democratic Opposition of Serbia (DOS) seeking reconstruction in cooperation with Europe. Milošević was unprepared for a unified opposition and the outright win scored by their candidate V. Koštunica in the Yugoslav presidential election and he was quickly forced from office by strikes and demonstrations. Despite a lack of charisma Koštunica evidently convinced the electorate that, with help largely restricted to China and Russia, there would be no economic improvement under Milošević who found himself lacking both an effective plan and army and police support as the results were declared. Immediately after his departure from office, Yugoslavia was admitted to the UN and the Stability Pact – subsequently the OSCE and IMF as well – and change was confirmed by a decisive victory by the coalition of 18 parties in the Serbian parliamentary elections later in the year. The coalition is holding and the astute premier Z. Djindjić has been able to deal with tension between his own Democratic Party and Koštunica's Democratic Party of Serbia. But the government cannot afford to disown nationalist adventures and defends Serbia's claim to sovereignty over Kosovo. There has been some economic improvement through increased rainfall since the end of 2000 which impacts positively on hydro-electricity and means shorter power cuts. But funding for reconstruction was crucial for growth in 2001 and the arrest of Milošević and his transfer to ICTY is a very significant step. It remains to be seen how Serb nationalism will be projected through the political process: can the presumption of a right to rule in former Yugoslavia – moderated by Tito but reinvigorated under Milošević – now be reconciled with European democratic norms? It will be helpful for the West to demonstrate the economic benefits of a European approach – e.g. through migration of workers to EU and easier access to EU for Serbian exporters – yet at the same time there must be an even-handed approach to all the countries in the region.

SANDZAK AND MONTENEGRO

Change in Belgrade should ease instability in the Sandzak region with its largely Muslim population. The local economy has suffered through pervasive police activity which has affected local industry (with its strong Turkish links) and the foreign exchange market. More significantly, there are powerful secessionist elements in Montenegro where, despite the imposition of federal law and the appearance of pro-Serbian nationalists in cultural institutions, the government has tried to frame its own policies. M. Bulatović and his Democratic Party of Socialists (formerly LCY) came to power in 1989 on a wave of anti-Albanian sentiment and the parliament in Podgorica subsequently decided that if the old federation was dissolved then Montenegro would also opt for independence. But in 1992 when dissolution was gaining momentum, Belgrade imposed a range of economic and political pressures to reverse the Montenegrin leadership's decision in favour of federation with Serbia, validated by a referendum that was widely boycotted. Nevertheless nationalism was fostered through restoration of the autocephalous church in 1993 (after abolition in 1921) and work began on a Montenegrin successor language to Serbo-Croat. Despite the 'Belgrade yoke' there has been more tolerance and less xenophobia in Montenegro, with a stronger opposition through the parliamentary alliance (based on shared anti-communist values) of the Liberal Alliance of Montenegro – showing strength in provincial centres like Bar, Cetinje and Nikšić, with some support from the Albanian minority – and the People's Party led by M. Djukanović who retained serious ambitions of Montenegrin independence. Montenegro took a more pragmatic stance during the civil war, for despite the territorial question of Croatia's control of the Prevlaka district which controls access to the Boka Bay naval base and the mouth of the Bay of Kotor, support for further fighting in support of Serbs fell away as a result of the casualties in the unsuccessful battle for Dubrovnik and subservience to the Serb-dominated federal army. Moreover, the economy was badly hit by sanctions as major installations like the Zeta Valley aluminium plant and the Nikšić steelworks ran way below capacity while the port of Bar was almost deserted and the tourist industry virtually eliminated.

The rift with Belgrade deepened when Djukanović came to power in 1996 and set about the promotion of Montenegro as an offshore zone to attract foreign investment. Assuming that the Serbian economy would not recover quickly, economic reform also included majority stakes for foreign investors in major companies like the Podgorica aluminium and tobacco industries and Budvanska Rivijera hotels. Although state assets did not attract good prices and rising unemployment provided ammunition for Bulatović, who occupied the presidency, enjoying much support from northern Montenegro as well as the trust of Milošević, his subsequent defeat and transfer to Belgrade has left Djukanović to continue his work for a more tangible assertion of Montenegrin autonomy, with independence the ultimate aim. But there is still a strong pro-Yugoslav element, especially among Serbs in the north who might break away if Montenegro became independent (thereby increasing the influence of the Albanian minority in the rest of the territory). And the presence of a substantial security force (especially a military police battalion known to recruit criminal elements) suggests that such an outcome – depriving Serbia of access to the sea – would be strenuously contested by armed conflict. In 1999 when Montenegro was pushed further towards secession by the Kosovo crisis, Milošević tried again to intimidate Montenegro whose government was seen as a tool of NATO 'aggressor' states, while despite a DM-based economy (to avoid the inflation of Serbia) Djukanović faced a deteriorating economic situation – including a Serbian blockade – and depended on a coalition which was divided over the timing of a referendum over Montenegrin independence. On the other hand Milošević was aware that any crackdown could bring calls for a Kosovo-style intervention and ensure Montenegro's secession. The issue remains delicately balanced. A majority in Montenegro still wishes to separate and pursue a distinct reform agenda, but the Liberals want a clean break with Serbia while Djukanović – although rejecting a confederal solution for Yugoslavia – would contemplate arrangements with Belgrade involving independent internationally-recognised states; accepting that Montenegro's population is only 650,000 and Serbia is Montenegro's largest trading partner and she desperately wants to retain access to the sea. Moreover, the West is keen to see violence contained and Montenegro is heavily dependent on foreign help. The election of 2001 returned Djukanović to power but without a clear mandate to renegotiate the federation.

The East European economy

This is not the place for an exhaustive review of the economic history of the region but the long-term function of the region as a supplier of raw materials must be emphasised, despite the success of mercantilist policies by the Habsburg and Prussian states in the late eighteenth century. The new Balkan nations set about increasing their primary exports to help create an industrial base to meet strategic requirements. Romania became one of the world's leading cereal exporters with rail as well as sea transport available. However, despite the proximity of the Danube the lack of irrigation made cereal farming vulnerable to severe summer drought. Feudal obligations were swept away but the peasants remained dependent on the large landowners who paid only low wages while imposing burdensome labour contracts. This prompted governments to combine industrialisation with a politically-motivated land reform programme to give peasants greater security and independence. Historic imbalances were also addressed by the governments of inter-war ECE and until 1945 there was some evolution along Western lines. But labour-intensive methods reflected the limited employment opportunities outside agriculture where the peasant character of farming was enhanced by share cropping arrangements to supplement the produce from the peasants' own land often arising out of land reforms implemented after the First World War. Thus the agricultural emphasis persisted amidst acute problems of rural overpopulation. A north–south gradient was very evident, for industries in the northern regions were expanding quite fast, albeit in the context of economic self-sufficiency.

It was in Stalin's interest to impose authority on the region to gain security and access to resources, including oil, timber and uranium initially. But the establishment of communist regimes also demonstrated 'the readiness of a determined minority of communists in each country to use coercion, deception and manipulation of their fellow countrymen in order to secure a monopoly of power'; taking advantage of 'the physical exhaustion and political disorientation of the war-ravaged population' (Batt 1991: 3). Elite penetration of all facets of government and international coordination made through Comecon legitimation difficult, but power was effectively consolidated and the propaganda war created an image of purposeful progress. Although his words were soon to be undermined by the flow of events, G. Schopflin (1988:147) had some justification for his assertion that the Stalinists 'actually succeeded in constructing a political system that was and remains superbly efficient at concentrating power and at ensuring that this is never seriously diluted'. It was even possible for Stalinist uniformity to be followed by growing national diversity and limited reform, provided the communist monopoly was not imperilled. Such rivalry is thought to have killed off any idea of a Balkan Federation that was discussed in the early post-war years.

Yet there was much grass-roots idealism for a new order in the region that would boost employment, accelerate the transition to an industrial economy and offer a better future for all sections of the population including the peasantry and the poorer people generally. The military–industrial establishment would provide, both nationally and internationally, a steady flow of orders, given sufficient raw materials within the bloc to maintain a high level of self-sufficiency without the need to take account of world market prices or the profits of foreign investors. A growing urban proletariat triggered a growing demand for food, and as rural infrastructure improved, factory industry spread through the central place hierarchy and took root in the more backward regions. However, objective Western appraisal of communist-inspired change was always complicated by the tension between the heroic voluntarism of traditional societies working flat-out for modernisation and the coercive state apparatus implementing a 'secret agenda' that had much more to do with the great power pretensions of the Soviet Union than the cause of prosperity for the masses. It was possible to herald a uniquely 'new' way forward as long as the environmental costs were minimal and the large population of forced-labourers could be explained away in terms of 're-education' for criminals and wartime collaborators. Some authors continued to applaud the social objectives and the spatial policies of equalisation well into the 1980s, despite the heavy social costs of the 'forced march' to modernisation through long hours of factory work and persistent shortages in housing, transport and consumer goods. And this included the reckless waste of human resource in terms of administrative/entrepreneurial skills and civic traditions and 'the destruction of minorities who had once been the spark plugs for economic progress but who are now missing' (Chirot 2000: 8).

Once installed, the system was difficult to change, given the importance of politically-reliable management and the prestige value of large enterprises. Factories were sometimes inefficiently located, incurring high transport costs, while poor organisation within the plant could result in production hold-ups, with knock-on effects in other industries dependent on monopoly suppliers of intermediates. High costs passed down the line boosted the 'value' of production on which wage increases might be based, but made exports less competitive. Meanwhile, the proletariat became dissatisfied with low incomes and reduced social mobility, while expectations of higher living standards that were satisfied by higher wages were not always supported by a commensurate improvement in productivity or the supply of consumer goods. Hungarian reforms associated with the 'New Economic Mechanism' (NEM) were always constrained by the communist monopoly on power which prevented significant decentralisation. There was also greater scope for individual initiative through the 'second economy' which was openly encouraged in some countries, like Hungary, and tolerated in others. But Gabor (1989) defined the second economy as a field of economic activity where income was gained legally or illegally outside the socialist sector. Almost by definition, the two-track strategy was nowhere adopted as official policy and the second economy was unable to grow to the point where it could act as a dynamic capital-accumulating stratum: it was not the primary source of the present economic elite recruited largely from the former socialist sector (ibid.: 9). Finally, socialist industrialisation remained wedded to the third 'Kondratieff' cycle, based on early twentieth-century engineering, until Western credits allowed the technology of the fourth cycle to be imported. But assimilation was only partial and as each national government reacted in its own way to the stagnation of the 1980s the basis for post-communist transformation was unwittingly created (Landesmann and Szekely 1996).

Transition to a market economy

The greatest failure of the communist regulation system was of course the lack of any contingency for its own demise. So hypothetical rescue strategies, such as retention of Comecon as a club for barter trade during the first phase of transition, were never tested. When the transition began it seemed that the region would quickly reintegrate with the Western world once the political constraints, emphasising self-sufficiency within the Comecon system, had been set aside. The 'radical change model' saw socialism collapsing 'as though all the blood [had] disappeared from the veins of the old body' (Andrusz *et al.* 1996: 233); any legacies 'are entirely negative since anything inherited from a system totally opposed to capitalism must be contradictory to a shift towards capitalism' (ibid.). But there was no precise modernising goal that would merit unqualified support for the notion of transition, for neither communism nor capitalism has agreed objectives that can be defined with any precision in operational terms. Following Heller (1998: 18–20), there has been an element of revolution because instead of a gradual evolution through modernisation of communist society ('system change'), there is 'system substitution' involving quite new elements and structures relevant to private enterprise and democracy instead of central planning and one-party rule. Overall, there has been movement towards a set of norms embraced by the EU which are more tangible than 'idealised generalisations' (Smith and Swain 1997: 32). But we have a multiplicity of possible development paths and regulation systems including a 'moderate change model' that would associate reforms under some communist regimes with the structures of capitalist and democratic societies. These innovations anticipated to an extent the 'coping mechanisms' developed to overcome the rigidities of central planning and were not far removed from entrepreneurship (Andrusz *et al.* 1996: 233). So system substitution has unleashed a complex range of adjustments at local, regional and national levels: all in various ways embedded in past experiences, as demonstrated by the different approaches to privatisation and land restitution.

Moreover, all commentators were unprepared for the complications that followed and the difficulties that attended a radical reorientation of trade, which sees Germany in a commanding position while major ties with Russia feature hydrocarbon imports (Table 1.5). Theoreticians denied any stereotype for a communist or a capitalist system (a 'pure' isolationist socialism was never achieved and there are many styles of market capitalism) and argued that pathways rooted in the specific networks and values of the past would indelibly shape the new order arising in different parts of the region, all in various ways embedded in past experiences, as demonstrated by the different approaches to privatisation and land restitution. Hybrid organisational forms involve 'recombinant' property relations blending state and private elements to spread the risks inherent in the transformation. Locally embedded networks and institutions appear to be guiding the transformation of the second economy into a self-sustaining sector able to foster regional economic growth (though some economic action remains locked into the rationality of the former system through the perpetuation of defensive overembedded networks). Thus, whole countries and individual regions are spontaneously adopting a wide variety of regulatory systems detracting from any stereotyped notion of transition. But equally there is a direction to change and arguably there is a transition (from 'real' socialism to European liberalism and a Western-style market economy) with paths of extrication which involve restitution/privatisation and the building of civil society and supporting institutions despite the somewhat incoherent transformations occurring under its umbrella.

There was no doubt about the direction of change, as major trade shocks concerned with German unification, war in Yugoslavia and the collapse of the Soviet Union accentuated the collapse of economic growth from which the region only recovered during the middle and later years of the 1990s (Table 1.6). Indeed in 1998 only Poland and Slovenia had moved ahead of the 1989 level (with 117 per cent and 104 per cent respectively) with Slovakia on 100,

Table 1.5 Major trading partners 1994–8

Country	Countries accounting for 10% or more of imports and exports	
	Imports	*Exports*
Albania	Ity 41.9; Ger 25.9	Ity 53.2; Gce 14.7
Bulgaria	Rus 27.2; Ger 12.4	Ity 10.0
Croatia	Ger 20.3; Ity 18.4	Ity 20.5; Ger 19.4; Sia 12.3; B&H 11.7
Czech Rep.	Ger 31.6; Ska 10.1	Ger 36.5; Ska 13.2
Hungary	Ger 25.8; Aus 10.7; Rus 10.1	Ger 32.9; Aus 10.8
Macedonia	Ger 15.5; Blg 10.0e	Ger 13.8; Yug 11.3; Blg 10.0e
Poland	Ger 25.9	Ger 35.5
Romania	Ger 17.4; Ity 14.7; Rus 11.9	Ger 17.8; Ity 17.4
Slovakia	Czh 24.3; Ger 17.5; Rus 15.3	Czh 29.7; Ger 21.8
Slovenia	Ger 22.0; Ity 16.9	Ger 29.8; Ity 13.9; Cro 10.1

Source: Eurostat 2000: 200–5

Notes
e Estimate

Aus	Austria	Czh	Czech Rep.	Rus	Russia
B&H	Bosnia & Hercegovina	Gce	Greece	Sia	Slovenia
Blg	Bulgaria	Ger	Germany	Ska	Slovakia
Cro	Croatia	Ity	Italy	Yug	Yugoslavia

the Czech Republic and Hungary on 95, followed by Croatia (78), Romania (76), Macedonia (69) and Bulgaria (69). And in 2009 it is calculated that Bulgaria will have only reached 100 per cent of 1989, with Romania at 108, Croatia 135, the Czech Republic 135, Slovakia 144, Hungary 150 and Poland 190. This reflects the EIU forecast of 4.6 per cent annual growth for Poland during 2000–9 compared with 4.3 for Hungary and between 3.0 and 4.0 for the other countries. There has been much disagreement over the speed of reform and the role of government. Policy has been framed with each national situation in mind. The IMF (which has come to play a crucial role in stabilisation programmes) takes a long-term strategic view, while governments must be aware of the short-term implications of change and doubt the capacity of private business to make up for jobs lost in the public sector. Again, the success of the neo-liberal strategy depends on the growth of native enterprise and foreign investment. In 1991–2 total FDI in the region amounted to only $6.13 bn. compared with 15.51 for China and 24.29 for Malaysia, Singapore, South Korea and Thailand combined.

At the same time, the investment in ECECs has been highly uneven with 88.1 per cent to the emerging Czech Republic, Hungary and Poland which accounted for only 47.7 per cent of the population. These countries received 1.8 times their population share and the others only 0.2. In 2001 these three states accounted for a total FDI stock of $74 bn. ($30.4 for 2000–1 alone, indicating recovery after the setback in Russia in 1998) compared with $21.8 bn. ($9.0–10.0) for the nine other ECECs (Table 1.7). Investment opportunities in the energy sector should help maintain the inflow of funds while early EU membership for the northern states will confirm their advantages. The disparities are serious when it is considered that

Table 1.6 Annual percentage change in gross domestic product 1990–2002

	1990	1991	1992	1993	1994	1995	1996	1997	1998	1999	2000	2001e	2002e
Albania	-10.0	-27.1	-7.2	9.5	9.4	8.9	9.1	-7.0	8.0	7.3	7.8	7.0	7.0
Bosnia & H.	na	na	na	na	na	33.0	28.0	15.0	20.0	10.0	5.0	7.0	6.5
Bulgaria	-9.3	-11.7	-7.3	-1.5	1.8	2.1	-10.9	-6.9	2.7	2.4	5.8	4.4	4.5
Croatia	-8.5	-20.9	-11.1	-8.0	5.9	6.8	6.0	6.5	2.5	-0.4	3.7	3.2	3.0
Czech Rep.	-1.2	-14.2	-6.4	0.5	3.4	6.4	3.9	1.0	-2.7	-0.8	3.1	3.1	3.3
Hungary	-3.5	-11.9	-3.1	-0.6	2.9	1.5	1.3	4.6	5.0	4.2	5.2	4.3	4.5
Macedonia	-9.9	-10.7	-7.9	-9.1	-1.8	-1.2	0.8	1.5	4.0	2.7	5.1	0.0	1.5
Poland	-12.1	-7.0	2.6	3.8	5.3	7.0	6.1	6.9	5.5	4.1	4.0	3.5	3.5
Romania	-5.6	-12.9	-8.8	1.5	3.9	7.1	4.1	-6.6	-7.3	-3.2	1.6	3.9	3.7
Slovakia	-2.5	-14.6	-5.4	-3.4	4.9	6.9	6.6	6.5	4.8	1.9	2.2	2.7	3.6
Slovenia	-4.7	-8.1	-5.5	2.8	5.3	4.1	3.5	4.6	3.9	5.2	4.6	3.7	4.0
Yugoslavia	-8.4	14.2	-26.2	-30.8	2.5	6.1	3.5	7.4	2.6	-21.9	7.0	4.0	4.0

Source: Economist Intelligence Unit

Note
e Estimate

Table 1.7 Foreign direct investment 1991–3 to 2000–2

Total investment over the period ($m.) (A) and annual investment ($ per capita) (B). Also total capital stock ($bn.) 2000 (C) and $ per capita (D)

	1991–3		1994–6		1997–9		2000–2e		2000	
	A	B	A	B	A	B	A	B	C	D
Albania	85	8.8	251	25.9	130	13.3	na	na	0.55	172
Bosnia and Herceg.	na	na	na	na	664	52.7	na	na	0.26	60
Bulgaria	138	5.4	287	11.4	1598	64.2	2000	122.0	3.00	366
Croatia	90[1]	9.9	707	51.2	1950	142.9	1950	203.1	4.10	854
Czech Republic	1600[1]	51.8	4600	148.9	7300	236.2	9000	436.9	23.00	2233
Hungary	5300	171.5	7600	248.4	4800	158.4	2900	142.2	21.00	2059
Macedonia	na	na	49	7.9	223	37.2	na	na	0.42	210
Poland	1000	8.7	4400	38.0	16100	138.7	18500	239.6	30.00	777
Romania	207	3.0	1021	15.1	4609	68.5	1000	22.2	6.00	267
Slovakia	350	22.0	703	43.6	1185	73.1	3500	324.1	3.50	648
Slovenia	265	44.2	479	79.8	449[2]	112.2	300	75.0	2.70	1350
Yugoslavia	na	na	na	na	na	na	1400	66.0	1.40	132
Total	9035	31.2	20097	62.4	39008	117.6	40550	180.1	95.93	779

Sources: EBRD 1999 (1991–9) and Economist Intelligence Unit (2000–2)

Notes
[1] 1991 not available [2] 1999 not available
e Estimate

FDI boosts exports (for companies with an element of foreign capital have increased their participation in Polish exports from 24 per cent in 1994 to 47 per cent in 1998 – with an even stronger showing claimed in exports of manufactured goods) as well as productive capacity and enterprise restructuring. This is most important given that all the ECECs have maintained a significant trade imbalance during the 1990s (Table 1.8). FDI has a positive impact on growth in the Czech Republic as measured by 'total factor productivity' or TFP (an indirect measure of technology transfer), such has been the transfer of technology and knowledge by foreign firms; much greater than is the case with joint ventures.

Hence neo-liberal strategies have been combined increasingly since *c*.1993 by 'neo-statist' approaches with change being led and controlled by the centre: either through the 'transformative model' of change formulated by technocrats being imposed by decree or a 'developmental' model with systematic intervention to help selected firms solve their problems with quality and productivity. In a sense the neo-statist approach is a pragmatic response to changing circumstances, including the growth of state budgets after several years of transition. But there is a change of emphasis philosophically now that a return to communism is impossible. Not that this makes for easy options. It is arguable that the region should follow Ireland's example with low corporate taxes – despite the pressure of debt repayments and the burden of education/health costs – on the grounds that spending could be reduced on large armies and support for non-viable banks and heavy industries. However, higher unemployment could be destabilising, given low welfare payments and a lack of social consensus on what constitutes a modern European state. Markets are not functioning well enough, with a low capacity for governance and cooperation and revenue depressed by criminalisation. Elements of conservatism in the SOEs and cooperatives may mean that change can be best accomplished in domains where socialist states performed badly.

Reform can in theory be postponed if the SOEs continue to serve as an elite vested interest with money from the export of primary produce to finance subsidies for loss-making firms. States with lucrative mineral deposits could conceivably earn enough from exports to support inefficient manufacturing industries based on the command system. But this strategy will impose continuing burdens and will be a source of discouragement for foreign investment outside the resource industries. None of the ECECs have the natural wealth to sustain such a conservative policy, so governments must intervene to stabilise the currency, control inflation and encourage investment. There were massive problems in the early years through inflation and monumental currency depreciation (Tables 1.9 and 1.10). The early stages of transition can be accomplished fairly quickly through 'shock therapy' involving macroeconomic stabilisation along with price and trade liberalisation and wage policy in the state sector to control inflation. Government must also discipline the state sector (by limiting borrowing and other forms of subsidy) and should balance the interests of unions and employers' organisations. Indeed, hasty elimination of the old institutions may cause excessive dislocation and prove counterproductive if the removal of controls allows the SOEs to operate unfairly and for private business to escape regulation. It must increase experimentation and train natural agents of capitalist development.

Table 1.8 Trade 1989–2002 ($ bn.)

Country	1989 Exp	1989 Imp	1990 Exp	1990 Imp	1991 Exp	1991 Imp	1992 Exp	1992 Imp	1993 Exp	1993 Imp	1994 Exp	1994 Imp	1995 Exp	1995 Imp	1996 Exp	1996 Imp	1997 Exp	1997 Imp	1998 Exp	1998 Imp	1999 Exp	1999 Imp	2000 Exp	2000 Imp	2001e Exp	2001e Imp	2002e Exp	2002e Imp
Albania[1]	133	224	155	262	82	3145	70	540	112	602	141	601	205	680	244	921	159	694	206	794	275	938	268	1083	325	1190	325	1225
Bosnia & H.	na	na	na	na	na	na	na	na	na	na	na	na	0.2	1.1	0.3	1.9	0.4	2.9	0.6	2.7	0.6	2.7	0.7	2.8	1.0	2.9	1.2	3.0
Bulgaria	3.1	4.3	2.5	3.3	3.7	3.8	4.0	4.2	3.7	4.6	3.9	4.0	5.3	5.2	4.9	3.7	4.9	4.5	4.2	4.4	4.0	5.1	4.8	6.0	5.4	7.2	6.2	8.0
Croatia	2.8	3.5	4.0	5.2	3.3	3.8	4.6	4.5	3.9	4.7	4.3	5.2	4.6	7.5	4.5	7.8	4.2	9.4	4.6	8.7	4.4	7.7	4.6	7.8	4.5	8.3	5.2	9.4
Czech Rep.	5.4	5.0	5.9	6.5	8.3	8.8	11.5	13.3	14.2	14.7	16.0	17.4	21.5	25.2	21.7	27.6	22.3	26.3	26.4	28.9	26.3	28.2	29.0	32.3	31.7	34.9	36.4	39.9
Hungary	6.4	5.9	6.3	6.0	9.3	9.1	10.1	10.1	8.1	12.1	7.6	11.4	12.9	15.3	14.2	16.8	19.6	21.4	20.7	22.9	21.8	24.0	25.3	27.4	28.6	31.2	33.2	36.4
Macedonia	na	na	1.1	1.5	1.1	1.4	1.2	1.2	1.1	1.0	1.1	1.3	1.2	1.4	1.1	1.5	1.2	1.6	1.3	1.7	1.2	1.6	1.3	1.9	1.2	1.7	1.2	1.8
Poland	8.3	8.4	11.3	9.9	13.8	14.6	13.9	14.1	13.6	17.1	17.1	18.9	23.5	26.7	24.4	32.6	27.2	38.5	30.1	43.8	26.3	40.7	28.3	41.4	31.7	44.0	37.9	51.2
Romania	10.5	8.4	5.9	9.1	4.3	5.4	4.4	5.8	4.9	6.0	6.2	6.6	7.9	9.5	8.1	10.6	8.4	10.4	8.3	10.9	8.5	9.6	10.4	12.0	11.4	13.5	13.3	15.8
Slovakia	na	na	5.9	6.5	4.7	5.5	6.3	6.8	5.4	6.4	6.7	6.6	8.6	8.8	8.8	10.9	8.8	10.3	10.1	11.9	10.2	11.3	11.8	12.7	13.2	14.4	14.9	16.3
Slovenia	3.4	3.2	4.1	4.7	3.9	4.1	6.7	5.9	6.1	6.5	6.8	7.3	8.3	9.5	8.3	9.4	8.4	9.4	9.1	9.9	8.6	9.9	8.8	9.9	9.2	10.1	10.7	11.8
Yugoslavia	na	na	5.6	6.8	4.7	5.5	2.5	3.9	2.9	3.0	1.5	1.9	1.4	2.4	1.8	4.1	2.4	4.8	2.9	4.9	1.7	3.0	1.9	3.3	2.0	3.9	2.1	4.1

Source: Economist Intelligence Unit

Notes
1 $m.
e Estimate

Table 1.9 Consumer retail price inflation 1990–2002: percentage change per annum

	1990	1991	1992	1993	1994	1995	1996	1997	1998	1999	2000	2001	2002
Albania	0.0	36.0	226.0	85.0	22.6	7.7	12.8	33.2	21.9	0.4	0.0	3.5	4.0
B&H–FB&H	na	na	na	na	780.3	4.0	−20.1	10.8	5.1	−0.9	1.2	2.0	3.0
B&H–RS	na	na	na	na	1061.0	118.1	17.2	−6.8	2.0	15.1	13.6	7.0	5.0
Bulgaria	21.6	333.5	82.0	72.9	96.2	62.0	123.1	1082.3	22.3	2.6	10.3	5.8	4.3
Croatia	609.0	123.0	665.0	1518.0	98.0	2.0	3.5	3.6	5.7	4.2	6.2	5.3	5.0
Czech Rep.	10.8	56.7	11.1	7.0	10.1	9.1	8.8	8.4	10.7	2.1	3.9	4.4	3.8
Hungary	28.9	35.0	23.0	22.5	18.8	28.2	23.6	18.3	14.3	10.0	9.8	9.8	8.0
Macedonia	608.0	115.0	1691.0	349.8	121.8	15.9	3.0	3.6	1.1	−1.1	10.6	6.5	6.5
Poland	585.8	76.7	43.0	36.9	33.3	26.8	20.1	15.9	11.7	7.3	10.1	6.3	6.0
Romania	50.1	174.5	210.9	256.1	136.8	32.3	38.8	154.8	59.0	45.9	45.6	34.0	23.0
Slovakia	10.4	58.3	10.0	23.2	13.4	9.9	5.8	6.1	7.0	10.5	12.0	7.5	6.7
Slovenia	548.7	117.7	201.3	32.9	21.0	13.4	9.9	8.4	8.0	6.1	8.9	8.8	6.3
Yugoslavia	593.0	121.0	9237.0	116.0	72.0	74.1	93.1	18.5	na	42.4	75.7	89.0	30.0

Source: Economist Intelligence Unit

Table 1.10 Currency exchange against the dollar 1990–2002

	1990	1991	1992	1993	1994	1995	1996	1997	1998	1999	2000	2001e	2002e
Albania (Lek)	5.6	14.6	75.0	102.1	94.6	92.7	104.5	148.9	150.6	137.7	143.7	148.0	154.0
B&H (Marka)[1]	na	na	na	na	na	100	100	100	na	1.83	2.12	2.23	2.03
Bulgaria (Lev)	2.4	18.4	23.3	27.6	54.2	67.1	177.9	1682	1776	1836	2123	na	na
Croatia (Kunar)[2]	11.2	19.7	258.0	3577	6.0	5.2	5.4	6.2	6.4	7.1	8.3	8.4	7.8
Czech Rep. (CzCrown)	17.9+	29.5+	28.3+	29.1	28.8	26.5	27.0	31.7	32.3	34.6	38.6	37.5	34.6
Hungary (Forint)	63.2	75.7	105.2	91.9	105.2	125.7	152.6	186.8	214.4	237.1	282.2	275.8	249.6
Macedonia (Denar)	na	0.2	43.2	23.6	43.2	38.0	40.0	50.0	54.5	56.9	65.9	69.6	69.1
Poland (Zloty)[3]	9500	10576	13626	18115	2.27	2.43	2.70	3.56	3.48	3.97	4.35	4.25	4.10
Romania (Leu)	22.4	76.4	307.9	760.1	1655	2033	3085	7168	8900	15333	21709	28810	33390
Slovakia (SkCrown)	17.9[4]	29.5[4]	28.3[4]	30.8	32.0	29.8	30.7	33.6	35.3	41.4	46.2	46.0	45.0
Slovenia (Tolar)	na	27.6	81.3	113.2	128.8	118.5	135.4	159.7	166.1	181.8	222.7	242.7	230.4
Yugoslavia (Dinar)	na	na	na	1053	1.6	4.7	5.1	5.9	10.0	11.0	37.7	67.0	69.0

Source: Economist Intelligence Unit

Notes
[1] Parity with DM
[2] Kunar replaced the Croatian Dinar in 1994 at a rate of 1:1000
[3] New Zloty introduced in 1995 at a rate of 1:1000 Old Zloty
[4] Situation in Former Czechoslovakia
e Estimate

Institutions for a market economy

For a market economy to operate efficiently a complex array of institutions are needed to encourage enterprise and investment while controlling abuses. The state must of course be proactive, yet too much effort to accelerate the transition can be counterproductive. If new institutions (such as an amended constitution, commercial codes, civil law procedures and banking regulations) are not in place very quickly to control large SOEs it may be necessary to introduce stringent monetary policy with restricted access to foreign currency that will impact heavily on the private sector. In other words, there is a danger of conflict between liberalisation of the economy and liberalisation of the state sector because SOEs may continue to attract disproportionate funding on the basis of political contacts with banks and parliamentary representatives. In the case of cultural institutions like universities, the collapse of party supervision without alternative arrangements can leave the management free to assert their own preferences (favouring particular subjects in the case of university rectors). So ideally there needs to be a trade-off between reform of the old state sector institutions and the creation of others relevant to the new private sector; yet there has often been a significant time gap between collapse and rebuilding.

On the domestic front, it is important to create an environment in which both large and small businesses can operate, because there is a danger that large firms in retailing, wholesale, distribution and transport may operate unfairly and stifle competition in what are the most promising sectors for small private enterprises. Large firms stifle competition and make it difficult for small enterprises to gain access to the banking system. Tight monetary policy will tend to favour a bank's largest customers and will delay repayment of trade credit advanced by small firms to large buyers. 'If there is a widespread belief that the state stands behind state firms and will honour their debts, then any state firm is in the position of being able to create credit' (Stiglitz 1992: 171). The difficult question is how to harden the budget constraint when the state has little choice but reconstruct on the basis of SOEs, given the time needed to create an indigenous capitalist class and a range of 'intermediary institutions' and the reluctance of foreign companies to invest. However, large state enterprises need to be broken down into smaller units and privatised wherever feasible. This will not be achieved quickly – despite the favours shown towards nomenklatura networks and the post-1989 political leadership – given the failure rates among small privatised businesses, leading to renationalisation through takeover by financial institutions holding deeds as collateral against loans. Thus privatisation does not simply connect old and new pathways but may describe loops back into state ownership (Williams and Balaz 1999: 731).

Privatisation reduces the state's involvement in economic management and increases the influence of people in close touch with the market: capital can be borrowed according to economic criteria (which helps to build a capital market and encourages personal saving); while management can organise production according to efficiency criteria (market discipline will then rub off on the remaining state enterprises). As ownership is extended enterprises become more accountable. Privatisation may also have political benefits in undercutting the entrenched positions of interest groups like the trade unions and the 'nomenklatura', thereby buttressing democracy. But foreign capitalists may not be interested in many of the assets and governments may in any case wish to place limits on their involvement. Considerable use has therefore been made of 'voucher' systems whereby a proportion of the shares in former SOEs are virtually given away to the citizens who can buy coupons or vouchers at a symbolic price and use them to obtain the shares in enterprises of their choice. However, governments are rarely rewarded electorally for their sponsorship of privatisation since most people do not warm to the principle of shareholding while the process often leads to higher unemployment and to corruption when well-connected individuals receive preference or gain economic assets in return for political favours.

At the same time new 'small- and medium-sized enterprises' (SMEs) must be encouraged, as was the case in some communist states in the early 1980s as part of a growing 'second economy'. Small businesses are prominent in the craft sector, as well as tourism and transport. They tend to suffer because of high interest rates and inadequate training for owners and managers and hence much new business is linked with people who gained middle-management experience under communism or accumulated wealth through the second economy protected by links with the communist elite or 'nomenklatura' (Harloe 1996). But small businesses are good for job creation, innovation and technological diffusion, filling market gaps and niches, while they provide a boost for tax revenue provided they do not disappear underground into the 'hidden', informal sector. They are also helpful in terms of decentralisation because they can operate in rural areas much more easily than large enterprises (though many small units spring up around the larger enterprises on the basis of sub-contracting). The encouragement of SMEs is therefore an important task of government and one that has attracted significant foreign funding and expertise. Innovation centres allow businesses to learn from each other and thanks to concessions by local authorities and large institutions, premises may be made available at low cost, while linkage with universities and polytechnics can add the benefit of technology transfer. In Romania the National Privatisation Agency introduced the concept of business and technology incubation centres in 1991 with the support of the Ministry of Research and Technology.

But whatever happens to industrial production a major structural shift in favour of the tertiary sector is already a reality. Services were typically under-developed given the endemic shortages in the planned economic systems of the region. There were also anomalies because although the cities were the main service centres the relationship between urban hierarchy and tertiary functions became increasingly irregular because of the priority for industrial developments. It is clear that job-creation during the transition will depend heavily on an expanding service sector. According to Bicanic and Skreb (1991: 228) 'a consensus has emerged that growth rates and the share of services are closely related' and while the nature of the linkage is not absolutely clear 'there is no doubt that inadequately developed services are a bottleneck for growth'. The market itself will dictate the extent of growth in consumer services (often provided by single-person operations with only limited capital requirements) which are no longer inhibited by administrative constraints and ideological stigma. In addition services like health and education, which already operate at quite a high capacity, need reorganisation and improved efficiency to provide a more even quality of service rather than major expansion. However, producer services, discussed in connection with transport and telecommunications, are less easy to stimulate though they are the critical engines of growth. They include information processing and both financial and legal services where the lack of an enterprise culture and appropriate management skills creates a large potential for Western business advice. 'For without a highly-skilled legal profession it is impossible to develop a contractual economy, without financial services it is impossible to operate capital markets or modern banking [and] without information processing it is impossible to organise a modern economy or participate in the world economy' (Bicanic and Skreb 1991: 229).

Regional variations

Each locality is itself uniquely embedded in the old socialist regional geography and must exploit its own perceived advantages in evolving a distinct pathway for future progress. But the potentials seem to vary between north and south. Ekiert (1998) paints a picture of a 'virtuous circle' which links economic restructuring with progress by civil society in control. Successful countries tend to be those making an early start to reform and liberalisation, with comprehensive macroeconomic stabilisation, a high level of privatisation and FDI, and tend to be those with the most secure and effective democratic systems (with former communists losing power in the first round of democratic elections) and with relatively close relationships with Western democracies, international organisations and the global economy in the past. By boosting their investment attractiveness, some countries are being 'disembedded' as they break out of the confines of the old behavioural environment by forging strong external links of a business or cultural nature. This is happening in the northern countries which are closer to Western Europe, with greater political stability and better prospects for consumer spending. By contrast, vicious circles prevail where there is a lack of foreign support and reforms affecting both the economy and civil society fail to gain momentum. Communist governments that resisted reform effectively 'destroyed the prerequisites of a creative adjustment' in less-developed countries where 'Annus Mirabilis' has been followed by 'Anni Miserabiles' (Berend 1996: 341).

These contrasting scenarios, which broadly fit the situation in the northern and southeastern countries respectively, tend to be self-reinforcing since reform is difficult to accomplish without the encouragement of foreign investment to provide cash and new jobs while for the successful economies investment provides the confidence for more wide-reaching reforms. Berend (1997: 27) argues that the countries closest to the EU might have a historic opportunity to break out of the vicious circle of peripheral status 'leaving [a] relatively declining humiliated and chaotic east and southeast Europe with its explosive ethnic and minority conflicts as well as an emerging nationalist extremism'; raising the possibility of a 'cordon sanitaire' between the new eastern border of the EU and crisis-ridden areas beyond where political instability can become a self-fulfilling prophecy. Hudson *et al.* (1997) think that progress in building institutions and inducing synergy between them is possible: 'global outposts' can be overcome by inward investments that create more 'embedded' branch plant investment, involving higher value added activities and greater linkages with the regional economy. But Ferge (1997: 107) argues that the collapse of state socialism came too late because 'at its advent the welfare consensus was over and the ideas of the "New Right" were gaining ground and respectability everywhere'; the situation for ECECs was made worse by the collapse of the Soviet Union and greater competition for investment (Kowalik 1994: 123). The poorer countries have no consumer boom to stimulate growth and must rely on exports from efficient and innovative companies. Hence the slow reformers have been disadvantaged despite the short-term employment benefits of supporting SOEs and the attempts of some economists to make a paradigm out of 'gradualism'. Governments may resort to crude nationalism and financial gain through connection with illegal trading activities and pyramid investments. Most of the slower reformers have been driven into action, usually as a result of electoral change, as in the case of Montenegro which sought the status of a low-cost manufacturing base for Italy.

The nationalists have found that there are no real alternatives. The authoritarian Mečiar government in Slovakia did not create serious complications for FDI which benefited from a high level of consumer spending, but poor external relations (complicated relations with Hungary and the EU) and corruption linked with privatisation created concern over the handover of power. The Slovak currency was also on the brink of devaluation in 1998, with rising interest rates, falling growth and rising inflation (partly due to fiscal laxity). Low levels of FDI can propel governments into high taxation: a top rate of 60 per cent in Romania (reduced to 45 per cent in 1998), with rising VAT (up from 18 to 22 per cent in 1998), placing a strain on poorer families. Governments may have to engage with the informal sector, including nominally illegal activities concerned with drug and tobacco smuggling and pyramid investments (in Albania it was calculated in 1998 that the five biggest pyramid firms controlled assets of $50 m. while owing $348 m. (not counting interest)). Mafia-ridden societies thrive in the Balkans where the bomb outrages of Shkoder and Tirana in early 1998 are an indicator of stress. Profits from drugs helped finance Albanian resistance in Kosovo while sanctions busting provided opportunities in all countries bordering on Serbia, though such illegality can create longer-term problems in getting business back 'above ground' and thereby building tax revenues.

Towards a critique of transition

The years since 1989 have seen a radical realignment of forces as the ECECs have aligned themselves with the West in the hope of gaining security and prosperity, at least over the longer term. Social scientists are however inclined to resist 'triumphalism' in preference for a critique of Western actions, given that

the West, under American leadership, was hardly neutral in its attitude towards
the Soviet position in the ECECs and was morally obliged to provide substantial assistance in 1989. At the very least there is arguably a new world order,
with the USA as the ultimate power above the regional alliances. But there is
still a need for a critique of global capitalism and its regional components, especially in transition countries which are arguably caught not without an ideology
– given the acceptance of globalisation and Europeanisation (allowing for distortions due to ethnic politics in the Balkans) – but without a blueprint for
implementation.

The West can hardly be blamed for the planned economies' failure to plan
their own demise and the resulting transition programme. The system collapsed
when the Soviet guarantee was withdrawn – given the pent-up urge for reform,
even by those who would not have opted for complete dismantling. Thus the decay
of the communist regulation system was largely spontaneous, though it was
undoubtedly assisted by the more radical reformers to ensure that there would be
no going back. Some critiques of neo-liberalism imply a 'wilful' and 'perverse'
rundown in the capacity of the state to regulate. Yet there was no consensus and
a vacuum seems to have been inevitable except where elements of the old system
were de facto preserved through gradualism. A new regulation system is needed
and this has been built on the basis of civil society. Here again, it is hardly the
fault of the West (which campaigned for human rights) that communism's destruction of the elites in the 1950s meant there was no 'lasting traditional order'
(Szakolczai 1996: 7) to ensure a consensus for orderly and purposeful change.
However, there has been an element of naivety, for while viable West German
institutions were introduced across East Germany, rising real wages unrelated to
productivity gains meant that the East's competitiveness declined sharply and it
required enormous welfare payments to maintain stability.

There has arguably been a conflict between helping ECE and expediting penetration by Western business interests and rather too uncritical a recommenda-
tion of shock therapy. While 'gradualism' can be a recipe for doing nothing, the
'machine of transition' (Smith and Pickles 1998: 4) is a highly complex mixture
of elements which cannot all be legislated for simultaneously and cannot be
achieved within a single time frame. Integration will not occur just when the institutions are in place. 'The provision of credit and innovation systems takes much
more time than is involved in the transfer of political and legal structures'
(Dunford 1998: 105). Transition is proving to be far more complicated and traumatic than anybody expected and while ECE has broadly regained 1989 levels
of output, the distribution of income has radically changed to produce 'winners'
and 'losers', while in the regional context pockets of development contrast with
areas of endemic poverty. Not surprisingly the sustainability of transition comes
into question, although another round of authoritarian redirection of capitalist
forces, at the expense of individual freedom, is probably unlikely while the international community can prevent war and stabilise national crises.

Part of the criticism also arises through rejection of a stereotyped transition
scenario. It is disingenuous to question ultimate destinations on the grounds that
neither communism nor capitalism are truly definitive conditions between which

there can be transition, since most people are simply attracted by Western institutions and wish to move closer to the EU. The power of the local which produces alternative strategies in the West is also prominent in the East given local resources and forms of governance embedded within wider national regimes. And while the local was largely invisible under communism it is now projected by the transparency of civil society: indeed the local was particularly strong – and unpredictable – in the early transition phase, given the temporary weakness of the centre combined with the withering of local communist networks. Thus local networks have their own ways of coping with stress with ethnic – or gender-based – groups to handle community welfare or organisation; small-firm cooperation in the face of volatile markets (Kuczi and Mako 1996); or strategic alliances between ailing state-owned industries and politically-influential local governments to press for state subsidies and protected markets (Pickles 1998: 190). 'Globalisation does not displace the properties of localities but makes them all the more salient' (Grabher and Stark 1998: 67).

2 Aspects of social geography

This chapter considers a range of social issues focusing on civil society and the more decentralised government which has replaced the authoritarianism of the communist years. A somewhat triumphalist note has been sounded by Fukuyama (1992), arguing that the end point of ideological evolution has been reached through the universalisation of Western liberal democracy. However, as society has struggled with the dilemmas of decommunisation, it is evident that full use is not being made of the region's substantial human resources and there is some nostalgia for the old system of full employment. Given the grave economic crisis, it is proving difficult to find broad agreement on the way forward and in countries with high unemployment nationalism may be essential to provide a sense of unity based on cautious reform to give some encouragement to potential foreign investors while avoiding the high unemployment associated with rapid restructuring. Yet a nationalist agenda is also dangerous on account of the ethnic tensions that are fermented when the leading nations discriminate against minority groups. The range of experiences encountered by evolving civil societies provide the context for a review of party politics. A further dimension concerns demography, with its implications for the labour market as well as governmental welfare obligations towards the young and old. Migration is also considered as a particularly sensitive indicator of economic and social wellbeing.

The communist era saw planning impinge on demography in order to control the labour market, while cultural policies emphasised homogeneity. But after the terror of the Stalin years, regimes tried to cultivate elements thought likely to render genuine support and priority was given to ensuring a degree of upward mobility, especially in the case of elites in large towns. Hence the isolation of social and economic actors under communism through the payment of salaries and welfare benefits largely unrelated to individual economic performance. This delayed the cultivation of attitudes associated with the normal functioning of a market economy and the formation of a consensus over 'expectations of behaviour' in terms of efficiency and productivity. There was some questioning of ideological principles although any trend towards pluralism before 1989 was seldom recognised because of Western infatuation with the totalitarian model and the homogenous nature of communist politics. However, Moran (1994) suggests that an explosive situation built up in the northern countries where there was no scope for either dissent or emigration. There was some economic change with the growth of a relatively informal second economy sometimes situated on the margins of legality, but reform was very limited and the system remained highly damaging in spiritual terms through regulations that had to be broken in order to survive.

New civil society

The end of authoritarian rule, with its aim of homogeneity and regional balance, gave scope for open debate about the nature of civil society and personal goals in the new situation. Privatisation and economic restructuring can be accomplished by authoritarian governments, but it is a tenet of philosophy in the Western world that a democratic regime is the best guarantor of a market economy – through the logic of free enterprise and the promise of opportunity without oppression – and the EU is taking advantage of its influence in the region to make democratic government and civil society a foundation for accession. The private domain did expand in the last years of communism through the second economy, but this was seen as a peripheral intellectual and spiritual domain entirely removed from an official system rooted in theft and other forms of illegality. By contrast, a partnership approach is needed today and the private domain thus becomes a critical sphere of activity with its own spontaneous and unpredictable development process. Communism also gave some scope for opposition since ecological protest became the most important form of expression of disagreement with the communist system. Such was the appeal of environmental issues as symbolic foci for wider anti-Soviet sentiment that Dawson (1997) has referred to 'movement surrogacy' where radicals could hide temporarily behind surrogate causes in order to target similar audiences. But there were limits to this outpouring of frustration until the breakthrough in 1989.

Transition dynamics

Communist party monopolies have now been overcome and as new social systems emerge there is scope for free civil society to exist apart from the state; owning property and taking initiatives in the transition to a democratic multi-party system and a market economy. Over the past decade there has been a codification of individual rights and press freedoms; renewal of the legal system and other state institutions in line with public opinion; and provision of standards of accountability for representatives in local and national government. Post-communist civil society is dynamic and it is evolving on a playing field that is based on family values and is essentially European. The search for a new form of citizenship arguably gives some sense of purpose to the ongoing trauma of the transition, with each civil society adjusting to European democratic values as the basis for legitimation and the 'socialisation' of new elites. Political pluralism is well established with so many parties that 5 per cent quotas have been widely imposed to restrict parliamentary representation to the stronger groups and increase the chances of single-party governments or stable coalitions. With environment as a central element in policy making, Tickle and Welsh (1998: 163) see the 'environment–society–economy nexus' as a rich source of empirical material relating the global with the local, with society taking its own mature decisions on environmental management, albeit in the context of EU norms and the support of international ENGOs. There has also been much change in the workplace with more efficient use of labour and wages more closely linked with productivity than before. Modern workplace culture involves job descriptions, incentives (such as share options), performance evaluation, representation and empowerment but not all Western practices can be applied immediately for there is a deeply-ingrained mechanical attitude to pay and promotion. Similarly foreign companies may adopt a stance of 'zero tolerance' towards alcohol and downsize by firing the drunks. But there is a balance to be struck between sobriety and the bonding promoted by liquor: change can be made more gradually by open-plan offices, reduced evening work and a family atmosphere for celebrations.

People seem to behave positively and look forward with some optimism, yet a successful struggle for democracy cannot be assumed. Given the legacies which exist in the public mind, as they also do in the landscape, stable democratic institutions may seem unlikely in the present climate of global disintegration. It will take much longer to lay the social foundations of democracy and the market economy than to achieve constitutional and economic reform. There is no easy consensus over foreign policy and domestic goals. Despite the services and opportunities arising out of transition, it is difficult to construct an ideology of legitimacy when there are not yet enough 'winners'. Conflict management calls for a careful balance to protect the 'losers' of the initial neo-liberal experiment to engineer a transition to 'normal' market economies while investing sufficient resources for growth.

There are still many conservative elements in political life and in public administration and 'persistent sociological realities' (Jowitt 1996: 5) are evident through the survival of some former bureaucratic practices and the retention of security police forces which – in the case of Romania – helped to bring the 'miners' to the streets of Bucharest in 1990 and 1991 and, it is widely believed, had a hand in the ethnic violence in Târgu Mureş in 1991. By contrast in East Germany there has been a determined drive against the communist elite and especially those who cooperated with the secret police: action against the 'Stasi' – once supported by 100,000 employees and 140,000 free-lancers out of a population of 17 m. – extended to people who profited through holding party membership, leading to a comprehensive clear-out of functionaries on the grounds of their unreliability. Elsewhere there has been no real confrontation with the crimes of the past. Some politicians have been imprisoned (like F. Nano in Albania who was released only after the collapse of the Democratic Party government early in 1997), while the Ceauşescus were executed during the December 1989 revolution in Romania, but most people seem content to forgo investigations that would reveal corruption and oppression but also high levels of complicity by people who had little choice but adapt to the system while maintaining a private domain as best they could.

Although any monist system will become unstable if it closes off the possibility of improvement, Fukuyama may be right in thinking that Western liberal democracy can assimilate many aspects of East European politics, especially if a sense of liberation and civic culture – pluralism combined with moderation and self-reliance linked with education in the rights and obligations of citizenship – wards off highly authoritarian systems, as advocated by communist and nationalist elements in Russia. But democracy in the region will take many different forms and the communist experience will inevitably remain a significant influence. But while the northern countries have now clearly embraced parliamentary democracy, there are areas where democratic traditions are not so well founded and a European approach to social change may encounter resistance. In the SEECs, historically prone to 'racial xenophobic geopolitics' (Heffernan 1998), pluralism has been constrained by dreams of a nation state secured by ethnic cleansing and reinforced in religious terms by 'myths and symbols spawned under the shadow of three great conflicting historical empires and civilisations . . . too salient to be overcome by modern national or socialist programmes of integration' (Khan 1995: 471). Tension and civil war in former Yugoslavia reflect ethnic divisions reinforced by a lack of a common approach to historiography arising from the varied interpretations of Balkan history by Catholic, Orthodox and Islamic (Albanian and Bosnian) professionals. Although the Patriarchate of Constantinople has tried to mediate, much Orthodox opinion is against ecumenical contacts with Catholic and Protestant churches at present. Serbian Orthodoxy contemplates a 'defiant self-imposed isolation' (Herbst 1998: 178) insulated from a World Council of Churches and ecumenists who are seen forging unity through compromise and hypocrisy.

Here in the Western Balkans, where national politics are only just being hitched to the EU bandwagon, local community values will continue to exercise a significant moderating influence. As Liolin (1997: 190) asserts for Albania, cultural and religious life will continue to make a difference. Nationalism must come to terms with regional identities grounded in the administrative arrangements of the Ottoman period. The schism between Ghegs and Tosks, complicated by the events of the Second World War, set the tone for Albania's communist era under Hoxha. 'A systematic championing of the Tosk identity and the destruction of the Gheg's, helped sustain this incestuous regime: it is the paradox of fragmentation, envy and fratricide which keeps Albania together' (Blumi 1997: 392). Yet Albania failed miserably under communism in material terms and even now the legacy of classic Balkanisation is not conducive to sustained FDI. Economic penetration has been led by mining enterprises concerned primarily with geological conditions and mine installations rather than location per se. Chronic capital shortage then pushed the Berisha presidency into pyramid funding although the resulting unrest was aggravated by strident regional prejudice. Fortunately, West European and North Atlantic institutions seem ready to confront the local where it loses all self-control (as exemplified by NATO intervention in Kosovo in 1999), but the good news is that Balkan behaviour patterns remain locally specific (as the Milošević ethnic cleansing project in Kosovo contrasts with a history of working through the political process in Macedonia) and 'apocalyptic scenarios' of regional conflict (Blumi 1998: 565) do not appear to be appropriate. Nevertheless, there remains the possibility that change may follow the Latin American model with sharper class distinctions and mafia politics with Peronist 'internalisation' based on xenophobia and protected domestic markets rather than free trade and international cooperation.

Arguably, communism was more of a revolution than the post-communist transition in view of its social destructiveness. While some Poles see communism as a force for modernisation – especially industrialisation – others highlight the destruction of pre-communist elites and the undermining of Catholic cultural identity and its regional expressions. Yet the transition has seen many communities weakened by the initial lack of an elite able to play a leadership role. Reconstruction is taking place in the shadow of communism which exerts an element of 'path dependency'. There has been a resurgence of media and social activities around religion, as part of a complex change in the cultural landscape. However, there is also a generation gap, for the over-30s are still speaking Russian and are cast in traditional ways to a much greater extent than the more open-minded under-30s. Meanwhile, corruption is understandable after a communist era that saw widespread theft of state property, all the more so in view of the dramatic increase in inequality in the SEECs with the change from communism to a highly stressful transition. But this has a corrosive effect on morality and affects the image of transition states in the eyes of the international community at a time when the attraction of the EU and the need to comply with IMF–WB homogenises party platforms after periods of 'retrogressive government' in Croatia, Slovakia and, above all, in Serbia. Indeed Lavigne (2000: 16) suggests that, despite a steady stream of advice pouring in to the region since 1989, 'the most effective incentive for institution building may have come as a side-effect of efforts to join the EU'.

Institutions: the churches

Civil society will show a partial dependency on institutions like the churches which are important for spiritual enlightenment, education and social functions such as the religious festivals that often persist in the countryside. Religion should exert a positive influence – although the coexistence of church and state may be far from ideal – while individual churches will be preoccupied with their institutional interests, including their security in the face of perceived threats of a political-ethnic nature. There is freedom to worship, but some churches like the Uniates in Romania have been holding services out of doors because of a delay in returning property, after the forced amalgamations of the communist period were cancelled after 1989. Furthermore, buildings need to be put into a state of repair and outside help is needed: like the Aromanian church in Korce (Albania) supported by members of the community in the USA. And churches are also under pressure from a shortage of priests, especially teacher-priests. Yet as the churches regain their establishment role they may lose the 'refuge' function enjoyed under communism. Arguably there is credibility in a middle ground where the church does not directly participate in policy making but where its presence and influence can be felt in terms of an ethical and moderating stance.

In this connection, 'ex-communist rulers have generally shown an unsympathetic but legally correct attitude towards religious questions' (Luxmoore 1997: 89). This ties in with the ideological split 'between confessional and secular visions of political order [which] is closely entwined with opposing stands on the issue of decommunisation' (Jasiewicz 1998: 3). However, while religion should arguably be 'a witness to the transcendent and a spiritual corrective to our faulty and imperfect world rather than as an instrument of politics' (Merdjanova 2000: 258), it is important to remember that especially for Orthodox peoples of the Ottoman Empire there was a rise of religious nationalism in the nineteenth century as the church came to represent their national aspirations. Under communism, the churches became a focus for political opposition as 'the only legal structures with buildings, leaders and (albeit limited) finances around which initiatives from below could be organised' (ibid.: 256), not to mention the 'international centre' of the Catholic Church (ibid.: 256). The church remains a force in the struggle against homogenisation, especially in the SEECs where religion has been linked with a level of fanaticism and cultural impoverishment that clearly conflicts with the ideal of European integration.

Institutions: local government

Hopes will naturally be pinned primarily on central leaders rather than local self-government. But there is now an enhanced role for local government with implicit support for decentralisation because 'a gap soon appeared between the centre's appetite for control and its ability to exercise it' (Surazska *et al.* 1997: 461). The state can no longer be in full control, for a centralised unitary regime needs a welfare system with centrally controlled redistribution of resources, a professional administrative class and a stable party system. And because these criteria cannot always be satisfied in the ECE capitals there is a case for decentralisation. 'Hollowing out' in the context of a financial crisis was reinforced by the ideology of Western neo-liberalism and a desire to destroy the infrastructure for communist control. The trend has been accelerated by demands for greater autonomy by ethnic minorities and regional cultures and also with the EU practice of developing direct links with regions with regard to structural weaknesses in the economy. With the weakening of integration – albeit state-centred over-integration – the early transition years saw the 'local' coming very much to the fore as political sentiments were projected very much in the raw, e.g. in agricultural restructuring, in response to national land restitution settlements (Stewart 1997), and in urban development when tensions arose between urban communities and the planners. While West European models exercised some influence, decentralisation and fragmentation – reversing the forced amalgamations of communism – were 'substantively shaped and carried endogenously

by a constitutional consensus which, as a result of Hungary's negotiated transition, was borne both by the opposition forces as well as by the reformist
communists' (Wollmann 1997: 477). But there is a case for some reintegration
because the local can only flourish in line with national government policies
and as legislation for a market economy has been consolidated the scope for
local deviation has been somewhat circumscribed.

Nevertheless local government is perhaps the most significant expression of
a restructured 'local'. Indeed Kolankiewicz (1993) sees authentic local self-
government (freely-elected and independent of the state administration, with its
own fiscal and property base) as a principal component of civil society as it
becomes more responsive to grass-roots opinion and fosters a sense of community instead of merely carrying out duties handed down from the central planning
machinery in prefectorial fashion. However, especially in the case of the more
immature democracies, antagonistic tensions can develop when local government is extensively controlled by parties that are in opposition nationally and
central government will be all the more reluctant to decentralise if local councils might back radical (even secessionist) parties as a lever for more state support.
Moreover, local authorities are left with scope for initiative which is out of step
with distinctly modest financial resources. There is much variation between local
authorities as regards tax yield and local government expenditure. Typically
there are high values in metropolitan and western regions, while the weakest
performances are registered in recently-industrialised areas caught in the trap
of incomplete urbanisation: being 'no longer rural . . . but not quite urban either'
(Surazska *et al.* 1997: 453). At the same time, cities seem to crystallise into poor
and rich neighbourhoods, with substantial inequalities in the provision of services.

Hungary has moved to directly-elected regional councils which provide 'the
most decentralist institutional design in all Central and East European (and even
West European) countries' (Wollmann 1997: 477). While the counties themselves
('megyek') have only limited powers (though they are valued as a traditional
institution from the pre-communist period), the 3,200 municipalities at the lower
level of the two-tier system have a key role in the new order. Two-tier structures in other countries generally give rather less power to the grass-roots and
relatively more to the districts or counties, but this still produces the reality of
a large number of municipalities with tiny populations and very limited resources.
Meanwhile, there is a tendency towards reform of the top-tier units to create
larger regions appropriate to EU structures of governance. These are now in
place in all the candidate countries although there are variations in their status.
In Romania the regions have a coordinating role above the two-tier system of
local government that remains based on the county ('judeţ') and commune. The
new systems often generate much debate on the trade-off between efficiency
and identity with larger regions widely seen as undemocratic as well as dangerous
in the context of autonomy sought within the Czech Republic by Moravia and
Silesia. In Poland it was left to the former centre-right government to complete
the work of the previous centre-left coalition which had been deadlocked over
the issue.

Institutions: the media

Education for good citizenship requires a better supply of information which the media has a major responsibility to provide. Newspapers and magazines have multiplied and there is wide choice especially in the towns, although most publications provide psychological satisfaction rather than objective information. There have been supply problems because in Romania state-controlled distribution was blamed in the early years for the limited availability of opposition newspapers which meant that opinions sharply critical of the governing Party for Social Democracy were effectively shut out of the rural areas where the regime attracted strongest support in the early 1990s. However, there is now real potential for a media bonanza. Several Western companies are active, including Axel-Springer Verlag, and Ringier. Hungary's daily tabloid *Mai Nap* and the weekly *Reform* have attracted $4.0 m. of investment from News International, but poor performance arose from an unsatisfactory compromise between editorial and advertising interests. Meanwhile, a considerable number of national-language editions of titles like *Cosmopolitan*, *Elle* and *Reader's Digest* are now available.

There has been a trend towards liberalisation in broadcasting, despite political resistance to independent institutions financed by advertising and the problems of finance: equipment is often 30–40 years old at a time when modern FM technology is in use in most other parts of the world. After major controversy, Hungary made conspicuous progress in keeping state television away from the direct control of the ruling party, but state broadcasting is still very strong with the Hungarian Broadcasting Corporation providing Hungary's first 24-hour TV channel in 1996. However, the breakup of state television monopolies is now evident and national commercial television stations have appeared. Foreign media involvement remains limited in some countries but the American Group Central European Media Enterprises (CEME) has been successful in developing commercial radio and television stations: 'TV Nova' started in Prague in 1994 after Nova's reporters had trained with staff of the Atlanta, USA-based CNN. The station gained increased advertising time (compared with what was allowed to the earlier competitors of state television) and developed links with the leading advertising agencies in Prague. Within a year it had gained over 70 per cent market share while the public broadcaster Czech Television faced decreasing viewing numbers due to limits on advertising below the 10 per cent figure available to TV Nova. In Bulgaria the Russian television station 'Ostankino' has been replaced by a private channel for which several companies have competed.

Meanwhile, the Italian-owned Premiera, a Prague-based regional broadcaster established in 1993 to serve central Bohemia, is developing a national network of regional stations. In Poland CEME joined forces with Neovision and lost out in a bid to broadcast nationally to Polsat (which now has only a small audience share), but then in 1996 it launched a network in competition with 'Nasza Telewizja' (run by business people linked with the Democratic Left Alliance) in the centre and north while complementing the Wisla network in the south. Hungary's first regional station opened in 1997 and local stations are also developing: Hungary has allocated over a hundred licences for TV and radio frequencies geared to 'local reach' (500,000 people in Budapest and 100,000 in other parts of the country) though it remains to be seen how effective this will be in terms of competition with the state MTV. Meanwhile, Belgium's 'FilmNet International' launched Poland's first subscription television film channel; while 'Canal Plus' of France operates Poland's subscription television broadcasting operation through a joint venture with Polish investors. E-mail is becoming much used in business as a means of getting round postal delays and uncertainties, while the expansion of Internet Cafés reflects the value of the web for job openings and educational programmes: even if servers are slow, web-searching is still faster than visiting the local library. *Polish World's* Business and Economics page functions like a business directory. And newspapers are joining in: daily editions of *Życie Warszawy* can be read and indexes consulted on articles for recent years. The Privatisation Ministry programme and parliamentary information can be found, while tourist information sites offer maps, pictures and contacts.

Quality of life and welfare

Communism clearly limited individual freedom through regimented societies numbed by the smouldering legacy of expropriation and the drabness of life in 'concrete barracks' in the larger cities. However, communities often worked together to maintain a caring environment, while there was full employment and welfare and, for many people, opportunity in areas of professional life where the state had a vested interest. Substantial resourcing went into the military-industrial establishment while party membership bestowed significant privileges and the importance attached to diplomacy, trade and overseas aid provided travel opportunities for young professionals and a temporary escape from the constraints on life at home. Reference may also be made to the conspicuous success of ECECs, as well as the FSU, in international sport. Heavy investment in sport helped to stimulate young people into acceptable areas of healthy activity, and good performances in the international arena helped to project communism in a positive light around the world. Despite some questionable use of drugs, there is little doubt that effective training and coaching methods were built in the 1960s.

Now there is the possibility of a better quality of life and opportunity to consider what the 'good life' means in terms of human rights (democracy and restitution), leisure activities and economic options through freedom to start businesses and run small farms independently. Variations in income, status – and stress – have widened, especially in the context of recession, although each person will have an individual balance sheet of pluses and minuses in the light of political processes in specific countries and municipalities. People have escaped from an economy of shortages and from the intimidation of the communist security services, but they have entered an altogether more competitive world that is full of uncertainty. Winners and losers are easiest to recognise in the economic sphere, for while voucher privatisation gave everyone an equal opportunity in the countries involved, other aspects of privatisation and restitution have enabled relatively few people – especially the politically well connected – to gain substantial industrial assets and real estate while the less fortunate may experience unemployment and homelessness through the same process. The new economic elites are clear winners over the majority of the working class, while the 'losers' are found among pensioners, unemployed, women and families with young children. The majority are probably part of a 'syndrome of distrust' as initial high expectations are constrained by a pervasive uncertainty and the stress of inflation, unemployment and crime.

Although not adjusted for purchasing power, Table 2.1 shows a pronounced north–south difference in salaries and pensions. Pensions seem to be falling back but remain high relative to wages in Poland. Meanwhile, wages have increased everywhere in absolute terms during 1994–8, but have grown by more than half in Croatia, the Czech Republic, Poland and Slovakia. Albania also shows such a rate of increase but remains way below the level of other SEECs. The highest per capita spending is registered by Slovenia, followed by Croatia and the Czech Republic: two countries which (along with Slovakia) have less than half the spending going on the basic items of housing, food and fuel. However, any recent progress should be seen in the context of the stress that was encountered almost universally through the decline in the real value of wages and pensions during the high inflation years of the early 1990s. While low incomes under communism were concealed by the limited range of goods available in the shops, poverty is now highlighted by the glare of Western consumerism. Although growth resumed by the mid-1990s, many families need 'multiple economies' in order to get by and poverty may exist even in families with more than one working person.

Table 2.1 Social indicators 1994–8

	A		B		C		D	
	1994	*1998*	*1994*	*1998*	*1994*	*1998*	*1994*	*1998*
Albania	40	62	23	25	35	na	66.5	na
Bulgaria	77	106	30	40	29	36	57.4	63.8
Croatia	302	579	92	174	na	196	na	47.2
Czech Rep.	202	322	95	150	147[1]	169	38.9	41.4
Hungary	271	282	112	108	98[1]	99	53.1	58.6
Poland	203	335	124	175	na	128	na	51.4
Romania	109[1]	136	31	40	41[1]	52	70.3	70.6
Slovakia	165	253	73	105	102[1]	127	42.4	40.6
Slovenia	619	850	321	365	267[1]	326[2]	na	na

Source: Eurostat 2000: 37–62

Notes
[1]1996 [2]1997
A Average monthly salary ECU
B Average monthly pension ECU
C Average per capita expenditure ECU
D Ditto: percentage spent on food, housing, fuel and power

Surveys of public opinion suggest that only one-fifth of households are content, while most expect to be better off only over the longer term. There are strong regional variations, for the highest level of car ownership in Hungary occurs in Budapest, with the western areas clearly ahead of the east, while the towns are better endowed than the rural areas, albeit by a smaller margin. In Poland there are relatively high living standards in urban areas (though the emphasis in investment is on the agglomerations) and former German territories. At the same time the more-successful regions are well integrated with the global economy while the less-successful regions remain 'embedded' (sometimes 'over-embedded') in the inherited structures. SEECs in general suffer from insecurity with a low level of economic development, inevitably reflected by limited progress in pluralism and civil society. 'Countries disappointed in the past, probably largely disappointed with the present and uncertain about the future are difficult to fit into a stable system' (Nowotny 1997: 87).

Low wages in the private sector have given rise to a 'new' group of poor people whose plight contrasts with the 'old' poverty associated with inefficient farming in remote rural areas. Poverty is also the result of unemployment which in itself is a traumatic experience for societies used to full employment with lack of a job construed as parasitism. Previously marginalised in former socialist countries by the traditional role of agriculture as the residual employer, unemployment soared in the early 1990s (1.7 m. in East Germany by 1991) and became an important line of geographical enquiry linked with alcoholism, crime and drugs. The region has a workforce that is highly skilled and motivated in relation to wage levels, but opportunity in the new economy seems to be by no means equal and there are major structural contrasts between and within countries. Hence the danger of transition from a 'low work-intensity, mass underemployment system to one characterized by highly intensive work practices and mass unemployment' (Hardy and Rainnie 1996: 258). Economic reforms should enable ECE to realise its labour market advantages, but losses of jobs have commonly occurred in SOEs after privatisation and there is an insufficient growth of new firms to compensate (with small business sometimes making a greater contribution in the small towns than in the large cities). In Poland, 3.2 m. left the public sector during 1990–3 – when there was little money to subsidise labour hoarding and low worker productivity – but the new private sector could only employ 600,000. Even FDI job guarantees may be undermined by voluntary redundancy packages. Thus in countries embarking on rapid reform there was particular concern over the private sector's ability to create employment on an adequate scale – with unemployment because of and not in spite of reform – and some countries still hesitate to reorganise too quickly because the social costs could be unacceptably high.

Poverty

According to Vuics (1992), more than a quarter of Hungary's population was then considered poor, according to studies of the 'new' poverty arising from closures of factories and mines. There are particular problems of poverty among pensioners and some ethnic minority groups such as the Roma. 'The poorest quartile is not only poor in relative and absolute terms but also more stressed, exhausted and anxious than the better-off parts of society' (Andorka 1997: 84). There is widespread alienation related to a sense of powerlessness leading to alcoholism and suicide. Housing is often a problem for poorer families given that there were 14 per cent more households than housing units in Hungary in 1989: 60 per cent of people aged under 30 with young families share with their parents, for housing loans are quite prohibitively expensive and while there is much construction for the new upper middle class, poor families are left without the possibility of acquiring a place of their own. People are displaced from apartment blocks when they cannot afford the rent and welfare workers say the homeless often become so demoralised that resocialisation is impossible. At one stage, many homeless people were sheltering at Warsaw's Centralna Station: a hundred in summer and many more in winter. A waiting room was laid out in an orderly manner with benches covered with cardboard and cloth; plastic bags contained personal belongings while socks dried on door knobs; and misguided travellers would be quickly forced to retreat by the acrid smell of unwashed bodies.

Hungary has the best record on unemployment with the Czech Republic the only other country that remains in single figures. The situation is worst in B&H, Croatia and Yugoslavia where rates exceed 20 per cent (Table 2.2). The unemployment among young people (below 25 years) is particularly disturbing with a rate 4.4 times greater than for the rest of the labour market. The ratio is high in Croatia (3.7 times), Slovenia (3.0 times), but in the best case – Hungary – the rate is double, followed by the Czech Republic and Slovakia (2.3 times), Macedonia and Poland (2.6 times) and Bulgaria (2.7 times). In Poland – where the situation is aggravated by the annual increase of 150,000 in the economically productive population – unemployment has been most persistent in the north and northeast: 41.1 per cent in Gołdap (Suwałki) compared with only 6.2 per cent in Warsaw (Weclawowicz 1996: 151). There is also high unemployment in Hungary's northeastern border areas while Budapest and the western regions are relatively prosperous. Some old industrial areas have suffered badly because their outdated plant is no longer competitive and there may be little experience over problems of diversification when enterprise managers previously discouraged competition within their labour catchments. Case studies have highlighted the plight of workers in the coal and uranium mines of the Pécs area of Hungary where social security and high salaries were enjoyed under communism. The case of the Hungarian metallurgical town of Ózd is also instructive. People who left the area before 1989 often found work elsewhere but now the rising unemployment levels limit the options available. Given the inadequate legislation for worker protection it is possible for dismissals to be disproportionately heavy among minority groups (especially the Roma), the poorly-educated and those who have frequently changed their employment in the past (easy to pick up through the Hungarian employment card system).

However, decreases in salaried employment do not necessarily generate official unemployment to the same extent. Much depends on the nature of

Table 2.2 Unemployment rates 1992–2002 (%)

	1992	1993	1994	1995	1996	1997	1998	1999	2000	2001e	2002e
Albania	27.0	22.0	18.0	12.9	12.3	14.9	17.6	na	na	na	na
Bosnia & Herceg.	na	na	na	na	na	39.0	38.5	39.1	na	n.a	na
Bulgaria	15.3	16.4	12.8	11.1	12.5	13.7	12.2	18.1	17.3	16.0	14.7
Croatia	17.8	16.6	17.3	17.6	15.9	17.6	18.6	19.1	21.3	21.1	20.5
Czech Republic	2.6	3.5	3.2	2.9	3.5	5.2	7.5	8.6	9.0	9.2	9.5
Hungary	12.3	12.1	10.9	10.4	10.5	10.4	9.1	7.0	6.4	5.9	5.7
Macedonia	26.2	27.7	30.0	35.6	38.8	41.7	41.4	47.0	na	na	na
Poland	14.3	16.4	16.0	14.9	13.2	10.3	10.4	12.3	14.4	16.1	16.3
Romania	8.2	10.4	10.9	9.5	6.6	8.8	10.3	11.4	11.1	9.9	9.1
Slovakia	10.4	14.4	14.8	13.1	12.8	12.5	15.6	19.2	17.8	18.0	17.5
Slovenia	13.3	15.5	14.2	14.5	14.4	14.8	14.6	13.6	12.2	11.5	11.0
Yugoslavia	24.6	24.0	23.9	24.7	26.1	25.6	27.2	27.4	na	na	na

Source: *Romanian Business Journal* and Economist Intelligence Unit

Note
e Estimate

registration formalities, while in the rural areas there is still a strong presumption that anybody without a salary will be doing some agricultural work. But small-scale family farming is by no means universal and where restructured cooperatives and private agribusiness looks after much of the land, surplus family labour may only find an outlet in gardening. Thus during the years 1990–2 there were much greater increases in unemployment in Hungary and Poland (over 14 per cent) compared with the Czech Republic (2.5 per cent) although the change in employment was between 11.5 and 12.6 per cent in all three cases. Yet in Poland, hidden unemployment accounts for 480,000–840,000 rural inhabitants, while the Council of Ministers' Social and Economic Strategy Council says there are 1.5–1.8 m. 'rural work force reserves' (a third of them under the age of 24). The rural situation is all the more difficult because politicians discriminate through 'reluctance to bankrupt large politically sensitive workplaces based in urban areas' compared with SMEs in rural areas (Hardy and Rainnie 1996: 62).

Welfare

Despite the very high cost of adjustment to European and world markets, especially in view of peripherality in relation to the core of Europe, civil society will have to provide greater mobility than was possible in the later years of communism when there was insufficient scope for people who achieved economic success. Society will also have to find ways whereby the market system can cope with increased inequalities and greater social differentiation; with more self-help – rather than state intervention – and more input by secular organisations to avoid unacceptable material contrasts and further polarisation. Advocating a better integration of social and economic policy, Heller (1998: 21) calls for 'reforms in political institutions so that as many people as possible are able to participate in economic growth and welfare'. People need to be involved through expertise over job creation and general support for transition policies. This requires that cheap labour, low taxation and the repatriation of profits be balanced by higher welfare spending and a more interventionist economic policy, especially as wealth increases.

The unemployment situation is even worse when social security systems are kept modest by IMF pressure to limit budget deficits, though early retirement and disability pensions have reduced the numbers of unemployed while state economic sectors have received considerable financial support to help maintain employment. Western workfare is hardly appropriate since the jobs are not available in sufficient numbers and welfare is not generous enough for Western-style welfare dependence to be a problem. But there is a dilemma between employment protection (perhaps through subsidies or bank credits to finance labour-hoarding) in the interest of social stability – for high unemployment could undermine democracy – and the encouragement of retraining and mobility for future economic efficiency. There should be a more comprehensive financial cushion, for there is only limited welfare provision outside the enterprise. But finding money to restructure and expand social services is a challenge throughout the region. There is also a need for housing benefit in the public housing sector to limit the fall in real living standards.

Meanwhile, the unions are struggling for legitimacy as former socialist brigade leaders seek a new role as shop floor representatives. Trade unions often retain some welfare functions particularly where recreation is concerned, but the tendency to enter politics through integration with centre-left parties could detract from their effectiveness in protecting working people and campaigning on behalf of the poorer sections of society. The case of Solidarity in Poland also suggests that unions cannot hold on to popular support if they start to share responsibility for macroeconomic management (Whitley 1995: 17). Despite a cooperative stance, some unions are excluded from bargaining productivity and flexibility in return for higher pay, partly because of the problem of individual pay contracts. The environment in the workplace should harmonise with the style of political life, but an authoritarian management culture is likely to threaten unemployment for poor work and marginalise critics as 'communists'. Fortunately, the well-established multinationals are setting high standards, but some other private businesses want quick access to the domestic market through cheap (especially female) labour with a short-term view in search of immediate profits. There is a need for compromise between aggressive marketing and worker disillusionment so that modern trade union organisations can appreciate market realities while at the same time protecting workers.

Job creation

A passive policy of providing job applicants with social benefits should be balanced by more active measures or job creation and also employability through higher motivation to work and measures to prepare people for the market needs through retraining and stimulation of enterprise. There is a change of attitude occurring away from the socialist perception that jobs should follow labour to the free-market attitude that labour should follow jobs (Weclawowicz 1996: 151). However, there are simply not enough jobs to go round. Management skills are limited (particularly in the financial domain) and entrepreneurship is in short supply after decades of propaganda portraying private enterprise as morally reprehensible – though former communists do not seem to be represented in management above the average and hence the 'nomenklatura model' of entrepreneurship may be questioned. Instead social capital seems to be a key factor through positive attitudes to risk taking: this reflects family background, occupation and connections. However, low spending power will tend to limit the growth of the tertiary sector and enterprise is most likely to blossom in the cities where demand is greatest and where infrastructure and business services are relatively good. But growth has been relatively slow throughout the SEECs due to the scale of the trade shock (with accompanying deflation, which discouraged business generally), low levels of foreign investment which helped to delay privatisation and high levels of poverty which make export markets crucial.

Employment may be stimulated through institutions such as business associations and loans to create new jobs. Broader projects to develop labour markets include the provision of International Development Association credits for this purpose in Albania, with reform of the social safety net at the same time. Employment promotion in Poland was assisted in 1994 by a World Bank loan implemented in part by consultants from Ohio State University through Small Business Assistance Centres to develop entrepreneurship, supply information and act as consultants. The PHARE Local Initiative Programme (completed in 1996) successfully fought unemployment in nine small areas of Poland, chosen by competition. With a subsidy of E0.7 m. each, the communes started training/ retraining programmes and created a capital access system for small businesses. PHARE's investments in infrastructure helped finance a new road and waste-disposal facility at Wicko near Słupsk in Poland, turning this coastal settlement into a bustling tourism/recreation centre. Special help for women has been provided through the International Women's Foundation, founded in Łódz in 1992: this has successfully imported ideas from the USA (including links with a fund to encourage small businesses) and a self-help booklet *The ABC of Small Business* has been financed by PHARE and the Stefan Batory Foundation. Poland also has a network of business incubators and local development agencies. More people are taking up self-employment, and while some are marginalised, others find they can become successful service providers.

Education and training

Under communism motivation was crucial for the acceptance of disciplined work routines and, more generally, compliance with party and government edicts: hence the emphasis given to propaganda along with the drive for homogeneity through educational and cultural programmes. Education is a priority to provide appropriate training: technical skills were needed for the anticipated explosion in factory employment and hence the emphasis on science and technology in the secondary schools, polytechnics and universities. But in Poland the communist education system now has a legacy in narrowly-specialised experts who have trouble in changing jobs, while new graduates don't have the skills – such as computer literacy – to find rewarding employment. The transition is a time for greater efforts in education and retraining so as to 'lower the barriers to entry' for participation in market processes (Landesmann and Szekely 1996: 19). However, by the standards of the EU where 42 per cent of 18–24 year olds are in education (23 per cent in tertiary education), only two ECECs come close according to Eurostat data (2001: 48): Poland with 41 per cent and 19 per cent respectively and Slovenia with 37 per cent and 25 per cent. Romania shows the largest discrepancy (with 21 per cent and 11 per cent) followed by the Czech Republic (28 per cent and 14 per cent) and Bulgaria (28 per cent and 22 per cent). It seems that social attitudes are an important aspect of skill, especially in service industries where much enhancement is needed. Indeed, early German investors in the region are now rediscovering Western work culture which may result in 'renewed interest in higher cost but also higher reliability locations such as Britain and Spain' (Cooke 1997: 697).

High unemployment goes hand in hand with shortages where new skills are concerned and this points to the importance of enhancing labour markets.

Assessment of the skills base in Polish industry reveals satisfactory recruitment of engineers for the electronics industry from higher education, while the quality of labour in branches of the food industry is well up to export standard. But education has to be adjusted to other sectors of the jobs market: vocational secondary schools and agricultural schools now train students for the so-called strategic professions (though there is a lack of money and workshops are poorly equipped). Post-secondary schools are being created with curricula adjusted to business sector demands (though more needs to be known about the jobs market). Poland has responded to the transition with a growth of academic as opposed to vocational education. While jobs in government and in the state-owned factories have in the past been secured with vocational qualifications, graduates from academic secondary schools have been doing better in the labour market recently. This is good news because academic courses are relatively cheap from the state's point of view. Meanwhile, market forces in higher education are encouraging private institutions: Poland has 72 private universities approved as degree-awarding institutions. There has also been a growth in higher education in Romania where a buoyant private sector, taking off in 1990, now complements the state-run institutions which are also expanding. There is also a need to reduce linguistic isolation arising from provision for minority languages and also for opportunities for mainstream communities to learn Western languages. The CoE launched a scheme in 1993 to twin each of Albania's 2,000 schools with similar institutions elsewhere in Europe or North America. This is a follow-up to the CoE emergency relief programme 'SOS Albanian Children' to tackle deprivation in Albanian schools.

Health

Health care was necessary under communism to maximise natural increase by reducing deaths, particularly among young children. However, under the transition the situation has deteriorated. In Romania, where AIDS is now a major problem, the living standard is the main factor influencing health and the situation will not improve in the near future. Infant mortality has risen, especially in areas with severe health problems in the northeast (Botoşani and Iaşi) and the southeast (Giurgiu, Călăraşi, Ialomiţa and Constanţa). Meanwhile, the funding of medical services has become problematic and there are inequalities within countries which may reach crisis proportions. Apart from problems of funding, access to health care in Poland seems to vary according to economic position and informal connections, and also with location, in the context of (a) a lack of choice between medical centres, when many local centres have only limited facilities and (b) inhibitions through the doctor's professional language. Low pay makes the recruitment of staff difficult and there are also deficiencies in Poland's western cities that do not provide health care commensurate with their pollution levels: part of a stressful situation – leading to psycho-social disorders, unhealthy lifestyles (smoking and excessive alcohol consumption) and increased suicidal behaviour – which makes a reliable health system all the more necessary. With health a crucial test for international policy-making in view of the scale of east–west differences much outside help has been provided. World Bank loans have assisted the health education (in respect of smoking, drinking and diet) necessary to improve performance over heart disease and life expectancy. The future would seem to lie with social insurance-based health care systems despite the immediate problem of affordability.

Crime

The rise of petty crime has become a serious matter and it is tempting to correlate this trend with stress in general and long-term unemployment in particular. Yet, organised crime (unrelated to poverty) seems to constitute a greater problem. In some countries it is now common for private security organisations to protect business premises. The reduced power of the state has provided opportunities for mafia activities, with the danger that local government may be vulnerable to the patronage of mafia organisations emerging out of contact between officials, businessmen and other influential local people. During the Yugoslav civil war oil deals for Serbia were struck in Szeged (Hungary's 'mafia city'). Meanwhile, organised car theft in Poland involves experienced thieves or 'fishermen' (often teenagers), bodyrepairmen who alter identification numbers, a 'printer' to provide forged papers (using stolen registration forms) and a 'dealer' who finds a buyer when gangs are not stealing vehicles to order. Central Warsaw is particularly dangerous (with car theft the greatest nightmare for American companies) while the CIS has been likened to a 'black hole' because there is little cooperation over stolen vehicles taken there from neighbouring ECECs. Pickpocket gangs are a further hazard, for example around Warsaw's Centralna railway station and Srodmiescie municipal bus station, while the tram lines in general are a pickpocket's paradise. Penalties are relatively light for people who are caught, given the low perceived level of 'social harmfulness' provided no violence is used. Again, the decrease in budgetary expenditure for education (and other social services) tends to limit access to good education only to children from better-off families and this creates an underclass society evident in youth subcultures linked with drugs and criminality which may include inter-communal violence in the form of skinhead outrages against the Roma and foreigners. It is said that police forces are too poorly paid and equipped to check crime and avoid becoming involved in bribery.

Illegal immigration provides further opportunity, especially for the Russian mafia using Polish guides. Travelling by air to Moscow, Sofia, Kyiv and Prague, migrants transfer to trains, cars, buses or TIR trucks to reach the Polish border; then they hide in Poland to consolidate into groups to reach Germany and 'the gate to heaven'. There are particularly high prices for Asians, who are easily recognisable, and a danger of prostitution where women are concerned; but those caught can still apply for asylum. The Slovak underworld uses illegal migrants for various criminal purposes. They are mainly from the FSU (squads of Ukrainians in particular) and former Yugoslavia but include others abandoned by guides who may have convinced them that they are already in Germany! The high profits from smuggling – $10,000 may be charged for a trip to Germany – may justify use of light aircraft to cross (say) the Lithuanian–Polish border. Local helpers – who may include Ukrainian military units – may receive $200 for a single assignment and will claim to be helping hitch-hikers in the event of interception. Although the EU is constantly pressing for better border security from its accession states, the police have not always given illegal immigration high priority since it is victimless crime. Further back down the line, it is reported that Africans, Chinese and Kurds cross into Romania accompanied by guides ('snakeheads') from Bulgaria or Moldavia.

Illegal trade in drugs consists largely of heroin smuggling through the Balkans, but war in Yugoslavia forced Turkish heroin smugglers, working with 'residents' (local people in financial difficulties), to find new routes through Ukraine (by ship from Turkey), then Poland (with considerable local-scale activity attracting police action against wholesalers), Germany, France, Britain and Spain. Meanwhile, cocaine supplies come from Latin America, which places Poland at a crossroads for the smuggling of the two drugs. Poland is also a major amphetamine producer: a major source of ecstasy for Western Europe (especially Germany and Sweden), though there is some trafficking through Poland for synthetic drugs and heroin produced in Russia and Latvia. Until 1989 heroin 'kompot' was produced domestically in Poland from poppy straw using simple kitchen utensils. But in the 1990s amphetamine (synthetic drug) laboratories snowballed and 80 were closed down during 1994–6 after the first was found at Tuszyn, south of Łódz. Some laboratories produce as much as 30kg/month and the Health Ministry estimates 20,000 drug addicts and the Narcotics Bureau thinks that at least 400,000 people use illegal drugs regularly.

Population: demography, gender and migration

Under communism population increased by almost 1 per cent per annum with Albania showing remarkable dynamism in contrast to Hungary, Bulgaria and FCS. By contrast, the transition has seen a slight decline (–0.07 per cent annually) with growth maintained only marginally in B&H, Macedonia, Poland, Slovakia and Yugoslavia (Table 2.3). In the past, despite some variations between northwest and southeast, there has been a distinct 'East European model' by virtue of early marriage and high fertility which has more than offset a relatively high death rate. But now, despite a further decline in infant mortality, the birth rate has been greatly depressed through the postponement of marriage (while the share of non-marital births has doubled in five years to 15 per cent, linked with the rise in divorce) and a general trend towards smaller families, despite the persistence of regional variations with some severe depopulation and ageing in rural areas in the west (Table 2.4). The ECECs are moving closer to the West European pattern. The Czech Republic's birth rate has dropped from 15 ptp in 1970 to only nine today, a decrease of 30 per cent in 30 years. Meanwhile, life expectancy is rising in the north but not in the SEECs (Table 2.5).

The removal of draconian measures to stimulate the birth rate is having a negative effect on natural increase, which may be compounded by the uncertainties of the transition. The peak reproductive age group for women in Poland is shifting from 20–4 to 25–9, reflecting changing social/professional aspirations: with

Table 2.3 Population 1950–2050

	Population (000)								Annual change (per cent)			
	¹1950	¹1970	¹1990	¹1995	¹2000	²2010	²2030	²2050	1950–90	1990–2000	2000–50	1950–2050
Albania	1230	2138	3289	3177	3113	3347	3957	4322	+6.68	-0.53	+0.78	+2.51
Bulgaria	7251	8490	8718	8499	8225	7753	6766	5673	+0.51	-0.56	-0.62	-0.21
FCS	12389	14336	15562	15680	15631	15522	14555	12665	+0.64	+0.04	-0.40	+0.02
Czech Rep.	na	na	na	10325	10244	10066	9229	7829	na	-0.16³	-0.47	na
Slovakia	na	na	na	5355	5387	5456	5326	4836	na	+0.12³	-0.20	na
Hungary	9338	10338	10365	10227	10036	9627	8627	7488	+0.27	-0.32	-0.51	-0.20
Poland	24824	32526	38119	38610	38765	39190	38680	36256	+1.34	+0.17	-0.13	+0.46
Romania	16311	20253	23207	22731	22327	21525	19335	16419	+1.06	+0.38	-0.53	*
Fmr. Yugoslavia	16345	19663	22808	22428	23095	23588	23227	21777	+0.99	+0.13	-0.11	+0.33
B&H	na	na	na	3415	3972	4330	4250	3767	na	+3.26³	-0.10	na
Croatia	na	na	na	4493	4473	4403	4098	3673	na	-0.09³	-0.36	na
Macedonia	na	na	na	1963	2024	2142	2286	2302	na	+0.62³	+0.27	na
Slovenia	na	na	na	1990	1986	1951	1760	1487	na	-0.04³	-0.50	na
Yugoslavia	na	na	na	10567	10640	10762	10833	10548	na	+0.14³	-0.02	na
Total	87688	107744	122068	121352	121192	120552	115147	104600	+0.98	-0.07	-0.27	+0.19

Source: FAO database <http://www.apps.fao.org>

Notes
¹ Estimate ² Forecast ³ 1995–2000 only
* Less than +0.1

Table 2.4 Population growth and migration 1994–5 and 1997–8

	1994–5						1997–8[1]					
	A	B	C	D	E	F	A	B	C	D	E	F
Albania	22.3	5.9	+16.4	−3.8	+12.6	na	18.9	5.8	+13.1	na	na	15.0
Bulgaria	9.0	13.4	−4.4	0.0	−4.4	15.5	7.8	14.5	−6.7	0.0	−6.7	16.0
Croatia	11.1	11.2	−0.1	+5.4	+5.3	9.6	11.3	11.5	−0.2	+8.6	+8.4	8.2
Czech Rep.	9.8	11.4	−1.6	+1.0	−0.6	7.8	8.8	10.8	−2.0	+1.1	−0.9	5.6
Hungary	11.1	14.2	−3.1	0.0	−3.1	11.1	9.8	13.8	−4.0	0.0	−4.0	9.8
Macedonia	16.8	8.2	+8.6	+1.1	+9.7	22.6	14.8	8.3	+6.5	−1.0	+5.5	15.7
Poland	11.8	10.0	+1.8	−0.5	+1.3	14.3	10.5	9.8	+0.7	−0.3	+0.4	9.9
Romania	10.6	11.8	−1.2	−0.8	−2.0	22.5	10.5	12.2	−1.7	−0.5	−2.2	21.3
Slovakia	11.9	9.7	+2.2	+0.6	+2.8	11.1	10.9	9.8	+1.1	+0.3	+1.4	8.8
Slovenia	9.6	9.6	0.0	+0.2	+0.2	6.0	9.1	9.6	−0.5	−1.7	−2.2	5.2

Source: Eurostat 2000: 13–29

Notes

A Births; B Deaths; C Natural increase; D Inferred migration; E Total change; F Infant mortality (per thousand live births)

[1] Albania 1998 only; Macedonia 1997 only

All figures are calculated per thousand of the total population except where otherwise stated

Table 2.5 Fertility, marriage, divorce and life expectancy 1993–8

| | Children[1] (per woman) | | Marriages[2] (per th. pop.) | | Divorces[2] (per th. pop.) | | Life expectancy at birth (yrs)[3] | | | |
| | | | | | | | Women | | Men | |
	1993	1998	1993	1998	1993	1998	1993	1998	1993	1998
Albania	2.7	2.6	8.2	8.8	0.7	0.6	na	75.4	na	68.5
Bulgaria	1.5	1.1	4.7	4.3	0.9	1.3	74.6	74.3	67.1	67.1
Croatia	1.5	1.5	5.3	5.4	1.0	0.9	na	77.0	na	70.2
Czech Rep.	1.7	1.2	6.4	5.3	2.9	3.1	76.4	78.1	69.2	71.1
Hungary	1.7	1.3	5.3	4.4	2.2	2.5	73.8	75.1	64.5	66.4
Macedonia	2.2	1.8	7.3	7.0	0.3	0.5	74.0	74.5	69.6	70.3
Poland	1.9	1.4	5.4	5.4	0.7	1.2	76.0	77.3	67.4	68.9
Romania	1.5	1.3	7.1	6.5	1.4	1.8	73.3	73.3	65.9	65.5
Slovakia	1.9	1.4	5.8	5.1	1.5	1.7	76.7	76.7	68.4	68.6
Slovenia	1.3	1.2	4.5	3.8	1.0	1.0	77.4	78.7	69.6	71.1

Source: Eurostat 2000: 13–29

Notes
[1] Albania 1995 and 1998
[2] Croatia 1994 and 1998; Macedonia 1993 and 1997
[3] Bulgaria 1995 and 1996; Croatia 1997; Hungary 1993 and 1997; Macedonia 1994 and 1997

a higher priority for housing/car ownership and better education. There has been 'demographic shock' in East Germany: a collapse of births and new marriages suggestive of a severe reduction in material well-being. Given very high female unemployment, a choice may have to be made between keeping a job and having a child. There is a need for new structures on the jobs market to combine the desire for work with the role of the mother. Meanwhile, the death rate is increasing, perhaps linked with transition and the downward pressure on wages and resulting hardship. Eberstadt (1993) sees falling rates of natural increase linked with inadequate health care and deficiencies in housing conditions and environmental quality. In the SEECs especially, death rates may be related in part to poverty and inadequate health care. Low incomes lead to the purchase of cheaper food products which means consuming more bread and low-quality cereals rather than meat, fish and fruit (necessary for an adequate intake of vitamins and mineral elements). This in turn leads to obesity, but many children are also anaemic, showing fatigue after minimum effort and low intellectual performance. Protein deficiency affects immunity and increases the risk of infection.

Migration is a significant factor, with negative natural increase accentuated by net out-migration in most countries, but rates have been generally moderate. Only Albania appears to have lost very heavily through a surge of emigration, according to the FAO database (cancelling out a high level of natural increase) although the movement is not quantified in the Eurostat data used in Table 2.4. Movements in B&H and Croatia presumably reflect adjustment following the end of the civil war. However, Albania is expected to resume overall growth over the next half century while all other countries (except Macedonia) are expected to decline – and by a more rapid rate where decline has already been evident during the 1990s. Albania certainly has the youngest population with about half aged under 30 years, followed by Macedonia with 40 per cent below 25 and Poland with 37.5 per cent in the same bracket (Table 2.6). The net result should see Albania increasing its share of the region's population from 1.40 per cent in 1950 to 4.13 per cent in 2050, while Poland is also destined to advance considerably from 28.31 per cent to 34.66 per cent. A 'realistic' forecast for Bulgaria anticipates decline by 5.8 per cent annually during 1995–2000 rising to 7.7 during 2015–20 when population will be 7.12 m. compared with 8.38 in 1995. FAO projections to 2050 bring out an average annual decrease of –0.62 per cent over a half century, followed by –0.53 per cent in Romania and –0.50 per cent in Hungary and Slovenia; which could create problems for the labour market. Croatia's population – squeezed by a lower birth rate and emigration that took place during the recent war – could fall to almost 3.5 m. by 2050, while the ageing process will affect Poland from 2005; accelerating after 2010: the percentage of population of retirement age will rise from 14 per cent in 1998 to 20 per cent in 2020 (although in 2010 Poland will still have 2.7 m. more working-age people than in 1991).

The picture varies considerably within individual states. Figure 2.1 provides profiles for the counties of Romania over a period of some two decades and the east–west split is quite clear with relatively high levels of natural increase in the east whereas the demographic transition in the west has reached the same

Table 2.6 Age structure 1993–8

	1993 % distribution						1998 % distribution					
	0–14	15–24	25–44	45–64	65–79	80+	0–14	15–24	25–44	45–64	65–79	80+
Bulgaria	19.0	14.5	27.3	24.9	11.8	2.4	16.8	14.8	27.3	25.5	15.6	2.1
Croatia[1]	20.0	14.0	30.0	24.0	12.0	2.0	20.0	14.0	30.0	24.0	12.0	2.0
Czech Republic	20.0	15.9	28.5	22.8	10.3	2.6	17.4	16.4	27.6	25.0	13.6	2.4
Hungary	19.0	15.4	28.3	23.5	10.9	2.8	17.5	15.8	27.8	24.5	14.4	2.5
Macedonia	25.1	16.3	29.9	20.2	7.1	1.3	23.5	16.4	29.8	20.9	9.2	1.1
Poland	24.1	14.8	30.7	19.9	8.4	2.1	21.1	16.4	29.1	21.7	11.7	2.0
Romania	22.1	17.1	27.3	22.3	9.3	2.0	19.2	16.8	28.6	22.7	12.7	1.8
Slovakia	24.1	16.0	30.1	19.3	8.4	2.1	21.0	17.2	29.5	21.0	11.2	1.9
Slovenia	19.6	14.6	31.5	22.9	8.9	2.5	17.0	14.9	30.9	23.9	13.2	2.3

Source: Eurostat 2000: 13–29

Note
[1] 1994 not 1993

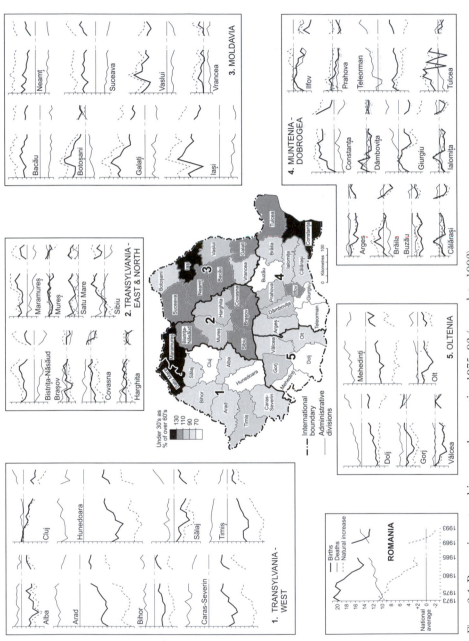

Figure 2.1 Romania: natural increase by counties 1975–93 (map data 1993)

The following labels appear within the figure:

1. TRANSYLVANIA - WEST
Alba, Arad, Bihor, Caraş-Severin, Cluj, Hunedoara, Sălaj, Timiş

2. TRANSYLVANIA - EAST & NORTH
Bistriţa-Năsăud, Braşov, Covasna, Harghita, Maramureş, Mureş, Satu Mare, Sibiu

3. MOLDAVIA
Bacău, Botoşani, Galaţi, Iaşi, Neamţ, Suceava, Vaslui, Vrancea

4. MUNTENIA - DOBROGEA
Argeş, Brăila, Buzău, Călăraşi, Constanţa, Dâmboviţa, Giurgiu, Ialomiţa, Ilfov, Prahova, Teleorman, Tulcea

5. OLTENIA
Dolj, Gorj, Mehedinţi, Olt, Vâlcea

ROMANIA
Births, Deaths, Natural increase
1973, 1975, 1980, 1985, 1989, 1993
National average

Under 30's as % of over 60's: 130, 110, 90, 70

International boundary, Administrative divisions

0 Kilometres 100

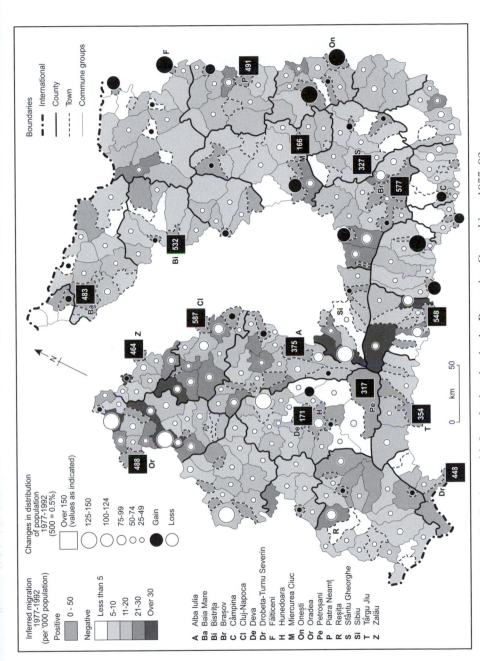

Figure 2.2 Population distribution and inferred migration in the Romanian Carpathians 1977–92

Source: Romanian Census 1977 and 1992

stage as in Hungary. While natural decrease is occurring in the west, there is still significant increase in Moldavia and the north Transylvanian county of Maramureş. Traditionally Moldavia has been unable to provide jobs for its expanding population and people from this province have therefore migrated to other parts of their country including Bucharest, Transylvania and the west-ernmost regions of Banat and Crişana. The situation can also be examined with reference to the Romanian Carpathians in Figure 2.2 which shows changes in the overall distribution between 1977 and 1992. Districts which have increased their share are all urban areas, especially the main polarising centres, whereas the rural districts have seen their shares fall through net out-migration compounded by falling natural increase. However, the losses are greater in the west whereas the situation is aggravated by high out-migration inferred from county-level data on natural increase. Whereas the east enjoys considerable local employment in mining and wood processing, the west and southwest have experienced annual out-migration exceeding 20 ptp, flowing mainly towards the county towns.

Gender

The gender balance is clearly tilted in favour of women (51.2 per cent) with only Albania and Macedonia registering a male majority (Table 2.7), though there are strong regional variations and the disposition of young women to migrate from remote rural areas of Poland produces male:female ratios of 7:1 in parts of the northeast. A historic imbalance arising from war losses is aggra-vated by selective international migration. Women have significantly higher life expectancy in all countries (Table 2.8), but while female activity rates are low

Table 2.7 Population by gender 1950–2050

	Percentage of the population which is male in:							
	1950[1]	*1970[1]*	*1990[1]*	*1995[1]*	*2000[1]*	*2010[2]*	*2030[2]*	*2050[2]*
Albania	50.3	50.5	51.3	51.2	51.1	50.9	50.5	50.1
Bulgaria	50.0	50.0	49.3	48.9	48.6	48.2	47.8	47.8
Fmr Czechos'via	48.6	48.7	48.7	48.7	48.7	48.6	48.4	48.3
Czech Rep.	na	na	na	48.7	48.7	48.6	48.3	48.2
Slovakia	na	na	na	48.8	48.7	48.7	48.5	48.4
Hungary	48.0	48.0	48.0	47.9	47.8	47.8	48.0	48.3
Poland	47.7	48.6	48.7	48.7	48.6	48.5	48.3	48.5
Romania	48.2	49.1	49.3	49.2	49.0	48.9	48.4	47.8
Fmr. Yugoslavia	48.3	49.1	49.3	49.3	49.3	49.3	49.4	49.5
Bosnia & H	na	na	na	49.5	49.5	49.5	49.4	49.2
Croatia	na	na	na	48.3	48.3	48.3	48.5	48.7
Macedonia	na	na	na	50.1	50.0	49.9	50.0	50.1
Slovenia	na	na	na	48.8	48.7	48.7	48.5	48.4
Yugoslavia	na	na	na	49.7	49.7	49.7	49.8	49.9
Total	48.3	48.9	49.0	48.9	48.8	48.7	48.6	48.6

Source: FAO database <http://www.apps.fao.org>
Notes
[1] Estimate [2] Forecast

Table 2.8 Gender 1998: demographic, social and economic aspects

	Female as a percentage of male:								Female employment share:				
	A	B	C	D	E	F	G	H	I	J	K	L
Albania	na	na	na	65	122	na	na	na	na	na	na	na
Bulgaria	105	na	na	82	194	99	101	95	36.0	na	na	na
Croatia	93	110[1]	na	77	na	122	126	110	48.8	na	10.6	na
Czech Rep.	106	110	126[1]	73	102	164	182	138	32.9	37.4	8.4	55.2
Hungary	109	113[1]	130[1]	73	134	82	85	78	24.0	39.1	8.3	54.1
Macedonia	na	106[1]	115[1]	63	133	115	121	106	24.0	39.1	8.3	54.1
Poland	106	112	128[1]	76	185	135	143	116	44.3	34.6	8.8	55.5
Romania	104	112	120[2]	79	114	94	87	111	50.0	40.4	12.7	48.9
Slovakia	106	112	127[1]	75	116	111	121	100	30.9	37.7	8.8	57.7
Slovenia	105	111	128[1]	80	130	104	103	108	46.7	na	9.9	na

Source: Eurostat 2000: 13–39

Notes

[1] 1997 [2] 1996

A Total population
B Life expectancy at birth
C Ditto: age 65
D Activity rate

E Graduates from tertiary education
F Unemployment rate
G Ditto: people aged 25 and over
H Ditto: people aged below 25

I Agriculture
J Industry
K Construction
L Services

in terms of formal employment, women experience higher unemployment, especially among the population aged 25 or over. Women are very prominent in participation in the tertiary sector of education, particularly in Bulgaria and Poland – but significantly in Hungary, Macedonia and Slovenia too. Women are particularly prominent in the service sector, while being strongly in the minority for agriculture and industry (albeit with considerable informal activity in the former case) and even more for construction.

As communist population policy enhanced the role of women in positions of leadership, they became more effective in the workplace, bargaining over pay when the organisations that were supposed to represent their interests were insufficiently supportive. Communism also gave women access to a variety of occupations traditionally reserved for men. In Poland, women became prominent in business, banking and finance; while the feminisation of professions like book-keeping gave women an advantage for entry into accountancy. In general communists went further than Western countries in providing abortion clinics (beginning in Bulgaria in 1956) and day nurseries. This anticipated progressive Western thinking, but abortion policy could change dramatically and in 1966 Romania introduced strict controls in order to halt a catastrophic decline in the birth rate. 'And changes came about because the ruling elite deemed they were necessary' to get women on to the labour market (De Silva 1993: 312) not because of pressure from the feminist movement. In Albania, the one country that found it unnecessary to introduce abortion, women were active in political movements though they did not achieve fundamental change towards equality. And while equality on the labour market gave women independence it also meant the forced imposition of unsuitable jobs and neglect of the family.

Gender issues are now highly relevant to the transition after half a century of dependence on the 'patriarchal state'. Curiously, while women have been prominent in some of the environmental movements, they have yet to surface in transition politics and even the strength of women's protest movements has subsided. Yet women have suffered disproportionately heavy unemployment: they have lost status and have been effectively neutralised by withdrawal into the private sphere of the family (Watson 1993). They bear the brunt of poor services, e.g. in rural areas where there may be just one nurse for every two to three villages; also where there is much overcrowding and malnutrition linked with alcoholism and domestic violence. And lower family incomes arising from unemployment tend to impact disproportionately on women in terms of domestic consumption through a 'tradition' of priority for children and male workers. Abortion has been an important issue, with each country searching for an appropriate contract that will safeguard population in the future. In Romania the very tough communist anti-abortion laws of the late 1960s (referred to above) were relaxed in 1989 and abortion is now available on request up to 12 weeks (and thereafter if medical evidence is supportive). French organisations have been helpful with information on contraception. However, regarding health care there is a need for clinics, media and NGOs fighting for women's rights to heighten women's awareness of their right to have access to health services: including modern contraception methods; sex education and prevention of sexually-transmitted diseases; cervical cancer screening; and mother and child care.

The reduction in jobs and the incidence of unemployment among women have been very noticeable in East Germany, for women seem to be unemployed for longer periods than men and face greater difficulties getting new jobs. Moreover, despite equal opportunity legislation, there is concern over the erosion of child care facilities which complicates female employment for young mothers. Under communism, women could be mothers and also enjoy independence in the context of single-parent families. They enjoyed easy access to nurseries and family planning assistance, although in the West the communist strategy was seen as tantamount to locking women in a traditional reproductive role. Now, with rising costs for nurseries in East Germany, it is difficult for unmarried mothers with more than two children to get a job. Forced to choose between family and career, some women are opting for sterilisation in order to compete for jobs. Families may become more stable if women resist the cyclical pattern of marriage, child-bearing and divorce. However, with the legacy of full integration into employment and economic independence they will be reluctant to accept the West German model of only partial labour market integration.

In Poland, too, women carry domestic burdens while bearing the brunt of unemployment, especially in the old industrial areas which have traditionally had high rates of female employment, while assuming responsibility for the family budget at a time of economic stringency with reductions in subsidies, social services and real wages. Initial passivity over the loss of many state-financed day-care facilities, deterioration in the health service and rising poverty gave way to a measure of organisation as the Catholic church became involved in debates on social issues. Women who desperately need to retain their salaries in the workplace have faced pressures to stay at home from 'traditional and highly reactionary images of the family' (Hardy and Rainnie 1996: 257) which emphasised female domesticity and downgraded the role of women in the labour market. In particular, women have tried to organise to fight legislation that would permit abortion only when the life or health of the mother was threatened and envisaged imprisonment for doctors and patients contravening the regulations. But while women's issues have been much more adequately articulated since a social democrat government came to power in 1993, the urban–rural divide inhibits consensus and most women seem to accept a prime domestic role should be in the home. But regrets have been expressed that even with the issue of women's right to abortion – a measure that many would see as progressive – the communist regime failed to convince the population as to its merits. But this may be part of a wider issue of equal opportunity which requires a change of attitude in both the home (with marriage a true partnership) and the workplace, especially in areas with a low level of female participation in local economic and political institutions. However, the situation varies between countries: according to Pailhe (2000) discrimination is stronger in Poland and Slovakia than in Hungary where the employment structure – with a large service sector – is more favourable to women.

Migration

Given the spatial imbalance between job losses and vacancies – and the inadequacies of public transport which complicate commuting – it might be expected that unemployment would stimulate migration both nationally and internationally. But despite some problem areas with former state farms and heavy industries (often in one-industry towns), the scale of movement is relatively low, which reflects 'severe inefficiencies in matching workers with jobs' (Blanchard 1994: 7). Quite apart from a general 'clannishness', housing is in short supply (factories no longer have their own 'workers' hotels' and renting apartments is very expensive) while many young people feel they have obligations towards dependent relatives: 'reverse dependency' means that young professionals often look after parents and grandparents while marrying young means that young executives tend to have family commitments. Moreover, 'the national housing shortage, which affects even the more developed urban agglomerations, serves to maintain the spatial structure of unemployment' with high levels in the north and northeast (Weclawowicz 1996: 149). The costs of immobility are especially high given the shortages in skilled positions. They also apply through constraints on factory relocation (say) from a city centre into a rural area or from the capital city to a provincial centre, since many employees may be lost unless there are better financial incentives and company housing.

Migration between regions shows a significant drift from peripheral to central areas – and also from country to town, though not as strongly as before. This is particularly evident where the exporting areas continue to maintain natural increase. Albanians have flooded into Tirana where the population has rocketed from 0.24 to 0.60 m. during the 1990s, while Durrës has grown substantially as the 'second city' from 85,000 to 130,000. Natural decrease now affects all Hungarian countries except Hajdu-Bihar, Borsod-Abaúj-Zemplén and Szabolcs-Szatmár-Bereg but there is heavy out-migration from these areas. In East Germany, too, rural agricultural areas have suffered the highest levels of out-migration, along with regions with poorest housing quality and fewest opportunities for finding subsidised employment, on-the-job training or places in higher education; rural agricultural regions; and regions with poorest housing quality. While there was an emphasis on urban concentration in the GDR period (outside the old industrial areas) there is now suburbanisation around large cities, but especially Berlin and Potsdam where the process extends further into Brandenburg for those who choose to commute to employment opportunities in the Berlin region rather than migrate to the West. Also the Land capitals are generally strong, especially Dresden. In other countries, rural–urban movement has continued during the 1990s and only in Albania and Macedonia is the rural population still increasing in absolute terms. Rising urban unemployment might be a disincentive to new potential migrants and a stimulus for return migration back to the countryside, but the predominant flow seems to be in the other direction. Emigration generates internal migration flows especially when the migrants leave the best economically developed zones and the Roma are attracted into former German villages. Counterurbanisation might also be expected in the case of professionals moving to suburban zones with good conditions for high-tech industry, although the scale of such movement seems to be rather small at present.

Emigration

The removal of bureaucratic controls on migration through simpler frontier formalities and the abolition of visas for many international journeys have also had an adverse effect on population growth in the region. The first dramatic upheaval occurred in East Germany and had immediate political consequences. While German unification certainly benefited from the cooperation of the allies in the 'Two plus Four' talks, alternative strategies were constrained by extremely high migration levels. Migration accelerated the unification process because it was only moderated by a decision to work for rapid and complete unification and a radical rebuilding of the East German economy. Elsewhere, massive outpourings of population seemed a probable response to the poverty problem at the onset of a transition from a 'low work intensity mass underemployment system to one characterised by highly intensive work practices and mass unemployment' (Hardy and Rainnie 1996: 258).

There was a surge of emigration by Albanians into Greece and Italy when emigration controls were relaxed in 1989, although apart from Greeks living in southern Albania who wished to move into Greece, many of those attracted by the lure of consumer goods have since returned (reportedly an estimated 100,000 under 'Operation Scoupa') and much of the movement now is concerned with temporary work. Nevertheless there are 230,000 Albanians in Greece legally (and roughly the same number illegally), with estimated figures of 150,000 and 50,000 respectively for Italy. Faster privatisation is needed to bring in the investment and technology to create more jobs. In Bulgaria the end to persecution of the Turks minority has made for stability. Nevertheless, economic recessions cause displacement, especially of skilled and educated people (potential administrators, entrepreneurs and professionals) who have left Albania, Bosnia & Hercegovina, Bulgaria and Romania in despair. The National Statistics Institute in Sofia calculates that, with emigration and the falling birth rate, only 6.8 m. people will reside in Bulgaria in 2020 (a reduction of some two million – just over one-fifth – of the present level).

But the potential for mass-migration has been reduced by the stabilising effect of land restitution. In the Podhale region of the Polish Carpathians, with a tradition in pluriactivity on the periphery of the former socialist economy, any contraction in local employment will immediately generate movement to take up opportunities in commerce in other places, including seasonal migration linked with family contacts as far afield as Chicago! But people in central Poland who have been 'highly integrated into the central economy' feel an acute sense of loss if unemployment forces them back into agriculture and will tend to accept any other employment that may be available in the local area (including sweatshops run by kinship networks), claiming the need to help on the farm as a reason for not migrating (Pine 1998: 121). There are also constraints over the lack of information and contacts relating to potential migration destinations, although movement could increase in the future as the opportunities become better known through official and kinship networks. Some groups may be demonised abroad: thus Greeks associate the influx of Albanians with an increase in petty crime, while the government in Athens is concerned at the cultural impact of Muslims in an Orthodox state. However, neighbourly relations have now been established, with Greece considering the longer-term economic bene-fits of dealing with Albania, while Albania appreciates the political problems in Greece arising from xenophobic opportunists in the regional centres of Ioannina and Konitsa near the Albanian border. Exaggerated fears about the risks of prostitution among women who migrate may also be alleviated with time. And the 'farming constraint' may also become less significant over time as the margin-ality of small-scale farming becomes clear.

Although the Iron Curtain has been swept aside, the EU attempts to regu-late migration across the 'Rio Grande' frontier which draws many migrants who wish to surmount the 'wage precipice' and enjoy higher living standards. However, relatively easy entry formalities in Germany in the early 1990s resulted in heavy in-migration from the ECECs with 127,500 from Bulgaria, Romania and former Yugoslavia in the first half of 1991; 30,100 in the second half and 107,600 in the first half of 1992 (compared with 32,000; 13,900; and 18,900 respectively for Turks and Vietnamese). There are now around 1.0 m. people from the region in Germany (excluding Germans from FGDR or Eastern Europe) comprising large Czech, Hungarian and Polish communities plus some people from FSU and the overwhelming proportion of Yugoslavs who were able to leave without difficulty to become Gastarbeiter. But, with a hidden labour reserve of 7.5 m., Germany is no longer a country of immigration and is revising agreements affording access to its labour market, and migrants from Angola, Cuba, Mozambique and Vietnam, introduced by the FGDR government into industrial centres like Chemnitz, Eisenhüttenstadt and Rostock, have been offered free flights home. So there is now much opposition to economic migrants who want to cross the 'Wohlslandsgrenze' (prosperity frontier). The arrival in Germany of over three million 'Aussiedler' since the 1950s constitutes a special case. During the 1990s many have arrived from the FSU but also from Romania, following the cancellation of the education tax imposed by the communist government. However, the EU will soon have to face the looming problem of free movement of labour in the context of association and ultimately enlarge-ment, expressed in one source as a stark choice 'between imports and immigrants' (Shepley and Wilmot 1995: 55). Even if the pent-up pressure is not over-whelming, Austrian and German growth areas may be affected at a time when there is considerable domestic opposition to 'Gastarbeiter'. On the other hand restrictions would have the benefit of avoiding the need for Schengen standards on the Czech–Slovak and Hungarian–Romanian and Polish–Ukrainian borders.

Meanwhile, a significant number of people have moved from West to East, including those returning to ECE after spending the communist years in the West as émigrés and are able to bring in capital and market behaviour. All the coun-tries have been affected but especially Hungary which lost 184,500 people in

1956: there was a trickle of below 1,000 return migrants per annum during the 1970s and 1980s, rising to 3,500 in 1991. There is some ambivalence towards those who return: they are acceptable if they work quietly (indeed there is a tradition of respect in rural areas for people who go abroad and prosper) but there has been some resentment against returnees who enter politics. For there is a sense of 'opting out' of the hardships of communism. In Poland such sentiments are tinged with suspicion that the returnees may have actually benefited under communism through the 'Polonia' companies of the 1980s which provided for joint ventures under favourable terms when the foreign partner was of Polish origin. However, an element of public disquiet has not discouraged some from entering politics, such as I. Raţiu in Romania (senator and presidential election candidate in 1990) and S. Tyminski in Poland (presidential election candidate in 1992).

Emigration: international movements within the region

Migrants to Slovenia from other parts of former Yugoslavia have generally stayed put despite poorer job opportunities. But fighting that occurred when the old federation broke up – and again during the Kosovo crisis of 1999 – brought substantial displacement which has largely been contained by visa formalities and the EU decision in 1992 that refugees could be sent back to a third country, through which they had travelled, if that country was safe (i.e. Croatia, Hungary or Slovenia in the case of refugees from Bosnia & Hercegovina). By the end of 1993 4.24 m. people were receiving some form of UN assistance, mostly in Bosnia & Hercegovina. Meanwhile, Kosovars have moved into Albania where they may mix rather uneasily with the local Albanians who have lived most or all their lives sheltered by the former Hoxha regime; though some north–south migration which has taken them to spare housing in the former Greek villages has been portrayed emotively as an exercise in ethnic cleansing. Since the war there has been heavy movement from Albania into Kosovo which is not restricted to Kosovars. Meanwhile, complex movements have occurred between the Czech Republic and Slovakia, including Slovaks formerly employed in Ostrava moving back to Čadca and Dolný Kubín, though the possibility of Ruthenians in Eastern Slovakia returning to Trans-Carpathian Ukraine has not yet been realised, given the poor economic prospects. Finally, Roma from Romania moving into Poland – encouraged by both the lack of visas and the tolerance of Poles – have made money through running pickpocketing gangs and spinning 'hard luck' stories of war in Bosnia and persecution in Slovakia, in order to finance new housing in Romania for extended families. Hungary has faced the possibility of return migration by Hungarians not only from Croatia due to the civil war, but also from Slovakia and Transylvania although in the event numbers have been small. Today much of the movement is temporary, in connection with training programmes and work contracts by people who normally move without their families, with FCS and Hungary drawing from Poland and the Balkans. Romanians take harvest work in Serbia, while suitcase traders sell goods obtained at subsidised prices in CIS shops and then return home with the proceeds converted from ECE currencies changed into dollars.

Most ECECs are expected to be fairly stable in migrational terms in future, although 'aggressive' migration pressure may be expected from the SEECs and FSU as people try to escape poverty and unemployment. Even so, movement is not on the scale feared in 1990–1 when some countries erected defences against an influx that never materialised. This transit traffic can result in some spending which is beneficial to the economy, but this is outweighed by the costs of the security operation – deportations involve the state in air tickets and hotel accommodation – as the EU presses associated countries to exercise firmer control. The recent annual rate of detected illegal crossings of the Polish–German border exceeds 10,000. And as a migration buffer zone develops around Western Europe, ECECs like the Czech Republic, Hungary and Poland may well have a positive migration balance on account of people from the FSU (especially Armenia and Ukraine), Africa and Asia (Afghanistan, Bangladesh, India, Iraq, Kurdistan, Sri Lanka and Vietnam) who wish to reach Western Europe but who may well settle in ECE in the interim. There were an estimated 150,000–200,000 illegal immigrants in FCS at the end of 1992 (1.5 per cent of the total population) and there are thought to be 10,000–30,000 at any one time in Bulgaria. Hungary has used barracks vacated by returning FSU soldiers to accommodate economic refugees, although many stay in ECE semi-permanently and find gainful employment.

Africans do not generally get involved in commerce but for other groups arriving in Romania from Bulgaria or Moldova, setting up a business may be a cheap way to get a residence permit. Businesses introduce cheap poor-quality goods (of Chinese, Indian, Middle Eastern or Turkish origin) in tune with the low spending power of the population. Arabs establish supermarkets, restaurants/fast food outlets, generally trade legally (apart from some black market currency dealings) and integrate well with Romanians who may be employed as bakers, drivers and shop assistants. The Chinese are into bulk buying/selling as well as restaurants and shops, but tend to live in isolation and employ only Chinese: most do not intend to stay (like the Indians who also intend to move on and maintain distance in their relationships) and make little effort to integrate, though they create no conflict with Romanians. However, it seems that Romanians show most tolerance towards Arabs (who are particularly interested in staying permanently) and Chinese, less so towards Africans. However, there is no proper monitoring and few organisations concern themselves with immigrants apart from an international organisation for migrants.

Ethnic minorities

Nowhere in the region is the nation state ideal achieved whereby compact nations are contained neatly within their respective territorial cradles. Rather, there is ethnic diversity, invariably over-dramatised as a source of stress rather than cultural enrichment, given the evidence for discrimination and genocide. But if human resources are to be effectively harnessed for growth and prosperity this constraint has to be overcome. Arguably, minorities are more likely to emerge in the European periphery in order to stiffen resistance to the debilitating effects of 'backwash' into the core. 'There is no question that the further we move to the east the greater the number of specific problems caused by nationalism that can be seen today' (Skubiszewski 1993: 25). This might include ethnic movements for autonomy or independence where the upper-middle-class elements (who tend to lead ethnic movements) have most to gain from a redistribution of power to reduce or eliminate dependence on the host country.

One may thus conclude that ethnic minority groups arise as a 'natural' component of the East European cultural landscape. They are an inevitable part of the evolving political process that cannot always be harmonious and well ordered. Some tensions are deep-seated, for the Roma have attracted hostility for their alternative lifestyles, most evident through casual employment, alleged criminality and 'contingency housing'. Attempted assimilation by the Habsburgs during the Enlightenment, to bring an end to nomadism and extended families, produced a situation in the pre-socialist period where 'they lived on the margins of society but had well-defined "niches" in the socio-occupational structure' as blacksmiths, brickmakers, horse dealers and musical entertainers (Andorka 1997: 81). The Jews came under pressure because of their perceived economic dominance. They were exploited by Polish landowners who oppressed the peasantry by using them as managers and tax collectors; also their language, dress and religion tended to 'distance' them from the mainstream communities.

In the context of nation states, mainstream groups in general have become less tolerant towards minorities, for people who are 'disappointed in their past, probably largely disappointed with their present and uncertain with their future will thus seek myths, most likely nationalistic and xenophobic myths that would enable them to blame someone else for their misfortunes' (Kochanowicz 1997: 68). Groups that may normally be respected and at least tolerated may thus be demonised on account of their identity, including cultural/religious minorities lacking territorial ambitions. This can be related in part to the heightened perceptions of fifth columnists manipulated by predator neighbours: real and potential threats are not difficult to construe bearing in mind complicated legacies whereby each group can create its own history and minorities instinstively identify with 'mother countries' elsewhere in the region. Potential subversion may lie dormant for long periods, and then emerge as a threat when there is a historical experience to cast that minority in an exploiting role. In the early transition years, neither Romania nor Slovakia enjoyed frontier guarantees; this led to sensitivity over activities by the Hungarian minorities who were supported by a neighbouring kin state which had (albeit temporarily) annexed parts of Romania and Slovakia as recently as 1938–40. However, the human rights of minorities are usually safe, although they may encounter problems when vested interests are eroded or violently contested. Again, the responsibility of nation-building in the Balkans by definition compelled a radical stance on the 'Turkish Question' and hence some animosity towards Turks. In Bulgaria, Slavic-speaking Muslims or 'Pomaks' (derived from the word 'pomachamedanci' which means Islamicised) have been treated as outcasts because they opted for the Islamic faith to retain land ownership in the Ottoman period. The communists tried to impose resettlement to the Rhodopes – because they crossed the border to maintain contact with relatives in Greece – and a change from Arabic-Islamic to Slavic-Christian names as a process of rebirth. By contrast, a small self-contained minority like the Sorbs in Lusatia (the Görlitz area of East Germany), posing no significant threat to state security and identifying closely with the German community, was not pressured by the majority group.

However, data are not entirely satisfactory because while most states publish information quite comprehensively, the Europa Yearbook does not provide an authoritative set of figures for Hungary or Poland (Table 2.9). Estimates have been used on the basis of the wide-ranging survey by Foucher (1994). Even so, figures for Roma minorities are particularly suspect since many members of these communities do not always declare themselves as such. And finally the figures for Bosnia & Hercegovina are misleading since the Muslims (Bosniaks) are regarded as the titular nationality although strictly speaking, given the state's confederal status, Croats and Serbs should be included as well. However, with these reservations in mind it is evident that (excluding Bosnia & Hercegovina) the minorities are most prominent in Yugoslavia (37.4 per cent) and Macedonia (33.4 per cent), followed by Croatia (21.9 per cent), the Czech Republic (18.8 per cent), Bulgaria (14.3 per cent), Slovakia (14.2 per cent), Slovenia (12.2 per cent), Romania (10.5 per cent) and Hungary (8.4 per cent); this leaves Albania and Poland as the most homogenous states (minorities only accounting for 2.0 per cent and 2.2 per cent respectively).

Immediately after the 1989 revolution local politics became potentially explosive as heightened national awareness clashed with pressures from ethnic minorities seeking a fairer economic distribution and enhanced political rights including regional autonomy: ethnic minorities became victims because of the involuntary opposition between nationalism and individual rights. And while such ethnic reawakening may be a progressive force in the new Europe, there is also a danger that memories of a perceived 'golden age' will inspire unattainable goals and legitimise aggression. As Germans manifested a yearning for unity based on the constitution of the Federal Republic, the German minority in Romania took advantage of simplified emigration procedures and in particular the withdrawal of the education tax. Again, where minorities felt themselves vulnerable the removal of bureaucratic controls on movement gave ethnic tensions considerable migration potential. Koulov (1999) raises the case of 'regionally concentrated minorities' which are especially significant where they form a majority in the areas involved: parts of Bulgaria, Romania and Slovakia as well as former Yugoslav states other than Slovenia. 'Comparisons with the former Yugoslav republics confirm the broad correlation between the share of geographically concentrated minorities and the level of social conflict' (ibid.: 201). Under such circumstances Bulgaria's Turks continued to suffer as former communists in the Bulgarian Socialist Party exploited the dilemmas in ethnically mixed regions where new tensions arose over restitution because former Bulgarian owners claimed land while the Roma and the Turks failed to qualify. Poles in the Grodno region were perceived as subversive by the Belarus government and others living at Salcininkai (Lithuania) faced persecution for a resolution framed in 1990 to seek autonomy. However, the most extreme cases of ethnic violence have arisen out of the breakup of the former Yugoslavia.

Table 2.9 Ethnic minorities c.1990

| | Year | Population (,000) | | | | | Other groups individually | | |
		Total	Titular group	%	Other groups	%	1st	2nd	3rd²
Albania	1989	3182.4	3117.6	98.0	64.8	2.0	Grk 58.8	Mac 4.7	Oth 1.3
Bosnia&H.¹	1991	4377.0	1905.8	43.5	2471.2	56.5	Srb 1369.3	Crt 755.9	Yug 239.8
Bulgaria	1992	8487.3	7271.2	85.7	1216.1	14.3	Trk 800.1	Rma 313.4	Arm 13.7
Croatia	1991	4784.3	3736.4	78.1	1047.9	21.9	Srb 581.9	Msl 43.5	Slo 22.4
Czech Rep.	1991	10302.2	8363.8	81.2	1938.4	18.8	Mor 1362.3	Slk 314.9	Pol 59.4
Hungary	1990e	10360.0	9490.5	91.6	869.5	8.4	Rma 450.0	Ger 175.0	Slk 110.0
M'donia	1994	1945.9	1296.0	66.6	649.9	33.4	Alb 441.1	Trk 78.0	Rma 43.7
Poland	1990e	38120.0	37265.0	97.8	855.0	2.2	Ger 325.0	Ukn 300.0	Brs 200.0
Romania	1992	22810.0	20408.5	89.5	2401.5	10.5	Hng 1625.0	Rma 401.1	Ger 119.5
Slovakia	1997	5378.6	4614.5	85.8	764.1	14.2	Hng 568.3	Rma 87.8	Czk 59.0
Slovenia	1991	1966.0	1727.0	87.8	239.0	12.2	Crt 54.2	Srb 47.9	Msl 26.8
Yugoslavia	1991	10394.0	6504.0	62.6	3890.0	37.4	Alb 1714.8	Mnt 519.8	Hng 344.1

Source: Europa Yearbook 1999 and Foucher 1994

Notes
¹ The Bosniaks (Bosnian Muslims) are regarded as the titular group while the other leading groups (Croats and Serbs) are treated as minorities
² Discrepancies between the totals for the three groups mentioned individually and the total for all non-titular groups constitute an Others/
Undeclared category
e Estimate
Abbreviations for ethnic minority groups:

Alb	Albanians	Grk	Greeks	Oth	Others/Unaccounted
Arm	Armenians	Hng	Hungarians	Pol	Poles
Brs	Belarussians	Itn	Italians	Rma	Roma (Gypsies)
Bul	Bulgarians	Lth	Lithuanians	Rom	Romanians
Crt	Croats	Mac	Macedonians	Rus	Russians (including
Czk	Czechs (including Moravians	Mnt	Montenegrins		Lipovans)
	and Silesians)	Mor	Moravians	Sil	Silesians
Ger	Germans	Msl	Muslims	Slk	Slovaks

Slo	Slovenes		
Srb	Serbs		
Trk	Turks		
Ttr	Tartars		
Ukn	Ukrainians (including		
	Ruthenians)		
Vlh	Vlachs		
Yug	Yugoslavs		

The European agenda

It is now evident that ECE is working increasingly towards a European agenda emanating from the Council of Europe (CoE), the Organisation for Security and Cooperation in Europe (OSCE) and the EU. As transition states need security in terms of their defence and their economic arrangements, European institutions offer the best way forward. Apart from Slovenia, war in the former Yugoslavia has tended to isolate the successor states from the EU accession process and it is most desirable that the Stability Pact (emerging late in 1999) should gradually overcome this discrimination. For the European institutions reject the thesis that minority rights are an internal matter and provide a check on the double standards of states that 'often seem to be very willing to offer protection to their minorities outside their territories while denying even the existence of minorities within their territories' (Christopulos 1994: 173). The OSCE deals with human rights, as well as security and technology, and a high commissioner has held office since 1992. And the EU's 'Copenhagen Principles' supporting pluralism and a market economy are also being imposed on candidate countries as a boost to the development of civil society. But good inter-ethnic relations based on mutual respect and appropriate facilities for free expression are also being assisted by the CoE 'framework convention' on the protection of minorities (now being adopted in friendship treaties) and greater confidence in frontier stability as potentially antagonistic states anticipate NATO enlargement. The year 1992 saw a European Charter on Regional and Minority Languages and language policy for the region should now stress the benefits of bilingual knowledge.

There is also a positive contribution from the strengthening of local government. This answers Verdery's (1993) suggestion that ethnic tension reflects a lack of institutions (except at the centre) and pent-up frustration by minorities in lacking leaders of their own, even where they are in the majority locally. This concept can be further extended to cross-border 'territorial alliances' (and formally-constituted Euroregions) which have a crucial role in solving the problem of ethnic minorities by both checking any minority rights' infringements and containing the escalation of tension. Such cooperation could allow 'both the national integrity and the development of international relations with bordering countries' (Cappellin 1992: 16). In terms of politics, therefore, each national and local government area can integrate the ethnic issue into wider deliberations over resource allocation and regional planning in order to focus on various aspects of socio-political development and not ethnic 'problems' specifically. Community leaders will have to accept compromise because while there has always been an underlying tolerance of ethnic rights, the fragility of the new democracies, and their greater risk of destabilisation than their Western Europe counterparts, places a premium on moderation. Equally, external observers should be sensitive to scenarios where minority groups make excessive demands through the political process in order to imply that their condition is unbearable. On this basis substantial progress has been made and the OSCE High Commissioner on National Minorities considers that minority politics in much of the region has moved into the arena of 'normal politics': for example, Poland and Lithuania have set up a commission to deal with mutual education problems. But the situation can be more difficult in countries like Croatia, Serbia and Slovakia where former dem-agogues 'have fashioned a new generation of political role models who would be extremely difficult to discredit' (Ingrao 1999: 314).

According to Table 2.9 the Hungarian minorities are among the largest in the region with a total of 2.53 m.: of these 1.62 m. are in Transylvania (Romania), 0.57 m. in Slovakia and 0.34 m. in the Serbian Vojvodina (Yugoslavia), besides oth-ers in the Ukraine. Numbers are fairly stable although there has been a sharp decline in Vojvodina where 0.44 m. Hungarians were enumerated in 1961; even so, the present figure may be too high because of recent emigration. In 1996, improvements were possible in Transylvania when Hungary and Romania – with CoE and EU support – agreed to 'adequate possibilities' for education in minor-ity languages. Later in the year a new government came to power in Romania with the main Hungarian party as a member of the ruling coalition. Romania provides guarantees of parliamentary representation for minorities. Demands for the separation of Cluj-Napoca's Babeş-Bolyai University into separate Hungarian and Romanian institutions (as occurred for a short time under communism) have been partially neutralised by the provision of more courses in German and Hungarian (a move supported by the OSCE High Commissioner). Meanwhile, reforms in local government require the use of bilingual signs where a minority accounts for more than a fifth of the population of a town or commune, although there is a tendency for these to be obliterated by local nationalist. It is also possi-ble for people to address the local authorities in the relevant language; also for information to be disseminated and debates conducted in the same medium. Hungarians would now like to see all statements of a national character removed from the constitution.

Meanwhile, in Slovakia a basic treaty negotiated in 1995 accepts the present frontiers while providing for European norms to protect national minorities, though falling short of regional autonomy. Most Slovaks living in Hungarian majority areas think positively about coexistence and accept the case for public signs in the Hungarian language. On balance CoE fact-finding missions have expressed satisfaction and there is promise of an era of genuine constitutionalism with reduced central authority and moderation of populist discourses previously projected by the confrontational style of most political parties. Sadly, however, the Hungarians in Vojvodina, politically divided and disturbed by the arrival of Serb refugees, cannot realistically expect restoration of autonomy (removed by Milošević) since their demographic strength is reduced and the precedent would not be welcome in Romania and Slovakia where the Hungarians are much stronger (both relatively and absolutely) in their respective majority areas.

Jews and Roma

Although the number of Jews in Eastern Europe is now extremely small, the former importance of this community ensures that it remains in the political spotlight, especially in view of the Holocaust. Meanwhile, the Roma remain a large ethnic group often exceeding official totals because of false ethnic identification or a failure to declare themselves. It is believed there are over 2.00 m. in Romania, 0.80 m. in Bulgaria and Hungary, 0.40 m. in Slovakia and 0.30 m. in the Czech Republic. Assimilation and conversion into a settled proletariat seemed inevitable under communism, following centuries of a nomadic lifestyle which involved seasonal employment. However, while some embraced mainstream community values and took steady employment, others maintained their independent lifestyle. Due to poor work discipline, transition has often cast the Roma as losers in the context of heavy redundancy among unskilled or semi-skilled workers. On occasions the Roma face further discrimination through their perceived association with alcoholism and petty crime. Many violent incidents have been reported and some extreme measures have been taken, with particular controversy arising over the wall – 65 metres long and 4 metres in height – built to marginalise Roma in the Neštěmice district of Ústí nad Labem in 1998. Czech Roma made an 'Appeal Against Romany Apartheid' assisted by Roma organisations in Germany and during the late 1990s considerable numbers gained asylum abroad (many in Canada and the UK) on the grounds of discrimination.

As the economic transition proceeds there is acknowledgement that job-creation schemes are needed, which could provide the Roma with private business opportunities. Over the longer term there will have to be more social work among local communities and a reduction in the birth rate through education. However, as already noted, some Roma elements have found resources abroad to build substantial houses for extended families and this may augur well for the future in stimulating a greater priority for higher living standards. In public life Roma representation is also improving: for example, in Hungary they are now represented through a National Gypsy Council. In Macedonia, significant progress has been made by the Roma in education, the media and politics. Although progress is slow many feel they are 'laying a foundation for the next generation who will be much better prepared to fight for the interests of Roma people' (Kovats 1998: 143).

Political parties and national profiles

Free elections – rather than coup d'état – are now universally accepted across the region as a legitimate means of transferring power and election turnout, though variable, has often been encouraging (Table 2.10). There is now a better balance between the 'supply' of political actors and an informed public articulating 'demands' to reflect its identities and interests. Some new parties have been too confrontational, lacking a willingness to compromise for the wider good, while party infighting may have more to do with personalities than with alternative ideological underpinnings for civil society. Fragmentation of parties in the early years seemed to be deliberately encouraged by interim regimes in order to splinter the opposition. However, parties have replaced movements; and they have come to rely on the steady support of sections of the electorate, including formal links with organisations representing major interest groups in the economy. Effective choices are being offered and in all states fundamental changes of government have occurred since 1989. However, while there are relatively few extremist parties and demagogues on the political scene – though there are charismatic leaders like Václav Klaus in the Czech Republic, the partisan nature of political life can still produce attitudes of intolerance towards opponents. Governments may allocate patronage overwhelmingly in favour of their supporters and isolate local governments which support opposition parties. New administrations may then conduct purges of a previous government's appointees in the police, judiciary, administration and universities. In Albania, the Socialist Party has removed Democratic Party appointees in the name of professionalism though the latter would argue that discredited communist offi-cials are being restored as a result. Where the new political elites were often not perceived as having voter interests at heart, parliaments and political parties inspired less trust than other institutions like the presidency, the military or the churches.

Table 2.10 Summary of general elections 1990–2001

	1990	1991	1992	1993	1994	1995	1996	1997	1998	1999	2000	2001
Albania		AC	Tr[1]			AC		Tl				AC
Bosnia & Herceg.	[1]						[2]		RC		Tc	
Bulgaria	AC	Tr[1]			Tl			Tr				Tc
Croatia	Tn[1]		AC			AC					Tc	
Czech Republic	Tc[1]		Tr				RC		Tl			
Hungary	Tr[1]				Tl				Tr			
Macedonia	Tn[1]				RC				Tc			
Poland	[1]		Tr		Tl				Tr			Tl
Romania[4]	RC		RC				Tr[1]				Tl	
Slovakia	Tc[1]		Tl				RC		Tc			
Slovenia	Tc[1]		Tl				RC				RC	
Yugoslavia			RC[3]				RC				Tc	
Montenegro	Tc[1]		RC				Tn					RC
Serbia[4]	RC		RC	RC				RC			Tc[1]	

Source: Mangott 1998, with modifications and updating, with kind permission from Kluwer Academic Publishers

Notes

[1] Loss of power by communist or neo-communist government

[2] Resumption of national parliamentary rule following the Dayton Agreement (1995) and the previous existence of separate assemblies in the Bosniak-, Croat- and Serb-held parts of the country arising out of the civil war following independence in 1992. Elections have been held in the two entities in the same years as for the confederation government

[3] New federal government based on pluralism

[4] Although elections were based on political pluralism from 1990 onwards the continued role of the secret police and former networks suggests neo-communism until the transfers of power indicated. Other countries show similar tendencies – e.g. Albania, Croatia and Slovakia – but arguably not to the same degree

T Transfer of power: (l) Left; (r) Right; (c) Centre; or (n) nationalist

RC Relative or restricted continuity (similar coalition or minority status)

AC Absolute continuity

These are matters to be placed in the context of remarks by Lagerspetz (2001: 11) that would see the transformation goal for civil society – at the same time as nation building and the creation of the institutions of state – as a 'modern' project in a 'post-modern world' where elites are prone to operate as 'resourceful and well-organized interest groups' forsaking universalist approaches expressing concern for all members of society. Those in power can resist universalism by dominating the less powerful networks (as under real socialism), for despite their tribal identity and a cultural form of 'resistance', the latter will not be able to challenge the elites in a way that is necessary to initiate social change. In this situation NGOs have an important role in providing a corrective force through the development of partnerships with governments and policy makers. It is crucial that electors retain confidence that politicians will eventually 'get it right' and do not offer support for alternatives to political pluralism. Dissidents now tend to be dismissed as naive dreamers without deep connections with the societies for whom they claim to speak. But to avoid opportunism the public may favour a candidate on the basis of good character and administrative capability rather than party affiliation. At times of crisis there has been significant support for government by technocrats.

Kitschelt (1992) sees the spectrum between pro-market/liberal (right) and pro-state/authoritarian (left) as fundamental to the development of political cleavages, although it is normal for coalitions to be preconditions for stable majority governments, including some elements which are themselves coalitions. Hence the simple choice between centre-right and centre-left; reflecting, respectively, the neo-liberal view that 'the state is still everywhere, markets are distorted, entrepreneurship is inadequately rewarded' and the leftist perception of the traditional middle class 'plundered [by] the rule of the few who are only tenuously accountable to the many' (Commisso 1997: 18). While the first view accepts the realities of globalisation and the need for foreign investment to create jobs and generate the wealth to pay for social services over the longer term, the latter insists that the state need not be so subservient to international finance (transnational companies and institutions like the World Bank and IMF) and that the economy can be regulated in such a way as to sustain welfare on a more generous scale through social services, protection of former SMEs and job creation. Such views, although dangerous for macroeconomic stabilisation (through inflation running out of control by printing too much money), are persuasive to people who feel that national culture is being invaded from abroad and that new civil rights cannot be appreciated by people struggling to make ends meet. Although there is no intellectually credible alternative to the market as the key to future prosperity, present difficulties offer a platform for populist forces exploring hybrid systems of economic management such as worker ownership of enterprises to enhance social protection.

However, the failure of socialist parties to offer truly viable and radical alternatives – as opposed to vague promises of amelioration – can give opportunities for regional parties to reflect the sense of alienation (a blend of aggression and helplessness) in depressed areas where people are particularly suspicious of politicians and sceptical of the whole political system. Regional contrasts arise as metropolitan areas struggle to approach EU average incomes, even on a 'purchasing power parity' basis, while the enterprise culture sponsored by the central government may well evoke strong resistance from regions that seek more interventionist policies. Indeed the peripheries of some countries constitute a political majority and restrict the modernising ethos to the capital city and some leading provincial centres. But unless there are export staples (quite rare in ECE) excessive protection of SOEs will not create the economic and political stability for foreign investment. It is an important matter for the political process to decide how far the state's limited resources should be used to stimulate the more dynamic areas or protect those areas least able to achieve a positive transformation. Foreign penetration is more readily accepted by the better-educated people (especially those who have done well in post-communist Poland) while those with a rural background are most distrustful. Few Poles want foreigners to invest in arable land, power stations, housing and medical services, but in Lublin in rural eastern Poland people have more reservations over FDI as a threat to employment and local business than in the more cosmopolitan Wrocław area in the southwest.

There are further options provided by ultra-nationalist parties rejecting the globalising agenda and EU accession. As people rediscover a history partially obscured by decades of totalitarian rule, nationalism may prove to be a divisive force opposed to reconstruction in the spirit of a wider 'Europe of the regions'. Nationalism has served in the past as a defensive mechanism against Western liberal capitalism and, having survived state socialism, it could now be exploited as a cheap political substitute for a radical transition in the SEECs where a successful struggle for democracy is compromised by hybrid nationalist–populist regimes. Evans and Whitefield (1998) argue for the primacy of ethnicity in Slovakia through much of the 1990s. This makes stability more difficult since compromises over identity issues are difficult when their emotive nature makes for demagoguery. But it is above all in the SEECs where hybrid nationalist–populist regimes have emerged, with ethnic tensions dominating the political agenda in some countries to the detriment of human rights. Pluralist mobilisation has been the catalyst for the breakup of Yugoslavia, but has the option of resisting pan-Serb ambitions, and indeed the initial acts of provocation by Serbia, been enhanced by the need for unity in the face of uncertain economic prospects? Where politics are rooted in ethnicity, change and consensus become more difficult. Meyer and Geschiere (1999) consider that globalisation and local identity can be an explosive combination which does not exclude communalist violence. Where reformist groups have failed to gain power 'unreconstructed former communists will use nationalist attacks on a large ethnic minority to establish their legitimacy and then to consolidate their support during difficult economic times' (Snyder and Vachudova 1997: 32). Ethnic minorities may well be further radicalised while the mainstream population is left ignorant of the main causes of economic stress. On the whole the trend seems to be away from crude nationalism. At a time when the Salvation Front/Party of Social Democracy government was making 'promises everywhere' it seemed appropriate to exploit nationalism and emphasise the originality of Romanian culture as an alternative to the 'spirit of Europe' (Marga 1993: 21). But since 1996 Romania has been breaking away from the 'protochronism' of the early 1990s (inherited from the later Ceauşescu years) when Western-style liberalism and pluralism were presented in some quarters as alien to 'indigenous' traditions (Tismaneanu and Pavel 1994: 405).

Following the revolution it was quite common for power to pass to centre-right governments promising radical market reforms by 'shock therapy' and a system of decentralised regulation that would be appropriate to both the state's reduced resources and the imperative of eroding the infrastructure on which totalitarian power had been based. But those who brokered revolution often became vulnerable, in their turn, because of deepening dissatisfaction over the short-term impact of radical reform and inadequate levels of welfare provision. Hence the opportunity for the communists to reprofile themselves as social democrats and gain power directly or through the formation of centre-left coalitions by linking up with agrarians and nationalists. This phenomenon was first evident in Hungary, Poland and Lithuania in the mid-1990s and was referred to as the 'LPH sydrome'. The results have been generally positive in demonstrating that former middle-ranking communists (like G. Horn, a former member of the Hungarian Politburo who was prime minister during 1994–8) 'are pragmatists anxious to occupy the centre-left political stage [and] in their quest for legitimacy, both domestic and external, they are willing to accept the rules of the game' (Karasimeonov 1998: 10). And since social democrat regimes (as in Hungary and Poland) can be as reform minded as right of centre parties, it has become evident that there are different strategies that may be adopted within the broad context of a market economy with support for global and European institutions. This helps to marginalise the more extreme parties which would combine an aggressive stance towards Europe with ethnic politics (on the right) or inflationary welfare deals (on the left). However, in the case of both centre-right and centre-left, coalitions must be kept together and this can be difficult on account of both policies and personalities. The Hungarian Democratic Forum (the senior partner in the Hungarian government of 1990–4) was weakened by defection of I. Csurka's Hungarian Justice and Life Party, while in Poland in 1998 Solidarity Election Alliance was weakened by defection of the Confederation of an Independent Poland and those linked with the Catholic Radio Maryja. Centre-left coalitions may be constrained if social democrats have problems over working with communists.

National profiles: Albania

Albania's communist government resisted reform and cast the country into a dangerous state of international isolation after the departure of Soviet and Chinese technicians in 1960 and 1978 respectively. Until May 1990 the country's constitution officially forbade any sort of foreign investment and even credits were ruled out. However, after the death in 1985 of E. Hoxha, the country's communist ruler from 1945 to 1985, the country began to borrow money to finance foreign trade deficits. The last communist government (under R. Alia) was very much constrained by conservative elements supporting the Stalinist policies of the Hoxha era and economic reforms had only a limited impact. But peasants were allowed to sell surplus produce from their 200 m² private plots on free markets, although the surplus produce available was very limited and the lack of consumer goods gave the farmers even less incentive. A new electoral law provided the opportunity for a Democratic Party (DP) to emerge with its own newspaper and electoral programme (including land reform). However, the DP failed to dislodge the communist Party of Labour of Albania (PLAb) in the 1991 election since it lacked transport to reach the villages where an overcrowded rural population was fearful of the consequences of privatisation. But a prolonged national strike in 1991 led to the removal of PLAb hardliners as half the country's industry came to a standstill. In 1992 the DP was able to govern in coalition with Social Democratic and Republican Parties; while the Socialist Party (SP: the former PLAb) maintained a strong opposition. Agriculture, minerals and tourism were identified as priority sectors and the centre-right government of S. Berisha did quite well until 1996, enjoying some-what uncritical Western support on the basis of commitment to free market democracy and opposition to neo-communists verging on the paranoid.

The government was returned with an increased majority but it was evident that there had been widespread corruption and intimidation while pyramid investment schemes were launched in order to raise campaign funds which were also dependent on money laundering, dubious arms sales and the Balkan drug trade. The inevitable collapse of these funds (administered by Vefa Holdings) in late 1996 and early 1997 reduced the country to chaos. The scale of ruin was extreme given that many families had even sold their homes in order to join the speculation. People in the south compensated themselves for losses by looting and towns were in effect taken over from central government control by their own populations (though Vlorë, with a tradition of criminality, had for some time been largely controlled by the narcotics gangs who now profit from the boom in locally-grown cannabis and marijuana). Most of the country was comprehensively ransacked and Berisha's fortunes collapsed in sympathy; a deci-sive reverse after the creation of a de facto one-party state by rigging elections and barring potential opposition MPs on the grounds that they were former communists or Sigurimi agents. It was assumed that he would set his sights on being re-elected president and would then establish a presidential dictatorship.

Berisha was replaced by R. Mejdani and returned to parliamentary life as DP leader but a Socialist Party victory in the new elections held in 1997 saw prime minister F. Nano replaced the following year due to corrupt and inef-fective government by a 'kitchen cabinet of friends' and Berisha's attempted coup also contributed to the country's political bankruptcy. The new Socialist Party premier P. Majko was considered incorrupt, with a European social demo-crat outlook, but he was considered a lightweight politician who would be unable to gain popular support and the president, R. Mejdani, had integrity while lacking the charisma of his predecessor whose continued presence in public life spoke volumes for the status of law and order. But there were grave problems in restoring order since every armoury had been ransacked and much of the country was in the hands of local militias (including the Kosovo Liberation Army – KLA – which controlled the northeast). Insanitary shanty towns around Tirana are inhabited by around 100,000 migrants from the north who left the mountains on the assumption that Berisha (a member of the northern Gheg community) would help them. While this influx has boosted the population of Tirana to some 450,000, the northern highlands are socially dislocated with farmland and forests degraded by soil erosion. Growth had already halted by 1997 because people were able to stop working while they lived off the interest of their pyramid funds. After taking savings and potential agricultural investments running into hundreds of millions of dollars, there were no agricultural banks with resources available to restart the economy even in a small way by overhauling irrigation systems or buying seed, fertiliser and basic machinery. Cheap imports from Greece and Macedonia mean there is no incentive even for local vegetable growers.

So the Majko cabinet (a compromise among the competing factions of the ruling SP and G. Pollo's small Democratic Alliance) had only nominal authority while most roads remained prone to armed attack. The population remains traumatised and at least 250,000 young Albanians have left the country, thereby creating a huge gap in terms of demographic reproduction and skills. Remittances from abroad are down to a quarter of what they were (reflecting a loss of confidence by Albanian émigrés) while foreign investment is virtually zero (although BP stays on in Vlorë in the hope of a future offshore deal) and it will need five years of stability to turn the tap on again. It needs law and order to make a start on production but this cannot be done until there is economic hope. A good start would be a working customs regime (especially on the transit route from Macedonia to Montenegro) instead of systematic criminal activity and large-scale smuggling to finance government when conventional tax revenue is not available. Crime and corruption have become a way of life, discouraging both reform programmes and foreign investment. The DP opposition sought to undermine the socialist government through a paralysing standoff by a prolonged parliamentary boycott; yet in any case the government lacked legitimacy in the eyes of the people in general. Thus Albania found itself weak just at the time when leadership is required in view of the Kosovo crisis, with the KLA as one of the country's leading economic actors through its arms purchases and property rental income in Tirana. There are strongly nationalist parties favouring a greater Albania, like the Bali Kombetar and the monarchist Legalitate, but Albania's present predicament as a de facto NATO protectorate does not advance any pan concept and hence independence remains the only option for the Kosovar Albanians.

The present socialist government of I. Meta has now restored some order and further stability arises through NATO and the need for permanent military facilities. Relationships are being built with Croatia, Macedonia and Montenegro. However, institutional infrastructure is lacking and there is difficulty over a northeastern border that is totally open to traders and traffickers. Illegality exists in various forms including the arms and drugs trades, car theft, money laundering and sanctions busting. There has been some growth from a low base since 1998:

Elbasan steel is being revived by Turkish managers, while a Greek–Norwegian consortium has acquired the country's GSM mobile operator and there is Italian interest in copper as well as Tirana's brewery and the idea of piping Albanian water across the Adriatic. Tourism showed some reivival in 2000, thanks to a marketing campaign and a large number of visits from Albanians in Kosovo and Macedonia. Hotels are being privatised and upgraded. However, the informal sector is believed to account for half of all economic activity. It is important that tax administration is improved and savings are mobilised for public purposes

Municipal elections passed off peacefully in 2000, which augured well for the general election in 2001. A Reconciliation Commission is helping to bring an end to blood feuding while a Transparency Commission is struggling with corruption. Yet the biggest issue here is recovery of money lost in 17 pyramid investment schemes, which at their height were promising interest at 8 per cent monthly and, at their height, had liabilities equal to half the country's GNP. But it seems that only $50 m. of the £1.0 bn. lost through fraud is recoverable, for lack of bilateral agreements in respect of foreign bank accounts in which up to $100 m. has been deposited. The government has had to retract promises of compensation and seeks to 'draw a line' under a thoroughly dishonest business which saw Vefa Holdings (the largest of the schemes) closely linked with the former Berisha government. The outlook remains poor and many Albanians embarrassed by their country's 'protected' status arising from the lack of a mature and responsible political class. However, recovery continues with another term of socialist government, although Berisha's Union for Victory Coalition gained some seats in the 2001 election and a Reformed Democratic Party, led by former ministers who have left the mainstream party, also contributes to stability. The government now claims a 'European' policy with rejection of 'Greater Albanian' aspirations and claims. On this basis an EU stabilisation and association agreement was negotiated in 2001. But the socialists need to do more to relieve poverty in the north and thereby draw more support from this part of the country – heavily dependent on mining – which is traditionally a Democratic Party stronghold. The long-running feud between Berisha and Nano continues to influence national politics and will doubtless come to the fore again if Nano seeks the presidency in 2002.

National profiles: Bulgaria

Bulgaria is a communist state which experienced a peaceful regime change although the transition has been quite bitterly contested given the need for radical restructuring for an effective market economy, yet the popular desire is to retain the welfare benefits of the old system. The long-serving communist leader T. Zhivkov was forced to resign in November 1998 and the opposition was legalised the day after the Berlin Wall was breached. Seven non-communist political groups entered into a Union of Democratic Forces (UDF) with the dissident leader Z. Zhelev as president. The level of public support for the UDF was such that the communists were obliged to bring the group into roundtable discussions, along with the Bulgarian Agrarian National Union (BANU) in March 1990. These opened the way for the election of a Grand National Assembly to serve, in effect, as both parliament and constitutional assembly. The revamped communist party, the Bulgarian Socialist Party (BSP), was formed in the spring of 1990 and immediately won the election by appealing to urban voters as a party of responsible change, while maintaining a hold over the rural population apprehensive over land reform. While the UDF advocated 'shock therapy', the BSP sought 'responsible' change that would (hopefully) achieve a relatively painless transition. This was electorally effective in 1990 but the programme could not be delivered. There were further falls in output, complemented by shortages of energy and food. The UDF took over but was seen as confrontational in style; taking initiatives haphazardly with inadequate communication. Following partial disintegration of the UDF in 1992 a government of technocrats then survived until 1994 when electors decisively rejected the UDF with its factional rivalries, in favour of the BSP under its young leader Z. Videnov.

At this point a reckless economic policy, boosting the money supply to an excessive degree, produced economic devastation in the form of a hyper-inflationary spiral leading to unprecedented street protests early in 1997. The UDF under I. Kostov was returned to power later in the year as the first post-communist majority government without socialist backing and participation. With experience in economic management and a relatively clean record, Kostov rebuilt the macroeconomic framework (achieving financial stability through a higher rate of tax collection) and the country's growth prospects improved. The government launched an anti-corruption campaign (also addressing organised crime and customs control) and embarked on a plan to privatise 80 per cent of state enterprises by the end of 1998. Despite low foreign investment the privatisation programme went ahead, Bulgaria joined CEFTA in 1999 and EU accession negotiations commenced, the nuclear issue now resolved by the promised phasing out of the four 440 MW nuclear reactors at Kozloduy power station (two by 2003 and the other two by 2010). The IMF provided support and there has been financial compensation for damage suffered through the NATO bombardment of Yugoslavia.

Unfortunately all this made little impact on incomes and so Bulgarians were disappointed with the UDF because living standards were not deemed to have recovered from the 1996–7 crash, while war in Macedonia scared away investors and caused much anxiety as the economy deteriorated in the face of closure of the Danube and trade routes through Serbia: hence the 4 per cent growth target could not be sustained. A lack of transparency over some privatisation deals further haunted the government. As the government gave priority to EU negotiations the danger of an 'accession trap' loomed, for approximation involves a bureaucratic-legal programme which can stifle development in the interim: a 'Bermuda triangle' absorbing reform efforts only for accession to become more problematic through economic malaise. There were still difficult problems to be resolved in the energy field for the decision over Kozloduy – along with cut-backs in the generation of electricity based on lignite (needed to alleviate pollution problems) – inevitably pointed to resumption of the highly controversial nuclear power project for Belene. There was also the threat of demographic catastrophe.

As the electorate yearned for mass popular capitalism but had no stomach for the return of the BSP it was left for the former king, Simeon Saxe-Coburg, to administer UDF policies through his 'National Movement Simeon II' which swept to power in 2001, offering an 800-day miracle cure in coalition with A. Dogan's Movement of Rights and Freedoms. The government has made a promising start with populist policies but there are tensions over social spending and a gentle acceleration of growth is needed. However, support has begun to shift back to the Socialists who now control the presidency.

National profiles: Hungary

Hungary was the most market-oriented socialist country, with a bond market in 1982, a bankruptcy law in 1986, a stock market in 1987, opening the way for joint stock companies in 1988 and also a modern taxation system and an unemployment compensation law in 1989. With a relatively low level of production for the military, there was considerable independence for state firms and a significant private sector developed. Radical reform by 'shock therapy' was hardly needed, yet given the emphasis on a social contract to protect the population during the communist years under the leadership of I. Kadar, the transition saw the Hungarian Democratic Forum (HDF) initially favouring a third way between communism and capitalism. However, the party soon transformed itself under J. Antall's leadership into a Western-style Christian Democrat party (albeit prone to nationalist sentiments) whereupon it was able to govern in 1990 allied with the Independent Smallholders Party (ISP). But continuing recession, linked with harsh emergency measures, caused reform to lose momentum and let in G. Horn's Socialist Party (SP: evolving from the reformed communists) in 1994, which sought to combine market forces with the security of the old regime.

Trade relations were diversified through business with the West for firms benefiting from new investment like Videoton and Tungsram producing TV sets and fluorescent lamps respectively. Meanwhile, Ikarus found new markets for buses in Iran, Taiwan and Turkey though it was some time before a foreign partner arrived and there is currently a Russian stake (reduced since 1997) alongside a large Hungarian government holding. Meanwhile a new bus company has arisen through a link between Rába of Györ (a well-known truck manufacturer) and an American company. The new firm – NABI – has developed alternative-fuelled buses (using compressed natural gas: CNG) and has attracted large orders from North America. Hungarian cigarettes found favour in Ukraine after Philip Morris acquired the Eger tobacco factory and set new standards for an industry capable of doing much better in export markets with new ideas for production and packaging. The Hungarian economy is now highly globalised, given a consensus over privatisation which is virtually complete with assets transferred overwhelmingly to foreign investors. Almost all recent GDP growth has come from transnational companies (two-thirds EU-based) although they employ only a fifth of the workforce. The 1989 production level was regained in 1999 and the economy continues to grow better than most others in the region. It can be deemed a success, like the agriculturally-based Polish economy, although there are severe regional contrasts.

Despite steady growth the socialists succumbed to their erstwhile coalition partner (V. Orban's Young Democrats–Hungarian Civic Party: HCP) which gradually consolidated the right-wing vote during the years 1994–8 after the HDF was further weakened by the hiving off of the nationalist right wing under I. Csurka (whose small Hungarian Welfare and Life Party is trying to launch a neo-fascist bandwagon similar to that of the Freedom Party in Austria). The HCP platform is one of faster growth through effective use of privatisation proceeds, combined with better public security, anti-corruption measures and equality of opportunity in education. There is also a green element through the decision against new dams on the Danube and rejection of negotiations with Slovakia to implement the 1997 International Court decision over the troubled Gabčikovo-Nagymaros hydropower and navigation project. The party received fewer votes than the Socialists but the electoral system, combining proportional representation with 'first past the post', gave them a small lead over their rivals. The government is stable and self-confident. The coalition of the HCP–ISP has avoided serious internal rifts with good relations between the respective leaders: Orban and his opposite number J. Torgyan.

Hungary joined NATO in 1999 and looks forward to becoming a full member of the EU around 2004 (being the first transition state to apply – back in 1994). Continuing the previous government's stabilisation programme, the economy has been doing well with good growth and strong exports and is largely free of structural deficiencies. With a new minister (G. Matolcsy) in charge of the economy – following friction between Orban and the national bank over high interest rates which were thought to be holding up the economy – a 5 per cent annual growth target has been set (reducing slightly to 4.5 over a 20-year period). There is a high level of indebtedness – $2,900 pc in 2000 – which is exceeded only by Slovenia (Table 2.11). More privatisation will be needed and there will be a boost for SMEs and stronger regional policy for the east; though the risk of higher inflation must be faced and waivers will be needed from the EU over environmental standards. Indeed, Orban may achieve his dream of a two-party system involving his own party and the Socialists who currently dominate the polls, for the Smallholders are in difficulty and the HDF – the country's first post-communist governing party – is under threat of extinction despite being part of the present governing coalition. But despite recent growth which has translated in higher living standards, FDI is now falling due to an appreciating real exchange rate, high inflation and a slowdown in productivity growth, for $-based unit labour costs are expected to exceed levels in the Czech Republic and Poland during 2001–2 (and by a bigger margin in the cases of Romania and Slovakia). The government's interventionist tendencies are also greeted with some concern by the business community.

Hungary is constantly preoccupied with the status of her minorities in the neighbouring countries. It is appreciated that world opinion will never accept the revision of boundaries, while Hungary cannot absorb her minorities in Romania, Slovakia and Yugoslavia in the manner of Germany. Hence the need for institutions to maximise 'collective national good' in the context of the present borders, a goal consistently sought by the Antall, Horn and Orban

Table 2.11 External debt ($bn.) 1988–2002

	1988	1989	1990	1991	1992	1993	1994	1995	1996	1997	1998	1999	2000	2001e	2002e	Rate*
Albania	na	na	na	0.5	0.6	0.8	0.9	0.7	0.7	n.a	na	1.0	1.1	na	na	0.39
Bosnia & H.	na	na	na	na	na	na	na	3.4	3.6	4.1	3.4	3.1	2.6	na	na	0.65
Bulgaria	7.2	10.2	10.9	12.1	13.9	12.2	9.8	10.3	9.8	10.1	na	9.9	10.4	10.7	10.9	1.34
Croatia	na	na	na	2.8	2.6	2.5	3.0	3.6	4.6	6.0	7.0	9.4	9.8	10.3	12.0	2.19
Fmr. Czech	7.3	7.9	8.1	9.3	9.5	na	na	na	na	na	na	na	na	na	na	na
Czech Rep.	na	na	6.4	7.2	7.6	9.2	10.7	16.3	20.1	21.6	22.3	22.6	20.4	21.7	24.0	1.99
Hungary	na	0.4	21.3	22.6	22.0	24.3	28.2	31.5	27.0	23.2	26.7	29.0	29.1	30.7	32.9	2.90
Macedonia	na	na	na	0.5	0.7	1.0	1.1	1.3	1.8	1.5	1.7	1.4	1.4	1.4	1.6	0.69
Poland	na	43.1	49.4	53.6	48.7	45.3	42.6	42.3	40.1	39.5	44.0	55.5	58.9	62.2	72.0	1.52
Romania	1.1	1.2	1.2	2.1	3.2	4.2	5.5	6.7	9.2	10.4	9.5	9.4	9.8	9.2	9.6	0.44
Slovakia	na	na	2.0	2.7	2.8	3.5	4.1	5.8	8.3	10.0	na	9.1	8.1	8.3	8.4	1.50
Slovenia	na	na	1.4	1.4	1.2	1.9	2.3	3.1	4.0	4.9	na	5.4	6.2	6.7	8.3	3.12
Yugoslavia	na	na	na	n.a	5.6	10.3	13.0	13.8	13.4	12.9	14.0	12.9	11.7	8.7	8.6	1.10

Source: Economist Intelligence Unit

Note
* Indebtedness 2000 $,000/pc
e Estimate

governments. Hungary's neighbours have been much preoccupied with securing their borders – as new states in the cases of Croatia, Slovakia, Slovenia and Yugoslavia – but the desire to join Europe (including Yugoslavia, following the political changes of 2000–1) increases readiness to engage in cross-border cooperation. Meanwhile, there is massive support for Europe in Hungary because it maximises the chances of softening the borders, though there will be some complications until more of her neighbours also become EU members and in the meantime Hungary should be careful not to use NATO membership in the cause of the diaspora. Meanwhile, Csurka's right-wing party shows customary impatience in calling for reannexation of pre-1920 territories – including the return of Vojvodina at the height of the Kosovo War (when a national debate on the status of the province saw the ISP advocating independence for this Yugoslav territory). While multinationals like IBM, Audi and Philips (jointly responsible for a quarter of all Hungary's exports) help to ensure steady growth, nationalists are resentful of multinationals buying up companies and markets.

National profiles: Poland

There was a surge of support for Solidarity when General Jaruzelski's military dictatorsip was replaced and a reform programme was launched. In defence to the Soviet Union an initial power-sharing agreement which emerged out of the 1989 Roundtable, but solidarity took control after the elections of 1991. Although the communists retained a grip on the economy, electoral extinction was a stimulus to them to adopt a social democratic stance and tolerate the shock therapy programme of 1990 which broke the monopoly of large state firms and encouraged the SMEs which have generated the growth of the 1990s. Despite regime changes Poland has persevered with market reforms and has steered clear of political extremism and criminal capitalism. The contest for the presidency in 1990 between the first democratically elected prime minister T. Mazowiecki and the charismatic union leader L. Wałesa precipitated a 'war at the top' which split Solidarity irrevocably into its intellectual and proletarian wings. After four short Solidarity governments, there was a chance of stability when a 5 per cent threshold was imposed so that many small parties were excluded from parliament. A centre-left coalition was then formed between Democratic Left Alliance (DLA) with the Polish Peasants Party (PPP) in 1993. Victory rested on several factors not the least important of which was the ability of the left to make common cause throughout the election and to stand as effective representatives to those who felt they had lost out in the reform process through unemployment or falling real wages. The shift also showed that the public wanted stronger welfare policies, although change was assisted by the failure of previous governments (especially the outgoing Suchocka government) to 'sell' their policies and Solidarity's inability to organise a coalition, aggravated by the new electoral law. Subsequently Wałesa lost the presidency in 1995 to the left-wing candidate A. Kwaśniewski.

Thus in Poland there is, typically, a socio-economic division between neo-liberal, free market/free enterprise policies and state interventionism in the economy with a welfare state-type social policy. This dichotomy runs broadly in parallel with the fact that 'an ideological split exists between confessional and secular visions of political order and is closely entwined with opposing stands on the issue of decommunisation' (Jasiewicz 1998: 3). In elections it all depends on how well the opposing blocs (Solidarity and the Social Democrats) can maintain cohesion. Poland's 'polarised pluralism' makes cooperation difficult. Different political entities are highly polarised, making it virtually impossible for them to join forces. This undermines the centre because collaboration with one faction obviates the possibility of establishing contact with others. It also makes for fragile coalitions since major differences tend to surface with left and right groups.

The socialists succeeded initially in 1993 because there was genuine co-existence between coalition partners as the urban-based intellectuals forged an alliance with the peasantry. It was expected that the new government would boost welfare (and also cause inflation to rise) rather than stick to a monetarist policy desired by the IMF. But privatisation proved a stumbling block: the government was unable to proceed with the 1995 mass privatisation covering 440 medium-sized companies, despite EBRD support for start-ups and other initial costs. A factor here has been the undermining of the pact between trade unions and government which amounted to a precondition for the privatisation deal. However, the coalition remained intact and the government ran its full length despite two changes in the premiership (J. Oleksy and N. Cimoszewicz), though the government was weakened by the PPP's 'no confidence' motion on Cimoszewicz on the grounds that little was being done to help the farmers. As the PPP veered away from the left the centre-right returned to power in 1997. In that year, M. Krzaklewski's Solidarity Election Action: SEA (bringing together right-wing parties that failed to get 5 per cent in 1993) formed a coalition with the Freedom Union led by the monetarist L. Balcerowicz insisting on fiscal reforms and financial rectitude. But prime minister J. Buzek of SEA (a member of Solidarity since 1980) had a torrid time holding the government together through a series of unsatisfactory compromises. Many SEA parliamentarians left the party and the coalition finally broke up in 2000 when the departure of the Freedom Union left Buzek with a minority SEA government.

Nevertheless reform accelerated in 1998 through privatisation and the reorganisation of heavy industry while local government was belatedly reorganised on the basis of 16 large regions. Although there are still Solidarity strongholds in the south and southeasts, tensions within SEA ensured that Kwaśniewski retained the presidency in 2000 (easily defeating the SEA candidate: Krzaklewski), especially since the public mood became somewhat less enamoured with the EU and showed a stronger nationalist tendency, with antipathy towards foreign investors. And parliamentary prospects for the centre-right deteriorated with the splintering of parties uneasily brought together under the SEA banner. Indeed Solidarity was annihilated in the elections of 2001 which returned the pragmatic DLA to power, with the Labour Union and PPP as coalition partners to secure a majority government at some cost in terms of political effectiveness. However, despite divisions in the former coalition over privatisation and tax reform – which precipitated a switch to minority government status in 2000 – the economy averaged well over 5 per cent annually and real GDP in 1998 was 17 per cent over 1989 (while Czech economy remained 10 per cent below), with inflation down to single figures. Indeed Buzek was the first post-communist premier to serve out a full term. Given this performance and the large domestic market, Poland should remain a magnet for FDI, despite a downturn in growth and employment which requires correction. Militarily the situation is less satisfactory because the large Polish forces are poor by NATO standards, especially in terms of mobility. And a strong pro-environment lobby is needed to boost environment as a major social problem. Legislative reform is delayed and implementation of existing legislation is sluggish.

3   Production

Industry and agriculture

Restructuring of industry

Manufacturing was the key sector under the command economy and great increases in output and employment were registered under Soviet leadership on the basis of low energy, transport and labour costs. The result was a dramatic structural change, particularly in the more backward economies of the SEECs. Heavy industry had priority because of its importance for the military-industrial establishment and for the engineering sector which produced the machinery needed for consumer goods. Particularly rapid growth occurred in the iron and steel industry and in non-ferrous metallurgy including aluminium. The priority for heavy industry meant exploitation of virtually all relevant raw materials. Hence the importance of the resource base, augmented by further mineral prospecting. Disparities in levels of production between countries and between regions narrowed but did not disappear. However, the generally large scale of production – frequently involving more than 5,000 workers – meant that towns could easily be dominated by just one or two industries. Moreover, the technological level was relatively low because the ECECs did not experience the scale of innovation which persistently marginalised the smoke-stack industries in the West. Socialist industry served relatively undemanding markets and there was little creative destruction: new capacities were almost invariably geared to net increases in output rather than the replacement of obsolete technology. When large enterprises did retool, their discarded machinery often cascaded down to small industries under the control of local authorities or cooperatives in the rural areas. Autarky was another distinct characteristic because, despite specialisation under Comecon auspices and substantial deliveries by ECECs to the Soviet market, the prime concern was the home market and few manufacturers achieved the efficiency necessary for a rewarding export business in the Western world. This was unimportant at first when almost complete separation was desired, but it became an embarrassment in the light of debts incurred in the West in connection with technological upgrading.

The transition left the inherited 'state-owned enterprises' (SOEs) without many of their traditional markets as East Germany became a hard currency trading area while the Soviet Union collapsed and Yugoslavia was isolated for a time by UN sanctions. Meanwhile, home markets were open to competition from Western imports. Finding new markets while boosting quality and efficiency at the same time was a major concern for enterprises which also had to cope with a much higher level of autonomy. Enterprises that were in effect production units of branch ministries could not easily function as firms making their own economic decisions. Since the revolutions were all about doing away with the command economy, restructuring was inevitable at a time when the state was no longer able to support all branches of production and viable private busi- nesses for the market economy had to be created. Although joint venture investments came from abroad, most of the restructuring cost – and the money needed to finance deficits – was initially provided by governments hard-pressed through low financial reserves and tax revenues. There was a heavy shake-out of labour in order to raise productivity and some units with obsolete plant and pollution problems had to shut down altogether. Yet the SOEs were crucially important for national survival and had to be safeguarded in order to build a new industrial economy and avoid the political instability that very high rates of unemployment would have produced. This does not deny the role of second economy entrepreneurs as building blocks of an emergent market economy but suggests that growth potential has been exaggerated, especially in the context of moonlighting and dependence on mafia support (in some countries) which limited the benefits to the state through taxation.

The state-owned industries

The restructuring process varied from one country to another and was inevitably 'path dependent', i.e. linked with the inherited socialist structures. In East Germany the communists organised their industry from the 1960s into large vertically-integrated 'Kombinate' of which there were 126 in 1989 (each comprising 20–40 plants with 20,000 employees on average). Even before the first free elections the GDR government set up Treuhand to hold the state prop- erties and dispose of them. The old networks were paralysed as 'the precondition for a smooth integration of the East German units into the Western corporate networks which proved to be the most promising route to capitalism' (Grabher 1995: 43). The large units were broken up in order to attract buyers, yet many isolated units had to be liquidated because they were left without their own administrative, distribution and R&D capacities. Plants that were sold were inte- grated into West German networks and – shorn of their former connections – were left stranded as 'cathedrals in the East German desert'. Von Hirschhausen (1995) portrays Unterwellenborn as a typical socialist Kombinat with a metal- lurgical industry (blast furnaces, oxygen converters and rolling mills); social functions (housing, hospital and kindergarten); and control through the party administration, union and prison. However, there was no interest in acquiring the complex until Treuhand broke it up and offered the generally old produc- tion facilities free to anyone willing to invest in modern steelmaking capacity. Arbed then became the core investor, while smaller departments were priva- tised by EMBO and the 'Community of Unterwellenborn' took over the housing and cultural/sporting facilities. Average employment per enterprise fell sharply and overall employment declined from 7,200 to 2,000.

On the other hand, Hungary set up 23 horizontally-integrated trusts during the 1970s, covering 350 state-owned firms, 3,000 plants and 600,000 workers in light industry alone. There was a higher degree of autonomy for the individual firm under the New Economic Mechanisn (NEM), which operated during the last two decades of communism along with some encouragement of the second economy and scope for 'Business Work Partnerships' within the enterprise during the 1980s. Restructuring in Hungary involved privatisation through a State Property Agency and other bodies, but the process was decentralised for small companies with the process 'basically driven by key actors in the old formal and informal networks which constituted probably the best organized social group in Hungary during the last decade' (Grabher 1995: 45). So while large enterprises were split into separate units they remained connected with the corporate headquarters, so that private, semi-private and state property might be regrouped into new businesses as 'recombinant property' without ownership being clearly private or statist. Outright fragmentation in East Germany thus contrasted with 'the Hungarian pattern of cross-ownership between satellites, corporate headquarters and banks, and above all the "social embeddedness" within clan-like coordination and management networks' (ibid.: 49). Nevertheless, it seems that small firms (employing 50 or fewer people) are becoming dominant everywhere, especially when account is taken of producer services catering for specific industries.

Crisis management within SOEs typically 'generates protective and cautious strategies, little actual technical or organisational restructuring, a severing of formerly paternalistic ties with smaller neighbouring communities and a careful deepening of strategic alliances with larger municipalities to mobilise support at the national level for state subsidies, guaranteed supplies and protected markets' (Pickles 1998: 190). However, these initial responses by SOEs have been gradually overtaken by the reality of privatisation which governments have – sooner or later – been forced to countenance. Not only has the drain of resources to support loss-making enterprises become politically insupportable, but global institutions have seen privatisation as a critical indicator of a reform agenda. Even firms that remain in state ownership have had to find ways of increasing efficiency. The Czech heavy engineering company Škoda Plzeň used to build everything that lay within their technical capacity without considering profitability. It has seen its business contract, but has retained some links with Russia in the context of pipeline building and the overhaul of locomotives. The programme is now to privatise the core business after selling off the other sections. Meanwhile, the Tesla company has been rather more specialised with a focus on radio and TV transmitters and has entered into joint ventures with Lucent Technologies (USA) and Siemens (Germany). In this way it is coming to terms with competition in a global world, though some diversification into engineering is desirable in order to retain its capacities in Prague-Hlocbětín, Mlada Vožice and Vimperk.

It is hard for firms to compete and maintain market share when they fail to find Western partners. The Hungarian state-owned cosmetic and household chemical producer Coala has dropped from a virtual monopoly position in the domestic market to a 40 per cent share because a lack of advertising at a time when Western products were making deep inroads (thanks to Colgate–Palmolive's promotional work through children's television and education programmes for the schools) left the company with just the lower end of the market where it could compete on price. Other businesses have been able to maintain quality and exports, like the Kowary carpet factory in Silesia which dates back to 1854 when the first employees trained at Smyrna in the then Ottoman Empire. Quality production – now mainly Axminsters for wall-to-wall carpeting for stores and institutions – is backed by keen marketing. On the positive side too, there is a growth of indigenous small businesses, first encouraged in some communist states in the 1980s as part of a growing 'second economy'. But while shops and bars can be set up with relatively little capital – often using space in dwelling houses – manufacturing is much more demanding; requiring capital and business skills, even in the case of activities tied to small farms like sawmills and craft workshops. Hence training is needed and also low-interest loans to avoid the extremely high commercial interest rates. Various forms of business incubation have been attempted and links with higher education institutions are developing. Poznań science and technology park was set up in 1995 to foster links between the Adam Mickiewicz University and local business. The results have been rather disappointing and more active involvement by local and regional government may be needed. The larger regional units introduced in Poland in 2000 may be a step in the right direction 'leading to an increasing social embeddedness of the required relational structures' (Muent 2000: 87).

Foreign direct investment

FDI is assuming a central role in many industries, although the level of investment has been generally disappointing, especially in the SEECs. Trans-National Corporations (TNCs) are operating on a global scale and so each municipality or region has to compete with other destinations which may have better growth prospects: and for the next few years growth in the region as a whole is predicted to be below the world average. Some countries are convenient for European markets but others are more remote and low labour costs may be offset by political instability, allied with unpredictable currency depreciation, poorly developed financial and banking sectors, inadequate regulation of financial markets and low spending power related to poverty. The devaluation crisis in Russia in 1997 spilled over into parts of the region and in some countries the fall in investment at the end of the decade has not yet been reversed. Consequently ECECs have revised their economic policies. There was an initial 'neo-liberal' phase to remove state controls in favour of market forces and prevent any return to totalitarianism. But because foreign investment has been generally limited, while domestic private capital resources are very small, the need for greater state involvement in industrial policy is being appreciated and the 'neo-statist' philosophy of intervention is now more in evidence. This requires an appropriate mix of general incentives (tax concessions, credit support, infrastructure, state-financed R&D, training and marketing) and 'creative destruction' to eliminate obsolete plants and create new high-tech capacities; rather than a 'reactive' policy to provide an ad hoc response to social and political pressures.

A further concern is the lack of linkages within the region and the consequent status of the factory as a unit 'disembedded' from the local economy apart from the use of labour and services. This is a reality of the global economy, especially for projects which exploit cheap labour for production geared entirely for export. But it is very much in contrast to the communist structures which generally forged linkages and networks within each region. The more successful regions do not necessarily have integrated industrial structures but larger numbers of industrial workers in enterprises which develop links that are almost exclusively external. It is clearly desirable that FDI should become more 'embedded' in the regions where it locates and this may well happen in the course of time. Meanwhile, factory regimes may be imposed which minimise the scope for the workforce to contribute to policy making. Writing about the motor industry in Hungary, Swain (1998) refers to hegemonic factory regimes which typify a fragmented industrial econ-omy which is difficult for local or regional government to regulate, while Pavlinek and Swain (1998) contrast 'defensive restructuring' with low wages and worker flexibility (typical of investments geared to export, which may disappear as soon as wage differentials are eroded) with 'offensive restructuring' involving the enrichment of skills, high wages and partnerships between economic and insti-tutional actors: most likely to be found in connection with projects geared to the build-up of domestic market share. As an extension of the latter scenario, Uhlir (1998: 683) argues 'institutional gradualism' through discussion and negotiation so that local elites can regain recognition and legitimation and thereby become more active in regional development.

Experts suggest that success depends largely on a strong national economy combined with social cohesion (Hudson *et al.* 1997) which could offer hope for regions which have achieved good inter-ethnic relations like Slovenia but not for regions experiencing stress. Trust and networking among companies may be conducive to 'particular forms of industrial organisation [which] are more impor-tant than the particular industrial sectors present in a region' (ibid.: 367). When measured against these standards the infrastructure often leaves much to be desired while the fall in living standards, particularly in SEE, impacts on public health. Local business seems slow to take advantage of opportunities to service foreign firms, given Hungarian experience that local suppliers fail to win busi-ness from foreign firms because they are unable to deliver consistent quality, on time, at the agreed price and over the long term. The situation is not helped when SMEs have difficulty in raising loans to finance expansion. However, firms are very sensitive to the lack of a level playing field. Ambivalence towards FDI from management of large state enterprises has been reported from Osijek (Croatia) where local managers and directors are apprehensive over foreign investment because their positions bring political power.

It is clearly desirable that political parties should be able to work together, as in Varna (Bulgaria), where the local UDF (Union of Democratic Forces) lead-ership can cross party lines. But the development of civil society may be retarded by highly polarised politics which threaten to exclude substantial elements of the population. It is evident that 'unreconstructed former communists will use nationalist attacks on a large ethnic minority to establish their legitimacy and then to consolidate their support during difficult economic times' (Snyder and Vachudova 1997: 32). This is an extreme form of tension quite often experienced where local affairs are in the hands of a party in opposition at national government level. In Slovakia in the mid-1990s, regional advisory and information centres were widely seen as state institutions representing the former Mečiar government and not necessarily responsive to local interests represented by district economic and social councils. There was considerable variation in the quality of business advice between the effectiveness achieved in Liptovský Mikuláš in contrast to the strained situation in Martin where local initiatives were considered 'non-sympathetic' by the Bratislava government and were not supported by state enterprises in the area.

Slovakia's 'crony capitalism' made the Košice VSŽ metallurgical complex (with 25,000 employees and a joint venture investment of $57 m. by US Steel of Pittsburgh) a Mečiar bastion in the mid-1990s. Meanwhile, the local mayor (R. Schuster) projected a less partisan approach to Slovak independence – appropriately in an ethnically complex city – and set up his own Party of Civic Understanding which was successful in the 1998 election as part of the anti-Mečiar coalition: indeed Schuster himself was subsequently elected the country's president. In Croatia during the last years of the Tudjman presidency, Split was governed by a coalition of opposition parties and such was the conflict with Zagreb over control of the local economy that national planners seem prepared to route the planned Zagreb–Dubrovnik motorway through Bihać and southern Hercegovina rather than Split. There were also strained relations in Dubrovnik where the activities of the Croatian Central Bank in support of the Dobrovačka Banka – regarding a hotel rehabilitation programme which brought control of local companies through 'debt for equity' swaps – were seen as an attempt by Zagreb to gain control of the city's economy.

It is however a central tenet of the global concept that regional diversity is maintained if not enhanced by direct involvement in the global economy. When communism collapsed in East Germany it was noticeable how the larger cities immediately began to project an identity on the strength of economic audits highlighting leading activities that could be promoted as cornerstones of their economic development strategies. Hudson *et al.* (1997: 369) believe that 'global outposts' can attract embedded branch plant investment that forges links with the regional economy, although chances are maximised where management and labour (and more generally the external and local interests) appreciate the reciprocal relationships between cohesion and competitiveness and where local institutions are well developed. Hampl and Muller (1999) also think that geographical unevenness can be put to effective use if, for example, sovereign states can opt to maintain low labour costs and a better environment for foreign executives to compensate for higher transport costs. It is therefore useful to look at some of the factors relevant to FDI location decisions which over the past decade have played a major role in the transition process. It takes note of recent scientific literature, including the business journals.

Foreign penetration typically raises efficiency and is a force for consolidation so that the remaining capacity (indigenously owned) can retain its markets. Such a process is occurring in the highly fragmented Polish meat industry, helped by a new veterinary law curbing illegal sales by farmers. In some cases there may be strong competition for market share, with the Czech brewing industry a case in point. The Anglo-Lebanese company 'Seament' now operates in Albania at Elbasan and Tirana (Fushe-Kruja) with the former complete and now undergoing reconstruction to quadruple production to 0.60 m.t. After takeover by foreign capital it is not unusual to see quite complex forms of specialisation taking place to achieve success in the global economy. A good example is the Slovak synthetic fibre manufacturer Chemlon Humenné which spawned a set of joint ventures after takeover by Rhône-Poulenc in 1993 (which subsequently merged with Hoechst in 1998): Humenné is involved in the Rhodia subsidiary producing industrial yarn for tyres and conveyors. Humenné is also tied up with Nylstar while a joint venture between Rhodia and Sina (Italy) secures nylon fibres for the textile industry. Through their foreign associations companies in the region can improve their performance and gain public recognition as 'national champions'. On the other hand rationalisation may be unwelcome and when the UK company United Biscuits sold on the famous Hungarian food producer Györi Keksz to Danone and the latter wanted to close down the long-established Györ factory in favour of consolidation at Székesfehérvár, there was very strong local opposition in Györ to the point where the government stepped in with help to keep the factory open.

Although reorganisation has been slow and much still remains to be done, the region's industry is very different from what it was in 1989. Production levels were down in the early 1990s, but there has been good growth since, especially in Hungary and Poland. Recent growth has not been continuous, given the poor years in Albania and Bulgaria (1997), Romania (1997–9) and Yugoslavia (1999), along with small declines in the Czech Republic, Slovakia and Slovenia in the latter year (Table 3.1). Despite poor statistical coverage, it is evident that businesses are now much smaller, but also more efficient and productive with a high input of private (often foreign) capital and a new range of markets. Typically the best companies are those which combine a foreign stake with local management like the VSŽ Košice steelworks in Slovakia (recently sold to US Steel), the Volkswagen car assembly plants in Bratislava (Slovakia) and Mladá Boleslav (Czech Republic), the Hungarian mobile phone company Westel and the Bulgarian distributor of lubricants Prista Oil. Foreign investment is considered highly desirable not only because domestic capital is very limited but because external influence can not only help with technological updating but deliver improved export market penetration through integration into new corporate structures. Experiences have been mixed. From the investor's point of view, some companies that went in quickly, like General Electric and General Motors in Hungary, have been disappointed in the short term because they underestimated the scale and duration of the recession. Guardian Glass at Orosháza found that float glass orders from ECE were making up only a quarter of the production instead of the half that had been anticipated. So more of the output has been

Table 3.1 Percentage change in industrial production per annum 1989–2002

	1989	1990	1991	1992	1993	1994	1995	1996	1997	1998	1999	2000e	2001e	2002
Albania	na	-8.2	-42.5	-20.4	10.0	-8.0	2.8	1.0	-5.6	9.3	5.0e	4.0e	na	na
Bosnia & H.	na	na	na	na	na	na	na	na	40.0	24.0	8.0	8.0	9.0	7.5
Bulgaria	-0.2	-17.5	-27.8	-21.9	-6.3	2.9	1.7	1.0	-13.1	4.0	2.3	2.0	3.0	4.0
Croatia	na	na	na	na	na	na	na	na	6.8	3.7	0.1	1.5	4.0	3.0
Czech Republic	1.5	-3.5	-22.3	-10.6	-5.3	2.3	9.2	9.8	4.0e	6.0	-3.1	5.4	6.1	5.9
Hungary	na	-8.5	-19.1	-9.8	0.6	5.0	4.8	3.4	11.1	12.6	10.4	18.3	8.0	6.2
Macedonia	na	na	na	-15.8	-13.9	-10.5	-10.7	3.1	1.6	4.5	-2.6	3.4	0.0	4.0
Poland	na	-24.2	-11.9	3.9	6.2	12.1	9.4	8.7	10.8	4.8	3.6	6.8	4.1	4.6
Romania	-2.1	-18.8	-19.6	-22.0	1.3	4.6	9.4	9.9	-7.4	-17.3	-8.0	8.2	5.1	8.1
Slovakia	1.1	-2.7	-21.6	-13.7	-13.5	7.5	7.5	3.0	2.4	4.0	-2.9	9.1	4.8	5.4
Slovenia	na	na	na	na	na	6.4	2.0	1.0	1.0	4.2	-0.6	6.2	3.5	3.2
Yugoslavia	na	na	na	na	na	1.3	6.0	7.5	9.6	3.4	-50.0	10.9	0.5	3.0

Source: Economist Intelligence Unit

Note
e Estimate

sold in the West, in spite of high transport costs arising from a factory location well beyond Hungary's frontier with Austria. The scale of trade liberalisation was sometimes underestimated, enabling much of the domestic demand to be met through imports. And lower import duties certainly facilitated the import of second-hand cars, thereby reducing the potential for home market sales by Western car makers setting up within the country. However, labour has performed better than was expected and the longer-term prospects are good.

Trends by industrial sectors

Steel

In readjusting to global markets cultural changes were needed such as the need for payment from customers in advance in a situation where firms had very limited financial services. However, the situation varied considerably between countries and industrial sectors. Heavy industry in particular has had to slim down and accept a significant reduction in capacity. The steel industry found itself grossly oversized, with much obsolete equipment, especially in older steel mills that were also too small to enjoy economies of scale (the Polish industry was particularly fragmented). High-cost production based on imported iron ore (and coking coal as well in some cases) cannot be justified, especially when the steel is used extravagantly to export ships and railway equipment at low prices. The Kremikovtsi complex was built on a low-grade Bulgarian orefield and fully integrated with iron, steel and rolling capacity in 1974. But the local ore and coke were inadequate, while the plant was badly located for imported ore which had to be railed up to 500 km overland after arriving by water at Lom on the Danube (following transfer to barges at Galaţi) and later at Varna where ore arrived off the 'Druzhba' ferry from Ukraine. 'Reckless investment in the machine tool, engineering and utilities section' was undertaken in order to consume the higher metal output (Palairet 1995: 503). It has proved impossible to maintain output in the context of a market economy with Western Europe extremely sensitive to dumping from ECECs.

The fall in steel production has been quite dramatic: from 2.94 to 1.55 m.t in Bulgaria between 1985 and 1992; 15.04 to 10.84 in FCS; 3.54 to 1.53 in Hungary; 15.36 to 9.83 in Poland; 13.79 to 5.38 in Romania – an overall reduction of 58.8 per cent. Recent trends are shown by monthly figures in Table 3.2. But given the national importance of heavy industry throughout ECE, attempts have been made to restructure the sector as a whole to adjust production to new market conditions while minimising employment decline and modernising individual plants to the point where they can be privatised. The VSŽ Košice steelworks in eastern Slovakia (opened during the late 1960s) has done relatively well, with a reduced workforce (partly redeployed in special steelmaking and in steel-using industries established in the vicinity), thanks to a low-cost product which benefits from efficient transport of raw materials (iron ore pellets from the new Dolinskaya plant in Ukraine and coke from Ostrava) delivered to the mill without the need for trans-shipment since standard- and broad-gauge rail access is available. When VSŽ needed a strategic partner to participate in privatisation, US Steel became a persistent suitor on the basis of a lucrative initial

Table 3.2 Monthly production of basic commodities 1982–2000

Cement (,000 t)

	1982	1986	1990	1994	1998	2000
Bulgaria	468	477	393	159	138[3]	166
Fmr. Czechosl'kia	860	858	864	na	na	na
Czech Republic	na	na	512[2]	438	384	300
Fmr. GDR	977	999	1022[1]	na	na	na
Germany	na	na	na	3355	3047	na
Hungary	120	186	328	233	234[3]	243
Poland	1336	1319	1043	1153	1243[3]	1148
Romania	1161	1088	789	500	608	529
Slovakia	na	na	281[2]	240	237[3]	359
Slovenia	na	na	131[2]	139	94e	105
Fmr. Yugoslavia	810	760	663	na	na	na
Yugoslavia	na	na	na	134	188	na

Steel (,000 t)

	1982	1986	1990	1994	1998	2000
Bulgaria	216	141	182	208	219[3]	166
Fmr. Czechosl'kia	1253	1259	1239	na	na	na
Czech Republic	na	na	610[2]	590	541	575
Fmr. GDR	597	664	na	na	na	na
Germany	na	na	3203	3403	3401	3881
Hungary	309	310	234	161	152[3]	149
Poland	1206	1429	1135	926	966[3]	790
Romania	1088	1189	518	469	533	398
Slovakia	na	na	375[2]	331	288[4]	na
Slovenia	na	na	33[2]	35	8	44
Fmr. Yugoslavia	168	175	127	na	na	na
Yugoslavia	na	na	8[2]	4	3	na

Electricity (m.kwh)

	1982	1986	1990	1994	1998	2000
Bulgaria	3370	3485	3511	3198	3475	3380
Fmr. Czechosl'kia	6229	7064	7060	na	na	na
Czech Republic	na	na	4941[2]	4892	5426	6210
Fmr. GDR	8576	9608	9915[1]	na	na	na
Germany	na	na	na	37786	45860	50137
Hungary	2065	2338	2361	2780	3080	2899
Poland	9797	11691	11311	11236	11931	11898
Romania	5742	5965	5345	4591	4374	4323
Slovakia	na	na	741[2]	726	863	915
Slovenia	na	na	na	na	na	na
Fmr. Yugoslavia	5194	6448	6871	na	na	na
Yugoslavia	na	na	3041[1]	2959	3388	na

Cars and Commercial Vehicles (,000)

	1982	1986	1990	1994	1998	2000
Bulgaria	na	na	na	na	na	na
Fmr. Czechosl'kia	na	na	na	na	na	na
Czech Republic	21.9	22.9	21.1	2.5[5]	3.9[3,5]	na
Fmr. GDR	18.4	22.0	21.4[5]	na	na	na
Germany	na	na	na	376.1	483.4	na
Hungary	na	na	na	na	na	na
Poland	0.9[5]	1.3[5]	0.7[5]	0.1[5]	0.2[5]	na
Romania	23.1	28.8	25.8	30.0	48.0	56.3
Slovakia	8.6[5]	10.7	10.4	8.3	10.9	5.6
Slovenia	na	na	na	0.7[5]	3.5[5]	16.0
Fmr. Yugoslavia	19.1	23.6	28.4	na	na	na
Yugoslavia	na	na	na	1.1	1.3	na

Source: United Nations Monthly Bulletin of Statistics

Notes
[1] 1989 [2] 1992 [3] 1997 [4] 1996 [5] excluding cars = excluding commercial vehicles

joint venture (1997) for tin-coated sheeting for the food industry and galvanised steel for the vehicle industry (though the latter had to be sidelined in 1998 as mismanagement pushed the firm towards bankruptcy). US Steel purchased the complex in 2000 and carried out a radical restructuring to dispose of all extraneous assets. VSŽ is now a major player as a high value-added steel producer, poised to supply the regional market for both products in competition with Thyssen–Krupp and Voest Alpine.

On a smaller scale, American capital has also been invested in electric steel-making at Reşiţa in western Romania where a late-eighteenth-century plant was allowed to expand under communism way beyond its local resource potential. Closure of the blast furnaces and coke ovens (and abandonment of a large coke-chemical plant constructed just before the revolution) left the works dependent on scrap metal charges for the Siemens Martin furnaces although a new electric furnace (supplied by the company's own hydro-electric installation) and continuous casting commissioned in 1999 provided a basis for privatisation which took place in 2000 when Noble Ventures took over. In this case it was fortunate that maximum expansion of the local hydro potential was embarked upon – to maximise energy self-sufficiency within large industrial complexes or 'centrals' – in the later communist years for the works would not be viable if power was needed from utilities. Even so the enterprise was plagued by strikes during 2001 due to disputes over the promised investment programme. Other privatised works face problems. Turkey's Kurum Steel acquired Albania's Elbasan works in 1999 and operates the mini-mill principle with an electric furnace – unfortunately affected by power shortages – to produce billets from which wire and other rolled products are derived. But markets in Kosovo and Macedonia will be needed to push production from 250 tpd towards the target of 1,000 tpd. And Kuwait Investment Agency has spent $90 m. to restart steel production at Zenica (B&H) but the venture is problematic now that the Yugoslav market has gone.

There is some unease in EU over the establishment of American companies in the 'backyard' where preferential tariffs have been granted. For ECE could become attractive as a peripheral location for blast furnaces to threaten those installations currently on the margins of Europe as presently constituted. At the same time, there is concern within ECE as to the long-term competitiveness of the remaining state-owned steel industries and heightened in the privatisation of such large complexes at Hungary's Dunaferr (Dunaújváros) and Romania's Sidex (Galaţi), though the latter seems destined for takeover by the European steel group LNM (in competition with Usinor of France) in a debt for equity swap that brings together a bold investor and a decisive government. Unable to contemplate liquidation of major economic installations – or limited takeovers by VSŽ! – other governments are doing what they can to boost efficiency in the hope that acceptable privatisation deals can be negotiated that will ensure investment from other sources and simultaneously address EU pressure to reduce the size of the state sector. Thus the Czech government is investing in the Moravian steelworks of Vitkovice (with the help of the Czech Revitalisation Agency), selling off non-core businesses like the hotel, sports stadium and reducing staff from 12,000 to 6,500. The plan was conceived as a short-term measure for the period up to the 2002 election – rather than a full-scale modernisation programme – but the government is now looking at the possibility of a wider grouping to create a Czech Steel Company ('České Ocelarsky Podnik') to combine blast furnaces in Ostrava with the Nova Hut and Vitkovice steelworks. Such a combination might be integrated with Třinecké Železarny, owned by Moravian Steel (with some US participation) since 1996 and including part of the Bohemian steel producer Poldi Kladno.

In Poland a joint venture for quality steel making was initiated – with EBRD support – between Lucchini of Italy (efficient operators of scrap-based steel mills) and Huta Warszawa (Warsaw) where a $200 m. modernisation should protect the workforce of 4,700 concentrating on steel strip for the vehicle industry. But Huta Katowice (HK) at Dąbrowa Górnicza (producing long products) and Huta Tadeusz Sendzimira (HTS) at Nowa Huta, Kraków (turning out flat products and tubes) face the prospect of 'slow death' as they lose market share and face severe financial problems. After the installation of a continuous casting line from Mannesman, HK attracted interest from the Netherlands/UK company Corus, Danieli of Italy and CVRD of Brazil with a view to splitting the complex. But this project has failed, while interest from Thyssen–Krupp (Germany) and Voest Alpine (Austria) has also faded, with the latter now envisaging a new coastal plant at Szczecin for tin sheet. Meanwhile, the southern mills supply metal to the small, specialised steel mills in the core of the Upper Silesian conurbation. Indeed privatisation hopes now rest with a deal that will combine HK and HTS with two small steelworks: Huta Cedler and Huta Florian. If interest from the Indian–Indonesian group ISPAT or the West European consortium Arbed–ThyssenKrupp is not sustained there may be no option other than to create a single company Polish Steel ('Polskie Huty Stali') under the Industrial Development Agency. But massive investment would be needed to meet demand for higher value sheet and the EU will not allow government 'bale-outs' unless a viable enterprise can be secured. Other Silesian steelworks have already been reprofiled like Huta Bankowa, which produces forged and rolled products from HK ingots, while Huta Batory in Chorzów has closed obsolete heavy production departments and is investing in a new electric mill. Zawiercie steelmill has boosted productivity by installing gas and oxygen burners (supported by an oxygen plant on the premises) and more steel is now produced from two electric furnaces than was previously produced from seven old furnaces (including open-hearth equipment now closed). Continuous casting has also been installed.

Non-ferrous metallurgy

This also faces severe challenges given the need for competitiveness and higher environmental standards. Many mines are not profitable while those with good reserves and long-term potential may be compromised by obsolete equipment and poor management (as well as disruption in the case of Albania where employment in mining has halved to around 10,000 after the insurgencies in 1991–2 and 1997). Albania's best prospects are in chrome (Dibra and Mat), copper (Mirdite and Puke) and iron-nickel (Pogradec). The Italian company Darfo controls the chrome mines of Bulqize, Pojske and Prrenjas – in addition to the ferro-chromium plant of Elbasan – while Germany's Preussag AG is also modernising mines in order to raise output. Elsewhere, there has been some modernisation in Bulgaria where the Pirdop copper smelter is undergoing a five-year investment programme after takeover by Union Minière in 1997, while other smelters at Kurdjali, Plovdiv and Sofia remain short of capital following privatisation by EMBO. Macedonia still experiences high pollution levels at Veles and Romania combines a dirty state-owned primary copper smelter at Zlatna with a relatively clean privately-owned electrolytic plant at Baia Mare. The aluminium industry, based on imported bauxite (apart from Romania and former Yugoslavia), has experienced similarly varied fortunes. While pollution brought about the closure of Poland's Skawina plant near Kraków in the Solidarity era (the premises are now used by SMEs) the similarly problematic Slovalco smelter (ZSNP) at Žiar nad Hronom in Slovakia has been rescued. Built in the early 1950s to comprise a power station, alumina factory and downstream activities offering employment for a poverty-stricken area of central Slovakia, the area became blighted by air pollution and the leaching of caustic soda from a red mud pile containing some 16 m.t of waste by 1985 when modernisation started on the basis of state-of-the-art technology introduced through collaboration with Norsk Hydro and completed after 1989 with EBRD assistance. Although profitability is not assured, efficiency has improved (with capacity doubled to 132,000 t/yr while energy consumption has increased by only a tenth since the old smelters closed during 1994–6) and both environmental and social conditions have been stabilised. There is a bentonite retaining wall enclosing the red mud pile, the alumina production closed during 1994–7 and the power station has switched from lignite to gas and Polish anthracite coal. Elsewhere, the French aluminium company Pechiney is poised to acquire the ailing Czech aluminium company HZB while Romania's relatively modern high-purity smelter at Slatina (and the alumina works Alum Tulcea) is attracting attention from an Alcoa-led consortium.

Engineering

It has experienced difficulties, given the need to increase productivity generally while slimming down capacity. Some of the problems may be appreciated with reference to shipbuilding in Poland. After a steep fall in orders during the 1970s and 1980s, the transition brought a collapse of trade with the FSU plus the failure of post-Soviet shipping companies to pay for some of the ships that had already been commissioned. Also, dollar contracts in the 1980s were set at unrealistically low prices in order to generate hard currency, while governments paid subsidies to the shipyards. But when these ceased in 1988 shipyards had to take out high-interest loans and huge debts were swelled by inflation. Poland's 'Stocznia Gdańska' was closed in 1988 for a time, while 'Stocznia Szczecinska' faced bankruptcy in the early 1990s. All Poland's five yards have now carried out debt-reduction and modernisation programmes although problems of low productivity and long lead times remain. However, 'Stocznia Szczecinska' is attractive to foreign companies because of low costs and short completion times (driven by the high cost of credit), while heavy-duty cranes have helped the yard to diversify. Moreover, there could be a spate of domestic orders within a decade when the aged 1970s fleet needs to be replaced.

High-tech industries were hit by the collapse of orders from the military-industrial complex since symbiosis between military and civilian production was generally lacking. Reorganising the arms industries of former Warsaw Pact countries has been a major challenge and there has been a great effort to retain employment while developing new products and markets. There has been a general preference for alternative links with Western companies, exemplified by the contract of 1999 between Poland's Huta Stalowa Wola and General Electric Marconi for the assembly of self-propelled 155 mm howitzers in Poland using the chassis of the modified F72 tank from Huta Łabędy (Upper Silesia) and gun barrels from GEC. The new equipment is needed by the Polish army in preparation for NATO membership while foreign markets will be sought with the help of GEC Marconi. Meanwhile, traditional eastern links have been valued by Slovakia where the new state inherited an extremely large armaments industry. Growth of the business in the communist period made sense in the context of Warsaw Pact forces stationed in Bohemia and Moravia, while locations in Slovakia contributed to regional development in this part of the country, with hydropower from the Váh Valley stations and close links with the metallurgical industry of Ostrava (subsequently at Košice in eastern Slovakia too).

Kiss (1993) refers to the Martin–Detva–Dubnica nad Váhom triangle of Slovakia where some 30,000 workers used to produce heavy weaponry under Soviet licence, compared with the largest concentration in Moravia amounting to some 10,000 workers in the Slavičín–Bojkovice–Uherský Brod area. Although there has been some labour-shedding, the relatively large workforce has been retained through short-time working. Some early conversion at Martin went ahead in the late 1980s thanks to contracts with Hannomag of Germany for tractors and construction machinery; also with Lombardini of Italy regarding the manufacture and sale of small diesel engines (to East European markets only). But the Russians are gaining considerable control through capital invested in 1993, although the Slovaks are tied to making RD35 aircraft engines and JAK130 aircraft from Považske Strojarne and navigation systems for MiG29 aircraft from Trenčin. There is an obvious Russian interest in cooperation with Slovakia; slowing the expansion of NATO and paying debts in military technology. But the Czech Republic has totally rejected this strategy, even though the amount owed by Russia was greater than in Slovakia's case, and while Hungary has imported Russian aircraft, she is not solely dependent on Russian technology. Slovakia could have developed some promising lines with Western assistance including mine-clearing devices and the T72 tank. However, her whole approach will now have to be modified in the context of a declared interest to join NATO.

Wood processing

This includes paper, board and furniture production, and benefits from good supplies of domestic timber over much of the region. Many new sawmills have opened up and the production of furniture is an expanding sector in some countries, notably Poland where Nowy Styl is a family business that set up in Krosno in 1992 exporting chairs around the world: after recovering from the Russian crash in 1998 it opened a new office furniture plant in Stalowa Wola SEZ. Wood processing is one of Poland's expanding industrial sectors, as is food processing. By contrast, ECE's leading sawmillers in the communist period, with large capacities and multiple-production lines using old technology, are unattractive to small Western companies. However, Italy's Calligaris purchased Croatia's Ravna Gora factory which produces beechwood chairs, attractive in terms of quality and price. Another Croatian unit, TVIN Virovitica, supplies the giant Swedish Ikea company with tabletops produced from particle board and German veneer (though Ikea has a very efficient partner in Slovakia and might also resume old contacts with hardwood manufacturers in Serbia). With its oakwoods in Slavonia, Croatia could do better when hardwoods come back into fashion, though high wood prices from state-owned forests could be a problem.

Building materials

These include cement and tiles, and are attractive to foreign investors encouraged by impending motorway projects and the growth in housebuilding. Moves are being made by Western building materials groups like Braas of Frankfurt and Redland of the UK who favour local production because long-distance transport is a burden while the setting-up costs in the region are acceptable because of low labour costs. Moreover, each local producer tends to create its own market with capacity for very rapid growth in the years ahead, while markets in the West are expected to be stable or declining. Pol-Float, the Polish subsidiary of French glass manufacturer Saint Gobain, has opened an environmentally-friendly glassworks at Dąbrowa Górnicza; while Pilkington's subsidiary in Sandomierz is also serving the domestic market (including the supply of windows for the car industry). American Building Products and Royal Plastics Group of Toronto are keen to get access to Mielec SEZ (offering a clean environment for high-tech firms) in order to produce virtually indestructible PVC construction materials. The Canadian company owns patent rights for making vinyl compounds and produces its own machinery for extrusion.

Light industry

Textiles and clothing production can be geared to Western markets on the grounds of cost although low-cost cotton from the FSU ceased to be available and Far Eastern competition is strong. In East Germany capital has been injected into the Lusatian textile industry which experienced drastic reductions in employment immediately after the revolution. High-quality textiles are now being produced in Forst, Guben and Spremberg using a new production concept rejecting mass production and concentrating on quality. Tiroler Laden of Innsbruck is building on the century-long weaving tradition of Forst through the activity of a subsidiary company (Brandenburgische Tuche) within the old 'Forster' factory. Meanwhile

Gubener Tuchmacher works closely with a Paris-based fashion designer and deals with orders from international fashion houses. There is the possibility that this business could extend to a network of Brandenburg textile companies (Kratke 1996: 23–4). Meanwhile, there have been some major changes at Poland's Podhale leather and footwear factory built in Novy Targ in the mid-1950s. Up to 10,000 were employed in producing shoes, but the factory collapsed under the shock of losing the Soviet market, not to mention the arrival of Western shoes with revolutionary designs. However, there are now more than 40 small businesses using the premises and those which still produce shoes – stimulated by a large domestic market and new opportunities in the east – now give greater attention to quality and design to compete with foreign imports. Meanwhile, the old firm's apartment dwellings have been purchased by the tenants while the local authority has taken over the skating rink and swimming pool. And foreign investment is coming into clothing and footwear in other parts of Poland.

In the food sector, meat processing and canning were hard hit while beverages – including brewing – has done well. There have been many takeovers throughout the region and new production capacities have been provided. The Czech breweries have attracted particular attention. The British company Bass controls Pražské Pivovary (Prague Breweries) with a 13 per cent share of the market and would like to control Radegast in order to gain a further 15 per cent; but Nomura (Japan) have been successful in combining Radegast with the market leader Plzeňsky Prazdroj (29 per cent) to form a very large organisation 'České Pivo'. However, in general ECE producers found themselves with a narrow product range and had difficulty meeting EU quality standards, not to mention a lack of good transport and logistics. There is good scope for competition in labour-intensive areas such as fruit and vegetables, but on the other hand the Polish sugar industry found itself handicapped by high energy consumption and high losses in production; needing an expensive upgrading of capital stock in order to compete in the West (Von Zon 1996: 94).

Growth in the vehicle industry

It would appear that ECE is the latest 'spatial fix' for the global automobile industry. There was already significant production in the 1980s following rapid expansion between 1965 and 1976 (from 0.25 m. cars to 0.76) through cooperation deals and licensing arrangements with Western companies, especially Citroën, Fiat and Renault. But growth was relatively slow during the 1980s despite attempts to produce specially for the Western market, such as the FCS Škoda Favorit. Factories were small by Western standards, yet all the ECECs except Albania and Hungary were involved. And the domestic markets showed some growth with 266 cars ptp in East Germany in 1989 and 242 in FCS, almost exactly doubling the figures for 1977: 131 and 121 respectively. But given the pent-up demand, underscored by the priority for rail transport under central planning, a prominent feature of the transition was bound to be a relative expansion in road traffic and a further increase in car ownership as the freeing of trade made it possible for both new and used cars to be imported. And despite a relatively slow pace of reform, car makers have remained impressed by the potential market. Firms are keen to get in behind tariff barriers, with the added attraction of low-cost labour suggesting the possibility of car production in the east for export, as was always the intention in Hungary.

Highly motivated and flexible workforces have been recruited with the minimum of trade union organisation, with wages above local averages to discourage moonlighting, but eminently affordable in the global context. Volkswagen invested in Škoda Auto (though without the Czech sourcing that the government wanted) with labour costs at Mladá Boleslav – previously using some Cuban and Vietnamese workers – only a tenth of the West German levels. However, this arrangement acted as the catalyst for some 50 joint ventures for the production of components (including six in Slovakia) and 20 greenfield component plants. The old Škoda component suppliers were encouraged to form joint ventures with Western companies as a means of increasing quality quickly: driven by the danger of losing orders and the stimulus of gaining new orders from other manufacturers in the Volkswagen group. As a result, 40 of Europe's top 100 car-parts suppliers now operate in the country (and some two-thirds of the components for Škoda cars now originate in the country). Moreover, several Czech parts manufacturers are linked up with larger, better-networked foreign manufacturers. One of the new enterprises is a fuel injector business set up by Robert Bosch with Motor Jirkov. Although Jirkov lost its contract with Škoda, Bosch increased its stake in the factory which now supplies its global network. Bosch have also transferred car headlight manufacturing from Reutlingen to their diesel subsidiary in Jihlava. Siemens (Germany) have invested in factories at Plzeň and Stříbro to produce cables for BMW and now employ some 1,500 workers (Pavlinek 1998b: 79).

Meanwhile, in East Germany the reorganisation of the vehicle combine ITA enabled General Motors to recruit a hand-picked workforce – albeit in the high unemployment area of Eisenach – for wages 40 per cent below the level prevailing at Russelsheim despite longer hours, while Volkswagen invested at Mosel near Zwickau. And despite having no stake in car production before 1989 Hungary attracted investment and restarted a tradition first established during the 1920s and 1930s. Magyar Suzuki have a project at Esztergom (and an alliance with General Motors for compact car production is mooted) while General Motors have located at Szentgotthárd in close contact with an existing engine plant just over the border in Austria. Toyota are reputedly planning an investment on the strength of the country's export base. The production of components is also expanding in view of the Ford factory at Székesfehérvár while locations include Gödöllo, Pásztó, Vezprém – and Györ where a Hungarian aluminium manufacturer has set up a cylinder head factory to supply Opel in Austria. Clarion (Japan) is investing in a factory for car audio equipment at Nagykáta 50 km east of Budapest while Harrison Thermal Systems (part of Delphi Automatic Systems) have embarked on a joint venture with Calsonic (Japan) involving a car air-conditioning factory at Balassagyarmat.

Most developments have occurred in Poland where a continuing relationship with Fiat is now geared to producing the Cinquecento at Tychy and the Siena at Bielsko-Biała. General Motors had plans to modernise the existing Warsaw factory and upgrade the Polonez (initially derived from the Fiat 125) while Peugeot embarked on a venture in Lublin and Volkswagen/Toyota invested in the farm vehicle factory near Poznań where the Felicia and Polo are now assembled. However, since 1995 the South Korean firm Daewoo have taken over the Warsaw and Lublin factories, effectively blocking the Peugeot deal while displacing General Motors to a new factory in Gliwice (1998) where the new Agila is being made for export to Europe. In Warsaw Daewoo offered far more than General Motors: an investment of $1.1 bn. for the assembly of South Korean cars and a production base for the manufacture of engines, gearboxes and components. Meanwhile, Lublin was used for the production of vans, some of which were exported in parts for assembly at a Daewoo plant in Ukraine.

The components industry is again dynamic and advantage is being taken of the opportunities to invest on favourable terms in a special economic zone (SEZ). Toyota has opened a $93.5 m. transmission plant for 250,000 gearboxes/yr in Wałbrzych SEZ with the intention of exporting to Europe, while Tychy SEZ is home to the Isuzu diesel engine plant, initially for supply to Opel in Gliwice but eventually General Motors plants throughout the EU as well. Delphi Automotive Systems (a subsidiary of General Motors) are buying up local parts manufacturers and are building a car seat factory near the assembly plant. Meanwhile, United Technologies Automotive, a subsidiary of United Technologies (USA), is building a factory at Mielec SEZ to make electric wiring elements for West European carmakers and Daewoo will manufacture automotive lighting equipment in the Suwałki SEZ. Thus foreign carmakers are using an increasing proportion of Polish parts: 75 per cent in the case of the Daewoo Matiz.

With so much foreign interest in Poland's vehicle industry the future is rendered highly uncertain for other established firms which have not been able to forge foreign partnerships. In this sense, a potentially important development was the creation in 1991 of Sobieslaw Zasada as an exclusive importer and service company for Mercedes-Benz products, an association which developed out of the owner's interest in car racing. The company now controls a range of Polish motor factories including the truck maker Star (Starachowice), the Autosan (Sanok) and Jelczanske Zaklady Samochodowe (Jelcz) bus factories and a producer of gearboxes for trucks at Tczew (Fabryka Przekladni Samochodowych). The aim has been to turn these flagging businesses around with the use of Mercedes technology (though not, as yet, their capital investment): the first Jelcz-built Mercedes-Benz bus was finished in 1995. Despite a cheap but skilled workforce, business has not been highly profitable – with falling demand in the early 1990s and competition from foreign suppliers, some of whom now have their own assembly facilities in Poland. The firm has produced some 4.0 m. vehicles since 1953 but annual output has fallen from a peak of 230,000 in 1989 to 14,000 in 2000. Although the company hopes to improve performance on the basis of exports – and the Polish commercial vehicle market is certainly a large one – a foreign partner is needed for a long-term investment (possibly by Peugeot or Volkswagen) to get the plant abreast of modern technology, for tax rates are now more favourable for importers.

Slovakia was involved in motor cycles (first produced at Považská Bystrica in 1927) and tractors began to emerge from Bánovce nad Bebravou ten years later. However, under communism Slovakia was mainly concerned with components for cars, buses and trucks assembled in the Czech Lands. Presently there are a few companies growing fast as part of the international supplier network (like Presskam Bratislava) but most of the 150 companies have limited capacity and obsolete technology. Now a 'National Programme for Vehicle Manufacture Support' seeks strategic foreign partners, working from manufacturing companies already present in Bratislava and Košice (the latter founded by firms in Bratislava, Puchov, Trenčin and Žiar nad Hronom as well as VSŽ Košice). The aim is for Slovakia also to expand vehicle assembly for cars, trucks and buses, particularly in central and eastern regions (Banská Bystrica, Prešov and Košice) where unemployment is relatively high.

There has been only limited development in the SEECs. Romania's two car plants at Craiova and Pitești-Colibași have been taken over by Daewoo and Renault respectively. But a small assembly project at Varna (Bulgaria) by Rover has folded. Zastava of Kragujevac, the ailing Yugoslav car maker with an over-manned and unfunded plant, desperately needs a new model (the Yugo Florida was designed in 1988). With only a slight recovery of output after NATO bombing, the company needs access to a home market protected by a temporary tariff barrier. There is also a strong desire for a foreign partnership – perhaps with Peugeot Citroën – following the restoration of links with Iveco to produce trucks for the local market and export. The workforce has been drastically cut, with 35 dependent companies hived off. Meanwhile, Volkswagen have resumed assembly work at Vogašcá near Sarajevo (where production had to be curtailed at the onset of war in 1992) with the aim of delivering Škoda Felicias to the Balkan market. In Slovenia, Alusuisse-Lonza (Switzerland) is involved in a joint venture with the Tomos Group to produce car components at Koper. Meanwhile, the truck and bus maker TAM of Maribor lost its markets when Yugoslavia broke up and the company has been split into 14 companies among which some survive as private companies, renting land and equipment from the bankruptcy administrator. A number of these have recently been sold to component supplier Cimos Koper which has markets in France.

The future of the car industry looks reasonably secure since costs are low by world standards. But much will depend on the growth of the ECE market and the labour cost advantages of different peripheral areas of Europe, for the region is very much in competition with Iberia. It is noticeable that, as demonstrated by the Hungarian experience, there are sharply contrasting approaches between projects which create classic branch plants ('enclaves dislocated from local market and regulatory upheaval') and 'relatively locally-embedded production facilities within a European manufacturing system' (Sadler *et al.* 1993: 348). The latter situation is obviously preferable from the ECE viewpoint, although the former does have its own attractions in terms of employment and technology transfer and establishes bases which could be enlarged in the light of continuing cost advantages and local market opportunities. Indeed, the region is now seen as a good export base. In the early years Poland was keen to attract car firms to cater for the home market with tax incentives as well as high tariffs against imported cars. But as EU membership gets closer these tactics are being watered down and tariffs are now 10 per cent compared with 35 per cent when protectionism was at its height. As the competition for the home market intensifies, Poland is being used more as a cheap export base with Toyota and Volkswagen both producing more cars and components for export than for domestic sale, while the balance for Fiat is just 50:50. As already noted, General Motors are exporting from Gliwice and while Škoda Auto holds on to half the Czech market, sales in Germany and Poland are also doing well. At the same time it cannot be assumed that capacity in Western Europe can be closed down to avoid global oversupply. Firms that set up low-cost assembly plants to beat import quotas by taking in kits may well reorganise in the light of overcapacity in the West like General Motors in Hungary which may switch from producing Astras to making components. And established vehicle builders which have not yet established joint ventures with Western firms will face greater problems in finding investment capital to maintain their domestic market share and develop an export business.

Component suppliers are also exporting. Italian-owned Magnetti Marelli, which started up in Poland in 1992 to supply Fiat, now serves most of the big car makers in Italy and started selling electrical components to China and India in 1999. The same trend is evident among the Daewoo joint ventures with 17 components factories and while Ford have closed their assembly operation at Plonsk, its components subsidiary Visteon is expanding its Duszniki Zdrój plant to supply foreign markets. In Romania both Daewoo at Craiova and Renault at Piteşti-Colibaşi are exporting more of their production now that home market sales are falling. However, high production standards appear crucial. While some Polish companies are exporting half their output of high-quality components, others that lack modern technology are on the verge of bankruptcy. Many Czech suppliers are oriented to Czech standards only and Czechinvest has launched a supplier-development programme with the focus on quality engineering and control. It is likely that there will be increasing discrimination as the car manufacturers impose stringent terms on their suppliers. To remain in the network, even for domestic markets, suppliers will need ISO 9000 quality certificates and some small firms will need to change their whole production system. This has to be seen in the context of modern assembly involving modules, for Škoda has only 150 direct suppliers (compared with 300 five years ago) and they must have up-to-date logistics in order to deliver on time and minimise warehousing. Thus transport is often being outsourced given the intense competition and the need for reliability.

Finally it should be noted that vehicle production includes trucks, buses and vans which are also attracting foreign firms and, in turn, requiring indigenous companies to improve quality with the help of FDI in order to retain their markets. The picture is highly complex and cannot be summarised comprehensively. While TAM of Maribor, Slovenia has gone out of business (as already noted), other established producers are trying to improve marketability with foreign help like Rába in Györ, Hungary, Roman in Braşov, Romania and Tatra in Kopřivnice, the Czech Republic. At the same time, foreign manufacturers are moving in. Notable here is the creation of the enterprise Volvo Truck Poland which started assembly at Jelcz, supported by a marketing agency in Warsaw and 11 service facilities throughout the country (with 15 more by the end of

1997). Given the large home market, there is also Volvo Bus Poland, which involves partnership with a Finnish vehicle body manufacturer, turning out buses in Wrocław. In both cases the possibility of exports is being explored. Also in Poland, Citroën are to assemble vans at Daewoo's Nysa factory. Meanwhile, the Hungarian government is looking for a foreign buyer for the Ikarus bus company but conditions are tough and Volvo withdrew in 1997. Motor-cycle production has a long history in the region but few manufacturers were still in business at the end of communism. In the Czech Republic ČZS of Strakonice initially decided to work closely with Cagiva of Italy but the arrangement broke down when the Czech brandname was phased out and the Strakonice factory was downgraded to the production of parts. It remains to be seen if ČZS can survive independently alongside Jawa which is another of the region's survivors.

Industrial location

Socialist industry: equity or efficiency?

Despite the articulation of 'socialist location principles', it would appear that (as in the West) the efficiency built into the Weberian 'least cost' rationale was of first importance. After all the central planners were looking for the highest possible growth rates in each country overall and inappropriate locations would harm this objective. Little is known of the decision-making processes which must have involved meetings and discussions within the various ministries with considerable input through the party hierarchy. No doubt the criteria taken into account were similar to those relevant in a Western situation although a different set of transport costs and commodity prices would have applied. It may be assumed that there was a good deal of lobbying within party circles so that influential local officials could attract industry which might otherwise have gone elsewhere. Yet the results were generally unsurprising. All Albania's 26 districts had some food-processing industry and most had building materials/wood processing – and some textiles, footwear and chemicals. Meanwhile, only six had large plants with export capacity: Durrës–Tirana–Elbasan plus Shkodër, Vlorë and Korçë.

Party leaders were known to intervene in favour of their home areas, especially when a location decision might involve a multiplicity of near-equal choices. Romania's two principal leaders of the socialist era (Gh. Gheorghiu-Dej and N. Ceauşescu) were well known for their support of Moldavia and Oltenia respectively, and especially their home districts of Bârlad and Scorniceşti. Such input was also significant in terms of both siting and implementation. Ceauşescu was especially disposed to interfere in decisions at all levels and his working visits would always include a good deal of instruction to local officials as to how developments should be carried out – fuelled by an acknowledged fascination with the built form of socialism, especially integrated industrial complexes. Thus the economic logic of siting the Anina power station in a river valley (with cooling water readily available), some distance from the bituminous schist quarried on the higher ground in limestone country, was firmly ruled out by the president who wanted the quarries, stockyard, power station and new town to form a single integrated complex despite pollution hazards and the need to pump cooling water to the top of a dry mountain! Yet in other instances Ceauşescu is reputed to have supported local interests, trying to resist the central-ising urges of regional authorities. The local mayor took advantage of a presidential visit to the Nehoiu sawmill in the Buzău Mountains to make successful representations against the county administration's plans to consoli-date the timber industry on the edge of Buzău City and accelerate the growth of a town which had previously been eclipsed by its inclusion within the larger Ploieşti region. Such valuable lobbying opportunities explain why local admin-istrations liked to provide 'dedicated' accommodation for party leaders even though it might only occasionally be used.

Importance certainly attached to an 'equity' principle to discourage concen-tration and achieve greater decentralisation than was normal under capitalism. This has led some writers in the West to emphasise a distinct ideological compo-nent in socialist decision-making to bring out a geography of production quite different from the capitalist counterpart. But the argument is complex and, in the first place, propagandist claims for the equity principle as a tenet of ideology are no proof of successful implementation. They merely point to the hope of improved opportunities and living standards everywhere by making factory employment more widely available (and thereby overcome one of the perceived contradictions of capitalism). Furthermore, it should be remembered that plan-ners worked in an environment that normally favoured very high levels of concentration. Given the central planning system there were powerful argu-ments in favour of large projects which were relatively easy to supervise and were impressive as national achievements. Moreover these large projects would naturally tend to gravitate towards the national capitals and the centres of administrative regions with the best infrastructure, including the communist party's supervisory capacity. Meanwhile, allowance must be made for the initial increase in concentration through the expropriation, consolidation and reloca-tion of small-scale industry (much of it in rural areas) through nationalisation and discriminatory fiscal policies. Hence the ideological pressure of equity as a counterweight to the bureaucratic instinct in favour of concentration. But in any case, there is no way in which the operation of the equity principle can be demonstrated in opposition to classical economic logic since the politicians did not present statements of comparative costs which might have revealed the rejection of a 'least cost' location in favour of an economically less attractive choice on the grounds of equity.

It is undoubtedly true that factory industry became more dispersed under socialism than had previously been the case under capitalism. 'All socialist soci-eties industrialised with less spatial concentration of population than market capitalist economies' (Szelenyi 1996: 299). But it is hardly appropriate to compare one system in the 1950s–1980s with another for the period 1900–50. The tech-nological environments were quite different for these two periods; not least in the sense that electricity became widely available (and the railway system more extensive), especially in SEECs, during the later (communist) period. Locations vindicating the equity principle could usually appear rational in efficiency terms at the same time. A diverse geography of extractive industries provided many options for raw material orientation, especially in the context of regional supply systems for the cement industry to limit transport costs. Linkages with Soviet raw materials and intermediates – and with Warsaw Pact troop deployments – also threw up a string of feasible rural locations well away from established industrial regions.

From the large enterprise point of view, sub-contracting in rural regions could be very beneficial in facilitating expansion where labour was relatively scarce. Serbia's 'Cervena Zastava' car factory is a case in point. The venture started in Kragujevac in 1954 after the Federal Office for Economic Planning selected the city's Military Technical Institute to take responsibility. There was an engineering tradition in the town (through 'Fabrika Motora 21 Maj' which produced engines for the new factory) and a decade of reconstruction after the war paved the way for the new enterprise which began with the assembly of the cars under licence from Fiat of Turin. Zastava expanded in other locations; so the 'Fabrica Automobila' and 'Komponente' organisations in Kragujevac were complemented by 'Heroj Toza Dragović' in Ohrid and 'Ramiz Sadiku' in Pec; to say nothing of 280 sub-contractors (over 100 supplying car parts) from 130 factories all over former Yugoslavia (and some abroad). The sub-contractors became responsible for all the components apart from transmission, suspension elements and car bodies which continued to be produced at Zastava's own plants. It was a pity that not all the producers achieved a high level of efficiency, for 'Ramiz Sadiku' (opened in 1968) was mismanaged on a grand scale with absenteeism, poor organisation and low-quality production which resulted in hold-ups on the Kragujevac production line. Hence while there was decentralisation some of the dangers of this process were highlighted. They seemed to reflect a lack of interest that may have been politically motivated by Kosovo's loss of republic status and its ultimate loss of autonomy within Serbia.

Urban locations retained their importance, and decentralisation in conurbations like Upper Silesia continued the emphasis on concentrated development. However, there was no simple correlation between city size and growth rate because all regional administrative centres were well supported while the other towns were overlooked and generally denied any significant international role. Moreover, despite expansion as regional administrative centres with large bureaucracies, the large cities did not grow in proportion to their functions. 'Underurbanisation' arose because enhanced job opportunities were not backed by adequate housing. Prague and Brno saw their population weight decrease during the 1960s and 1970s (15.2 to 15.1 per cent) while their share of jobs increased (16.8 to 18.5 per cent) (Dostal and Hampl 1994: 208). At the same time the settlement system was distorted with jobs in industry generated only in the 200 centres of over 6,000 population and jobs in services in less than 100 centres of 15,000 inhabitants. Thus the medium-size towns – like the new county administrative centres of Romania identified under the 1968 administrative reform – did particularly well under extensive socialist development, but restructurning since 1989 has been very damaging, leading to out-migration and a deterioration in living standards in recent years.

The transition: capital cities

Over the last decade there has been relatively little new greenfield manufacturing, apart from light industry and warehousing on the edge of cities, and the prime concern has been connected with restructuring in order to retain as much of the inherited plant as possible. New enterprises with domestic capital have been relatively small and have emerged in and around the more dynamic urban centres, although a rural, raw material orientation is evident in the case of small sawmills, distilleries and meat-packing plants which complement the large state-owned units which are now extensively privatised. Discussion focuses primarily on trends evident over FDI which has discriminated locally as well as nationally. Where existing installations have been taken over decisions may reflect overall efficiency whereas greenfield investments do involve conscious locational choice. FDI tends to concentrate in capitals – also leading provincial cities and ports, with their immediate rural hinterlands – due to a favourable mix of comparative advantages: existing manufacturing amenable to restructuring; labour with diversified skills; good infrastructure; and conditions conducive to seedbeds and incubators for SMEs. Firms entering an ECEC for the first time will typically establish a sales office in the capital from which distribution will be organised (Domanski 2000). FDI will tend to favour the capital initially with the possibility of dispersal at a later stage as the firm becomes more adventurous.

The northern capitals especially have seen a massive transformation, triggered in Prague by what Sykora (1993) has called the 'rent gap' – between the social use and potential best use of sites – given property restitution and subsequent sale to developers who have encountered attractive acquisition and refurbishment costs. Berlin has been strengthened by the return of the German government in 1999 and authorities who have supported the internationalisation of the economy (Kratke 1992: 229). The Sony European Headquarters in Potsdamer Platz indicates the confidence of business – as does the decision of Bombardier (Canada) to service Eastern Europe from Berlin – while many German firms have moved in from cities in the west where they established themselves at the end of the war. However, with low-cost labour just over the border in the Czech Republic and Poland, organisational networks are more likely to be rooted in Berlin rather than factory production. Equally, there can be no doubt about the potential of Budapest in the context of the European metropolitan system (Barta 1992). Previously stranded on the margins of Western Europe as the large centre of a small country, Budapest has attracted heavy inward investment and combines proximity to Vienna with a business community with much experience of East European markets. PepsiCo have moved from Vienna to Budapest, identified as a stepping stone into SEE, while Opel set up a sales and marketing operation in Budapest to cover the same region and overcome delays in working through Austrian and German offices. A 'stratum of early entrepreneurs [was] established during the era of the liberalised command economy' of the NEM (Csefalvay 1994: 358), followed by joint ventures and a swing towards the tertiary sector. Budapest was also sustained as an intellectual centre with major educational capacities and a reputation for social innovations which contributed to a smooth transition.

However, when it comes to hub status across ECE as a whole, Budapest is in competition with Warsaw which attracts big companies interested in the large Polish market: promoting local offices de facto to the regional level. Furthermore, a Warsaw base may also coordinate activities more widely over the northern countries (the Czech Republic, Hungary and Slovakia), as in the case of Colgate, and also in Russia and the CIS (Procter & Gamble). But there has been a surge of office building (e.g. Warsaw Financial Centre) especially in Centrum borough where land values are extremely high; so that development is now moving into other parts of the city. To enhance attractiveness in a city that is perceived as expensive, with problems over pollution, telecoms and road traffic – despite a cheap and well-developed labour market – the Polish government is supporting the city through a 'Contract for Warsaw' which will boost government expenditure in grants and subsidies from 12.7 per cent of what the city pays into the central budget through taxation to 21.4 per cent. This will be done through the metro and the inner and outer beltways: the latter are long overdue with Warsaw almost unique among European capitals through the lack of such facilities (due to a failure to agree over the proportion of private financing). A factor in the shift away from the centre is traffic congestion, aggravated by the slow pace of metro and ring-road construction.

Finally it is worth noting that the capitals act as centres for dynamic regions containing provincial cities attractive to investors partly in their own right and partly because their image is enhanced by the proximity of an international airport and the fullest range of economic services and cultural facilities. Budapest is the centre of a dynamic region, for northwest Hungary in general has proved to be an area of innovation (Nemes Nagy 1994) where the vehicle industry has been particularly successful. This area includes the Baroque city of Székesfehérvár, in easy motorway contact with Budapest, which has attracted $1.3 bn. of investments and reduced unemployment from 30 per cent in 1990 to 7.8 in 1998. Companies like Nokia (Finland), Stollwerck (Germany), Denzo (Japan), Philips (The Netherlands) and IBM and Ford (USA) are all present and there is now a substantial cluster of expertise in the electrical field (IBM and Nokia both set up in 1995, producing disk drives and loudspeakers respectively). A new facility for Videoton (established in the communist period and currently Hungary's fifth biggest employer with 16,000 workers after the national railway, post, oil and telephone companies: MAV, Magyar Posta, MOL and Matav) is supplying parts to US electronics manufacturer Texas Instruments, IBM, Alcoa-Fujicura and 13 other TNC end-users. Tatabanya is also experiencing growth. A subsidiary of France's Framatome Group (FCI Connectors Hungary) has an electronic appliance factory at the local industrial park, Győr home furnishings maker Graboplast opened a PVC floor tile factory at Tatabánya in 2000 with round-the-clock production to meet demand from Europe. Elsewhere, Novi Sad (Serbia's second city) could emerge as an overspill centre dependent on Belgrade's services; with lower rentals making it a potential distribution and warehousing hub based on road, rail and river links. There are already computer and hardware distributors and local enterprises ('Novkabel', producing electric cables and the oil monopoly 'Naftna Industrija Srbije') are restructuring with the help of local banks. Ploieşti in Romania is only 75 km from Bucharest, benefiting from an excellent range of inter-city rail services and nearness to the international airport of Otopeni.

Transition: entrepreneurial cities

Again, there is the benefit of experience for in referring to Upper Silesia, and specifically Gliwice, Domanski (2000: 49) refers to a historical trajectory affecting the town's future given the record of success based on favourable location and 'the interdependent local economic, social and spatial qualities of the town itself'. The larger cities accommodated major socialist enterprises which have contributed to a relatively rapid recovery during transition. It seems that the status of regions in the state socialist economy has had a distinct bearing on their fortunes during the transition; e.g. involvement in Poznań by Alcatel and Glaxo-Wellcome dates back to business in the communist period. Bachtler (1992b: 136–7) emphasises the FDI potential of 'diversified industrial regions with a relatively good material and technical base and experienced personnel, good infrastructure and international links', especially areas close to EU borders and areas of tourism potential with an unspoiled environment. There are clear attractions in Poland's 'second cities' (Katowice, Kraków, Łódz, Poznań, Szczecin and Wrocław) through business-friendly authorities and state enterprises with the potential to become well-connected local partners, though each may have specific qualities: Kraków offers tax incentives at a new high-tech park while Wrocław is close to the western frontier and has a good position on the main road network (Kobierzyce, where Bielany Industrial Park is available).

Asea-Brown Boveri (ABB) are prominent in Wrocław (where they were active in the communist period) through control of ABB-Dolmel and Dolmel-Drives (generating equipment and machine drives) and Pafawag (producing rolling stock – including 'Adtranz' joint venture with Daimler-Benz of Germany). And businesses remaining in Polish hands have succeeded in the global market place. Szczecin has had success with shipbuilding (concentrating on container ships and improved quality despite EMBO and debt-equity swap with local banks rather than a strategic investor) and has negotiated downtown infrastructure deals – including a civic and business centre. The hope is that Szczecin will become the main port for Berlin after EU accession. There is good potential for an airport (former military base) where a cargo terminal is needed and NATO has decided to locate an international allied command headquarters in the town despite a 'thuggish reputation' arising from black market/mafia activity.

In the GDR era, Leipzig retained only its spring and autumn trade fairs but this helped to generate a clear identity, adding substance to such promotional gambits as 'Leipzig means Business'. The 'Media City Leipzig' project – with public and private capital for a media centre, office block and publishing houses (involving greenfield developments to avoid property ownership problems) – also builds on an established reputation and a mixed area near the city centre is in the process of transformation from the printing trade to office location. Industrial traditions are also important in Győr, where the established base in household textiles has attracted Amoco Fabrics Europe (a subsidiary of Amoco US). And Györ's rich agricultural area has initiated local specialisation in high-level processing which foreign capital is now extending. United Biscuits invested in 'Györi Keksz' in 1991 and in the same city a modern potato crisp factory ('Croky Chips') has developed a greenfield site in the local industrial park. Meanwhile in Debrecen, Reemtsma (Germany) have acquired a local cigarette factory and boosted productivity. Even in the car industry where a great deal of new investment has been undertaken the plants are generally rooted into existing large enterprises or sites with important infrastructural advantages relating to the communist military-industrial complex. Downsizing at Rába (the Györ truck builder) has provided an opportunity for the German carmaker Audi who have been using part of Rába's premises since 1993.

Immediately after the events of 1989 when deindustrialisation first arose, it seemed that old industrial areas with social and environmental problems might well be marginalised. But these established industrial centres have shown considerable resilience and have promoted themselves through a marketing effort undertaken by leading social and economic groups as well as the city authorities. In Łódz, foreign investors have found both skilled labour and an entrepreneurial outlook, evident in both the business community and the local authority. The workforce is well qualified and land prices are relatively low. Coats decided to set up their own cotton-spinning business in Łódz after expanding their exports from the UK to Poland in the 1970s and 1980s (employing over 100 in 1995) and Wrangler have also set up in the city. But Łódz is also home to Bosch und Siemens Hausgerate washing machine plant occupying a former stocking factory. Łódz voivodeship is active in stimulating the development of the economy, supporting entrepreneurship and attracting foreign capital; while Łódz City has its own departments for municipal investment and property management. The Łódz Chamber of Commerce and Industry (CCI), set up in 1990 and financed by local membership fees and various EU sources including PHARE, brings together private companies, state enterprises and cooperatives to provide information and stimulate development and trade. The real estate market is strong and in 1999 a start was made on the Łódz Diamond Business Park on Manewrona Street to yield 20,000 m² of space for warehousing, light industry and offices.

With a good chance of direct access to the future Berlin–Moscow railway, new businesses like Barry Callebaut Polska, Intercell and Lincoln Polska continue to reduce the dominance of textiles. But it is also impressive to see the many ways in which mixed private/public companies have emerged. Locations are being considered in terms of their living environment and here it is relevant to consider not only the overhaul of urban transport (referred to in the next chapter) but the consumer revolution that is well under way. Gdańsk is experiencing investment in soft drinks (Coca-Cola and Pepsi-Cola); fast food (McDonalds); petrol stations, supermarkets, hotels and entertainments. Access to city centres for foreign companies has been complicated by the delayed introduction of parking charges and the continued presence of former state shops paying 'historic' concessionary rents that have not been updated: in 2003 legal changes will force Budapest to equalise rents for all shops and many existing businesses may have to close. Meanwhile, shopping in the provincial centres has improved greatly with the arrival of the Western supermarkets. Many cities are short on hotels – Leipzig had hardly more than 3,000 beds when reunification occurred – but new hotels are opening and older ones are being refurbished. A complex including hotel and sporting facilities (sports centre, swimming, yachting marina) with underground parking complex is under way at Wyspa Spichrzow in Gdańsk by US Hardage Group; while a 15 ha site beside Głowny railway station in the same city is earmarked for a hotel and shopping/service centre with multi-level parking space. In Dubrovnik hotels are being renovated through income from the selling off of assets: note the Dubrovnik Hotels restructuring plan for 24 hotels which should draw foreign investment into tourism (already Club Méditerranée and Hilton are established) and make the town more attractive for tour operators.

Differentiation is also evident over environmental problems including relatively high levels of air and water pollution and uncertainty over the quality of drinking water. The pollution is linked closely with rapid industrialisation and it became a serious problem in areas involved in coal mining and electricity generation. But substantial improvements have been made; partly by deindustrialisation and partly by investment in installing air filters, waste-water-treatment plants, sewage plants and improved drinking water provision: Gdańsk completed its Wschod (East) sewage plant by 1998 and Warsaw is currently implementing a series of major environmental improvements. With an accelerating pace of urban reconstruction it becomes realistic for FDI to rediscover the small Baroque towns of Central Europe. Pécs is an attractive place to live in, with good business and education/training services and a supportive local government. There is also a large middle class linked with services (including higher education) which provides the purchasing power to maintain the shopping and entertainment facilities, though much of the surrounding area is poor following the collapse of coal and uranium mining. Several Finnish companies (Elcoteq, Ensto Plastic and Nokia) are present while UK BAT bought the local tobacco plant in 1992. These urban values are being enhanced by city centre facelifts in across ECE because, despite some architectural impoverishment through excessive redevelopment during the communist period, most have retained a good stock of buildings which can now be refurbished; not to mention the psychological value for the community of a city centre facelift such as has occurred recently in Banská Bystrica. An expensive facelift in Košice, strongly supported by the former mayor, has included restoration of the fourteenth-century cathedral which should be of major benefit for the tourist industry. In Łódz the main street has been upgraded and the role of tourism is being stressed through the local environmental resources: woodlands, parks, a spa and some fine architecture which stimulate a concern for 'green issues'. However, a strong emphasis on conservation can complicate planning procedures and some developers have experienced difficulty in adapting historic buildings for modern commercial uses.

Transition: transport

For companies exporting all or part of their production, transport must be considered in an international context with the main road, rail and river routes extended as EU transport corridors and supplemented by provision of 'missing links' through formerly closed border regions. There are clear advantages for cities well situated on the main transport arteries, especially when the distances to the frontiers of the region are relatively short. A 15,500 kg container costs less than E1,000 to transport to Vienna from Croatia, the Czech Republic, Hungary, Poland, Slovakia and Slovenia, but upwards to E2,000 for the other ECECs (excluding Albania) along with the Baltics, Belarus, Moldova, Ukraine and western Russia – and over E4,000 for the Caucasus and Central Asia. Győr in Hungary stands out as a rapidly growing regional centre close to the Austrian and Slovak borders and with long-standing cultural links with the West. The role of international and regional airports should be emphasised. Meanwhile, the region also offers good links for firms wishing to export to the FSU, though there are problems in the short term due to recession and collapse of purchasing power in Russia: broad-gauge railway links run eastwards from Košice in Slovakia and from the eastern edge (Sławków near Dąbrowa Górnicza) of Upper Silesia (Poland); while Varna retains ferry links with Ilichevsk (Odessa) in Ukraine and Poti in Georgia. Some cities in peripheral areas are enjoying considerable cross-border custom given today's more permeable frontiers with relatively simple formalities, especially where Euroregions have been created. Cross-border perspectives on supermarket shopping would see Hungary well placed to draw customers from Croatia, Romania, Ukraine and Yugoslavia.

The matter has a wider significance both in terms of spreading investment and modernisation into SEE bearing in mind the importance of simplified frontier administration for freight movement to Western Europe and also the scope for cross-border cooperation where reference could be made to the model of German investments in western Poland and the Czech Republic: taking advantage of lower wage costs in the contexts of permeable frontiers created through Euroregions and the minimal transport costs of returning the production to Germany. Schmidt and Naujoks (1996) have underlined the scope for German–Czech and German–Polish 'workbenches' which achieve a division of labour between high and low wages which could be a basis for innovative networking in automobile, chemical and metallurgical industries. Kontavill (France) produces electrical equipment in Szeged and exports to former Yugoslavia; likewise Villeroy and Boch from Germany, who make sanitary products, the Italian company Sole who supply milk plant and the Hungarian enterprise Pick salami. During the civil war in Yugoslavia, Szeged made money from petrol smuggling to Serbia although tourism suffered, along with the textile and food industries, but under the normal conditions of the present, Hungarian firms in the southeast of the country are distributing modern Western equipment into neighbouring countries and contributing to a technological upgrade. Collaboration between Chambers of Commerce is also a force for modernisation. Meanwhile, Split has links with B&H and Montenegro: the UK RMC Group has a majority stake in the local cement producer Dalmacijacement which in turn has acquired 60 per cent of the Montenegrin cement maker Servis. Osijek also has potential for trade with B&H, Hungary and Serbia, while Zagreb could act as a regional base for former Yugoslavia: it has the infrastructure to act as a business hub, with advertising, auditing, banking and legal services, while Croats understand all the languages once grouped under Serbo-Croat. However, trade barriers necessitate manufacturing and distribution centres in each country.

Transition: industrial estates and special economic zones

Entrepreneurially-minded local authorities now appreciate that an important facet of the region's excursion into globalisation is the provision of industrial zones combined with centres of scientific research and development. 'Technological parks' are a primary condition for the development of a national economy based on the restructuring of obsolete industrial regions. According to Lorber (1997), arguing the case for Slovenia, ECE must catch up with the EU in this respect and combine industrial estates with appropriate support services and fiscal incentives. Some cities can offer such a package in addition to stability and centrality combined with good human resources and infrastructure. Hungary has been quick to appreciate the advantages of the business park ('Uzleti Park'), for most of the new industries in Székesfehérvár have been built on a former Soviet military base run by the American Loranger Industrial Development Company. And Videoton (the old Hungarian electronics company which started manufacturing for Western firms) has made part of its site available as an industrial park. Meanwhile, Györ has an industrial park with joint Austrian–Hungarian management. This has led to a surge of greenfield development (mostly Austrian and German) in a town offering good education and training. Plzeň in the Czech Republic has much to offer in terms of stability, accessibility and human capital. But the city has also profited from a clever land deal struck in 1993 when the former army barracks of Borska Polje was turned into an industrial park, offering space to foreign investors at attractive prices. The site is close to the Škoda complex as well as the university and technology and business innovation facilities. During 2001 Slovakia began to apply legislation which provides financial support for estate projects which secure an investor in the automotive industry, electrical engineering, telecommunications or processing. An estate is at Bratislava (Lozorno) while Jarovce and Záhorie (a high-tech 'Euro Valley') operate nearby. Gabčikovo, Levice, Lučenec and Rožňava are all poised for investment from Hungary while Námestovo is close to the Czech and Polish frontiers. Martin, Liptovský Mikuláš, Prešov, Spišská Nová Ves and Humenné are other estate projects for the central and eastern regions. This activity provides a follow-up to Slovakia's earlier strategy which relied on a network of state 'Regional Advisory and Innovation Centres' and 'Business Innovation Centres' – and locally-organised social and economic councils – which picked out many of the towns where industrial estates are now being provided (Figure 3.1).

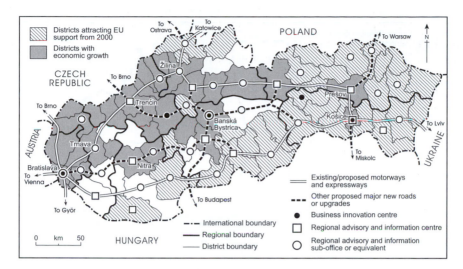

Figure 3.1 Slovakia's communications and business innovation centres

Source: Constantine the Philosopher University, Nitra

More controversially, however, special economic zones (SEZs) have been created in Poland under a law of 1994 to generate new jobs and make better use of assets and infrastructure in areas not coping well with the transition. The best chances are thought to exist in cities dominated by one troubled industry where restructuring requires massive layoffs. The first project – Euro-Park Mielec (1995) – was modelled on Ireland's Shannon Airport estate and comprised 575 ha in the grounds of an aviation factory. Within the zone 7,000 new jobs were anticipated with another 3,500 in the area round about. Ten years' exemption from income tax was given, with 50 per cent exemption for another ten years, and complete exemption from real estate tax and possible higher rates for fixed asset depreciation. Most were created in 1996–7 and a zone was established in the northeast at Suwałki and another in Upper Silesia since the coal industry has been restructured into seven state-owned holding companies and mine closures have resulted in considerable unemployment. At the end of 1998 there were 6,000 ha of SEZs with $3.2 bn. invested. Bazydko (1999: 302) lists a total of 13 zones responsible for 25,600 jobs with Katowice accounting for 9,700 followed by Mielec, Suwałki and Legnica with 3,000–4,000 each.

In Katowice there have been major social changes as miners' wives look for jobs (never necessary before), but alternative jobs for men have been provided at Gliwice, where General Motors have built an assembly line to take advantage of booming car sales in Poland. Gliwice offered fiscal incentives as well as 'fast track' clearance for land permits (Figure 3.2). Other sectors of the Katowice SEZ are located at Dąbrowa Górnicza (where the French glassmaker Saint Gobain has opened a factory), Jastrzębie Zdrój, Tychy and Żory (site of the ABB tube factory). Meanwhile, Katowice itself is being promoted as a national centre of finance, manufacturing services, culture and science; shifting from a traditional

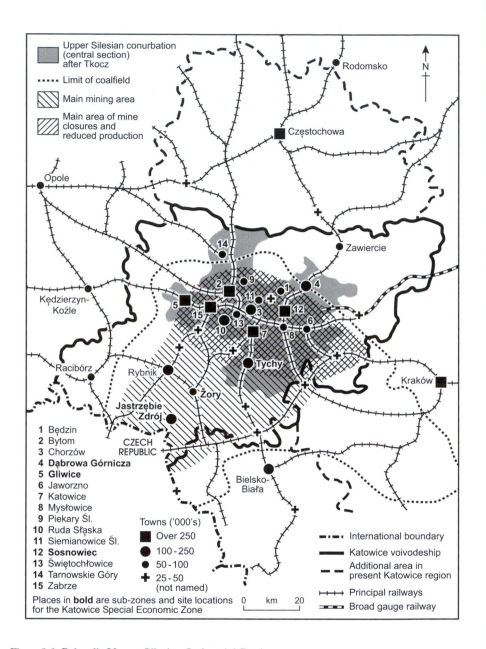

Figure 3.2 Poland's Upper Silesian Industrial Region
Source: Katowice Special Economic Zone

role linked with coal and metallurgy towards ecologically sound industries. These include IT which has been developing since the 1970s. The first private computer company arrived in the 1980s and momentum is maintained thanks to backup from Silesia University, Silesia College of Technology and the Economic Academy (though the SEZ discriminates against computer companies by not extending the usual taxation benefits). However, automotive companies have taken full advantage in a region that scores highly for banking, insurance, education and training, as well as transport including a road–rail–canal interchange at Gliwice, a railfreight node in Tarnowskie Góry, a broad-gauge link at Sławków (Dąbrowa Górnicza) and Pyrzowice international airport.

Relatively dynamic cities hardly need additional stimulation yet in 1998 Poland established Technology Parks at Modlin on the edge of Warsaw as well as at Kraków, followed by the Okęcie Airport Zone in 1999. The aim is to encourage clusters of high-tech industries in circumstances where large investors still expect some concessions for a firm decision. The Czech Republic was forced to provide incentives in 1998 when investors started going elsewhere and Romania is finding it difficult to refuse large investors like Daewoo, Continental and Renault. Kraków therefore offers 12-year tax breaks to firms who make arrangements with Kraków High Technology Centre: the enterprise managing the Academic Technology Park in Czyżyny and Pychowice (offering laboratory, office and production space) and also the Kraków Industrial Park which adjoins the Sendzimira steelworks (formerly Nowa Huta). Yet Kraków is an outstanding provincial centre of enterprise, which has seen a dramatic growth in private sector business. In addition to high ratings for transport, labour, real estate costs, the local authority is highly accessibile to prospective investors. The city also benefits from the revamped Balice airport work (renamed after Pope John Paul II) which can now accommodate the larger jets used on international services. 'Cultural specificity' comprises a town square of unique quality (recognised by UNESCO in 1978 as being of world importance in terms of cultural heritage) and a rich cultural life based on the very large Jagiellonian university (Hardy and Rainnie 1996: 180). SEZs are becoming unpopular with the EU on the grounds of 'unfair competition', despite the principle of support for poorer regions. Thus it seems likely that existing zones will be retained while non-functioning zones will be closed, scaled down or amalgamated and no new zones will be created. The issue is a complicated one. Thus, the future would appear to lie with special favours for 'less favoured areas' (LFAs) defined carefully according to criteria acceptable to the EU. Romania can now offer attractive deals through the designation of disadvantaged areas suffering from high unemployment as a result of mining closures. The greatest needs arise in Petroşani, centre of the Jiu Valley coalfield, but there are also problems linked with the mining of coal, lignite and metalliferous ores in Banat, Maramureş and Oltenia. The LFAs are being administered by regional development authorities set up in 1998 to handle cohesion funding emanating from the EU. They also take over ad hoc promotional work started in some areas at an earlier stage in the transition process. Some significant investments are being made but it is evident that the local authorities need to be fully committed to supporting the initiative, especially in areas where little outside investment has previously been generated.

Transition: gateways to rural areas

Large metropolitan cores may be able to transform themselves but arguably at the expense of rural peripheries that have difficulty in projecting a positive image through poverty among villagers excluded from the labour market after the better-educated people have left. However, a more positive scenario would anticipate opportunity arising from the progressive development of civil society in regions where dynamic entrepreneurial cities provide gateways to unlock the rural work resources. Some small towns are expanding their business functions and food-processing industries which have in many cases been acquired by Western capital are often located away from the conurbations. Furthermore, rural districts are featuring in many mining ventures, a significant example being the joint venture between Minvest Deva and Gabriel Resources of Canada which has found up to 50 m.t of relatively rich gold- and silver-bearing ore in the Roşia Montană area of Romania's Alba County. Rural areas are also attractive to émigrés who wish to support their home areas. Hence the activities of Romanian entrepreneurs who have returned from Sweden to establish the 'European Drinks' complex in the villages of Rieni and Sudrigiu in the Beius Depression of the Western Carpathians. And a special case concerns those investments occurring on greenfield sites in rural areas on the edge of large towns that are subject to a suburbanisation process of the kind recently discussed in Zagreb where intense 'satellisation', as at Lučko on the south side of the city, reflects strong commuting flows on the major transport routes.

Rural areas and small towns in the western parts of the Czech Republic, Hungary and Poland are obvious candidates for FDI since these locations are convenient for export with cross-border cohesion enhanced by Euroregions. However, some other small towns in Poland have proved attractive in terms of labour catchment, social climate and leisure facilities, technical and business infrastructure and road access. In Poland, 260 towns were examined by Gdańsk Institute for Market Economics in terms of marketing, labour catchment, social climate, technical and business infrastructure, road access, economic transformation and leisure. Size proved to be important, but also the relationships with Warsaw and the western frontier. Piła has done best with 11 foreign investments of over $1.0 m., followed by Gniezno with eight and Ostrów Wielkopolski with seven. Motorway intersections offer particular centrality: as at Stryków near Łódz (where the planned intersection of the east–west and north–south motorways will reinforce the city's strength as a regional centre in the heart of the country) and Kobiercyze near Wrocław. Firms wishing to set up as quickly as possible will be much influenced by local government leaders who are able to assist in providing land while expediting various other formalities. Another example is Mszczonów, formerly a dormitory for Grodzisk Mazowiecki, Warsaw and Zyrardów, but potentially the 'pearl of Mazovia'. For it has good rail links with Warsaw, Łódz and the southern part of the country generally, while it is also close to the intersection of the main east–west and north–south motorways already referred to.

The local authority in Mszczonów has cooperated with the British developer 'Europa Distribution Center' and many well-known companies have moved in. Rapid processing of development proposals has boosted interest from distribution firms like FM Logistic (France), Maersk (Denmark) as well as Bauernfeind, Knauf and Mostva (Germany), the latter having completed a central warehouse for imported cars with an East–West Duty-Free Center to follow. Poland's first true national distribution and light industrial centre has now achieved a 'critical mass' with $500 m. invested over the five years to 1999. The estate extends over 130 ha and offers fully-serviced, custom-built facilities for purchase or lease, geothermal energy for electricity (7.5 MW) and domestic heating. Commune income has risen (allowing investment in a heating system and power station – 7.5 MW – based on geothermal water resources) and new jobs are being created. Population growth is expected and planners are working to a target of 30,000. Elsewhere, the potential of rural areas is likely to hinge considerably on their relative accessibility. The planners in Cluj County (Romania) have identified a number of polarising communes where land has been made available for small-scale industrial development (Figure 3.3). This initiative is linked with a desire to contain growth in cities and reduce commuting by stabilising population within local cells. However, the local roads will need improvement if new

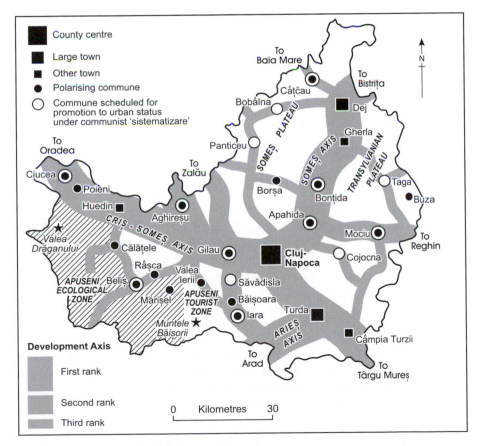

Figure 3.3 Transport axes in Cluj County, Romania

Source: Cluj County Council

businesses are to start up and transport will be needed to connect the villages within each cluster. It is evident that most of the centres lie along the main transport axes and were also scheduled for growth under communism.

Some rural industrial enterprises have grown from roots established in the communist era. Although most industry was developed in large towns, various decentralisation exercises produced patterns of dispersal which are being perpetuated in some degree under the transition. These 'islets of development' may well be developed by joint ventures linking former SOEs with foreign companies supplying investment capital, new technology and a marketing system; and state governments may be persuaded to assist with low-interest credits and tax exemptions. A good example is the decision of American Standard (sanitary ware) to develop in the small Bulgarian town of Sevlievo initially through privatisation (including purchase of employee shares) and subsequently with a boost in employment to 1,650 through a new vitreous china plant in the town – a 90,000 m² green field project completed in 11 months – with a further fittings plant at Gradnitsa nearby (850 workers) and an air-conditioner factory to follow. With most workers bused in from Gabrovo, Lovech and Veliko Turnovo, growth seems to owe much to Sevlievo's pragmatic and business-friendly mayor. A land deal was quickly brokered and negotiations expedited for fibre-optic cables and gas pipelines to improve infrastructure (partly at company expense), while a language school now trains proficient English speakers and a new hotel has been opened. The total amount of rural industry is modest but physical plans are tending to highlight district growth centres which may be endowed with investment in infrastructure to boost their attractiveness to outside investors.

Transition: shifting eastwards

It is evident that the ECECs do not provide a level playing field for the foreign investor. Some states do not offer a stable environment and must therefore enhance their 'capacity' through inclusive social policies, stimulative fiscal regimes and transparency in decision-making. This will be difficult in those parts of SEE that have not yet formulated restructuring programmes fully supported by the global institutions. Within countries it is clear that urban locations enjoy the balance of advantage although counterurbanising trends have occurred where former SOEs in rural locations prove irresistible and local authorities are able to deliver on all matters of concern to a foreign developer. Meanwhile, family connections with specific rural areas should not be overlooked given the substantial diaspora elements in Western Europe. But the majority of investors will be looking for urban locations which are convenient for distribution within the relevant domestic market (sometimes in other countries as well) and/or the shipment of production westwards. And given the relatively poor transport system this will tend to favour the northwestern part of the region.

Attractiveness is rarely assured on all fronts and in a highly dynamic situation investors must be prepared for trade-offs. The attractive cities of today may not retain all their advantages. As the capitals become overcrowded special skills exceed supply and wage levels rise. So investors are looking to the east and the

French chemical company Chinoin has moved beyond Budapest to set up its headquarters in Miskolc where a new industrial park has been provided by the Hungarian Investment Bank and the EBRD. Towns with an attractive environment but a poor infrastructure may therefore gain more FDI when established locations overheat and the lack of direct motorway links there may well be balanced by lower operating costs and regional incentives that have steered vehicle component firms to small towns like Balassagyarmet. However, the improvement in transport is becoming more and more apparent in the poorer regions. Debrecen has been disadvantaged by its situation at the end of a long single carriageway road from Budapest, but the Szolnok bypass has been provided and there is a link with the Budapest–Kecskemét motorway at Albertirsa. Meanwhile, another motorway is advancing northeastwards from Budapest to Miskolc and should reach the Ukrainian border in 2004. Peripheral areas like Białystok and Lublin in eastern Poland, which are potential launching pads for FSU trade, can offer environmental advantages. Białystok is blessed by a special ecological zone 'Green Lungs of Poland' embracing the Warmińsko-Mazurskie region and including Białowieza national park; while Lublin has benefited from a long-overdue renovation of the old town which has just been completed. The attraction of such cities should increase as they are drawn into the web of modern transport facilities, for there is the possibility of a motorway through Białystok to the Baltic States and another from Warsaw to Lviv which would pass through Lublin.

A final, crucial issue is whether all parts of the region will find a rewarding role in the global economy despite a long history of backwardness on the European periphery. Spatial unevenness must certainly be exploited to the full, through labour costs, amenity and other advantages, but even the stronger regions may find it difficult to throw away all the legacies of the periphery if FDI falters and if indigenous SMEs are unable to absorb more of the unemployed. Making a virtue out of peripherality involves a formula which is being operationalised to a greater extent than before and reconstituted civil societies are showing some skill in packaging the 'local' to catch the 'global'. But the harvest is still awaited in much of the region. Now that private investment accounts for the bulk of the capital flow, the transition states need an active industrial policy involving an appropriate mix of general incentives (tax concessions, credit support, infrastructure, state-financed R&D, training and marketing) as well as commitment to 'creative destruction' to eliminate obsolete plants and create new high-tech capacities. This has to be placed in the context of economic management to secure faster growth (for the region as a whole is predicted to grow below the world average over the next decade), a more stable currency and better regulation of financial markets. There is also a widespread belief that larger municipalities are needed to provide greater resources for units of local government and improve their capacity to promote their areas and support industrial projects. Perhaps the region can act as a proving ground for this aspect of industrial policy to see how far national and local governments can stimulate the growth of networks and programmes which can explore the potential for greater local input in terms of individual plant requirements and match these with existing enterprises with help over technological updating where necessary.

Transformation in agriculture

Socialist farming

The most substantial agricultural resources – which included areas like Hungary's Great Plain, offering good soils and a warm climate during the growing season – were fully exploited because communist industrialisation rested on an assured supply of cheap food and raw materials. But the business was reorganised in order to proletarianise the peasantry and control production at prices determined by the state. State farms comprised former experimental units and crown domains while cooperatives (or collectives as they were usually labelled in the early days) arose through coercion to pool peasant land, livestock and equipment without compensation apart from a right to work for the new organisation and retain a small private plot. This enabled the state, through the party hierarchy, to secure food deliveries from communally-farmed land while leaving each peasant with the obligation of additional work to secure a measure of self-sufficiency for the family and to create small surpluses for sale on local markets. In theory the old system could have achieved similar results through compulsory deliveries on a quota basis and the disruption experienced in the FSU in the 1930s (and on a less dramatic scale in ECE in the 1950s and 1960s) could have been avoided. But the ethos of revolution prevailed – with the so-called rich peasants being assumed to be enemies of the state – supported by the economic case for the planning of production and investment over large areas without the complications of individual peasant decision-making.

There was also an element of welfare because, although the profits of the more successful cooperatives were checked by loan repayments and the charges levied by the machine and tractor stations, the weaker cooperatives could usually generate some income for their members through the periodic writing-off of debts. But the strong peasant links with the land through private ownership were broken and most young people left for careers in industry: a useful by-product even though it was not intended at the outset. As collective farm members, parents encouraged their children to leave whereas as farm owners they would have been keen to ensure a succession. Moreover, the highly formalised cooperative system (going way beyond traditional forms of mutual assistance) generated a small corps of salaried officials who, as party members as well as farm managers, became part of the village elite providing what was in effect a local government institution. The scope for security police activity among this group (and other sections of rural leadership concerned with commerce, education and even the church) effectively neutralised the villages politically.

Despite some decreases in the amount of agricultural land, evident in the Sudetenland through a decline in arable farming and an increase in woodland after the Germans were expelled, communist agriculture achieved significant intensification. Farm production increased but as a result of the growth of population (with a demand for higher living standards) and the need for more agricultural raw materials in industry, there was a change from net exports to net imports. However, there were variations in the level of socialist farming with cooperatives playing a relatively minor role in Poland and former Yugoslavia. The cooperative programme was abandoned in Poland in 1956 and the private sector was tolerated as a short-term phase in the longer-term plan of socialisation: even large private farms could be accepted in the wider context of state regulation of agriculture. And the structure of agricultural production retained considerable diversity under socialism, with many variations on the basic logic of combining cereal farming with livestock. The mountain districts certainly specialised most heavily on stock rearing (cattle, sheep and pigs) and much of the cereal production (especially rye and maize) was undertaken in the interest of local self-sufficiency. But the main arable zones in the lowlands also expanded their herds and raised whole complexes of buildings for large-scale meat and milk production. The arable sector provided the fodder and the finished animals were conveniently situated in relation to the domestic food market and the international trading system.

There was a danger that such a strategy pursued by the state farms and cooperatives would ignore the full potential of the mountain grazings and it was appropriate that individual farmers should be left with the incentive to maximise output (all the more in view of the logistical difficulties of cooperation in areas of dispersed settlement). Meanwhile, although there was a growth in production through attention to plant and animal breeding, irrigation, mechanisation and fertiliser production, intense specialisation was discouraged by the inefficiency of inter-regional trade which perpetuated traditional emphasis on self-sufficiency. But there were some outstanding specialities (some of long standing). Bulgaria retained its reputation for tobacco and also for oleaginous roses yielding rose oil or attar of roses for the manufacture of perfumes (the latter specialism prominent in the Kazanlik Valley). Specialisation was also noticeable in connection with local market demand: flowers at Turany near Brno and Jabłonna near Warsaw. There were also anomalies of the kind described at Jina near Sibiu (Romania) where intense specialisation on sheep – wintered on cooperative farmland in lowland regions at nominal cost – produced very high incomes through the relatively high prices paid by the state for wool. However, high rural incomes were the exception rather than the rule.

Reforming the Stalinist model

Regular wages and improved welfare benefits were introduced and the importance of economic rather than ideological incentives led to the allocation of specific tasks to individuals and groups in return for payment by results. Hungary took a relatively pragmatic stance from the 1950s for after the setbacks in 1953 (through de-Stalinisation after several years of coercion to establish cooperatives) and in 1956 through revolution, the Kadar government rebuilt the socialist system through drawing in the middle peasants (thereby introducing a social hierarchy) and increasing investment in response to the challenge by the Chinese communists in favour of a fair balance between agriculture and industry. During the 1980s reforms (which were particularly far-reaching in Hungary) encouraged individuals to expand output from their gardens and plots through various forms of collaboration with the state sector. Improved coordination measures brought together cooperatives, state farms and private plotholders so as to make the fullest use of land and labour. The cooperative at Ócsa near Budapest allocated private plots according to the number of cows each member agreed to look after (0.75 ha for each animal) while at Pásztó (Nógrád County) vineyards were leased to responsible individuals in units of 1,500 m^2. Another cooperative near Sopron collaborated with several hundred individual gardeners and smallholders to produce grapes, vegetables and poultry for sale in Austria under 'small border traffic' arrangements.

Meanwhile, individual farmers, although inevitably tied up with the state marketing systems, were able to survive through pluriactivity. Pine and Bogdanowicz (1982) have shown how Carpathian farming communities in southern Poland were able to diversify through light industry, tourism and even temporary migration to the USA on the basis of family ties with migrants who left the area at the beginning of the century. But mainstream thought was more conservative and tended to rely on 'gigantism' through larger socialist farms (both state farms and cooperatives), coordinated production from groups of farms and the integration of farms and processors, as in the case of Bulgaria's Industrial–Agricultural Complexes and Romania's associations of state and cooperative farms. Here the peasants were already disillusioned by the experience of the early 1960s as state farms were allowed to take the best land while deteriorating rewards for farmwork legitimated theft and gave the peasants primary interests in retaining their private plot and securing waged employment in the towns for their children.

Transition: the political imperative of restitution

Agriculture has adapted more readily to the transition than industry, through decline in output in the early 1990s and new farm structure. Basically it has to be said that farming 'had no possibility to avoid pressures from reform' (Kraus 1994: 125) because there were very strong political pressures for restitution and governments had few reasons for not relinquishing responsibility, given the relatively straightforward procedure of land subdivision; unlike the situation in industry where the developments of the communist period made the expropriated property more difficult to identify. Former owners could have received shares as part of a privatisation process but this raised the sensitive issue of employment and downsizing which was approached with great caution in all the ECECs except East Germany where the agenda was driven by unification. Naturally there was a concern over food supplies and a desire not to lose all the opportunities for large-scale farming but these concerns were met by retaining control over state farms and a limited procurement role with some produce offered at subsidised prices for an interim period. Moreover, restitution for former landowners did not rule out the maintenance of cooperative structures. More generally, it has been argued that governments have little money to plough into agriculture and that – in any case – state intervention should be resisted, despite the temptations arising from the recent history of central planning and the elaborate arrangements for agriculture in the EU. The problems of agriculture are not unique and financial assistance should be provided by banks on a normal commercial basis, though it could be argued that when joining the EU each country will want the highest quotas it can get and so there will be a political incentive to maintain agricultural production as far as possible.

Some state farms have been subject to piecemeal transfer as part of the restitution process, but in most countries they were initially regarded as a quite separate category with the national interest uppermost. Some farms had long been run along the lines of agri-business, while others were created under communism by seizure of assets considered most suitable for efficient, large-scale production with a relatively advanced technical base. They provided a measure of food security while leaving the cooperatives to make up the deficit through more labour-intensive systems which ensured that the available rural workforce would be fully engaged on the cooperative estate and on private plots. After 1989, 24 of the largest and most technically-advanced state farms in Hungary – like the famous Babolna farm which dates back to a Habsburg initiative of 1796 – were initially kept intact to ensure that expertise in key areas (seed production and livestock breeding) would not be dissipated. This was a wise precaution in the light of Bulgaria's experience, where the breakup of state farms happened so quickly that stock had to be slaughtered when animals could not be readily absorbed on to small peasant farms. However, selling and leasing has now become widespread, and foreign capital has also been attracted to large holdings on particularly fertile land with a tradition of progressive farming where the state is unable to provide the investment needed. In the case of historic estates this will leave a residue of buildings to be sold off separately for restoration as hotels and conference centres. Poland has set up an Agricultural Property Agency for this purpose.

Transition: the emerging farm structure

Radical land reforms after the First World War established peasant proprietorship on such a scale as to create a vested interest in favour of restitution which the political process could not ignore. Conceptually there has been a sweeping process of 'reagrarisation' which has reconstituted peasant farming. Instead of 'collectivisation' to neutralise the rich peasants ('kulaks') and simplify the command structure, the evolutionary process towards a democracy based on property ownership ('embourgeoisement') is being resumed after an interruption caused by the ideological aberration of communism in the interest of the Soviet war economy. However, a choice arises between favouring former farm workers (through land 'distribution') or the pre-communist land owners (through land 'restitution'). Generally emphasis has been placed on the latter option, with Albania as the main exception among ECECs. However, many beneficiaries have received only tiny holdings of less than five hectares.

Vasary (1990) raises the question of 'competing paradigms' in respect of these restitution holdings, for they can be organised in many different ways to create farm enterprises. The land can be leased, wholly or in part, to a private farmer or to a cooperative which may be a reconstituted version of the farm operating in the communist period. Thus while Hungary has 5 m.ha owned by some 2 m. separate owners, two-thirds of this area is actually worked in the context of large farm structures. The cooperatives in turn may work the land themselves or lease land and buildings to separate businesses run by farming companies or individuals who wish to opt out as Kobayashi (1996) has noted for Ludersdorf in East Germany. Cooperation certainly makes sense where restitution holdings are too small to be viable and most landowners have full-time jobs outside agriculture. However, the new cooperatives are inevitably different from their communist predecessors. Although they may replicate some old communist structures, this reflects pragmatism in using methods that are successful. Fundamentally, the new cooperatives will be different, without the same level of solidarity and mutual support previously evident.

If the new-generation cooperatives are well managed and the markets for farm produce (both domestic and international) are stimulative, there could be a powerful trend towards a modified cooperative system (led by Hungary with its existing economies of scale and limited traditions of entrepreneurship). Yet newly-formed cooperatives have frequently run at a loss, even with a drastic reduction in administrative costs. In such a situation, support for cooperatives might be anchored in the 'egalitarian principle' and a belief that 'special favoured treatment of the sector is justified as it is the most sensitive industry producing basic foodstuffs, providing employment for the rural population and creating non-market values such as the landscape. The objective of this advocacy is quite explicitly to keep and raise the level of social welfare in rural society' (Ratinger and Rabinowicz 1997: 99). Failing cooperatives might also attract support, as in Slovakia in the early 1990s, by a financial regime of relatively soft budget constraints; enabling banks to lend to cooperatives to prevent liquidation and safeguard fertiliser stocks, machines and livestock. Other privileged options available to cooperative managers, but not private farmers, include payment in kind and liquidation of debts with transfer of assets to a new company (Kabat and Hagedorn 1997: 271). However, hard budget constraints are now very much the rule and cooperatives must be financially viable if they are to survive.

Private family farms versus reconstituted cooperatives

At a time when cooperatives enjoyed the privilege of soft budget constraints through the persistence of communist networks, there was scepticism over ability of family farms to compete. There might be high transaction costs to exit from a cooperative if it was necessary to take a share of the debt, while in the early years there was always a chance that a new government might change the rules to the disadvantage of private farmers. There was the prospect of having to mortgage property to buy a tractor and rent additional land at a time when land markets are poorly developed. Meanwhile, 'aggressive' renting of the kind seen in East Germany created tension and individual farmers might be ostracised by neighbours who preferred to work with the new cooperatives (the GmbH companies formed out of the former communist LPGs) and who perceived private farmers as capitalists hungry for profits. Traditionally-minded villagers are also uneasy about the prospect of increasing amounts of land being farmed by people who might have no links with their community: incomers might unwittingly upset local practices and conventions and tensions could be aggravated by 'historical resentments' linked with pre-communist landholding which might complicate thinking about the most desirable arrangements for the future. Despite official encouragement of private farms in East Germany, reformed cooperatives comprise the largest single category of land management over the agricultural land area cooperativised in the 1950s. However, given the leadership of a number of families prominent before communism, land at Singatin (Apoldu commune near Sibiu in Romania) is effectively worked as viable family farms through leasing on the holdings of the agriculturally inactive owners. 'Their activity reflects a civic consciousness that slowly and piecemeal creates trust-generating practices' (Stewart 1997: 78).

Yet it could be hypothesised that '(a) cooperative production has only a poor chance in competition with other organisational forms and (b) in the long run, family farms (and perhaps other organisational forms with similar economic structures) will be the predominant outcome of the transformation process' (Beckmann and Hagedorn 1997: 147). This is based on high transaction costs (for cooperatives) due to collective decision making and inappropriate incentive structures and, over the long term, the scope for minimisation of transaction and production costs (taken together) in the context of the family farm, since there seem to be no overwhelming economic advantages in favour of really large holdings. In the Czech Republic some private farms may reach 900 ha (with owning and leasing) while others may fall into the 50–100 ha range, with the smallest below five hectares. Farms over 100 ha (some comprising small estates recovered through restitution in FCS) are particularly rational for the use of machines, but 30–70 ha farms can also be viable, specialising in grain and sunflowers. Even mixed 'nostalgia farming' on 15–30 ha restitution holdings could be satisfactory in the long term although it may be most suitable for middle-aged people with income from other jobs and access to basic machines. Holdings of ten hectares can be rewarding where machinery can be acquired cheaply (second-hand from Germany with loans obtained through local contacts) and where high-value crops such as hops can be grown. Below this level are the subsistence holdings which have no hope of viability. Effectively therefore it is a choice between the large farming company ('agri-business') and the medium-sized family farm. Cochrane (1994: 335) expects to see family farms of 50–100 ha emerging in Romania, Bulgaria and Poland to exploit potential that remained suppressed under socialism: e.g. intensive pig rearing systems based on farm-grown potatoes, whey and vitamin supplements. Meanwhile, the bigger farms will certainly succeed with good management in regions like the Poznań area of Poland with its successful orchards and greenhouses linked with food-processing plants (with leasing and cooperative arrangements). At the other extreme, very small holdings below five hectares are linked by Pavlin (1991) with 'hobby farming' or 'post-deagrarisation': a recreational approach rooted in weekend cottages and far from the reality of subsistence farming imposed by poverty which applies in most of the SEECs and accounts for the majority of individual farms in the region so far. It is suggested that hobby farming may recruit talented young people from the towns who may be able to take over holdings owned by fathers or even grandfathers. But young people who are brought up amidst the realities of subsistence farming driven by poverty have no such pretensions.

The north–south contrast is confirmed by Swinnen and Mathijs (1997) who have calculated an 'Individual Farming Index' which relates share of individual farms in terms of total agricultural land in 1995 compared with the 1989 situation. Albania (94.2), Romania (55.2) and Bulgaria (45.4) are way ahead of the Czech Republic (22.1), Hungary (17.3) and Slovakia (3.1): demonstrating a relationship – through the 'way of life' syndrome – with the level of agricultural employment, even where holdings are fragmented, and a close inverse relationship with cooperative farm productivity in 1989. Yet it is likely that private farming will expand because of 'generational turnover' in the countryside and the arrival of younger more individualistic people without memory of work under collectivised agriculture. A more stable macroeconomic environment should reduce risk and offer the prospect of profitability in the context of better marketing and agricultural services which will make the 'shelter' of cooperatives less important. There will have to be a supply of land, which presupposes a high level of employment to reduce interest in subsistence cropping, and the land will need to be fertile, for marginal land – of which there is plenty – would not justify investment. However, there is no doubt that the economic factors will be crucial: the availability of land and credit, along with stimulative markets to encourage high productivity. Also important is a supportive environment with institutions concerned with information and training. Poland's 'Agricultural Circles' can help with marketing, specialisation and pluriactivity. And demonstration farms have been set up under PHARE auspices in Bulgaria, Hungary and Romania with the help of Dutch organisations.

The new private farmers: peasants or professionals?

A critical issue for the future is the emergence of skilled full-time farmers working viable holdings. The new private farmers may be professional people who played a leading role on the cooperatives and now take full advantage of their skills and contacts. But there are also cases of partnerships between peasants with complementary skills arising to lease state farm land in Poland, with the possibility of buying it at a later stage when restitution issues have been resolved. They may also rent machines – with the intention to buy later if they can – or may purchase outright through down payments of one-fifth and instalments for the rest over seven years. Good relations between local peasant families will accelerate the process of putting together small units of land ownership in larger functional units with a pooling of machines and labour. This may happen irrespective of whether the units of ownership derive from existing farms, restitution holdings, land taken out of cooperatives or acquisitions arising through vouchers.

Several factors seem likely to tip the scales in favour of the professionals or alternatively the peasantry. Cooperative leaders are in a strong position when farms are declared bankrupt because of superior knowledge about markets and other aspects of the farming business. They can exploit their monopoly of power and information to obtain the best property to launch their own private farm and some might even hasten the bankruptcy scenario. Where cooperatives are very large (including several village communities) the leaders have great informational advantages over the dispersed membership. On the other hand, family solidarity can strengthen the peasants' position and result is farms based on kinship ties or informal networks. Skilled rural workers may have connections through workshop industries which can be beneficial in joint farming and encourage group withdrawal from the cooperative, taking with them an appropriate mix of assets. Tractor drivers can work together on a private farm with the skills to drive and maintain machines. Expertise in agricultural engineering (coping with old Soviet machines, making spare parts and handling fertilisers and pesticides) may be complemented by wives' knowledge in finance and accounting. People who did well in farming under the communist reforms (in poultry rearing for example) may have capital to buy land.

All these scenarios are more likely to emerge in a well-integrated village community, without too much commuting to divert younger people away from agriculture. It is also important to have a village environment approving of private enterprise. Thus while it is difficult to deny that the 'green barons' are most likely to succeed because they have modest stocks of capital, managerial experience and agribusiness contacts, there are many people with a peasant background who could be successful in family farming. Suli-Zakar (1996) sees long-term promise in family farms which have traditionally been adaptable and sensitive to innovation, despite technical deficiencies and heavy dependence on female workers and family friends. Accordingly they should meet new challenges when there is an upswing in the market, having always lived in symbiosis with the very large farms. Indeed he sees this as an opportunity arising from incomplete restructuring under socialism. However, this will only be possible if smallholders forced back to the land through unemployment can find places in an expanding service sector and provide scope for farm amalgamation.

Agricultural production

As in the case of manufacturing, agricultural production has fallen sharply. Recovery was not very evident until 1995 and since then there have been some major drought problems, especially in Romania and Yugoslavia in 2000 (Table 3.3). The decline in livestock occurred everywhere except Albania in the early 1990s and continued in the later years, except for B&H and Slovenia. However, decline had already started in the late 1980s, except in Albania and FCS (Table 3.4). Thus while Albania had more units of livestock in 2000 than in 1985, numbers have declined by more than half in all the other countries except FCS and former Yugoslavia. These trends cannot simply be correlated with changes in farm structure. The restitution process certainly disrupted some forms of production when large livestock rearing and fattening complexes inherited from communism found themselves cut off from fodder supplies. The ones that have survived have required investment to rebuild the feeding and marketing systems, e.g. by integrating with a fodder factory, slaughterhouse and meat-packing plant. The disruption affecting both cooperative and state farms complicated the working of large irrigation projects, while machinery depots lacked adequate finance to modernise and maintain an efficient service. But there were other problems which affected production. Under the communist system farm production went to state warehouses for distribution to food markets or processors, sometimes with excessive transport hauls. These networks collapsed with the rest of the central planning system and had to be replaced by new systems which were built up only slowly by private enterprise. Industrial crops like cotton and tobacco were affected by the breakdown of the former producer–processor links, while the sale of fresh farm produce encountered problems, for while it is relatively easy for farmers to distribute dairy products, meat, fruit and vegetables in the local area in order to earn profits which can rejuvenate private agriculture, dependence on local markets means that farmers are exposed to price variations when localised deficits and surpluses cannot easily be ironed out because of inadequate intelligence and logistical support. Private dealers have not yet got the expertise to operate through wider networks that would balance out the surpluses and deficits. They seem to prefer a small turnover with prices as high as possible; so farmers may be left with unsold produce that is nevertheless in short supply in areas just beyond the immediate locality.

Even though governments often found it necessary to get involved to ensure that the urban areas would be supplied, there was a big reduction in demand and output fell in sympathy. People did not buy as much as before. Hungary reported a reduction in consumption of meat and dairy products of 30–40 per cent at the onset of transition, with bread becoming a more prominent element in the diet. Demand was also reduced in part by the withdrawal of the Red Army which generated a substantial market in East Germany, FCS and Poland. Meanwhile, domestic producers also became subject to competition from imports: lower import tariffs in the ECECs compared with the EU are good for eastern consumers but not for producers. Romanian millers want higher tariffs against Hungarian and Turkish cereal and flour imports until the domestic industry can boost efficiency and compete internationally: in 1998 the domestic market price for wheat was $115/t compared with the world price of $90/t. Meanwhile, although Austrian, French and German companies have acquired Romanian sugar factories and improved efficiency, some state processors have been forced out of business due to cheap imports, including duty-free imports from Moldova (which include some re-exports from Ukraine). There have been calls for a ceiling on duty-free imports and reduced VAT on sugar.

Given the relatively unrewarding markets, the cost of farm inputs such as fertiliser and machinery rose in relation to market prices for agricultural commodities. This price 'scissors' meant reduced intensification with less use of fertiliser and irrigation water, which in turn created difficulty for the chemical industry and for water management to the point where irrigation systems broke down. Fertiliser use in Poland decreased from 200 to 70 kg/ha, with a ten-fold decrease in agrochemicals as part of 'a tendency for the extensification of farming

	1990	1991	1992	1993	1994	1995	1996	1997	1998	1999	2000	2001e	2002e
Albania	na	na	na	na	6.8	13.0	15.0	-2.0e	5.0	5.0	4.5	4.0	3.5
Bulgaria	-8.6	-6.4	na	-18.1	10.8	16.3	2.0	26.2	2.5	1.0e	2.0e	na	na
Czech Rep.	na	na	na	-0.8	-2.5	4.2	3.5	3.0e	1.0	5.0	-2.0	4.0	2.0
Hungary	-3.8	-3.0	na	-14.7	3.0	0.0	5.0	-1.0	2.5	2.0e	2.0e	na	na
Poland	-2.2	-2.0	na	2.4	-10.7	16.1	1.1	1.5	3.5	2.1e	2.0e	na	na
Romania	-2.9	1.2	na	12.4	3.0	4.9	-3.1	1.4	-7.6	4.7	-15.8	-1.0	2.0
Slovakia	na	na	na	-12.0	-3.0	-9.4	-1.0	0.7	1.0	-1.0e	2.0e	na	na
Yugoslavia	na	na	na	na	3.7	4.0	-1.0	5.0	3.0	4.0e	-19.7	10.0	2.0

Source: United Nations World Economic Survey and Economist Intelligence Unit

Note e Estimate

Table 3.4 Livestock numbers 1985–2000

	Livestock numbers (,000) 2000				Livestock units[2]			
	Cows	Sheep[1]	Pigs	Poultry	2000	1995	1990	1985
Albania	720	2458	81	4000	1122	1292	1177	1009
Bulgaria	682	3595	1512	13919	1935	2233	4793	5381
Fmr. Czechoslovakia	2239	493	5278	42617	4710	5333	7832	7695
Czech Republic	1574	116	3688	29500	3256	3529	na	na
Slovakia	665	377	1590	13117	1454	1804	na	na
Hungary	857	1094	5335	25890	2976	3138	5279	6153
Poland	6200	392	18224	49526	7608	12176	15426	15993
Romania	3155	8526	5951	72000	7914	8912	14480	16541
Fmr. Yugoslavia	3382	5206	6563	40521	6863	6573	9191	9792
Bosnia & Herceg.	350	285	80	3870	864	459	na	na
Croatia	440	566	1362	440	739	747	na	na
Macedonia	290	1550	197	3339	634	811	na	na
Slovenia	471	87	552	8550	860	719	na	na
Yugoslavia	1831	2718	4372	24322	3766	3837	na	na
Total (ECE)	17235	21764	42944	248473	33128	39657	58178	62564
Germany	14574	2404	27049	103000	22109	22733	na	na

Source: FAO database <http://www.apps.fao.org>

Notes

[1] Includes goats but not for B&H, Macedonia, Poland and Yugoslavia

[2] Calculated using the Romanian conversion system 0.84 per cow, 0.20 per pig, 0.14 for sheep or goat and 0.04 for poultry

operations in the face of declining profitability and market uncertainty' (Morgan 1992: 147). On top of all this there were the usual natural hazards arising through flood, drought and disease; also the negative consequences of air and water pollution linked with industrialisation and soil infertility – along with compaction and erosion – arising from excessive use of chemical fertilisers under communism.

Figure 3.4 examines cereal production in Romania during the early transition years with the south and southwest showing above average yields when calculated in terms of percentage deviation from the national average for the period (far greater than the variations in the share of the sown area) and the average deviation (tonnes per hectare) year by year. However, while the variations in yield in individual counties considered in isolation are generally within 20 per cent of the average, they exceed 25 per cent in Tulcea; 30 per cent in Galaţi, Mehedinţi and Teleorman; and 40 per cent in Dolj and Olt where the consequences of dry years are particularly severe given that the irrigation systems were not working properly at this time (Figure 3.4a). In the later years of communism (1970–85: the late 1980s are excluded because data were not published for this period) the average annual variations in yield as deviations from the national average were greater than for the early 1990s, yet the percentage deviations from the individual county averages were quite small, especially in those areas highlighted by the first analysis. With irrigation systems in place very high yields were secured consistently – pushing up the national average – and this meant small variations year by year in individual counties – but major negative deviations in northern counties which applied fertiliser heavily but had no defence against poor summer growing conditions (Figure 3.4b). Olt is a particularly good barometer showing progressively stronger annual performances against the national average from the early 1970s to the early 1980s. During the early 1990s the margin has fallen quite noticeably while annual variations within the county have doubled with the deterioration in the farming infrastructure.

A further factor in production decline has been the removal of domestic food subsidies and consequent reduction in demand which, in turn, has discouraged investment. In Hungary, direct subsidies to producers dropped from $2.6 bn. in 1987 to $265 m. in 1992; and to consumers from $1.4 bn. to $200 m. Hungary then subsidised only 8 per cent of the value of agricultural output compared with 44 per cent for the G24 countries and 49 per cent for the EU (Berend 1996: 344). However, most help is typically given to the marginal areas, for the reduction in the level of intensification is particularly evident in areas with poor soils like western Bohemia and northwest Poland. In Slovakia, agricultural work is most important for the local economy in the Danube lowlands and least important in north central districts (Figure 3.5F). However, most government support for agriculture (Figure 3.5E) is provided for the Carpathians while assistance to the southern basins and eastern lowlands is relatively modest and the highly productive Danube valley receives virtually nothing. This reflects the relatively high value of output (Figure 3.5D) and also of the agricultural land (Figure 3.5C). More than 80 per cent of the land is used for agriculture (Figure 3.5B) and three-quarters of the farmland is arable. Thus an important strand of agricultural policy emerges through support of fragile rural communities that need

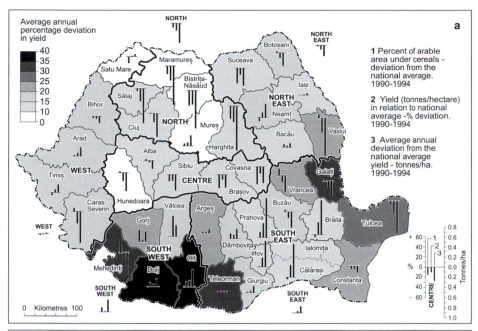

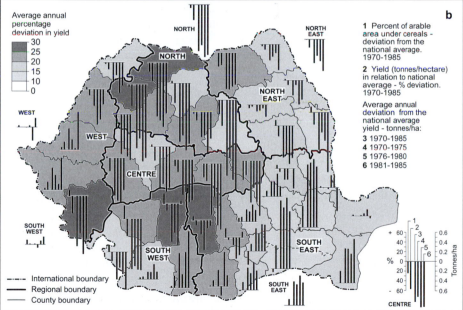

Figure 3.4 Romanian cereal yields by county 1970–94 (a) 1990–94 (b) 1970–84

Source: *Romanian Statistical Yearbooks*

to diversify. It will be shown in Chapter 5 how help is given to rural tourism (e.g. through SAPARD in Poland) while Albania helps small farmers in the mountains with a Fund for Financing Mountainous Zones and the policy study of Romania (below) highlights action to cope with rural poverty.

Land use change

Land use at the end of the communist period is shown for the Polish Sudetes using data discussed by Czetwertynski-Sytnik *et al.* (2000). Figure 3.6a shows that forests exist mainly in the high mountains and also in some lowland areas (surrounding Węgliniec), where there are poor soils developed on sandy glacial outwash material. Most lightly forested was the Sudetic Foreland, with fertile clay and loess soils as well as major urban developments (Jelenia Góra and Wałbrzych) and industrial zones related to lignite mining and electricity generation (e.g. the 2,000 MW Bogatynia power station). State farm ownership of land in Lower Silesia is mainly limited to the depressions with the best conditions for cropping, while private farms predominate in the mountains proper and individual farms typically comprise many scattered pieces of land, reflecting both the topography and rapid changes

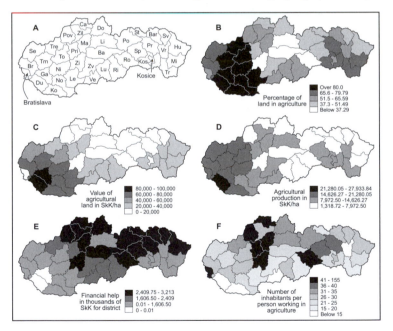

Figure 3.5 Agriculture in Slovakia 1994 (a) administrative units; (b) percentage of land in agriculture; (c) market price for agricultural land; (d) agricultural production; (d) state financial assistance; (e) employment. Key to administrative units: Ba Banská Bystrica; Bar Bardejov; Br Bratislava; Ca Čadca; Do Dolný Kubín; Du Dunajská Streda; Ga Galanta; Hu Humenné; Ko Komarno; Kos Košice; Le Levice; Li Liptovský Mikuláš; Lu Lučenec; Ma Martin; Mi Michalovce; Ni Nitra; No Nové Zámky; Po Poprad; Pov Považská Bystrica; Pr Prešov; Pri Prievidza; Ri Rimavská Sobota; Ro Rožňava; Se Senica; Sp Spišská Nová Ves; St Stará Lubovňa; Sv Svidnik; To Topolčany; Tr Trebišov; Tre Trenčin; Trn Trnava; Ve Velky Krtíš; Vr Vranov nad Toplou; Zi Žiar nad Hronom; Zil Žilina; Zv Zvolen.

Source: Drgona *et al.* 1998 pp. 252–3

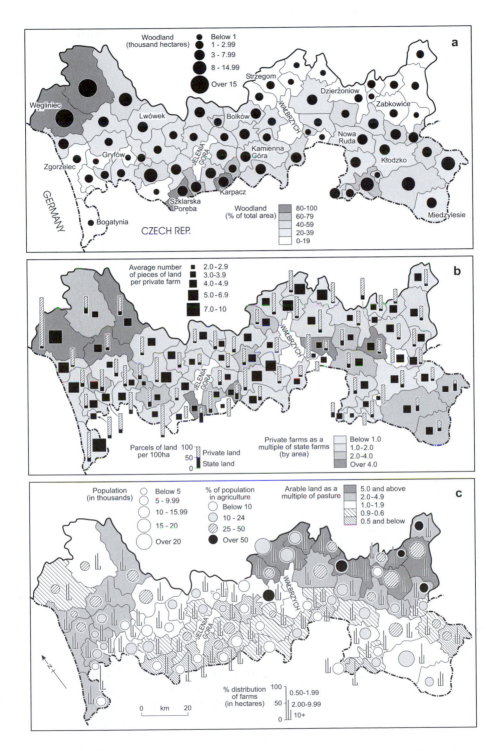

Figure 3.6 Land use in the Polish Sudetes in the 1980s (a) woodlands (b) farming
(c) employment

Source: Czetwertynski-stytnik *et al.* (2000: 78)

in ownership (Figure 3.6b). This is very noticeable in the neighbourhood of the main urban and industrial centres where rural dwellers were interested in farming as a source of ancillary income and domestic food supply. Generally, the fragmented holdings result in a loss of efficiency.

The Sudetic foreland shows arable lands are more important than pasture (Figure 3.6c), with a roughly equal balance in the mountain depressions. As natural conditions become more severe with altitude, pasture becomes relatively more important. While state farms took 37 per cent of the land and private farms 63 per cent, the latter averaged only 3.2 ha and were highly fragmented. Most farms fall into the 2.00–9.99 ha band (the average for Poland being 8.00 ha), but there were many under 2.00 ha near the towns and industrial centres where farm work was complementary to salaried employment. Agriculture still employed 10–25 per cent of the active population; exceeding 25 per cent (and sometimes reaching 50 per cent) in areas remote from the larger towns with especially good soils (e.g. Dzierżoniów, Kamieniec Ząbkowicki and Strzelin). But the index fell below 10 per cent in heavily industrialised areas (e.g. Bogatynia), tourist centres like Karpacz, Kudowa Zdrój and Szklarska Poręba, and also in forested zones like Miedzylesie where the forest administration and the wood processors were large employers.

It is now apparent that, for all its failures to meet plan targets, socialist agriculture was well resourced and relatively intensive compared with the situation today. Under present conditions the former levels of production cannot be maintained and smaller harvests may become the norm. For ecological reasons alone some relaxation may well be desirable. The effects of reduced intensification and market-orientation can be seen in areas where small family farms regained through restitution are giving greater emphasis to maize and potatoes than to sugar beet, fruit growing and viticulture. This may involve cutting down fruit trees to free land for maize which supplies human food, animal fodder and domestic fuel. Meanwhile, arable land that can no longer be farmed profitably to supply the market is being converted to pasture while breakdown of extensive grazing systems in some mountain areas is leading to the invasion of scrub and an organised programme of afforestation. Some changes have occurred in the Czech Republic in areas bordering Germany because of the disintegration of state farms, some of which were working whole districts like Cheb and Tachov. In East Germany, as in Slovakia, it is intended that agricultural land should become more highly compartmentalised through the restoration of some of the 'biolines' removed during the communist period. But greater ecological consciousness will require re-education for all users of the landscape as well as more adequate land-use documentation to ensure that business intentions are properly harmonised with the landscape potentials of the area.

Land use changes presented by FAO reflect these trends (Table 3.5), although it is difficult to interpret the figures with regard to divergent trends over agriculture under communism when an overall decline in farmland might be expected in view of the large scale of socialist urban-industrial development. The overall trend is to be expected and cases of expansion in agricultural land may reflect the reclamation of wasteland (other than woodland). The more rapid pace of change since 1989 reflects the marginalising of much land

Table 3.5 Land use change 1950–99

	Agriculture			Arable and Perm. Crops			Pasture			Woodland*		
	A	B	C	A	B	C	A	B	C	A	B	C
Albania	1128	−2.2	+1.8	699	+8.5	−0.8	429	−10.7	+2.6	1048	−7.9	0.0
Bulgaria	6203	+10.3	+3.5	4511	−12.3	+36.5	1692	+22.6	−33.0	3590	−3.3	+42.5
Fmr. Czechosl'kia	6724	−19.4	−2.5	4926	−7.8	−18.2	1798	−11.6	+15.7	4646	+10.2	+9.5
Czech Rep.	4282	na	na	3332	na	−12.8	950	na	+12.8	2630	na	+0.5
Slovakia	2442	na	na	1594	na	−2.2	848	na	+2.2	2016	na	+9.0
Hungary	6186	−24.9	−29.8	5039	−17.3	−24.8	1147	−7.6	−5.0	1811	+13.6	+17.6
Poland	18435	−42.3	−37.2	14401	−38.8	−35.8	4034	−3.5	−1.4	8942	+46.7	+38.6
Romania	14781	+11.4	+2.3	9845	+14.7	−20.8	4936	−3.3	+23.1	6680	−1.9	+0.4
Fmr. Yugoslavia	13127	+7.5	−97.9	7414	−2.6	−34.0	5713	+10.1	−63.9	8674	+38.5	na
Bosnia&H	2140	na	na	940	na	na	1200	na	na	2710	na	na
Croatia	3151	na	na	1590	na	na	1561	na	na	2076	na	na
Macedonia	1291	na	−2.3	635	na	−3.9	656	na	+1.6	1020	na	na
Slovenia	500	na	−9.1	202	na	−4.9	298	na	−4.2	1099	na	+5.5
Yugoslavia	6045	na	−23.8	4047	na	−4.8	1998	na	−19.0	1769	na	na
Total	66584	−59.6	−159.8	46835	−55.6	−97.9	19749	−4.0	−61.9	35391	+95.9	na

Source: FAO production yearbooks, FAO database <http://www.apps.fao.org> and Eurostat 2000: 136–7

Notes
A Land use in 1999 (th.ha) (Yugoslavia 1998)
B Change (th.ha/yr) 1950–89
C Change (th.ha/yr) 1989–99 (FCS successor states 1993–9; former Yugoslavia 1992–9)
*Uses Eurostat data for 1995–6 (except for Albania and former Yugoslav states other than Slovenia). Trends apply to 1996 only

(including some high mountain pastures reverting to scrub). The figures also show a continuing gain in woodland, as part of a long-term trend in some areas. Nationalisation of woodlands under communism gave an opportunity for restoration after centuries of deforestation. In Hungary there was an increase from 1.1 to 1.6 m.ha over three decades, although the first drive during 1947–54 was not very successful because only 42 per cent of the trees planted over 300,000 ha survived. There was an increase in the self-sufficiency level for timber from 54.7 to 70.5 per cent during 1950–85 (Keresztesi 1993: 67) but heavy cutting for an expanded wood-processing sector meant that the most accessible forests across the region tended to become thinner and younger.

In Hungary there was a prognosis even in 1988 which saw that higher crop yields would allow 1 m.ha of farmland to be abandoned and that 850,000 ha could be used for forestry: hence a 'great green' programme to plant 150,000 ha during the 1990s and a total of 700,000 ha over the six decades to 2050. However, there is now a larger private sector in Hungary and in the other ECECs and despite a forest code the need to secure profits from processing has led some enterprises to fell their best stands without replanting. Overall it seems that woodland is being used less intensively than before. Eurostat (2000) data give a net annual increment of 147.4 m.m^3 of timber growth across the region against cutting of 88.5 m.m^3 which accounts for only 60 per cent of the growth; though ranging from 99 per cent in Macedonia, 79 per cent in the Czech Republic and 74 per cent in Albania and Poland, to 62 per cent in Croatia, 53 per cent in Slovakia, 48 per cent in Hungary, 43 per cent in Romania, 41 per cent in Bulgaria and 36 per cent in Slovenia. Meanwhile, the lack of integration between silviculture and hunting means that game numbers can increase and inhibit woodland regeneration. Extension of woodlands is still desirable but the forests need to be worked sustainably, with systems of timber certification to ensure this, and management should respect all legitimate interests including grazing, recreation, fruit and berry picking, biodiversity conservation and hunting as well as wood production.

Measures to stimulate production

To stabalise farm incomes the Czech Republic decided to intervene in the restructuring of supply and demand through a market fund for the buying and selling of beef, cereals and milk. The system operates when market prices sink below quoted floor prices. There is also limited market intervention through Hungary's Agricultural Market Office in respect of five commodities (maize, wheat, beef, pigmeat and milk) when overproduction occurs. Intervention prices are now available in Poland for pork, butter, milk powder and grain. Meanwhile, PHARE has been helping to modernise wholesale marketing in large cities through collecting centres where produce can be sorted and packaged. The Bucharest Wholesale Market project is a prominent example. International trade is also an important issue, for the East could be Europe's granary. Another approach is to improve other aspects of the infrastructure to improve efficiency. Thus a Kuwaiti fund is contributing to the rehabilitation of irrigation systems in Albania. This is a costly business (though gravity systems should have priority since they have economic advantages over sprinkler systems) and the work needs to be linked with incentives to ensure the use of the land for high-value crops like vegetables, fruit and industrial crops. Similar work is needed elsewhere in the SEECs and since they have been subject to severe drought in recent years non-conventional sources may have to be tapped: sea water, drainage water and mineralised groundwater.

As processing is restructured the demand for quality raw materials will impose a pressure on farmers to consolidate, modernise and close the gap between agricultural and food industry standards. Otherwise the processors will go into farming themselves, as some have done already: e.g. the Agros company owns 12,000 ha of farmland. Foreign investors are taking a lead and French 'Celia' cheesemaker, located near Poznań in Poland, has negotiated contracts with farmers for 25,000 l of milk daily, which provides the stimulus to invest and improve quality through new equipment that could not have been afforded otherwise, even with bank loans. The same is happening with the vegetable processor Bonduelle which has enabled irrigation systems to be introduced in the Gniewkowo region of Poland; with training provided as well as the leasing of equipment. French penetration of the Polish distilling, poultry and sugar beet industries is also expected. Higher-quality production and processing opens the way to rising food exports and for contacts to process farm products from other countries: Arad's sugar factory ('Zaharul') is processing beets from the adjacent Békeés area of Hungary.

Indeed, it may well be considered desirable to stimulate exports to boost farm incomes and improve the balance of payments. The Czech Republic considers it is uneconomic to try and produce more for export because of the need to compete with subsidised exports from other countries including the EU. However, in Hungary, where agriculture is export-led (with a surplus of $1.5 bn. in agricultural trade in 1995), the government sees the ability to export as crucial for the balance of payments and has launched a National Agricultural Programme to increase agricultural exports with money for export subsidies. Over the long term Hungary believes that competitiveness should be stimulated not by subsidies but by comparative advantages (good arable land and skilled farmers). Hungary is trying to boost the export of organic apples while Romania is finding opportunities for livestock exports within CEFTA, with particular scope for pigs since a quarter of the herd is still in large units (in contrast to cattle and sheep). Cereal exports through Constanţa are facilitated through the new terminal opened in 1998: this is beneficial not only for Romanian wheat and maize but also for commodities from Bulgaria, Hungary, Moldova, Ukraine and Yugoslavia using the Danube navigation and the Danube–Black Sea Canal.

The picture emerging from Table 3.6 is mixed. Cereal production in the 1990s was generally lower than the 1980s, which shows gains on late 1970s driven by the need to reduce exports after Soviet supplies proved inadequate to meet all deficits. Thus Polish imports were down in the 1980s with reduced use of cereals for livestock which contributed to the reduced animal numbers already noted. There has been some resurgence of cereal imports since 1989

Table 3.6 Cereal production and net trade 1975–99 (m.t)

	1975–9		1980–4		1985–9		1990–4[1]		1995–9	
Albania	3.76	+1.47	5.02	–*	5.20	+0.15	3.14	+1.87	2.93	+1.53
Bulgaria	39.36	+1.53	43.22	+0.64	38.09	+4.98	32.13	–0.36	26.51	–1.94
Fmr. Czechosl'kia	49.03	+7.58	53.43	+4.46	58.31	+1.82	54.91	+0.31	50.88	–2.01
Czech Rep.	na	na	na	na	na	na	1.33	+0.54	3.40	–0.66
Slovakia	na	na	na	na	na	na	6.86	+0.02	1.69	–1.35
Hungary	61.53	–4.12	71.22	–6.46	44.51	–8.50	58.68	–7.27	61.23	–14.06
Poland	119.36	+31.25	111.12	+25.06	126.17	+12.65	120.96	+4.92	129.52	+13.99
Romania	92.01	+1.02	106.60	+2.48	113.68	+0.09	82.44	+7.70	88.67	–3.21
Fmr. Yugoslavia	77.38	+2.17	83.66	+0.21	80.41	+3.61	70.70	–0.19	70.30	+3.28
Bosnia&H	na	na	na	na	na	na	3.53	+0.28	5.39	+1.68
Croatia	na	na	na	na	na	na	7.69	–0.49	14.79	–0.22
Macedonia	na	na	na	na	na	na	1.75	+0.46	3.32	+0.98
Slovenia	na	na	na	na	na	na	1.46	+1.52	2.51	+2.47
Yugoslavia	na	na	na	na	na	na	22.84	–1.53	44.29	–1.63
Total	442.43	+40.90	474.27	+26.39	466.37	+14.80	422.96	+6.98	430.04	–2.42

Source: FAO production and trade yearbooks

Note
[1] Yugoslav successor states 1992–5 only; Czechoslovak successor states 1993–5 only (hence the totals for FCS and former Yugoslavia will not tally with the returns for the successor states for the 1990–4 period)

in Albania, but not in Bulgaria and FCS which suggests that livestock numbers are down to more sustainable levels. Hungary has managed rising exports and there is promise in Romania with more exports in the late 1990s than in previous years. Meanwhile, trade in all agricultural and food commodities shows a trend from export to import in Albania, with falling net exports from Bulgaria balanced by a strong export performance in Hungary. Many countries are net importers – e.g. Romania after net exports in the 1980s (paying off debts) – while net imports are falling in Poland, where the food industry has developed, but rising in Yugoslavia (Table 3.7).The indices presented in Tables 3.8 and 3.9 show that recovery in Albania is balanced between crops and livestock, whereas in B&H and Macedonia the improvement is more evident for crops than livestock (especially B&H) and there is a big difference in Croatia. Crops have shown a better capacity for recovery in 1999–2000 than livestock, but in Slovenia and Yugoslavia the situation is reversed (also in Poland slightly) while the figures for Hungary, Romania and Slovakia in 2000 reflect the serious drought, which also affected Yugoslavia.

The social problem

FAO data indicate the continuing prominence of agriculture in some of the ECECs. A high proportion of the rural population is working in agriculture, particularly in Albania, Poland and Yugoslavia, and the proportion is even higher in Albania as a proportion of the total active population (Table 3.10). The trend

Table 3.7 Trade in agricultural commodities ($bn.) 1965–99

	1965–9[1]	1970–4	1975–9	1980–4	1985–9	1990–4[2]	1995–9
Albania	−0.22	−0.22	−0.53	−0.06	−0.04	+0.92	+1.01
B&H	na	na	na	na	na	+0.57	+2.39
Bulgaria	−5.10	−0.86	−2.54	−3.21	−2.76	−2.85	−1.96
Croatia	na	na	na	na	na	+0.18	+1.60
Czech Rep.	na	na	na	na	na	+0.33	+3.88
FCS	+2.91	+3.80	+7.68	+5.92	+6.41	+1.86	+5.74
Hungary	−0.49	−1.66	−2.96	−6.09	−5.88	−7.88	−8.02
Macedonia	na	na	na	na	na	+0.13	−0.23
Poland	+0.77	+1.34	+6.67	+6.85	+1.89	+0.17	+4.58
Romania	+0.86	−0.97	−0.38	−0.09	−0.90	+3.78	+1.65
Slovakia	na	na	na	na	na	+0.48	+1.86
Slovenia	na	na	na	na	na	+0.84	+2.29
Yugoslavia	na	na	na	na	na	−0.03	+0.71
Fmr. Yugoslavia	+0.09	+1.28	+3.22	+0.24	+1.03	+3.68	+6.76
Total	+11.18	+2.71	+11.16	+3.56	−0.25	−0.32	+9.76

Source: FAO trade yearbooks

Notes
[1] Calculated on the basis of 1969 only for Albania, Bulgaria and Romania
[2] Yugoslav successor states 1992–5 only; Czechoslovak successor states 1993–5 only (hence figures for FCS and former Yugoslavia will not tally with the overall picture for the successor states for 1990–4)

Table 3.8 Livestock production indexes 1988–2000 (= 100)

	1988	1989	1990	1991	1992	1993	1994	1995	1996	1997	1998	1999	2000
Albania	89.7	94.5	104.4	101.1	113.1	122.0	140.5	163.7	165.6	146.7	148.8	157.2	168.5
Bosnia & H	na	na	na	na	50.1	47.7	36.4	32.5	20.5	31.2	37.6	42.5	24.5
Bulgaria	108.4	198.2	104.0	87.8	85.4	72.8	62.4	62.2	64.4	59.2	62.4	61.8	59.2
Croatia	na	na	na	n.a	56.1	51.5	47.6	45.3	44.6	43.8	48.7	48.9	48.0
Czech Republic	na	na	na	na	na	89.9	75.2	76.4	74.8	73.3	75.0	73.1	59.5
Hungary	104.6	104.8	102.4	92.8	81.6	73.4	68.3	67.4	70.6	67.8	69.2	70.6	70.2
Macedonia	na	na	na	na	88.7	94.4	96.7	93.6	87.7	87.7	85.1	86.1	86.2
Poland	100.4	101.5	102.6	95.8	89.6	83.0	77.6	79.7	83.2	83.5	87.8	86.8	82.4
Romania	102.4	96.8	99.8	103.4	95.0	97.0	98.4	93.0	90.5	90.5	88.4	89.9	90.7
Slovakia	na	na	na	na	na	76.4	70.2	68.1	67.0	69.2	66.9	69.8	67.7
Slovenia	na	na	na	na	77.4	92.6	98.0	96.0	103.7	104.7	104.2	102.2	98.2
Yugoslavia	na	na	na	na	101.0	97.9	97.8	108.4	115.9	108.8	97.6	90.4	92.1

Source: FAO database <http://www.apps.fao.org>

Table 3.9 Crop production indexes 1988–2000 (= 100)

	1988	1989	1990	1991	1992	1993	1994	1995	1996	1997	1998	1999	2000
Albania	104.8	124.1	106.2	69.7	69.2	86.7	82.5	94.3	96.2	95.1	100.5	95.6	103.6
Bosnia & H	na	na	na	na	98.7	88.1	74.5	85.5	81.1	102.6	107.7	105.3	107.6
Bulgaria	99.2	108.2	96.4	95.4	82.0	60.2	70.8	81.1	55.3	70.1	65.2	63.8	62.2
Croatia	na	na	na	n.a	68.7	75.2	72.6	77.1	80.4	84.8	94.6	91.2	75.8
Czech Republic	na	na	na	na	na	91.0	82.4	86.6	88.1	85.3	86.4	92.2	87.9
Hungary	105.2	102.4	93.0	104.6	72.2	65.0	76.3	74.5	81.0	85.8	85.0	79.2	65.5
Macedonia	na	na	na	na	112.6	80.9	93.5	96.5	97.8	101.9	107.9	110.3	105.7
Poland	98.2	101.9	103.1	95.0	75.7	99.1	77.1	89.7	90.9	83.2	94.1	83.5	81.9
Romania	115.7	112.2	93.2	94.6	76.3	98.7	93.9	103.6	93.6	109.9	90.4	101.4	77.6
Slovakia	na	na	na	na	na	84.9	88.4	85.4	89.0	91.3	94.2	78.6	62.9
Slovenia	na	na	na	na	81.0	84.6	99.4	94.3	100.8	91.8	96.9	85.8	87.8
Yugoslavia	na	na	na	na	77.2	76.3	83.6	86.4	87.3	98.9	90.0	82.6	62.0

Source: FAO database <http://www.apps.fao.org>

Table 3.10 Rural population and agricultural population 1990–2000

	1990						2000					
	A	B	C	D	E	F	A	B	C	D	E	F
Albania	3289	2113	64.2	753	35.6	48.3	3113	1895	60.8	757	39.9	48.9
Bulgaria	8718	2922	33.5	542	18.5	12.1	8225	2462	29.9	321	13.0	7.5
Fmr. Czechosl'kia	15562	5901	37.9	774	13.1	9.3	15631	5552	35.5	757	13.6	8.9
Czech Rep.	na	na	na	na	na	na	10244	3457	33.7	484	14.0	8.4
Slovakia	na	na	na	na	na	na	5387	2095	38.8	273	13.0	9.3
Hungary	10365	3933	37.9	596	15.1	11.5	10036	3317	36.5	532	16.0	11.0
Poland	38119	14549	38.1	4037	27.7	20.7	38765	13351	33.0	4423	33.1	22.2
Romania	23207	10765	46.3	2368	21.9	20.2	22327	10032	45.3	1709	17.0	13.8
Fmr Yugoslavia	22808	11200	49.1	2341	20.9	21.6	23095	10129	44.9	1492	14.7	13.6
Bosnia&H.	na	na	na	na	na	na	3675	2258	61.4	100	4.4	5.5
Croatia	na	na	na	na	na	na	4481	1891	42.2	192	10.1	9.0
Macedonia	na	na	na	na	na	na	1999	768	38.4	124	16.1	13.5
Slovenia	na	na	na	na	na	na	1993	941	47.2	22	2.3	2.1
Yugoslavia	na	na	na	na	na	na	10635	4271	40.1	1054	24.6	20.8
Total	122068	51383	42.0	11411	22.2	18.5	121192	46036	37.9	8499	18.4	13.9

Sources: FAO Agricultural production yearbooks and database <http://www.apps.fao.org>

Notes
A Total population (,000)
B Rural population (,000)
C Percentage of population rural
D Population active in agriculture (1990 and 1999)
E D as a percentage of the rural population
F D as a percentage of the total active population

during the 1990s has brought out a reduction in the rural population share and in the shares of the agriculturally active population, considered in relation to both the rural population and the total active population. But the change has been quite small in Albania, Hungary and Poland. Indeed when the three values are averaged as an index of rurality the value has risen from 49.4 in 1990 to 49.9 in 1999–2000 and is now more than double the ECE average of 23.4 (down from 27.6 in 1990). Poland also shows a rise from 28.8 to 29.4 while in Romania – the other country well above the average – the figure has been reduced from 29.5 to 25.4, surprisingly, given the high level of dependence on small farms which the statistics do not adequately represent. Bulgaria is well below average (21.4 to 16.8) as is Hungary (21.5 to 21.2). The other countries, for which 1990 data are not available, have values ranging from 17.2 in Slovenia, 18.7 in the Czech Republic, 20.4 in Croatia and Slovakia, 22.7 in Macedonia, 23.8 in B&H and 28.5 in Yugoslavia.

Policy for industry and agriculture: the case of Romania

The 1989–96 period

Approaches to the restructuring of industry and agriculture are considered here in the context of one of the ECECs, with treatment divided between the first phase of transition when government by the National Salvation Front (NSF), renamed the Party of Social Democracy (PSDR) – during the presidency of I. Iliescu – adopted a conservative approach; and a second phase (1996–2000) when the Democratic Convention (DCR) and its coalition partners attempted a faster rate of reform. It is relevant to point out that Romania consistently failed to implement significant economic reform prior to 1989. Despite strikes in Brasov in 1987 and some discontent within the party in the run-up to the 14th Communist Party Congress held in November 1989, change was ruled out and the liberalising trends in other ECECs were deplored. Meanwhile, the population was also battered by coercive demographic policies and a radical resettlement programme which sought to marginalise small villages (under long-term threat of total abandonment and demolition), while accelerating redevelopment in growth centres from Bucharest downwards. The communist leadership under N. Ceauşescu extolled a 'pure' life divorced from Western materialism and not entirely at odds with the calls of the Orthodox Church for fortitude in the face of economic stress. That the 'mămăligă' (maize pudding: a traditional food) did not 'explode' says much for the survival mentality adopted by the mass of the population and by the lower echelons of the party hierarchy who showed flexibility wherever they had the confidence to do so.

Revolution occurred after demonstrations in Timişoara after which Ceauşescu lost the initiative in Bucharest and the army refused to support him. A provisional government moved quickly from the wings and had the Ceauşescu executed before disbanding the Communist Party and gaining power in the 1990 elections (as the National Salvation Front) with solid control of the rural areas of Moldavia and Wallachia. The NSF was able to offer a home to former communists, including the security police ('Securitate'), and gather an effective constituency among the managers and workers of the state sector who had most to fear from the introduction of market reforms. This reinforced the need for stability after the trauma of the later communist years, marked by the most severe austerity combined with extravagant and socially-disruptive public works, and ensured that an end to central planning and toleration of private enterprise would go hand-in-hand with state subventions for SOEs and a very gradual dismantling of price controls which prevented a 'level playing field' for all forms of business. Ministerial changes during the early 1990s signalled a highly cautious approach, responding to a series of violent demonstrations in Bucharest by miners from Petroşani in the Jiu Valley, but the burden of supporting loss-making enterprises soaked up potential investment capital and generated budget deficits financed by credits from the financial institutions.

There was certainly popular apprehension over a market economy: following the great effort demanded under communism, people would not 'sacrifice themselves [again] for the sake of future generations' (Nicolaescu 1993: 102). While the principle of free competition might be acceptable as an ideal, most people would hesitate to 'replace government control with more seriously binding controls that are inflicted by market-related institutions' (ibid.: 103–4). For radical reorganisation of all the SOEs would almost certainly generate unacceptable levels of unemployment at a time of very limited social security. Caution was reinforced by the major dislocation of Romania's commerce through the 'trade shock' of German unification which meant that East Germany started trading in DM. There followed the collapse of the USSR, bringing many long-standing exchanges to an abrupt conclusion; while civil war in Yugoslavia led to a UN trade embargo against Serbia: another important Romanian trading partner. The result was a transition marked economically and socially by extreme 'gradualism', despite preliminary moves to join Europe.

The reform process was highly inconsistent. A stabilisation plan was formulated in 1991 with IMF support, but momentum was lost through expensive price controls (retained until 1997) and an overvalued exchange rate that constrained exports while encouraging imports in the run-up to the 1992 election. An attempt was made to establish a transparent, functioning foreign exchange market late in 1993, but there was only a low priority for privatisation that could have brought an inflow of private capital. Instead there was some progress through self-financing and (possibly) unwarranted bank lending: a populist macroeconomic strategy leading to rapid growth in 1995 – in time for the 1996 election – but also a growing trade gap and a budget deficit that maintained high inflation and rising external debt. Low growth limited welfare payments and nourished a culture of industrial unrest, while the lack of consensus over economic reform strained relations between the PSDR and its coalition partners and eroded public confidence in political parties generally.

When government failed to deliver on its promises it had to play the nationalist card and work with parties which had emerged through regional movements like 'Vatra Românească' in Transylvania, reflecting a sense of conflict against the indigenous Hungarian population by Romanian migrants from Moldavia who could not qualify for land restitution. The originality of Romanian culture was therefore invoked to rationalise isolationism as an alternative to synchronism and the 'spirit of Europe'(Marga 1993: 21). It was argued that the country could not withstand radical economic reform and needed to be protected by an interventionist system to support a centralised economy (through subsidies and social benefits) and counter perceived risks of subversion on the part of some ethnic minorities and neighbouring governments. The 'protochronism' of Ilie Badescu and his 'Movement for Romania' seemed reminiscent of the Legionary/Iron Guard Movement of the 1930s and equivalent to the 'propaganda mythology of Ceauşescu's providential role in the country's history' (Tismaneanu and Pavel 1994: 405).

Policy for industry

In advance of privatisation the SOEs became autonomous in 1990, as joint stock limited liability companies. But the state remained responsible and was obliged to cover losses. New trade relations developed gradually among EU and other countries, but production fell sharply in the meantime. Urban industrial employment decreased by a million and some towns witnessed a decline of over 60 per cent, especially those industrialised only late in the communist period. Rationalisation of industry, with closure of the least efficient plant and investment in the more viable units, was much discussed. But politicians believed that restructuring would incur heavy social costs and increase instability. The result was a compromise involving modest lay-offs and retention of core establishments, sometimes with compulsory 'holidays' taken in rotation. Further pressure on the workforce arose from delayed wage rises to keep in line with inflation. Hence major strikes and demonstrations occurred from time to time, as in the heavy industrial centre of Reşiţa in 1994 as workers were forced to leave the town in the face of redundancies and low living standards.

Heavy industry was in great need of restructuring because of low productivity and the inefficiency of using the large steel capacities built under communism (dependent on imported ore and energy) to turn out ships and railway rolling stock for export at competitive prices. Because only a few bulk steel producers will be needed in the future, the smaller and less well-located plants will have to reduce output and concentrate on quality with the use of scrap in electric furnaces. Production also declined sharply in the chemical industry since few farmers could afford fertiliser. It is possible that much of the capacity in oil refining and petrochemistry may be needed when Caspian oil starts flowing westwards to Europe. However, the great problem was securing the funds needed for modernisation to increase efficiency and reduce both energy consumption and pollution. The advantage of low wages can only be realised with higher productivity and this issue, combined with good design and short delivery times, constrained progress in some major industries (like furniture) where Romania has advantages through long experience, ample capacity, skilled labour and access to raw materials. Meanwhile, effluent and concentrations of dust, sulphur dioxide and other gases in the vicinity of the main industrial centres resulted in deterioration in the quality of river water and the health of the forests. Investment was also needed to overhaul the substantial tourist industry. In addition to climate, scenery and natural monuments, Romania has much to offer in terms of cultural and historic sites. Folklore has its importance and the Dracula legend has paved the way for the promotion of Bran Castle as the model for Dracula's castle in the Bram Stoker novel. But foreign investment has been slow to arrive and Romania still suffers from the poor image acquired in the last communist years.

Unfortunately, privatisation was delayed by a lack of purchasing power within the country, not to mention the public's unfamiliarity with the concept of share ownership and the state's desire to retain a large measure of control. But there was also a reluctance to sell to foreigners at low prices, for some political mileage was won from brave declarations against selling the national patrimony. Privatisation moved only slowly for several years although a law was passed in 1991 to sell off 55 per cent of state equity, including some direct sales to foreign buyers. The government tried again in 1994–5 to deliver a public voucher privatisation under World Bank pressure. The SOF eventually announced that 658 companies passed into private hands in 1995; less than half the target figure. A new programme for 1995–6 covered 2,200 large and medium-size enterprises representing all branches of the economy and drawn from all parts of the country. After much initial confusion, over 2.3 m. citizens invested their coupons through computers installed in post offices. At the same time the government saw a continuing role for autonomous state corporations ('regii autonome') operating in strategic areas like power, transport and telecommunications. As late as 1996, a decree provided for full state control of the oil and petrochemical industries, including the subsoil and subsea resources. Compania Română de Petrol (Petrom) sought to integrate the petrochemical industry with pipeline operators, distributors and fuel retailers. The opposition deplored such a return to gigantism and state intervention (including holdings of at least 51 per cent in 62 companies). Shareholdings by foreign companies were limited to 49 per cent.

It will therefore take time for a new generation of technically advanced industries capable of sustaining human-capital-intensive exports to replace some of the older branches which have only limited growth potential because of a poor economic infrastructure, obsolete capital stock, high imported raw material and energy costs and exports that involve only limited value-added. However, progress was not helped by resistance on the part of the state bureaucracy, state enterprises and trade unions. Investment stayed low, although the Romanian Development Agency recognised a succession of investment waves. The first in 1990–1 involved small and medium projects, followed in 1991–4 by large-scale projects by transnational corporations: ABB, Alcatel, Amoco, Coca-Cola (with its network of bottling plants), Colgate-Palmolive, Shell and Siemens. From 1995 some strategic investments by transnational firms arrived in the context of an improved international environment. In regional terms, the investment was highly uneven with the bulk of the foreign capital going to Bucharest and a handful of other urban centres. Transylvanian cities were among the most dynamic in the country, notably Cluj-Napoca which has recovered from its association with the Caritas pyramid investment scandal. Cluj has a local development association and seeks to build on its successful industries like the privatised Ursus brewery (with a majority stake by the German brewery Brau und Brünnen) and Porcelaine Manufacturers, a Romanian–German joint venture which began production in 1990.

However, investment in the car industry has been a very positive element. Pride of place goes to the Daewoo (South Korea) investment involving more than $150 m. in the former Oltcit factory in Craiova. Indeed, Romania has tried to create a favourable environment for investment and entrepreneurship, with low wages as a powerful incentive. A liberal foreign investment law in 1991 was followed by the 'Daewoo Law' (July 1994), designed expressly to attract a company, well known as a hard bargainer over tax incentives. Now, 100 per cent foreign ownership of businesses is allowed, with repatriation of profits, tax holidays and exemptions from customs duties. The 'Dacia' car works near Piteşti began looking for a foreign partner and other attractive investment targets appeared. Hydrocarbon exploration was opened up to foreign companies in 1990 through production-sharing agreements. UK Enterprise Oil has found gas in the Black Sea while Shell has drilled for gas in Transylvania beneath a salt layer at depths of 1,500 m which is beyond the capacity of Romgaz, operating in shallower fields. Amoco have also been prospecting onshore.

To stimulate new enterprise the Romanian Ministry of Research and Technology coordinated a programme to provide low-cost space (often by arrangement with local universities and other institutions) for fledgling businesses to develop technological or industrial ideas. Enterprise was assisted by the scientific support of the EU PHARE programme, which has backed the Institute of Technology's Business Incubation Centres, and similar initiatives by individual states (like the UK Know How Fund) have also facilitated technology transfer. Furthermore, technological centres with an educational and training role (as well as a remit to develop foreign contacts) were supported by partners such as Washington University. Business has been fostered by new institutions like the Romanian Chamber of Commerce and the Romanian Development Agency with responsibilities for stimulating business.

Policy for agriculture

Change occurred relatively quickly in agriculture because the cooperatives were extremely unpopular in many areas and there was a strong desire for land restitution to restore family farming. The NSF had little option but to respect the near-universal desire to reclaim family land but there was a priority to retain the structures of cooperation if possible. The study of restitution shows that people often did not wait for legislation but simply took their land back, because they knew precisely where each piece was situated and there was an overwhelming sense of justification for direct action. It seems that even unilateral seizures of land from state farms were tacitly accepted by the authorities. Although some associations were formed as successors to the communist cooperatives, the local authorities were unable to prevent the de facto collapse of the system in many areas where attacks by the peasants gave rise to minifundia.

Unilateral seizures were then legalised through legislation allowing restitution holdings of 10 ha arable equivalent. This decision left the vast majority of restitutions on the right side of the law, while allowing a large section of the peasantry to own land: at the very least through the symbolic 1.0 ha plots (0.5 ha in the case of families who were not members of cooperative farms). Where there was insufficient land in the former cooperative sector to satisfy restitution, claims could be transferred to state farms by allocating appropriate shares of farm production, making the financial share-out increasingly fragmented. Thus between 1991 and 1994 Romanian state farms 'became veritable rubber sacks, their capacities stretched in some cases well beyond those implied by the farms' actual surfaces' (Verdery 1994: 1081).

These actions were very important for the NSF politically. When the centre-right opposition – gradually consolidating as the DCR – advocated restitution on a more generous scale that would also benefit the former owners of small estates, the government was able to hold out the spectre of further expropriation of smallholders (or alternatively their bankruptcy in the face of radical market reforms). Peasant values, trivialised under communism to provide propaganda against Western consumerism, make for sustainability through unrivalled experience reinforced by cultural-religious values that emphasise the permanent stake of the family in a specific parcel of land. There is, however, considerable waste of investment through the redundant buildings of former cooperative farms and peasant individualism is currently holding sway without the necessary complementary forces of either modernisation or ecological coordination. Nevertheless, there was a presumption (fed to no small extent by government propaganda) that unfettered market forces would drive many small farms out of business and result in much larger capitalist holdings, perhaps controlled by the landowners who were expropriated after the First World War and whose claims might be accepted under the more generous restitution terms promised by the opposition. Hence the NSF and its successors controlled the rural areas – almost solidly in Moldavia and Wallachia – and this was the key to its electoral successes until 1996. But at the same time it became clear that commercial farming would have to emerge through consolidation of small plots and an extension of the farm associations. But little was done to accelerate this process. There were many constraints on efficiency including the lack of a land market that could enable successful farmers to enlarge their holdings.

Romanian agriculture was competitive enough for Fructexport to send tomato juice to Heinz in the UK for making ketchup. But for many more deals of this kind an improved infrastructure was needed, with modern processing to generate competitive exports. But agricultural credit was in short supply and many of the new owners who benefited from land restitution lacked the know-how and technology to produce for the market. Government was primarily concerned in keeping food prices low, so Romanian farmers had to compete with cheap imports. Hence the market price of sugar might remove the incentive for some peasant farmers to produce sugar beet. Price controls were also imposed in the interest of consumers but private farmers refused to sell to the processors when the fixed prices were below production costs. Moreover, subventions and price controls – and ambivalence over cheap imports – placed the currency under pressure and discouraged investment in agriculture and food processing. Thus yoghurt was imported from Turkey while the milk price was controlled because money could be profitably invested in milk processing. Meat processing was also depressed (only breweries did well since beer prices were not controlled). The control of egg prices plunged Romania's 43 state poultry farms into crisis in 1996 when prices for barley and protein supplements rose sharply. And the tobacco monopoly initially impeded the manufacture of Western cigarettes in Romania; whereupon illegal imports – under mafia control – were sucked in.

Meanwhile, private farmers, especially those in the hill and mountain zones, placed subsistence at the heart of their coping strategies. They had limited cash for inputs and minimised purchases of fertiliser and hire of machinery, while maximising the use of available family labour in pastoral farming (for which machinery is not required) despite some local increases in grazing pressure: there was a sharp overall decline in livestock numbers but virtually no fodder was brought into the mountains following the end of communist production plans. The peasant farmers would seek a greater maize output by cutting down fruit trees on restituted land and restricting off-farm sales to livestock, using either the state marketing system or direct sale to a local abattoir. However, Romania can never be a peasant society in the way it was before communism and family farms will only attract part-time work when waged employment was available. The approach is one of pluriactivity wherever possible, with government helping to sustain this approach by picking up the 'mountainology' agenda of alpine regions through a Commission – later Agency – for Mountain Regions dedicated to intensification (linked with research on Carpathian farming and encourage-ment of local processing and diversification) combined with the growth of agrotourism and production of handicrafts. This was a significant initiative which spawned a number of farming associations and twinning arrangements with the West, though it failed through insufficient promotion, expertise and market stimulus.

The 1996–2002 period

There was a change in government to a DCR-led coalition in 1996. Despite engineering an economic boom through heavy borrowing (capitalising on the low debt levels of the early 1990s), the election took place under conditions of inflation, devaluation and industrial slowdown related to a widening trade deficit and growing stocks of manufactured goods. At the same time the value of the minimum wage continued to fall and agriculture suffered a poor harvest after a difficult planting season beset with delayed provision of credit. Also the PSDR was hit by scandal through party links with the state-owned foreign trade bank Bancorex and misuse of bank funds with respect to unsecured loans. Meanwhile, the DCR offered a vigorous reform programme which paved the way for a new governing coalition, while their presidential candidate (E. Constantinescu, running for the second time against Iliescu) was also effective in promoting his own programmes for educational reform, cultural transformation and moral leadership, including a resolution of ethnic tensions following the treaty with Hungary entered into during the last months of the outgoing government.

The new government was aware of the fact that, unlike the ECECs to the north, Romania had a weaker capacity for export to the West and a low level of consumer spending (despite a large total population): hence weaker FDI. Romania needed to gain 'policy credibility' after six years of inconsistency (grounded in a lack of pre-1989 reform experience) had failed to build momentum. FDI had a crucial role for restructuring – without which intense strain and instability would continue – and needed stimulation through an improved business environment and privatisation opportunities. At the time, financial discipline with hard budget constraints was a precondition for IMF support and investor confidence. The new government had a mandate for reform and yet remained acutely sensitive to the risks of rising unemployment arising from hasty reorganisation of the SOEs. The balancing act was played out by a sequence of three prime ministers (V. Ciorbea, R. Vasile and finally the National Bank chief M. Isarescu) who experienced considerable pressure from Democratic Party members of the coalition who pressed strongly for reform while tending to work to personal and party agenda.

An initial surge in FDI evident in 1996 and 1997 was not maintained. The year 1996 saw liberalisation of foreign exchange, but greater inflation and currency devaluation than was expected, and also a fall in GDP since growth was stopped by very tight credit conditions: Romania suffered a second trans-formational recession. Then there was turbulence arising from Russia's economic crisis which also impacted negatively on the country's prospects bringing further GDP decline and rising unemployment, and also increased sensitivity over external payment default, the adequacy of central bank exchange reserves and the danger of banking crisis brought on by bad loans. Following these setbacks there was a resumption of reform in the second half of the government's term with resources to implement a radical restructuring of the mining industry. And while some weakness had previously been shown in dealing with the unions, it confronted the miners effectively in 1999 and loss-making mining companies – like Pitcoal Jiu and the National Coal Company Ploieşti – were restructured, despite strong opposition by PSDR and nationalist parties on the grounds of deprivation and unemployment. Growth resumed in 2000 with the prospect of the much-needed switch in fortunes from a 'vicious' to a 'virtuous' circle, with lower deficits and interest rates stimulating more FDI and more competitive exchange rate linked with privatisation. But the upward trajectory was not enough to prevent a disaster for the DCR which not only lost power to the PSDR but failed to regain significant parliamentary representation. The attempted reforms of 1996–2000 had delivered growth only belatedly although the situation was not helped by a 'failure of coalition'.

Meanwhile, the government made considerable strides in Europe, having taken over from its predecessor a position of initial ambivalence which gradually evolved to the point where integration with the West reached the top of the polit-ical agenda by 1995. Following the improved relations with Hungary, outstand-ing disputes with Ukraine (going back to the Soviet annexations of 1940) were settled, thereby demonstrating that 'as Romania reaches maturity as a distinctly post-communist country, national values are being defined less in the nationalistic and chauvinistic terms which dominated the early 1990s and more in line with Western expectations and the Western "civic" model of national identity' (Light and Phinnemore 2001: 290). However, the EU 'Copenhagen Criteria' constitute a set of tests of a candidate country's fitness for entry. These include the issue of judicial and administrative capacity and human rights applied across civil society as a whole and with respect to minority groups – especially the Hungarians and Roma – but they are primarily economic: a functioning market economy able to cope with competitive pressures and the process of economic and monetary union. The EU's review in 2000 concluded that Romania could not be regarded as a functioning market economy able to cope with the competitive pressures of the European market. Despite some progress with regard to a competitive private sector and a functioning land market, the overall economic prospects were not improving substantially. However, approximation of the 'acquis' led to the formulation of a regional policy and the establishment of eight large regions (groups of counties) as a basis for disbursement of EU and state funds and the coordination of local development initiatives (Figure 3.7).

However, Romania's role in the Black Sea Economic Cooperation demonstrates her interest in European institutions, while involvement with the Balkan Stability Pact is seen as a way of scoring points in Brussels (though ties with Central Europe – of which Romania claims to be a part – are considered far more significant for the country's continued modernisation). Romania is now actively negotiating for membership – though she will almost certainly not feature in the first wave of eastern enlargement – and the massive task of approximation could be eased by a 'European' focus in every ministry – rather than a European super-ministry which does not seem to work well. Meanwhile, Romania has also made great strides to join NATO; so much so that in 1996 reforms were implemented very steadily to avoid major social problems that might compromise the bid for membership that was made at the time. Although this bid did not succeed the door has been left open and strong support for NATO actions in Kosovo was intended to enhance membership chances. Romania's position is problematic. There is considerable poverty and very limited purchasing power to justify investment except on the basis of production for export. The indigenous sector is not making much headway and income is much more dependent on remittances (e.g. from Israel) and FDI than the ECECs to the north. The population is however disillusioned because living standards have fallen and GDP decreased in three successive years (1997–9) to finish the decade at only 74 per cent of 1989 level (80 per cent in the case of GDP per capita). More people have become alienated from politics (election turnout was lower in 2000 than 1996) and wish to leave the country. It could be that there is still a tension between external pressure for reform and the domestic problems that inhibit change: a 'flagrant contradiction between the economic and social realities of Romania and economic policies arising through pressures exerted by the IMF and other international financial bodies' (Iancu 2000: 6). However, nationalists have only a small stable vote and there is no immediate risk of extremism.

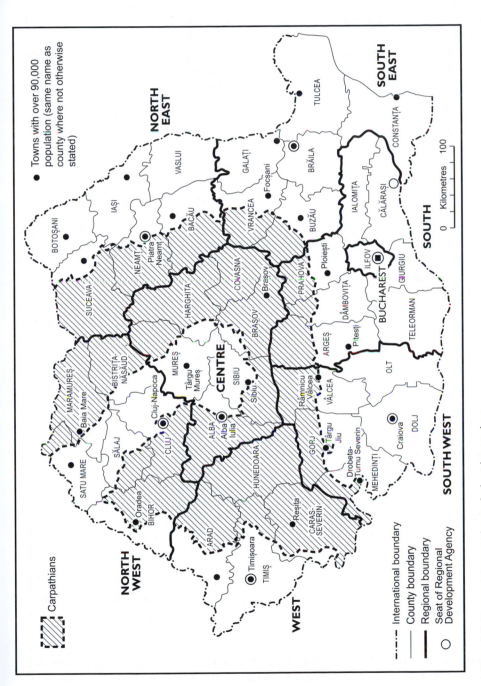

Figure 3.7 Romania's counties and planning regions

Source: Regional Development Authority, Bucharest

As already noted the DCR lost decisively at the election of 2000 and the leading party, the National Peasant Christian Democrats, were excluded from the present parliament and have only just begun to regroup under A. Marga, though a union with V. Ciorbea's Christian Democrat National Alliance is a useful beginning. The return to power of the PSDR under a further Iliescu presidency raised the spectre of a brake on privatisation and a reassertion of 'Securitate' economic control (for this was the party in which most of the 70,000 former secret police officers – and their hundreds of thousands of collaborators – found a new political home), yet the party also has its modernising wing – against continued state ownership indefinitely – and a deal with the unions will improve workplace safety, reduce the grey economy and stop the appointment of managers to SOEs on the basis of political connections. And if the appointment of a new leader – T. Basescu – to the Democratic Party improves relations with the PSDR this could strengthen the modernisers in government. In return for job creation to keep unemployment below 10 per cent, unions accept that wage rises must be linked with productivity. Reforms relating to land restitution and the conduct of local government business in the languages of ethnic minorities are going ahead, although there is some strain in Romania over a Status Law passed in Hungary which gives all Hungarians rights to employment, education and medical facilities: a measure considered divisive in Romania and Slovakia but enacted to head off an emigration stampede in the wake of EU membership from the east by offering short-term access. Meanwhile, negotiations with Brussels continue and the defence and foreign ministries are doing their best to inspire confidence abroad in the hope of gaining admission to NATO at the Prague Summit in 2002.

Policy for industry

The later 1990s saw much greater concern for privatisation that would enhance FDI and stimulate exports. The SOF has now privatised over 5,500 former SOEs and the private sector now accounts for over 55 per cent of GNP. This is a significant achievement given the Russian crisis and the fact that many enterprises were unattractive to members of the Romanian diaspora. Important deals include the sale of 51 per cent of the Romanian Development Bank to France's 'Société Générale', while the massive Daewoo investment at the former Oltcit factory in Craiova found an echo in the Dacia car factory at Piteşti-Colibaşi which was able to launch a new model thanks to SOF funding and subsequently benefited from a major shareholding by Renault. Siemens took over Elcaro Slatina, a producer of electric cables, so that exports could start after technological updating. The Greek company OTE bought 35 per cent of Romtelecom and restructuring has been carried out. On the other hand the electricity company has been burdened by customer debts and an inefficient distribution system which consumes half the power generated (while industry suffers from high energy and imported raw material costs). And while the SOF should have sold all their holdings by the end of 2000 there were still 3,100 enterprises early in the year. Indeed, there are few takers for most SOEs and the money paid takes the form of guaranteed investment rather than payment for the assets as such. Thus 51 per cent of the equity in the Mangalia '2 Mai' shipyard was sold to Daewoo for $51 m., though the deal really amounted to $20 m. in cash plus Korean components of ships to be assembled locally using Romanian labour.

Investment is often frustrated not only by bureaucratic delays but by union resistance to a high level of redundancy needed to raise productivity. Some privatisation deals may be delayed for years while gradual lay-offs reduce the workforce to a level that private owners can broadly accept. Privatisation of tourism has been surprisingly slow, but the industry consists of large inefficient units. Much of the progress has been by EMBO which helps to motivate the people who run the business but does not facilitate investment. With heavy losses among many of the remaining SOEs, the government was driven to contemplate the liquidation of some 50 prominent loss-making companies if they could not be privatised in 1998–9. There was external assistance for this radical stance and also for the parallel drive to reduce the losses sustained by the state mining companies based in Baia Mare, Deva, Petroşani and Ploieşti. The reduction in the workforce has been balanced by special fiscal measures to stimulate investment in LFAs that were set up in specific mining areas (Figure 3.8). This has stimulated some development among SMEs but it has also influenced FDI decisions and therefore constitutes a timely intervention to moderate the spatial polarisation evident hitherto and still quite highly concentrated on Bucharest and the western provincial city of Timişoara. There could be further intervention through the designation of industrial restructuring areas, especially in Moldavia which has been particularly neglected over the past decade. Investment might again be guided by fiscal incentives and the promotion of individual areas on the basis of raw materials, skills and infrastructure for specific industrial sectors.

There has also been renewed interest in joint ventures as a response to the compromising of traditional Romanian light industries that have been choked by imports of confectionery, soft drinks, textiles, plastics and construction materials. Also the limitations of Western 'storehouse industries' – such as Coca-Cola, Colgate-Palmolive and McDonalds – which meet internal demand but without a commensurate effort to develop production for export. Although the trend was evident before 1996, increasingly ways have been found of introducing new technologies that will generate competitive exports. The Câmpulung 'Land Rover' builder ARO is updating its engines through collaboration with Austria, while Romaero of Bucharest has been making subassemblies for a Canadian aircraft and IAR Brasov has a cooperation agreement with Franco-German Eurocopter (Aerospatiale and Daimler-Benz Aerospace) for the assembly of Ecureuil helicopters. McDonnell-Douglas has explored a deal with Avioane Craiova for the production of components and subassemblies which could be exported to pay for the purchase of aircraft by the Romanian airline TAROM. The Turnu Măgurele chemical fertiliser complex forged a link with Columna Bank and Euro Trading Chemicals to provide capital for essential equipment to eliminate sulphur dioxide pollution and secure new export markets. Meanwhile, Romanian shipbuilding has benefited from partnerships with foreign

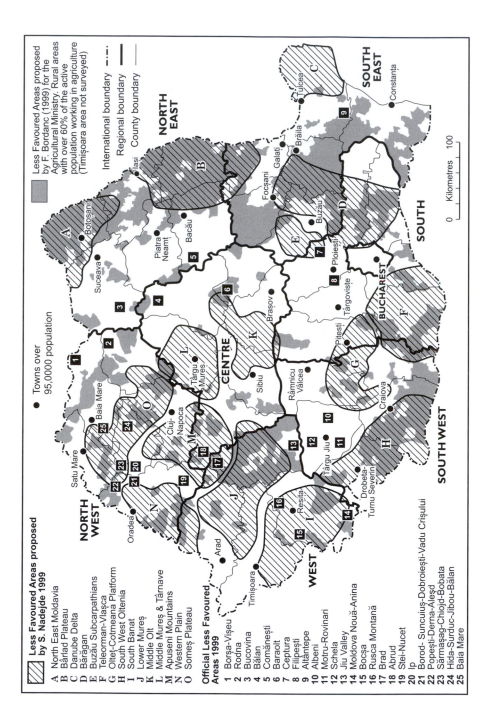

Less Favoured Areas proposed by S. Nadejde 1999

A North East Moldavia
B Bârlad Plateau
C Danube Delta
D Bărăgan
E Buzău Subcarpathians
F Teleorman-Vlaşca
G Olteţ-Cotmeana Platform
H South West Oltenia
I South Banat
J Lower Mureş
K Middle Olt
L Middle Mureş & Târnave
M Apuseni Mountains
N Western Plain
O Someş Plateau

Official Less Favoured Areas 1999

1 Borşa-Vişeu
2 Rodna
3 Bucovina
4 Bălan
5 Comăneşti
6 Baraolt
7 Ceptura
8 Filipeşti
9 Altântepe
10 Albeni
11 Motru-Rovinari
12 Schela
13 Jiu Valley
14 Moldova Nouă-Anina
15 Bocşa
16 Rusca Montană
17 Brad
18 Abrud
19 Stei-Nucet
20 Ip
21 Borod- Suncuius-Dobroieşti-Vadu Crişului
22 Popeşti-Derna-Aleşd
23 Sărmaşag-Chiojd-Bobata
24 Hida-Surduc-Jibou-Bălan
25 Baia Mare

Less Favoured Areas proposed by F. Bordanc (1999) for the Agricultural Ministry. Rural areas with over 60% of the active population working in agriculture (Timişoara area not surveyed)

International boundary –··–
Regional boundary ▬
County boundary ▬

● Towns over 95,0000 population

0   100
Kilometres

Figure 3.8 Romania's Less Favoured Areas

Source: Regional Development Authority, Bucharest

shipowners, resulting in export opportunities for ship bodies especially to Belgium, Germany and Norway. Braila shipyard is building Arctic trawlers for Norway using the German 'Westfalia'-type engine. Installation of high-performance equipment by the clothing factory Tricotton of Panciu has enabled the plant to regain traditional links with Germany, Italy, Sweden and USA, paving the way for further modernisation. SOF credits are helping Electroprecizia Săcele to undertake the manufacturing of electrical subunits for Daewoo Cielo cars, while Posada of Curtea de Argeş has a joint project with Gantner Electronics of Austria producing security cards with imbedded electronic chips.

However, investment has been very disappointing by comparison with the northern states which form the core of CEFTA and have made greater progress with economic reform. It is possible that the effect of Romania's economic weakness has been reinforced by the complexities of the bureaucracy and frequent changes in regulations which potential investors can find highly discouraging. A more stable regime for entrepreneurs combined with a single 'one stop shop' for FDI would certainly be an improvement. Meanwhile, exports to the EU show a growth in the share of labour-intensive exports, with growth heavily dependent on textiles and footwear with little value-added and very sensitive to competition (when wages rise above $100/month). Moreover, Romania may well be involved in only one section of the production process: sewing uppers for Western shoe manufacturers and making up items of clothing. In turn, cheap goods from Turkey have eroded the home market for Romanian textile manufactures, as have imports of secondhand clothing from the West. But it is through exports, rather than consumer spending, that Romania has potential for rapid growth over the next decade if the economic infrastructure can be improved, including international railfreight, and a new generation of technically advanced industries can evolve. There is, for example, buoyant foreign demand for Romanian furniture, based on strength and price, which could be increased still further by assimilation of new finishing technologies and greater emphasis on highly processed goods. In this connection several factories are now obtaining funds from EU or SOF sources. The markets of Russia and the CIS also provide opportunity for exports on the top of the revenues that accrue from the Russia–Turkey pipeline (which crosses Romania) and future projects associated with Caspian oil.

Despite some fears that the return of the PSDR would signal a slow-down on privatisation, the new government has made good progress in utilising the World Bank Private Sector Adjustment Loan (PSAL) negotiated by its predecessor in 2000. Recent successes have included the Sidex iron and steel complex in Galaţi and the Oltchim chemical complex in Râmnicu Vâlcea. The latter has been able to increase production after a technological overhaul but is embarrassed by heavy debts. The agricultural business will be hived off while the new owner, Exall Resources of Canada with a 53 per cent stake, will need to negotiate a price for the major input – ethylene – supplied by the state-owned Petrom refinery in Pitesti. With substantial debt write-offs on offer, there is every intention of finalising the sale of 25–30 large companies by the end of 2001 including Electroputere, the electrical engineering firm in Craiova, and the drugs company

Antibiotice in Iaşi. Indeed the government is now working towards a second PSAL to cover another 30 major companies including the Commercial Bank. At the same time there are attempts to simplify bureaucracy for large investments exceeding $1m. Indeed, projects involving more than $10 m. will attract the support of a new Department for Relations with Foreign Investors providing a 'one stop shop' service. And with investment dipping in 2001 there is a new law providing for fiscal concessions (including exemption from customs duties and delayed payment of VAT) for direct investments with significant impact on the economy.

Policy for agriculture

There has been a greater emphasis on improved competitiveness. To overcome the high cost of credit all farmers have been given coupons (in proportion to the area of their holdings) to help pay for inputs. Foreign loans have been secured to extend mechanisation and irrigation, while the transfer of expertise through pilot projects has been accelerated, e.g. through a dairy cow breeding project at Bestepe (Tulcea). An exportable cereal surplus of 5 m.t (in the context of a total harvest of some 20 m.t) is considered feasible and has been expedited by the construction of a grain terminal at Constanţa. Attention has been given to new schemes of cooperation linked with more efficient wholesaling with contracts to stimulate investment. Improvements in processing have occurred, for the abandonment of price controls enabled distribution firms to bring milk to Bucharest from the Carpathians where it had previously been fed to pigs. And a higher level of business confidence has contributed to growth in other fields. For example, Sucmerom of Reghin, producing natural apple juice from local orchards, has financed new equipment from its own revenues and natural juice concentrate is now sold in Austria and Germany at advantageous prices. An EBRD credit for the Mureş dairy in Târgu Mureş has helped to upgrade production (with modern packaging for butter, powder milk and fluid products) and create an ice cream factory complete with sales vehicles and stands. Links between agriculture and food processing show some promise. The indigenous Tofan Group has demonstrated its skills in the field of privatisation and restructuring through a successful takeover of tyre factories which have been sold on to Michelin. The proceeds could well be invested in building up a private food empire from state farms and bakeries. The 20-year lease of Brăila's Insula Mare to the Romanian company Trei Brazi, conditional on a seven-year investment programme and delivery of up to a fifth of the production, is another indication of private capital finding its way into agri-business. This is particularly significant as evidence of a continued entrepreneurial approach by the PSDR government after its election victory in 2000.

However, the great majority of farms continue to be small family units geared to subsistence and structural change is very slow. Consolidation hinges on the balance between subsistence value of land and value for efficient market production (associated with more waged employment and stimulative agricultural markets attracting more people to set up SMEs). The present position is unstable but current poverty levels mean that there is no chance of rapid change in the immediate future. Indeed the 2000 restitution law allows for the recovery of holdings of up to 50 ha arable equivalent and 10 ha of woodland. Instead a broader rural policy has evolved. Rural employment – totalling 5,904,000 in 1998 – is dominated by the primary sector: 4,289,200 people work in agriculture, forestry and fishing (of which 4,132,900 are self-employed, including family members) with 677,200 in mining, manufacturing, energy and water; 138,200 in construction and 799,400 in the tertiary sector. And especially since 1998 the imbalance has been aggravated by redundancies in the mining sector through the restructuring programme. Rural services are relatively poor and access to education is described as 'precarious' through reduced demand and high share of non-qualified teaching staff (Ionete and Dinculescu 2000: 81). The falling birth rate is linked with a falling marriage rate, but the low level of social protection means that children increase the risk of a family being pushed below the poverty threshold. Low incomes lead to the purchase of cheaper food products which means consuming more bread and low-quality cereals. Sociologists have therefore raised the problem of exclusion of rural people from the market economy and the sharp reduction in the number of salaried employees. They have also highlighted the problem of preserving the traditional work ethic in the countryside, rooted in the needs of agriculture. There are problems of leadership: although there were almost 33,000 NGOs registered in 1997, only 400 were nationally effective and very few of these were based in the least-developed rural areas where initiative depends on influential families who are 'liderii de opinie' and spiritual mentors for their villages.

The former emphasis on diversification in the Carpathians has been modified because the rapid run-down in factory employment in lowland areas of the northeast and southwest has reinforced extremely high levels of dependence on agriculture, and made some mountain regions appear privileged by comparison. Thus the agency (referred to above) has been disbanded and all that remains (apart from the local research and marketing organisations) are the special programmes for some mountain areas (e.g. the Apuseni Mountains) and the initiative for agrotourism which has been taken up by Romanian and international NGOs and the Tourism Ministry but without close links with agriculture. At the same time increasing involvement with the EU as a candidate country made for the formulation of an overall rural planning concept. Multidisciplinary research on rural problems through the 1998 PHARE-sponsored 'Rural Development Project' brought together two Bucharest institutes and four universities for an analysis of strengths, weaknesses, opportunities and threats (SWOT) to distinguish between areas with good potential for development and other areas with restrictions. This provided the basis of an official 'green paper' in 1999, followed by a national programme for rural development. This aims at a better quality of life in rural areas, with more entrepreneurship and cooperation between sectors, which will attract cohesion funding especially through the EU pre-accession funding (ISPA and SAPARD) which will bring in up to E540 m. annually during 2000–6. Progress is to be made in part through a network of some 500 advice centres which will provide agricultural information and help with entrepreneurial skills and limited credit for farming inputs. Funds will also be directed into improvements in rural households, with a stimulus to marketing and food processing outlined in the green paper. However, the money on offer is quite small when set against the scale of improvements needed and it remains to be seen what can be achieved both in the short term and also after EU accession when it is unlikely that the CAP will be sustainable in its present form. But the rural centres (small towns and key villages) are showing signs of dynamism through the growth of shops and other small businesses.

In any case, it will take time for results to show up in terms of enhanced infrastructure and investment attractiveness and in the meantime policy must consider 'firefighting' in areas of greatest need. These can be defined in terms of high level of dependence on agriculture (exceeding 60 per cent of all employment) and/or 'underdeveloped areas' (identified on the basis of geography, demography, economy and social issues) to reveal 15 rural regions where four or more criteria apply (with the great majority lying outside the mountain zone) (Figure 3.8). Although the commercial development of agriculture will help some of these areas – notably through grain exports from impoverished lowland areas which paradoxically have the best land resources – such high levels of dependence are plainly unsustainable in the modern world. Further work has been done in these areas by the United Nations Development Programme, cooperating with the Romanian government during 1997–9 with regard to a national poverty-alleviation strategy which included regional gender-empowerment projects in rural areas with experiments in setting up pilot food-production units operated by women. There has also been World Bank support for a $45 m. social development fund to help with rural housing, infrastructure (electricity, roads and water) and community centres, with almost half the money earmarked for disadvantaged rural areas. However, the problem remains to select a methodology which may be used to identify 'poor' or 'under-developed' areas in which a special regime of support can be devised that may apply over the longer term.

4 Tertiary sector geographies

Transport, energy and tourism

The tertiary sector

This chapter deals with transport and tourism, as well as energy which is often seen as an extension of the secondary sector, while retailing is covered in Chapter 5 as part of the urban scene. But, by way of introduction, it is worth emphasising that probably no transformation has been as dramatic as the growth of the tertiary sector, especially if construction is discounted. It has to be said that different procedures for calculating contributions to national output (national income in 1989 and gross value-added for the 1990s) create problems of comparability while the occupation figures have to allow for the substantial contraction in the total employment through the scale of de-industrialisation and the difficulty of showing agriculture realistically when there is so much non-salaried and part-time work (thus the 1989 figure for Romania is probably too low in comparison with 1994 although some increase might be expected). Nevertheless the increases registered by services do point to a significant economic trend which brings ECE much closer to West European profiles. In terms of employment as well as value-added, the service is now the largest of the four shown – with Albania the only exception, although in Romania the employment share is still well below the 50 per cent threshold which virtually all the other countries have crossed (Table 4.1). It is worth emphasising that the tertiary sector tends to be strongest in large cities where agriculture is virtually non-existent and where certain administrative, commercial, educational and financial services are heavily concentrated.

The regional breakdown for Romania (Table 4.2), which fits the areas shown in Figure 3.7), shows the sector (including construction) accounting for 53 per cent of both GDP and salaried employment nationally, but in Bucharest this proportion rises to 64 per cent for salaries and 71 per cent for GDP. By contrast in the industrially-strong Central region only 46 per cent and 48 per cent respectively are recorded. Figure 4.1, covering the Romanian Carpathians, shows that tertiary employment was still weak in rural areas at the time of the 1992 census, accounting for less than a fifth of the total in some districts. By contrast the level rose above 40 per cent in the towns, but rarely exceeded 60 per cent. The map also shows a major shift in the balance between agriculture and industry and mining between 1966 and 1992, although there was still a majority in agriculture in the latter year in most rural areas. Meanwhile, the urban areas show a massive industrial lead.

Table 4.1 Major economic sectors (%) (a) by national income (1989), gross value added (1994–9) and (b) by employment

(a)	Agriculture			Industry			Construction			Services		
	1989	1994	1999	1989	1994	1999	1989	1994	1999	1989	1994	1999
Albania	na	na	na	na	na	na	na	na	na	na	na	na
Bulgaria[1]	13.1	15.4	17.3	59.5	25.9	23.1	9.4	4.3	3.7	18.0	54.5	55.9
Croatia[1]	11.4[2]	10.7	8.9	50.4[2]	28.4	25.4	5.8[2]	5.9	7.1	32.4[2]	57.9	63.2
Czech Rep.	7.1[3]	4.9	3.7	60.6[3]	33.6	34.3	10.9[3]	7.4	7.5	21.4[3]	54.1	54.5
Hungary	13.8	6.7	4.9	50.1	25.3	28.0	12.6	5.1	4.6	33.5	62.9	62.6
Poland	14.0	6.9	3.8	48.2	31.1	27.7	12.9	7.4	8.9	24.9	54.6	59.6
Romania	15.7	20.6	15.5	58.1	37.6	30.9	7.2	6.8	5.4	19.0	35.0	48.2
Slovakia	na	5.3	4.5	na	34.7	29.3	na	7.5	5.8	na	52.5	60.4
Slovenia	na	4.5	3.6	na	34.7	31.4	na	4.7	6.1	na	56.1	58.9

Source: Eurostat 2001: 509

Notes
[1] Bulgaria 1996–9 and Croatia 1995–8; [2] former Yugoslavia; [3] FCS

(b)	Agriculture			Industry			Construction			Services		
	1989	1994	1998	1989	1994	1998	1989	1994	1998	1989	1994	1998
Albania	na	67.2	70.8	na	9.5	7.8	na	1.6	1.0	na	21.8	20.4
Bulgaria	19.3	23.2	25.7	38.0	29.1	26.4	8.3	5.7	4.4	34.4	42.0	43.5
Croatia[1]	28.7[2]	19.9	16.5	23.6[2]	22.9	23.5	7.3[2]	6.2	6.7	40.4[2]	51.0	53.3
Czech Rep.	10.8[3]	6.9	5.5	37.3[3]	33.4	31.5	9.9[3]	9.3	9.8	42.0[3]	50.4	53.1
Hungary	20.0	8.7	7.5	30.4	27.6	28.0	7.0	5.4	6.2	42.6	58.3	58.2
Macedonia	na	8.7	7.5	na	27.6	28.0	na	5.4	6.2	na	58.3	58.2
Poland	27.7	24.0	19.1	29.0	25.8	25.0	7.8	6.2	7.0	35.5	44.1	48.9
Romania	28.9	39.0	40.0	38.1	28.7	25.4	7.4	4.1	4.0	25.6	28.1	30.6
Slovakia	na	10.1	8.2	na	30.7	30.2	na	8.9	9.3	na	50.2	52.3
Slovenia	na	11.5	11.5	na	36.6	33.7	na	5.6	5.6	na	46.2	49.3

Source: Bachtler 1992b; Eurostat 2000: 63–7; Romanian Statistic Yearbook 1990

Notes
[1] Croatia 1996–8; [2] former Yugoslavia; [3] FCS

Table 4.2 Romania: the service sector by regions 1998–9 (a) by salaried employment 1999 and (b) by gross domestic product 1998

(a)

Region	Sector: Primary	Secondary	Services							
			A	B	C	D	E	F	G	H
North East	4.97	42.08	5.84	14.67	6.00	1.12	2.52	10.85	7.50	4.45
South East	7.63	35.62	6.94	15.41	10.82	1.25	2.68	7.90	6.45	5.30
South	7.09	45.49	6.02	13.28	6.63	1.18	2.84	7.70	4.80	4.97
South West	5.02	41.04	7.61	15.05	7.56	1.79	3.37	9.02	5.55	2.96
West	5.74	42.31	7.05	13.90	7.52	1.27	3.05	8.25	5.82	5.09
North West	3.68	44.53	6.07	13.95	7.17	1.37	2.62	9.99	6.82	3.80
Central	4.88	48.76	5.81	13.94	6.21	1.22	2.33	8.22	5.01	3.62
Bucharest	1.36	34.19	7.05	12.63	10.96	2.89	4.46	7.83	5.56	13.07
Romania	5.04	41.82	6.49	14.07	7.86	1.45	2.96	8.72	5.94	5.65

(b)

Region	Sector: Primary	Secondary	Services							
			A	B	C	D	E	F	G	H
North East	22.25	27.51	4.92	12.96	8.65	1.45	5.06	3.81	2.81	10.58
South East	16.93	31.25	6.38	14.12	9.21	1.43	3.85	2.37	2.02	12.44
South	20.41	33.02	4.87	11.16	8.68	1.45	4.53	2.69	2.33	10.86
South West	20.58	32.15	5.80	11.44	8.31	2.11	4.32	3.27	2.39	9.63
West	18.45	25.25	5.45	13.99	10.51	1.47	3.72	2.71	2.56	15.89
North West	17.53	28.96	4.72	12.64	10.13	1.64	4.10	3.82	2.75	13.71
Central	14.95	37.07	4.93	13.99	9.00	1.48	3.61	2.83	2.15	9.99
Bucharest	1.46	26.79	6.70	22.89	14.98	3.76	2.25	2.28	1.88	17.01
Romania	15.81	30.30	5.51	14.55	11.13	1.92	3.89	2.93	2.33	12.63

Source: Institutul National de Statistica 2001, Anuarul statistic al Romaniei 2000 (Bucharest; INS): 656–9, 718–19

Notes

A   Construction   E   Administration
B   Trade and Hotels   F   Education
C   Transport and Telecommunications   G   Health
D   Finance   H   Real Estate and Others

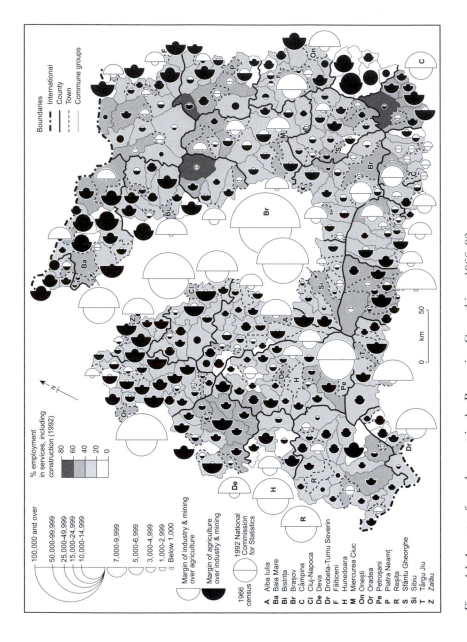

Figure 4.1 Aspects of employment in the Romanian Carpathians 1966–92

Source: Romanian Census 1977 and 1992

Transport

The communist period

Transport was crucial for communist industrial development but, with 'tight' central planning, capacity increased only slowly in the wake of increased demands exaggerated by a high level of transport absorptiveness caused by a policy of industrial specialisation and by long hauls for many raw materials and intermediates that were not realistically costed. Meanwhile, there was enhanced passenger mobility, especially in urban areas and their commuter zones. There was a relative increase in the importance of pipelines, roads and shipping – along with air transport, geared largely to the needs of the administrators – but the outstanding characteristic, in contrast to Western Europe, was the emphasis on rail transport, especially for freight. Railway lobbies were politically powerful; so the hoarding of labour kept productivity low and there was little 'creative destruction' so characteristic of market economies. Hungary was quite exceptional in eliminating some 1,500 km of lightly-used rural lines (many narrow gauge), comprising some 15 per cent of the total network, between 1965 and 1985.

Railways

The northern countries inherited a dense railway network including many industrial lines. There was investment in new equipment, including electrification and double-tracking, including some restoration of second tracks on lines initially singled to meet Soviet demands for war reparations in East Germany. The importance of the main east–west line through Warsaw and Berlin was reflected in two projects advanced during the 1950s: the 'Aussenring', avoiding West Berlin, and a 160 km transit route running to the south of Warsaw. Other major works in Poland included the 225 km high-speed 'Central Trunk Line' (CMK) completed in 1977 to connect Upper Silesia with Warsaw (225 km) and the 410 km broad-gauge 'Steel/Sulphur' line opened in 1979 from the Soviet border to Sławków near Dąbrowa Górnicza on the edge of the Upper Silesian Industrial Region. In the SEECs there was a greater element of new construction, especially in Albania where all the country's present railways originated in the post-1945 period. The country's first external rail link was built between Shkodër and Titograd and was opened in 1985–6. The main projects in former Yugoslavia were the Belgrade–Bar line and Bosnian narrow-gauge conversions: the latter came first with the Šamac–Sarajevo line of 1948 followed by the extension to Mostar and Ploče in the 1960s. In Bulgaria a new, more direct link was provided from Sofia to Burgas and Varna while Romania completed more direct routes from Bucharest to Craiova and Iaşi and built a number of branches, especially in the Oltenian lignite field.

Road transport

Meanwhile, road transport gained ground throughout ECE with the surfacing of the main roads and a substantial increase in freight and passenger services.

There was heavy use of public transport while private car ownership remained low by Western standards, although the 1970s and 1980s did see considerable increases, especially with the production of relatively inexpensive vehicles like the Trabant and Wartburg cars manufactured in East Germany. However, while some duplication of road and rail did arise, integration was the overriding principle. So bus services were prominent in urban areas and in more backward rural districts, while road freight services were provided for short hauls to feed the railways. And while most national road systems had good surfaces by the 1980s there was little dual carriageway or motorway available: in East Germany the autobahn from Berlin to Rostock was the only substantial addition to the inherited system. Likewise there were few purpose-built urban ring roads and bypasses.

Shipping, inland waterways, pipelines and airlines

There was more shipping business for the Adriatic, Baltic and Black Sea ports and particular expansion at Rostock when the GDR decided in 1957 to minimise its dependence on West German facilities. Some of the traffic was linked with the Soviet Union because of ferry services to reduce pressure on the railway routes through Poland and Romania. In the Baltic the Klaipeda–Rostock freight service was replaced by the Klaipeda–Mukran (near Sassnitz) train ferry which was a precaution against mounting political instability in Poland in the Solidarity era. And in the Black Sea theatre the Ilichevsk (Odessa)–Varna ferry connected the USSR directly with Bulgaria. Fleets were built up through the construction of new ships, usually built in ECE yards. There was much discussion about the scope for greater use of the waterways which were grossly neglected in some countries. But the huge cost of connecting the Danube with the northern rivers (Elbe, Oder and Vistula) was off-putting as it had been to previous governments. However, there were improvements on the Danube when Romania and Yugoslavia collaborated over hydropower and navigation facilities at the Iron Gates, although a similar project planned for Gabčikovo–Nagymaros was incomplete in 1989 and remains a bone of contention between Hungary and Slovakia. Romania also proceeded with the Danube–Black Sea Canal (Cernavodă–Constanţa) during the 1980s, at huge expense, and with the completion of the Rhine–Main–Danube Canal in Germany there is now a waterway for 1,000 t barges between the Black and North Seas. Finally, reference should be made to the pipelines transporting Soviet crude oil and natural gas to ECECs (especially the northern group) and the network of air services connecting the capital cities with Moscow and Western Europe and domestic services to cities more than 200 km from the capitals.

Comecon coordination

On the whole, transport in ECE was characterised by poor infrastructural maintenance, especially for the road network with its inadequate load capacity, poor surfacing and deficient roadside servicing. It is also striking that Comecon did not make a greater impact, considering the crucial importance of transport for the bloc as a whole. In theory the programmes introduced by the Standing Committee for Transport should have produced a rational system, but investment strategies were too modest in scale and inappropriately organised, with insufficient emphasis on the reduction of costs and improvement in the quality of service. Comecon member nations were inured to isolation and the emphasis on economic autarky introduced into central planning in the early post-war years tended to marginalise the international dimension and restrict functionaries to domestic perspectives orchestrated by the propaganda of each national regime. There were ambitious plans for north–south trans-European railways and motorways but these were never finalised, although some national transport programmes did involve ad hoc contributions to these grand designs. Once again a critical breakthrough was delayed until after 1989.

There was Comecon coordination of the production of locomotives, trams, motor vehicles, aircraft and ships. But inferior technology was most evident in the case of Soviet-built aircraft with high fuel consumption and heavy maintenance demands; not to mention obsolete motor vehicles, continued heavy dependence on animal power and overloaded telecommunications. Western firms became involved in the production of motor vehicles in the 1970s and 1980s but urban pollution problems, aggravated by poor vehicle maintenance, was only belatedly relieved by the construction of metro systems in the capital cities. Most significant perhaps were the mobility constraints arising partly from the inadequacy of the transport system but also from bureaucratic measures. While internal movements became largely free of regulation (apart from Albania), the stringent controls on the issue of passports, visas and foreign currency continued to impede foreign travel, even between socialist countries. Special 'local border traffic regimes' might operate in some frontier districts, but frontiers were generally closed and tightly guarded, with the most elaborate precautions along East Germany's fortified border with the West.

Transition

Since 1989 the situation has been changing through the erosion of the old state monopolies through privatisation – leading to new management and improvements in technology, facilitated by foreign involvement – and the emergence of new private sector airlines and road transport businesses (both passenger and freight) which has provided competition and opened up new routes. The railways remain state-owned but private capital is finding its way into the inherited state airlines and road transport companies and where state control continues through a majority shareholding the change 'can result in economic behaviour which is better oriented towards profit, efficiency and subsidy minimisation' (Hall and Kowalski 1993: 29). Improvements are very evident in such areas as ticketing, catering and publicity, often aided by joint ventures with Western companies. Polish shipping companies have been divided into more flexible units and the old monopolistic structures (such as those embracing all aspects of trade and maintenance at seaports) are being reformed. Loss-making ships (large oil tankers and bulk carriers) have been sold in preference for small modern vessels. Meanwhile, new private airlines and road transport companies have emerged to handle the boom in international traffic. However, while individual travel is now quite free of bureaucratic obstacles (with relatively few journeys in Europe now requiring visas), the 'softness' of ECE currencies means that international travel is extremely expensive. Most impressive however has been the modal shift from rail to road exemplified by the shift in the railway share of East Germany from 75 per cent to 40 per cent between 1989 and 1992 brought about by declining industrial production and increased use of modern Western trucks. The rising cost of public transport combined with the relative ease of acquiring second-hand cars initiated another major shift throughout the region.

This section will review the various transport modes, noting that the extensive railway network is contracting slowly, by 2.3 per cent during 1994–8, while motorways and related highways have increased by 24.0 per cent, and the pipeline system by 12.3 per cent to exceed 16,200 km (Table 4.3). Navigable waterways appear to be in modest decline (by 3.4 per cent) although there are major projects of long-term significance. Planning is placing more emphasis on the international dimension than was the case under Comecon. There was a initial sense of competition between ECECs to develop main lines of communication that would stimulate commerce, including transit movements and the international conference trade; while Western neighbours sought to reclaim their former influence. However, with the EU-enlargement process dominating European politics, transport coordination is being taken in hand and Pan-European Corridors ('Eurocorridors') have now been agreed as priority investments for high-speed combined transport, though some modifications have been made to accommodate the warring factions in former Yugoslavia who were left out of the initial decision-making: Corridor 9 has been extended and a new Corridor 10 added (Figure 4.2). Frontier posts on the main road routes are now major commercial and administrative complexes especially on the borders with the EU, like Hegyeshalom – Hungary's main border crossing with Austria, with 15 m. passenger movements in addition to freight – where there is now a large shopping complex in addition to catering and banking services. There are also large 'dry ports' on ECE's borders with the CIS: Čierna nad Tisou in Slovakia, Záhony in Hungary, and, in the case of Poland, Małaszewicze near Terespol can handle 23,000 t of transhipment freight and offers a 166 ha duty-free business zone on the main road nearby. Great efforts are made to keep these corridors open and minimise delays. War in former Yugoslavia caused many problems for transit traffic and prompted the development of an alternative route between Greece and Austria through Bulgaria, Romania and Hungary with a ferry over the Danube at Calafat-Vidin where a bridge is now under construction.

Transition: railways

Railways remain in state ownership, but new management structures are in place to increase efficiency and reduce losses, with privatisation in prospect: for example in Poland where a substantial part of the inherited narrow-gauge system

Table 4.3 (a) Transport systems and (b) Road vehicles 1994–8

(a)	Length of system (km) and density 1998 (systems km/,000 km² of territory)							
	Rail		Motorway		Waterway		Pipeline	
Albania	1194	43.6	na	na	na	na	189	6.9
Bulgaria	4290	38.6	319	2.9	470	4.2	578	5.2
Croatia	2726	48.2	330	5.8	933	16.5	601	10.6
Czech Rep.	9430	119.5	498	6.3	664	8.4	736	9.3
Hungary	7642	82.2	448	4.8	1373	14.8	7201	77.4
Macedonia	699	27.2	144	5.6	na	na	na	na
Poland	23210	74.2	268	0.9	3812	12.2	2278	7.3
Romania	11010	46.2	113	0.5	1779	7.5	4629	19.4
Slovakia	3665	91.6	288	7.2	172	4.3	na	na
Slovenia	1194	58.8	325	16.0	na	na	na	na
Total	64586	64.3	2733	2.7	8977	8.9	16212	8.9

(b) *Number of vehicles (,000) and rate ptp*

Cars				Buses/Lorries[1]			
1994		1998		1994		1998	
67.9	21	90.8	27	35.8[2]	11	46.4	14
1587.9	188	1809.4	219	296.6	35	325.6	39
698.4	150	1000.1	222	72.5	16	120.6	26
2923.9	283	3493.0	339	218.7	21	300.3	29
2176.9	212	2218.0	219	318.8	31	355.7	35
263.2	134	288.7	145	22.4	11	24.5	12
7153.1	185	8890.8	230	1140.6	30	1644.4	42
2020.0	89	2822.3	125	362.5	16	455.7	20
994.0	186	1196.1	222	161.6	30	168.0	31
667.3	335	813.4	410	36.6	18	46.3	23

Source: Eurostat 2000: 165–87

Notes
[1] Includes road tractors [2] 1995

has been closed down. The Hungarian State Railways (MAV) have an ambitious plan for a market-oriented railway with reductions in subsidies. There will also be some privatisation, along with transfers to local authority management. There is much improvement on international routes thanks to new track, signalling and rolling stock for speeds of up to 200 km/hour. Supra-national information systems will be an important innovation, especially for freight with the monitoring of wagons to announce the arrival of consignments and the introduction of new rolling stock including low-deck trucks to carry TIR lorries. Frontier crossings are being upgraded, with customs formalities at main stations rather than the frontier itself. Investments are being made, especially on the principal international routes, including new trains facilitated by EIB and EBRD loans. Rail traffic is still declining for both freight and passenger traffic (Table

Figure 4.2 Eurocorridors and some additional Balkan routes: 1 Tallinn–Riga–Warsaw;
Riga–Kaliningrad–Gdańs; 2 Berlin–Warsaw–Minsk–Moscow; 3 Berlin/
Dresden–Wrocław–Lviv–Kyiv; 4 Berlin/Nurnberg–Prague–Budapest–
Arad–Bucharest–Constanţa; Arad–Craiova–Sofia–Istanbul/Thessaloniki;
5 Trieste/Koper/Rijeka–Ljubljana–Budapest–Uzgorod–Lviv; Bratislava–
Žilina–Košice–Uzgorod; Rijeka–Zagreb–Budapest; Ploče–Sarajevo–Osijek–
Budapest; 6 Gdańsk–Łódz/Warsaw–Katowice–Ostrava–Žilina; Toruń–Poznań;
7 Danube River Corridor with the connecting Rhine–Main–Danube Canal;
8 Durrës–Tirana–Skopje–Sofia–Burgas/Varna; 9 Helsinki–St Petersburg–
Kyiv–Chişinău–Bucharest–Plovdiv; Kyiv–Moscow; Kyiv–Minsk–Vilnius–
Kaunas–Kaliningrad/Klaipeda; Ljubasivka–Odessa; 10 Salzburg–
Ljubljana–Zagreb–Belgrade–Slopje–Veles–Thessaloniki; Graz–Maribor–
Zagreb; Budapest–Novi Sad–Belgrade; Niš–Sofia; Veles–Bitola–Florina

Source: European Commission, European Union

4.4). Only Slovenia shows a slight increase in freight while Bulgaria's rising share
for rail is to be seen in the context of a sharp decline in the total amount of
freight handled by all modes combined (Table 4.5). The latter situation also
arises for passenger traffic in Bulgaria, along with Romania and Slovenia and

the Czech Republic marginally (Table 4.6). Only Macedonia shows a rising share for rail in the context of a stable level of usage where rail and bus travel are combined.

High-speed services

On the main lines domestic services will achieve higher speeds of 120–200 km/ hour and this has major implications for track which is currently defective, especially in those areas of the Czech Republic, East Germany and Poland that have handled heavy coal traffic. The sleepers have started to disintegrate due to the excessive alkaline content of the concrete formula applied in the 1970s. Modern track maintenance machines are needed in Poland where coal dust (arising from the milling of coal in Silesia for power station use) falls from leaking wagons to the tune of 1.0 m.t each year. Meanwhile, track improvement is also needed in East Germany because only a third of the route length has double or multiple track compared with half in the West. Many sections of single-track railway are having a second track restored following its removal – usually for war reparations – in the communist period. Moreover, only a quarter of the route length in the ECE is electrified compared with 40 per cent in the West. Modernisation also raises problems of safety, for there are many collisions at level crossings because vehicle drivers do not observe the rules. Crossings will have to be eliminated by bridges or better protected by automatic control systems. Much of the older rolling stock has become redundant with the decline in freight traffic and there are also passenger coaches in the SEECs that provide only Third World standards.

In East Germany higher speeds will be possible on the radial routes around Berlin, including the line to Cottbus via Frankfurt/Oder and the semicircular route from the present Hamburg line through Magdeburg, Halle, Leipzig and Cottbus. In Poland fast trains run on the CMK between Warsaw and Kraków and there will be further benefits when the link to Wrocław is complete and the line is extended northwards from Grodzisk to Gdańsk. The Czech Republic's Prague–Brno–Bratislava route needs improvement and a new high-speed line may be the best solution to overcome the bottleneck posed by the tunnels between České Trebova and Brno. In Hungary, electrification is to be extended to additional frontier routes. There will eventually be a radial network of high-speed double-track electrified lines from Budapest to Debrecen, Györ and Miskolc (including the frontier crossings beyond), with single-track facilities on the domestic lines to Pécs and Szeged and the international routes to Arad (Romania) and Belgrade (Serbia). Inter-city services will also benefit from the introduction of high-speed 200 km/h coaches (ordered from Deutsche Waggonbau and CAF of Spain). Transit through Bulgaria requires electrification from Sofia to both Dimitrovgrad and Slivengrad in cooperation with both neighbouring countries (Yugoslavia and Turkey respectively). However, over the longer term magnetic levitation is seen as the best technical basis for a new high-speed network (superseding the classic 'wheel to rail' system) although only the Berlin–Hamburg line is currently contemplated.

Table 4.4 Railway traffic, monthly average 1986–2000

	Freight (m.t-km)								Passenger (m.passenger-km)						
	1986	1988	1990	1992²	1994	1996	1998	2000³	1988	1990	1992²	1994	1996	1998	2000³
Albania¹	na	na	na	4499	4389	3466	2117	2491	na	na	18562	17949	14022	9636	10854
Bulgaria	1527	1465	1177	647	648	624	513	442	679	649	449	422	485	398	294
Croatia	na	na	na	148	127	143	164	143	na	na	82	76	86	76	81
Fmr. Czechosl'	6263	6274	5360	na	na	na	na	na	1617	1612	na	na	na	na	na
Czech Rep.	na	na	na	3998	2033	2015	1627	1480	na	na	1404	707	676	583	597
Germany	na	na	na	5849	5915	5645	6133	na	na	na	4828	5111	5444	5219	na
Fmr. GDR	5746	5030	na	na	na	na	na	na	1922	na	na	na	na	na	na
Hungary	1841	1714	1398	835	642	636	675	609	897	950	765	709	715	740	756
Macedonia	na	na	na	48	13	23	34	43	na	na	9	6	5	12	15
Poland	10148	10184	6961	4814	5482	5694	5247	4469	4344	4198	2714	2301	2214	2139	2051
Romania	6591	6717	4771	2315	2059	2240	1642	1413	2887	2549	2023	1526	1530	1110	845
Slovakia	na	na	na	1391	1020	1001	887	892	na	na	454	379	314	260	243
Slovenia	na	na	na	210	204	213	257	227	na	na	46	49	51	54	53
Yugoslavia	na	na	na	480	111	172	220	138	na	na	245	210	153	143	98
Fmr. Yugoslavia	2298	2118	2037	na	na	na	na	na	964	921	na	na	na	na	na

Source: United Nations Monthly Bulletin of Statistics

Notes
¹ Data in thousands
² 1993 data for 1992
³ For 2000 calculations are made on the basis of the first six months only

Table 4.5 Freight transport 1994–8

	1994	% distribution:				1998	% distribution			
	m.t-km	Rail	Road	Pipe*	Wtwy	m.t-km	Rail	Road	Pipe*	Wtwy
Albania	na	na	na	na	na	1863	1.3	98.3	0.0	0.4
Bulgaria	40858	18.5	79.8	0.8	0.9	29293	20.4	76.9	1.9	0.8
Croatia	2910	53.8	43.7	0.1	2.4	5570	36.3	45.7	0.7	17.3
Czech Rep.	49750	45.8	47.4	2.4	4.4	55562	33.8	61.0	1.5	3.7
Hungary	15227	50.5	17.4	5.1	27.0	27102	30.1	46.4	5.8	17.7
Macedonia	1649	8.4	91.6	0.0	0.0	1302	31.3	68.7	0.0	0.0
Poland	125175	51.8	36.2	0.6	11.4	150019	40.6	46.4	0.7	12.3
Romania	47722	51.7	38.4	4.0	5.9	41973	47.0	37.6	10.0	5.4
Slovakia	17992	68.0	27.3	4.7	0.0	18031	65.2	26.3	8.5	0.0
Slovenia	4383	55.9	46.1	0.0	0.0	4762	60.0	40.0	0.0	0.0
Total	305666	47.1	43.2	1.9	7.8	335417	38.9	49.6	2.9	8.6

Source: Eurostat 2000: 165–87

Note
* Oil pipelines only

Table 4.6 Long-distance passenger transport 1994–8

| | 1994 | | 1996 | | 1998 | |
	A	B	A	B	A	B
Albania	412	52.2	391	43.0	306	37.9
Bulgaria	13000	38.9	10334	49.0	8581	55.2
Croatia	5002	19.2	5581	18.4	5137	17.9
Czech Rep.	20004	42.4	17846	45.4	15699	44.7
Hungary	17149	49.6	18346	46.8	19506	45.5
Macedonia	1100	6.1	1008	11.9	1014	14.8
Poland	61872	44.6	60553	43.9	59699	43.0
Romania	32371	56.6	31198	58.8	22384	60.0
Slovakia	15122	30.1	14866	25.4	11932	25.9
Slovenia	3185	18.5	2961	20.7	2743	23.5
Total	196217	37.9	163084	44.4	147011	44.0

Source: Eurostat 2000: 165–87

Notes
A Million passenger-km by rail and bus
B Percentage for rail

Heritage services

Steam railway enthusiasts generate a considerable demand for 'Plandampf' events in the Meiningen and Suhl areas of Thuringia. Poland retains steam at Wolsztyn for regular services to Leszno and Poznań – the fastest steam passenger service in the world! – in connection with income from Western heritage steam organisations who use the facilities for driver training. But tourist use of narrow-gauge lines is most significant. In East Germany DR operate heritage narrow-gauge services in the Dresden area between Radebeul, Moritzburg and Radeburg and also between Freital and Kipsdorf, while other lines have been privatised, such as the 'Mecklenburgische Baderbahn' from Bad Doberan to Kühlungsborn on the Baltic coast, the 'Rügenscher Kleinbahn' (Putbus–Göhren) on the island of Rügen and the 'Sächsische Oberlausitzer Eisenbahn Gesellschaft' which connects Zittau with Kurort Jonsdorf and Kurort Oybin. Most outstanding however is the privatised 'Harzer Schmalspurbahnen' in the Harz Mountains, combining the 'Harzquerbahn' (Nordhausen–Wernigerode) and the Gernrode connection via the 'Selketelbahn'. The network includes the Brocken Mountain branch, which was closed on account of the GDR's heightened border controls of the 1960s but restored after reunification; also the Harzquerbahn–Selketelbahn link of 14 km (Stiege–Güntersberge) removed in 1945 but restored by the GDR in 1983. Poland retains narrow-gauge feeder lines, particularly the Pomeranian system ('Pomorska Kolej Dojazdowa') in the northwest based on Gryfice, Koszalin and Stargard with the integration of such lines as Gryfice–Trzebiatów, and Stargard Szczeciński–Dobra Nowogard. There are forest railways in the Carpathians and lakes of Poland (Bieszczady and Ełk); also Romania (Comandău in Covasna and Vişeu in Maramureş) and Slovakia (Hronec). The Polish Bieszczady system dates to 1895 and was eventually developed over 80 km (Cisna–Łupków–Rzepedź). Closure in 1994 was followed three years later by the reopening of a 10 km section at Majdan.

Meanwhile, the Ełk lines to Turowo and Zawady were renovated in 1994. Sochaczew – 50 km from Warsaw – offers a narrow-gauge museum with trains running to Wilcze Tułowskie in the southwestern part of Kampinoska National Park.

Railway workshops

Rationalisation will affect employment, particularly in workshops that are tooling up to handle modern rolling stock. In East Germany the inherited workshop network, involving 22 separate locations, remained little changed from the pre-communist era. But in view of the collapse of rail traffic and the introduction of new equipment requiring less maintenance, there is now a ten-year reorganisation to reduce the locations to 11 (each specialising in either motive power, coaches or wagons) and bring the workforce down from 34,400 to 8,900 in the process. All ECE railway manufacturers are now participating in the modernisation process, but they cannot afford to restrict themselves to the domestic market because they come under increasing pressure from competitors in Western Europe. Firms in the region can offer low costs, but they also need greater efficiency and Western partners to inject funds for modernisation. It is also necessary to acquire skills in contracting directly with customers, a matter previously handled by state trading companies. The Hennigsdorf installation north of Berlin is now AEG Schienenfahrzeuge, working on export orders (including trainsets for Shanghai) as well as electric locomotives for domestic use. In future more powerful electric locomotives will be needed for the faster passenger services now envisaged and Soviet-built diesel locomotives, which have been the mainstay on non-electrified routes, will also have to be replaced along with diesel multiple-unit trains.

The Budapest firm Ganz-Mavag faced bankruptcy even in 1987 and after continuing to operate at a reduced level (with a metro train prototype) entered into a partnership with a subsidiary of British Telfos to create the new company Ganz-Hunslet. This new company is now building electric trains for the Hungarian Railways (MAV) and trams for Budapest City Transport. In Hungary there is also a joint project between MAV and Voest Alpine Eisenbahnsysteme for pointwork manufacture at Gyöngyös. In East Germany 'Deutsche Waggonbau', which continues to export passenger coaches and refrigerated trains to Russia as well as to Amtrak in the USA, has the prospect of a joint venture with the Russian rolling stock company at Tver. In Poland, Adtranz (formed out of the transport sections of ABB Zürich and Daimler-Benz) have acquired a majority shareholding in what is now 'Adtranz Pafawag' of Wrocław where modern technology has attracted orders for electric locomotives for the Polish State Railways (PKP) including dual-system locomotives for the Berlin service. In connection with the Polish–German joint company Interlok, Piła Works in Poland specialises in the overhaul of steam locomotives for heritage workings in Western Europe and several coachworks are in the same line of business. Meanwhile, in the Czech Republic ČKD of Prague has been able to negotiate a joint venture with Daimler-Benz subsidiary AEG to manufacture rail systems, while in Romania GEC-Alsthom provides a Franco-British stake in the Bucharest locomotive and rolling stock company Faur which is working on Romanian and third country requirements.

Network change: Germany

The unification of Germany has given rise to major improvements on the old 'inter-zonal' routes. Since 1993 Germany's high-speed inter-city (ICE) trains have been operating from Berlin to Frankfurt, Hamburg, Cologne, Munich and Stuttgart; running at regular intervals with intermediate stops at stations such as Halle, Magdeburg, Potsdam and Wittenberge which were not available before unification. Of the 27 lines affected by the inter-zonal frontier imposed in 1945, only a small number have been reinstated under a special gap-closing 'Lückenschlussprogramm'. The first new line over the old frontier (supplementing the eight available before unification) opened in 1990 between Arenshausen and Eichenberg on the Kassel–Nordhausen–Halle route, but four others quickly followed: Wieren–Salzwedel on the Hamburg–Berlin route via Uelzen and Stendal; an entirely new line between Bad Harzburg and Stapelburg for the Hildesheim–Halberstadt route; Mellrichstadt–Rentwertshausen between Würzburg and Erfurt; and Neustadt–Sonneberg for Nürnberg–Erfurt trains. Between Walkenried and Ellrich (Northeim–Nordhausen) a railcar service was introduced as a local initiative immediately after the Berlin Wall collapsed and the arrangement was later formalised. However, this line was already in use for occasional freight transfers. The gap-closing programme was followed by the 'Verkehrsprojekte Deutsche Einheit': VDE (German Unity Transport Plan) involving a special rail construction company which restored double track and electrification, while improving track for faster speeds (since few lines in East Germany could handle speeds of over 100 km/h). In 1993 a through electrified line (15 Kv, permitting speeds of 160 km/h) became available between Hannover and Berlin via Helmstedt–Marienborn: one of the most important border-crossing points prior to unification. This allowed Germany's high-speed ICE trains to reach Berlin – although by an indirect route into the city until the direct route through Potsdam was available two years later. Another early improvement was the restoration of the original alignment at Gerstungen (between Bebra and Eisenach), replaced during the GDR era by a more direct but badly-graded route which simplified frontier administration.

The German network will eventually cater for 250 km/h speeds on lines that will be either entirely new ('Neubaustrecken': NBS) or substantially upgraded ('Ausbaustrecken': ABS). One 'Neubaustrecke' has been opened between Berlin and Hannover via Stendal (parallel to the existing single line) and this is now used by the principal trains travelling west from Berlin. It is also used as a new line to Hamburg (and Bremen) because it connects with the new 'gap-closing' route from Salzwedel to Wieren and Uelzen; reducing the journey time to 1 h 46 m, although the older route through Wittenberge was rebuilt with electrification by 1997. In 1994 the 'TransRapid' project was unveiled: a Berlin–Hamburg line (noted above) working on magnetic levitation which could achieve a speed of 500 km/hr. However, although still 'alive', this project has not yet been started. Meanwhile, the services between Berlin, Dresden and Prague along the Elbe valley have been improved and the local link that has been restored at Weipert/Vejprty between Annaberg–Bucholz in Germany and Chomutov in the Czech Republic could be used for trains between Chemnitz and Plzeň via Saaz. ABS projects also include the routes from Berlin to Dresden and also to Frankfurt a.Main via Leipzig, Jena and Coburg; plus the line from Dresden to Leipzig and Eichenberg near Kassel which will carry the Dresden–Cologne ICE. Tilting diesel railcar sets based on the ICE design are being used on winding routes – such as the 'Sachsenmagistrale' which provides a Dresden–Hof–Nürnberg service – where very high cost new lines would not be justified. Average speed has been raised from 60 to 100 km/h (though this means a reduction in line capacity – due to greater speed differential between Inter-City and local/freight services – calling for additional overtaking loops).

Network change: Poland

The ECECs will become more important for transit using the west–east routes through Germany and Poland, especially the main route through Berlin and Warsaw. The Euro-City 'Berolina' service was introduced in 1992 to connect Berlin with Warsaw in 6 h 10 m with German dual-system locomotives in use between Berlin and Rzepin to cut out the locomotive change previously needed at Frankfurt/Oder on account of the transition between the German-Austrian 15 Kv system and 25 Kv in use in Poland. Meanwhile, the new political geography in the East means that a range of routes into the FSU can now be developed, including the historic link between Berlin and Kaliningrad (formerly Königsberg) via Poznań, Toruń and Olsztyn. Furthermore, a broad-gauge single-track railway is to be built to link Gdańsk with Kaliningrad and facilitate forwarding of traffic throughout the FSU. This could make Gdańsk the best handling place for Russian companies in general and the Kaliningrad SEZ in particular. The growth of traffic between Poland and the Baltics will require improvement of the 'Białystoka' (Warsaw–Białystok–Grodno) to serve the Baltic states and St Petersburg. However, the possibility of a Belarus blockade has prompted the suggestion of a new line from Białystok to Kaunas. Meanwhile, the independence of Ukraine has resulted in a growth of traffic from Warsaw and other Polish cities to Lviv, Kyiv and Odessa. This highlights the axial route from Dresden across southern Poland via Wrocław and Kraków. Local trains have restarted over the German–Polish border between Strausberg and Gorzów; also at one point on the Czech–Polish border: Mezimĕstí–Mieroszów on the line from Teplice to Wałbrzych.

Finally there has been a reappraisal of the 'Sulphur/Steel Railway' (LHS): a broad-gauge railway from Upper Silesia to the eastern border which seemed destined for oblivion because of its origins as a colonial project to bind Polish industry with the Soviet Union (taking iron ore westwards and sulphur eastwards) and facilitate the penetration of the Red Army in an emergency through loading platforms and warehouses along the 400 km route. But although the military rarely used the line and freight declined from 8.0 m.t in the 1980s to 4.5 m.t in 1993, there is now growing appreciation of its value for trade with the east since transhipment at the border is eliminated. West European firms are making considerable use of the facility which is managed by Ewita, representing a range of companies based in Poland, Russia and Ukraine, as well as Germany, Hong Kong and the Netherlands. Loading and forwarding terminals are being opened along the line and at Sędziszow (Kielce) a modern facility has been opened to adjust the axles of Russian cars to European tracks. Although the Polish steel industry may obtain more ore in future from Brazil and South Africa, there is an expanding market for Polish coal in Ukraine. There are now eight trains a day in each direction on the LHS. In addition, passenger trains are running thrice weekly, including through passenger services from Moscow to Olkusz which is the point where the broad gauge ends as a separate formation (though it does continue along the old alignment to an iron ore stockpile at Sławków close to the Katowice steelworks in Dąbrowa Górnicza). With rail links mooted between Russia and Korea (with Czech involvement in construction) and between Uzbekistan and China, the LHS could become part of a rail link between Europe and the Pacific via Poland, Ukraine and either Central Asia or Siberia.

Network change: Hungary

Major improvements have been made on the Budapest–Vienna line, where new Austrian locomotives avoid the change previously needed at Hegyeshalom on account of the difference in voltage. Hungary is supplying new coaches while the improvements will also benefit Austro-Hungarian freight, 80 per cent of which uses this route. Hungary will also have an interest in the growth of north–south traffic originating in Scandinavia and running through the ECECs en route to the Middle East: such flows would certainly be encouraged by a Bosporus Tunnel in Turkey and faster transit through Bulgaria. However, freight from Austria and Italy will be able to move more easily through Hungary if a new west–east transit railway can be created so as to avoid Budapest. A possible route would pass through Székesfehérvár, Kecskemét and Cegled en route to the Ukrainian frontier at Záhony. This would require a new 35 km central section, including a bridge over the Danube north of Dunaföldvar (between Sárbogárd and Fülöpszállás). Alternatively, a more northerly route across the Danube at Adony would provide a more direct link between Székesfehérvár to Cegled but would require some 70 km of new construction. Sections of the line could also be used by freights moving eastwards from the Adriatic ports of Koper, Rijeka and Trieste. Meanwhile, Hungary provides an example of local realignment on the northeastern side of Debrecen in the interest of urban planning.

Network change: Romania

The railway to Constanţa now connects with Black Sea ferries to the Caucasus which will become more important in the light of oil production from the Caspian region: oil could flow through Constanţa either by rail or by means of a pipeline to Trieste which is projected. Meanwhile, new connections between Poland and Romania are using routes through Košice in Slovakia and thence through either Hungary (Debrecen) or Ukraine (Beregovo) to reach Oradea as alternatives to the line through Lviv which involves transit through only one country (Ukraine). But planning is overshadowed by the historic problem of capacity on the Carpathian route between Ploieşti and Braşov. Improvements were made by doubling the track and subsequently by dieselisation and electrification. However, in the 1980s this was deemed insufficient and the decision was taken to provide a new route from Bucharest to Sibiu through Piteşti and Râmnicu Vâlcea, requiring much new construction between the last named places. Although virtually complete when the revolution occurred, there was no money to finish the project – through country prone to landslides – and since the reduction in rail traffic has now made this project less essential the installations are currently abandoned. Meanwhile an improved link between Bucharest and Odessa has been proposed but the pro-gramme of local railways which brought Siret and Târgu Neamţ into the net-work during the 1980s has now been discontinued and only commuter lines in the Bucharest area (e.g. to Bolintin) remain a possibility.

Network change: former Yugoslavia

There have been major upheavals, most notably through war in Bosnia & Hercegovina, which restricted trains to local services in the Sarajevo, Tuzla and Žepče areas. Much of the damage has now been repaired and the route between Sarajevo and the port of Ploče is now operating again after the line was cut for several years north of Mostar. Montenegro is to modernise and electrify the 60 km rail link between the Nikšić bauxite mines and aluminium smelter at Podgorica, while the need for a direct rail connection between Hungary and Slovenia will require the reopening of the Zalalövő–Hodoš–Murska–Sobota route: some 20 km of new construction is needed to avoid the need for transit through Croatia. However, Croatia faces the greatest challenge to create a coherent national network. A project is under way to unify the network in Istria where the Buzet–Pula section is currently linked with Slovenia yet has no direct connection with Croatia. Planned in the communist period, a 23 km link from Jurdani on the Pivka–Rijeka line to a new junction station at Novi Lupoglav was started in 1993; including the 1,350 m Brecca viaduct and the Brgud and Čičarija tunnels. Croatia will also double the electrified Zagreb–Rijeka railway, with a new alignment through the Kopa Valley and a tunnel through Gorski Kotar to reduce the dis-tance. Croatia also plans a new railway along the Adriatic coast to Montenegro and Albania; seen potentially as a major trunk line carrying lorries from Greece towards the motorway system of Central Europe, this could lead to a series of improved railways from the Adriatic ports to the interior including a connection between Zadar and Zagreb via Oštarije. A more direct line from Austria passing down the Sava valley could be achieved by linking Pragersko in Slovenia with Krapina in Croatia and also by connecting Sisak in a straight line with Novska.

Network change: Albania

While Albania desperately needs to refurbish the existing railway track, the country is also at the centre of some of the most ambitious projects in the SEECs by way of possible international links with Macedonia and Bulgaria by means of a line from Durrës to Sofia via Skopje. Linkage with the Bulgarian network in Sofia would open a through route across the Balkan Peninsula and offer parallel advantages to an Adriatic–Black Sea road which has also been proposed. It would be of particular interest to Italy whose role in the region could be significantly enhanced. But there is no hope of early completion, although the Bulgaria–Macedonia rail link between Kjustendil and Kumanovo was started in response to the Greek blockade of Macedonia and the increasing importance of trade with Turkey. The line has now reached the frontier station on the Bulgarian side but little has been done to advance the line up the Kriva valley from Kumanovo, while the long tunnel on the border itself has not been started. The line will in any case be negotiable only at low speed. Meanwhile, Albania's only existing rail link with the outside world – the 11 km line from Han i Hotit to the Montenegrin border – is presently destroyed, although it is potentially recoverable as part of the Adriatic coast line already referred to. Albania also hopes to extend its Pogradec branch a short distance to the border from where it would require 68 km of new construction through Resen to reach the Macedonian network at Bitola and 111 km to the present Greek railhead of Florina, which would allow Albania to export non-ferrous ores through Thessaloniki. Albania would also like to have a rail link with Kosovo by means of a 100 km line from Rrëshen to Prizren. However, all this new construction – some 300 km in all – would cost $565 m., which is hardly justified given the recent drastic fall in Albanian freight traffic. However, a decision has been taken to retain the country's rail system, despite the damage which has occurred through insurrection, and branch projects currently discontinued – such as the Milot–Klos branch to handle copper and chromite, which was in an advanced state in 1989 – could be revived when economic conditions allow. However, trackwork is in a poor condition everywhere and train speeds are restricted to some 40 km/h for passanger trains and 25 km/h for freight.

Transition: road transport

A sharp growth in the relative importance of road transport has taken place throughout the region. There are many more vehicles on the roads and Western products – cars, trucks and to a lesser extent buses – are now very prominent as a result of imports or new assembly plants located within the ECECs (referred to in the previous chapter). Car ownership at 250–300 vehicles ptp was predicted for the end of the century by ECMT (1991: 210) compared with 1985 figures of 200 in East Germany, 160 in Czechoslovakia, 135 in Hungary and 100 in Poland. In 1998 there were more than 300 cars ptp in the Czech Republic and over 400 in Slovenia, though only 125 in Romania and 27 in Albania (Table 4.3). There has also been a modest motorcycle renaissance, mainly by enthusiasts who can afford Japanese machines (some German and Italian) as luxury items: there is unlikely to be any significant revival of production within the region after the boom in the interwar years and the early communist period. However, the average age of vehicles may not have changed much because of the many used cars that have been imported, while the wholesale renewal of bus fleets in East Germany has displaced the old Hungarian Ikarus buses to SEECs like Romania where they are gladly accepted at very low prices and where the old German markings are proudly retained. Road surfaces have improved, especially in East Germany (though there is a substantial maintenance backlog in the SEECs – leading to more frequent breakdowns and accelerated vehicle wear) but roads are becoming more crowded while agricultural traffic (tractors and carts) and slow-moving lorries keep average speeds low. Border delays, especially for truckers, may extend to between two and four days in the most notorious cases. With greater use of heavy trucks weight restrictions cause many problems especially in the case of the Vistula bridges in Poland.

Road development has brought about substantial improvements in accessibility and it is difficult to provide the highway capacity and frontier administration to keep pace, despite PHARE support for better border-crossing facilities, exemplified by the Koroszczyn station on Poland's frontier with Belarus (under construction during 1996–9). The economic impact is problematic because while road corridors may be welcomed as a boost to national and regional development, penetration by foreign goods could also grow. Yet roads seem destined to increase their relative importance over rail and it is predicted that in 2010 the roads will carry 87.5 per cent of the traffic in East Germany compared with 8.5 for rail. Much of the region's industry has developed on the basis of good rail access through private sidings, many of which are now closing even where lineside sites are still occupied. The road haulage business has seen striking modernisation with private firms acquiring modern Western trucks. Thus the Hungarian state company 'Hungarocamion' has been acquired by Ventura (Deutsche Bank and Bank of America) and, with its fleet rationalised, the company is now a modern, IT-based logistics provider able to transport Shell's lubrication products from Austria, Germany and Poland for warehousing and distribution in Hungary. The shape of things to come is being mapped out in a pioneer venture by Rynart Transport of the Netherlands in opening a 30,000-pallet warehouse at Biatorbagy in Budapest (adjacent to the MO) but shortcomings are highlighted when 'just-in-time' delivery is complicated by poor regional roads and many customers cannot receive loads outside normal working hours (while lorry traffic is restricted on Sundays). Improved standards in the Czech Republic are being driven by the trend towards outsourcing which is dependent on good logistics and capacity improvements by local haulage firms. The Western-owned supermarket chains like Makro and Tesco are setting an example, with Ahold of the Netherlands operating its own distribution centre north of Prague.

Network development

In addition to all-round improvement there will have to be a greater emphasis on new expressways and motorways. The East German 'autobahnen' are being thoroughly renovated and integrated with those of the West and extensions are planned, e.g. from Dresden and Bautzen to Görlitz (where a new frontier post is now operational); thence to Wrocław, Katowice and the Ukrainian frontier (intersecting in Upper Silesia the dual carriageway from Warsaw to Bielsko Biała). A new coastal motorway is planned to connect Lübeck with Szczecin –

although this is unlikely to run further east along the Baltic coast for environmental reasons – while Halle could be linked westwards with Braunschweig and northwards with Magdeburg, Stendal and the Berlin–Rostock motorway. Jena could also be connected westwards through Meiningen to Würzburg and a branch from the Leipzig–Dresden motorway to Chemnitz might continue southwards into the Czech Republic with branches to both Prague and Plzeň. However, East Germany is building on a substantial existing network with the resources of the whole German state. It is more difficult for the other ECECs to achieve the same standards starting with a relatively limited network and with greater financial constraints. However, Visegrád countries seem determined to improve their roads as quickly as possible.

In Poland motorways will run southwards to the Czech frontier from Gdańsk through Katowice and from Szczecin through Zielona Góra, while east–west routes will extend through Poznań and Warsaw to Belarus and through Wrocław and Katowice to Ukraine as already noted. Łódz will benefit from proximity to the intersection of the two principal routes, not to mention the new link road planned between Łódz and Wrocław. Over the longer term a 'Via Baltica' may link Warsaw with Tallinn and Helsinki: a project launched in 1996 for the renovation of 930 km of road. Also, a route from Warsaw to Lublin could continue into Ukraine and possibly the Black Sea coast. There has also been a Russian proposal (dating to 1996) to link Kaliningrad with Belarus through northeast Poland in order to avoid high transit dues charged by Lithuania. But the plan has been rejected by Poland because of the threat to the Suwałki area where 'special environmental protection' is needed (although the local population think the new road would stimulate their economy). All these roads will have international significance and links through the ECECs will eventually create an 'Amber Route' in the shape of an Oslo–Athens highway (taking the Karlskrona–Gdynia ferry into account) while the east–west axis will connect Moscow with Western Europe.

In the Czech Republic, the Prague–Plzeň motorway, opened in 1995, will continue into Germany and there is also a series of loops around the edge of Prague to connect with the motorway to Brno and Bratislava built during the communist period. Bratislava provides a link with Hungary where the orbital expressway in Budapest (MO) connects a developing radial system of motorways which will eventually reach the frontiers of Croatia through Nagykanizsa, Slovakia through Miskolc, Ukraine through Debrecen and Záhony, and Romania and Yugoslavia through Szeged (the Austrian link through Győr already exists). Hungary is also involved in a Milan–Kyiv motorway project which is referred to internally as the 'Deli Autopalya Projekt' or DAP (southern motorway project) avoiding Budapest, although Hungarian interests might be better served by a semicircular route closer to Budapest on the route Székesfehérvár–Dunaújváros–Kecskemét–Szolnok–Eger. Hungary is also upgrading regional roads to integrate better with the motorways. Thus the Budapest–Szeged motorway has good connections with Debrecen via Albertirsa and with Békéscsaba and Arad via Kecskemét. Complex choices also arise for Slovenia in connection with proposed east–west (Hungary–Italy) and north–south (Austria–Croatia) motorway links, for there is a case for neutralising the pull of the main intersecting routes through complementary regional roads that would connect provincial towns without going through Ljubljana.

In the SEECs the Eurocorridors provide the blueprint, updated by the 'Motorway Project for the Europe of Tomorrow' which covers the Sava Valley (Ljubljana–Zagreb–Belgrade) and two complementary roads: Budapest–Zagreb–Rijeka and Mohács–Osijek–Sarajevo–Ploče; but nothing tangible yet for the coastal corridor, although this cannot be discounted as indicated below. However, it will be many years before such a motorway network is realised. For Romania the main priority is the Bucharest ring and links with Constanţa and Giurgiu to complement the existing motorway from Bucharest to Piteşti (to be extended eventually to Sibiu and Oradea; also to Timişoara and the Hungarian frontier at Nădlac). Romania will however proceed with the Calafat–Vidin bridge in cooperation with Bulgaria and this could accelerate road improvements northwards through Drobeta-Turnu Severin and Timişoara; also through the Jiu Valley to Petroşani and Deva. A new regional road along the Danube between Orşova and Moldova Nouă could integrate with this concept. In Bulgaria a road tunnel under Mount Shipka would shorten the north–south route through the country and avoid seasonal delays. Meanwhile, Yugoslavia will need to improve the Belgrade–Bar link because the present road through Moraca Canyon is dangerous and the final 65 km stage from Podgorica needs upgrading with 10 km of bridges and tunnels (one 8 km tunnel, also for use by rail and oil and water pipelines).

Contested issues

There is considerable crime and prostitution connected with the business. Theft remains embedded in the corporate culture, with cargoes of cigarettes and alcohol a particular risk on the Berlin–Moscow road due to hold-ups. Mafia gangs are known to pose at destinations as the official recipients: hence the practice of posting half a torn dollar bill to the customer while the driver takes the other portion and matches the two before unloading. There has been a rise in organised car theft, with the CIS likened to a 'black hole' because there is little police cooperation over stolen vehicles. Resource and environmental issues should also be mentioned. Heavy expenditure on motorways diverts spending from other roads: the 'Via Baltica' has seen disproportionate resources invested in renovating 333 km and building 110 km of new road (so far) while other roads in Poland and the Baltic States are in dire need of attention. Cash for expanding the network is leading to concessions and toll arrangements which now affect many motorways in FCS, Hungary and Slovenia. For example, the French builder Bouygues is working on the toll motorway from Budapest to the Yugoslav frontier, while Autostrada Wielkopolska has received a concession for Poland's first toll freeway from Łódz to the German frontier and Astaldi (Italy) is building the motorway from Zagreb to the Hungarian frontier at Gorican. However, because toll roads deter local use there is a desire to negotiate 'build, operate and transfer' (BOT) arrangements that avoid such charges and this system will be used in respect of work by the Israeli 'Housing and Construction' company in completing the motorway through Moravia linking Austria and Poland. Furthermore, route selection involves sensitive environmental issues. The projected motorway from Sofia to Greece is held up by the controversial 19 km section through Kresna Gorge where options for the avoidance of this ecological zone were to be considered during 2001. Motorways along the Baltic coast and also through the protected Středohoří Park in northern Bohemia are facing opposition which in the latter case may require more tunnel sections. From several viewpoints there is concern over the priority for new roads when more money might be spent on other services such as education and health. But there is strong encouragement coming from the EU and from the motor industry.

ALBANIA'S DILEMMA

Finally, although the Eurocorridors reflect a degree of consensus it is evident that conflicting pressures remain over priority routes. While Albanian national roads are being repaired there is a fundamental split between international schemes which either follow the coast (supported by Croatia and Greece) or run across the Balkan Peninsula (advocated by Italy and the EU). Albania has a dilemma because there is a commitment to Corridor Eight to link Durrës and Vlorë with the Bulgarian ports of Burgas and Varna. Such a link would connect with ferries across the Adriatic to Italy and over the Black Sea to Ukraine, Turkey and the Middle East. Furthermore, in 1994 a protocol was signed in Sofia by Albania, Bulgaria, Macedonia and Turkey to develop a Trans-Balkan road route linking Istanbul with Tirana via Skopje. The trans-Balkan project also has American support – dating back to 1996 – for a Southern Balkans Development Initiative bringing together a network of US businesses and the three corridor countries (Albania, Bulgaria and Macedonia). However, Greece is opposed to such a route because of its own project for an 'Odhos Egnatia' – between Igumenitsa, Thessaloniki and Istanbul – and would much prefer to see road improvements along the Adriatic coast underpinned by an entente between Greece, Albania and Croatia. There are many ways in which Greece can signal its displeasure including deportation of illegal Albanian immigrants and demands for Greek military units to operate within Albania to enhance border security.

There is no plan at the moment for a coast road, although it was contemplated in 1988 and abandoned when conflict arose within Yugoslavia. However, the Bay of Kotor ferry needs replacement and Albania already has a World Bank loan to improve road links between Tirana and northern Albania. Currently the Lezhë–Shkodër section is in desperate need of rehabitation to help reduce the high costs for overland transport to Albania from Central Europe and Croatia. Further north, Croatia is keen to improve routes to Rijeka and Trieste that will connect with the European system through northern Italy. Yet, if the project were to re-emerge it would certainly be opposed by Italy with its own options over pressing for the Macedonian road corridor instead. It seems that Corridor Eight is proceeding by stealth as foreign engineering companies (Albanian firms are ruled by the need for ten years' experience) work to a poor standard in widening national roads – while using gravel from unlicensed pits in the Shkumbini Valley. Meanwhile, the Albanian government has received $66 m. for national roads from various sources (including EBRD, EU PHARE and the World Bank) but is not controlling operations adequately and denies that Corridor Eight is being developed officially: in other words the work can be passed off as a national programme of rehabilitation rather than an international scheme to complement work being done in Macedonia on a new six-lane highway (a standard that applies in Albania only between Durrës and Tirana). In short, Albania sees the Eurocorridors at their most problematic with conflicting international tensions on top of the basic problem of creating modern European highways well in advance of demand when a conventional national road network would seem a luxury.

Transition: ports and inland waterways

The seaports are entering into a new phase of competition as German and Italian ports struggle to regain their former influence. With the move towards free trade, ports are likely to regain their pre-communist hinterlands and planning must take transit functions into account. The big gainers seem likely to be the German ports of Bremerhaven and Hamburg (also Cuxhaven) which can offer better facilities (for container ships for example) and shorter steaming times than Baltic ports. Hamburg is well placed to draw traffic from FCS as well as East Germany which formerly traded through Rostock. Bremerhaven is equally accessible for the southern part of East Germany and Hungarocamion is now trucking freight there from Hungary. The leading German container carrier Hapag-Lloyd has therefore opened offices in the Czech Republic, Hungary, Poland and Slovakia as well as in East Germany; containers are taken to port by road although improvements to the Elbe could alter the picture. Meanwhile, Rostock (promoted heavily in the GDR era and used by Poland for the import of oil products) lacks the capacity to handle the larger container ships and will have to fall back on its Scandinavian Ro-Ro business through Warnemünde. Sassnitz has problems of road access through the town and also with the 'Rügendamm' connection to the mainland. But Mukran, where the Klaipeda ferry opened in 1988, could be retained as a ferry base for a range of Scandinavian destinations although there will obviously be competition from Polish ports and from the Puttgarden–Rødby route with its planned 20 km tunnel which will accommodate a car-carrying shuttle as well as through trains to Copenhagen.

The Baltic: Poland

The privatised ferry company 'Polferries' has a competitive programme of operations from Gdańsk to Stockholm and from Szczecin to Copenhagen, Malmö and Ystad (seasonally to Rønne in Bornholm). Yet the Polish ports have suffered a sharp fall in traffic (from 70 to 45 m.t of freight during 1979–93), despite a potential for handling goods for the CIS, including Russia, Ukraine and

Kazakhstan. Polish ports clearly lie much further east than their West European competitors and are less efficient, but they have much lower distribution costs and can compete for bulk commodities, if not for containers where Hamburg has the edge. Both the Gdańsk–Gdynia and Szczecin–Świnoujście complexes have the prospect of motorway links (although in the short term better local access is needed) and hence they can expect to handle traffic moving between Scandinavia and the SEECs. Gdańsk is the most northern ice-free Baltic port and can accommodate ships of 150,000 t (equalling the capacity of the Danish Straits). The oil facility at Port Północny has been joined by a liquified gas terminal which offers alternative supplies of both hydrocarbon fuels to the CIS pipelines. The Baltic Container Terminal is linked with the rail 'container bridge' to Sosnowiec opened in 1994 while in 1995 the port accepted an investment proposal by the American–Canadian consortium 'Europort' for a fodder- and grain-processing terminal (the most technologically advanced in the region).

Iron ore is another bulk cargo since the steel company Huta Katowice decided to open a terminal with a 5.0 m.t annual capacity (which will meet half the plant's requirements). Hungary and Slovakia are also interested in this facility which will accommodate 100,000-t ships bringing ore from Australia, Brazil and South Africa; so reducing dependence on Russia and Ukraine. Ironically an iron ore terminal was built in the 1970s for Soviet iron ore exports but never used: it was then taken over by the Europort grain terminal. Finally, Gdynia handles general cargo and has developed ferry links with Sweden, beginning with the Karlskrona service in 1991. Total traffic at Gdańsk–Gdynia is expected to exceed 50 m.t in 2005 compared with barely 20 m. in 1991. To improve access, the Vistula Bay project envisages a 3 km canal (20 m wide and 5 m deep) through the Vistula Spit (Mierzeja Wiślana) at its narrowest point north of Elbląg. This will greatly shorten the journey to the Tricity which at present involves a detour past Baltisk where the entry to the Baltic is Russian controlled. The project would also give a boost to the tourist industry since the canal would be available for pleasure boats and yachts and flooding problems caused by the build-up of water near Żuławy Elbląskie would be overcome. However, the cost is beyond the capacities of the local authorities and, although the canal might pay for itself after 15 years, central funding will not be available in the immediate future. Meanwhile, to the west Szczecin handles general cargo including timber and grain while bulk commodities are handled at Police (liquids and chemicals) and Świnoujście. The port has a major advantage in lying close to Berlin: aided by a bus shuttle there is a booming passenger trade (with regular ferries to Copenhagen, Malmö, Rønne and Ystad) and the possibility of a canal link with lifts. The complex should ideally be planned as a single unit, but Świnoujście fears that its profits will be used for the benefit of Szczecin.

Adriatic and Black Sea ports

Competition is also becoming sharper on the Adriatic and Black Sea coasts where much will depend on the rate of modernisation of the ports including warehousing, overland transport facilities and the development of agencies throughout the region. A small country like Slovenia would like to provide transit facilities through its Adriatic port of Koper which – assisted by the Baltic ferries and the Danube facilities – could in theory handle goods from ECE, Russia and Scandinavia en route to Africa, Asia and Australia. Koper has been developed since 1957, following the partition of the disputed Trieste territory, and can handle general and bulk cargoes, with special storage and the possibility of a free port environment which could stimulate the growth of semi-manufacturing facilities. But Croatia and Italy have similar ambitions in respect of Rijeka and Trieste respectively. Rijeka – the largest Croatian port in a group which includes Dubrovnik, Ploče, Split, Zadar and others – consists of general cargo and shipbuilding facilities in the town itself, complemented by bulk-cargo handling at Bakar, the petrochemical port of Sepen and the Omišalj oil terminal with capacity for 350,000-t supertankers and pipeline connections inland. Trieste offers an expanding container terminal, business innovation centre and science park. Albania's ports are coming into the picture, for Durrës has Ro-Ro facilities while Sarandë is being developed in the south and Shëngjin in the north may follow. Finally, the Black Sea theatre presents competition between the Romanian port of Constanţa and the Bulgarian rivals: Burgas and Varna. Constanţa has the canal link with the Danube and has new facilities through the grain and LPG terminals, but Bulgarian ports will benefit if the land bridge from Albania is developed and this would compensate for the uncertain future of the ferry service from Varna to Ilichevsk (Odessa) inaugurated in 1978. However, Odessa overshadows all the ECE ports and Caspian oil for Western Europe could be piped from the Odessa area as easily as from Constanţa.

Inland waterways

Although of minor importance for the region as a whole, after years of communist neglect, waterways have the advantage of operating costs of only one-third of the level for rail and one-sixth for road. The Danube is of great significance (despite temporary blockage through NATO bombing of the bridges in Novi Sad) and has enormous potential in the context of river–sea integration and the completion of the canal link with the Main and Rhine. A new container terminal is being provided at Bratislava (assisted by Austrian cooperation) and Hungary has opened a new international cargo terminal at Csepel (Budapest) in a joint venture with a German forwarding company. From Constanţa there are now ferry links to Istanbul, Samsun (Turkey) and Poti (Georgia). However, since finance is difficult to come by, Romania has stopped work indefinitely on the Bucharest–Danube Canal and is taking no active interest in canalising the Olt to Râmnicu Vâlcea or the Prut as far as the Stânca-Costeşti hydropower plant, although the project to reopen the Bega Canal and thereby reconnect Timişoara with the Danube system appears more soundly based. In the middle of the Danube basin the completion of the Gabčikovo-Nagymaros hydropower and navigation complex is deadlocked, despite the ruling by the International Court of Justice that – notwithstanding Hungary's unilateral withdrawal from the project – the original agreement with FCS in the communist period should be honoured. Meanwhile, a Danube–Tisza Canal is still a possibility in Hungary and Croatia has an interest in navigation on the Sava, at least in the context of a Vukovar–Šamac Canal even if the grandiose notion of a Sava–Adriatic link is abandoned. A link between the Danube and the Aegean is also a possibility, noted most recently by ECMT (1991: 500). Unfortunately the Danube was seriously disrupted due to bombing of the bridges at Novi Sad in connection with the Kosovo War. While boats were trapped – especially the Romanian and Ukrainian fleets – until clearance during 2002, rail and road transport was used more extensively for bulk cargoes, despite higher costs.

The Elbe is now facing a more assured future now that the port of Hamburg is able to reclaim its former hinterland in East Germany and Bohemia, lost through the redirection of freight by rail to Rostock and Szczecin during the communist years. In addition to its direct waterway connection, Hamburg has more modern facilities, while ocean-going ships require less sailing time compared with Baltic ports. The Port of Hamburg now has offices in Berlin, Dresden, Leipzig and Magdeburg. This raises the possibility of links between the Elbe and Danube, considered appropriate by ECMT (1991: 484–90) in conjunction with an Oder–Danube Canal (Koźle to Bratislava) that has historically been regarded as the more feasible: considered by Czechoslovakia before the split (with a completion date of 2002!) and previously advocated by the United Nations Economic Commission for Europe in 1959 (which also looked at the Danube–Rhine and Oder–Vistula–Dnieper links) not to mention its attraction for both Hitler's 'Fortress Europe' and the Habsburg Empire. The new canal would extend Bratislava's hinterland into southern Poland and provide Poland with a waterway link with Western Europe superior to the low-capacity canals from the Oder to Berlin (by the Havel and Spree) and the German 'Mittelland' system. But in addition to the financial implications there are formidable environmental objections.

Transition: air transport

This is becoming a more important facet of the transport scene, for transition has brought about an intensive interchange among professional people and the leading Western airlines have started to compete with the ECECs' 'flag-carrying' airlines on the key routes. There is also a substantial tourist traffic which has led to the formation of some new airlines within the region. In Romania the Jaro Airline with a US majority shareholding operates in Europe, North Africa and North America, while Air Budapest was set up in 1991 to operate refurbished Russian aircraft on excursions to the Canary Islands and Malta. Finally, airfreight is an alternative to road transport when there are concerns over security. Services have been modified to reflect a global world rather than the narrower confines of the communist club pre-1989, although Cuba still features as a holiday destination in the programme of Dresden regional airport. New orientations within the region – notably the Tirana–Skopje–Sofia axis – will be easier to achieve by setting up an air corridor than by other means. Since the national airlines of the ECECs have been retained and reorganised, some have progressed with Western partners like Malev which has Alitalia as a minority shareholder and Delta Airlines as joint operator of two Boeing 767s acquired in 1994 for the Budapest–New York route which Malev had difficulty in filling.

On the other hand, Československé Aerolinie (CSA) obtained support from domestic banks and has done well, especially with its Prague–London service. Losses are being reduced as load factors increase. More cargo is being carried and the airline's activity is being diversified into travel agencies. Air France took a modest stake in the company but the arrangement was a failure and the minority interest has been transferred back to the Czechs. With partition in 1993, CSA was retained as a Czech interest while the Slovaks have gradually established their own Slovak Air Transport System with aircraft secured under a 'jets for debts' deal with Russia in 1996. Polish Airlines (LOT) also decided not to make a strategic alliance with a Western airline and only after programme reorganisation to achieve profitability (with services to Chicago, New York and Toronto) have American Airlines been chosen as the code-sharing partner. New routes to Japan, Korea and South Africa, as well as North America, are being evaluated, although medium-range European routes give the best return. The SEECs' flag carriers including Balkan (Bulgaria) and TAROM (Romania) have encountered more serious problems in seeking profitability as a basis for privatisation. The situation in former Yugoslavia is highly complex for the old Yugoslav airline (JAT) remains intact with full ownership transferred to the Serbian government while separate airlines operate from the other successor states, with Adria (Slovenia) and Croatian Airlines the most developed. However, even Bosnia & Hercegovina's Republika Srpska has its own airline (flying to Belgrade, Budapest, Moscow and Timişoara) and Montenegro's Oki Airways links Podgorica with Skopje and Thessaloniki.

The region's airlines are starting to replace fleets of Soviet aircraft (Antonovs, Ilyushins and Tupolevs) in order to increase efficiency in the absence of Soviet oil – at concessionary prices – to compensate for high fuel consumption. It is also a matter complicated by insurance since the old arrangements for the former communist states broke down in 1990. However, fleet restructuring is a major undertaking given the number of aircraft involved, because in 1992 the nine national airlines (including Adria and Croatian Airlines) were operating some 200 aircraft altogether, including only a few of Western design at the time: mainly Boeings and BAC One-Elevens. CSA along with Malev, LOT and TAROM have ordered Boeing 737s (though LOT has also purchased aircraft from McDonnell–Douglas), while Balkan has purchased from Airbus (as have CSA and TAROM). There has been considerable financial and technical help from the West while relatively weak world demand for new aircraft in the aftermath of the Gulf War has meant that favourable credit deals have been negotiated with the large manufacturers. Malev is using Boeing 767–200s, financed by a loan from the UK banking group Barclays, for flights to New York via Rome. There has been some recourse to the hire (or second-hand purchase) of Western planes as an alternative to the ordering of new aircraft which would involve both great expense and delay. Croatia has obtained Boeing 737s from Lufthansa while LOT has obtained some aircraft on lease. Although

Poland has sold some Tupolevs to Ukraine, the inherited Russian fleets will not disappear quickly and indeed Malev intends to keep its Tupolevs after privatisation for use on charter flights. SEECs are generally retaining Russian aircraft but enlarging their fleets with Western designs. Slovakia is flying new Russian aircraft while some other small operators have purely Russian fleets.

Domestic services have been extensively modernised and the once ubiquitous Antonovs are now comparatively rare: their demise began with the withdrawal of some internal flights during the immediate post-1989 depression years. Alternative aircraft include the Fokker SAAB 2000 which can be used for both domestic and international flights and Canadian Bombardier CRJ and DASH 8–400. It is worth noting that the ECECs are themselves involved in aircraft building though usually only light aircraft are assembled. However, there is some scope for manufacturing components for large passenger aircraft under licence to defray expenses. In the case of Romania, Romaero has a contract to build upgraded BAC 1–11 aircraft for the new US airline Kiwi International and an agreement has been signed with Boeing that will enable the Romanian company to produce parts for 737 and 757 aircraft. Romaero also makes subassemblies for Bombardier CL-415 aircraft while the Craiova aircraft manufacturing plant makes spare parts. This ties up with the operation of Bombardier aircraft by the new private company DAC Air Romania on domestic routes to the Black Sea (and some neighbouring countries) in competition with TAROM. The Brasov-based IAR ('Intreprinderea Aeronautică Română') manufactures AH-IF Cobra helicopters under licence from Bell Helicopters of the USA, with the possibility of export to former Warsaw Pact countries. Meanwhile, Romania's 100 MiG 21s will be modernised at Bacău by Aerostar, with electronic equipment from El Bit (Israel) who are also cooperating over a new fighter plane called the Lancer.

Airports,   Those especially in the capital cities are being modernised and extended with new arrangements for the supply of aviation fuel and the management of duty-free shops which in the past have often been functions of the national airlines. There are special problems in Berlin because of disagreement between the federal government and both the Berlin and Brandenburg authorities over the suitability of the former East German terminal at Schönefeld as the future Berlin international airport. Both Tegel and Tempelhof airports are very close to the city centre and are also too small, but former Soviet airfields to the south of the city could be considered. In the other countries such choices do not arise, apart from the option of overspill to former military airfields. In Hungary, enlargement of Ferihegy was planned in conjunction with the now-abandoned World Expo in 1996: Terminal One (opened in 1950) has been refurbished for cargo and charter flights while Terminal Two (dating to 1985), along with two new sections currently under construction, will handle all commercial passenger flights. There is also modernisation of air traffic control with Siemens-Plessey equipment and better city links are being provided, notably through the highly efficient and affordable Airport Minibus service charging a standard fare to all city destinations. Ferihegy will be joined by new cargo facilities at Kiskunlacháza (a former Soviet airfield south of Budapest) and it is intended that seven first-class airfields left by the Russians will be used for feeder services.

In Warsaw a single modern terminal at Okęcie caters for international and domestic flights with the help of Western expertise over in-flight catering (by Scandinavian Airlines), while the US company AMR Services deals with ticketing, luggage handling and aircraft maintenance and an Australian group has the franchise for duty-free shops. A feature of the Polish domestic scene is the keen competition between air and road with 'Polski Express' coaches operating from Okęcie to the main provincial cities. However, there is also a project, attracting the attention of an international consortium from Germany, Italy, UK and USA, for the former military airfield of Modlin in the Mazowsze region which could become a major airport and transit centre, with an associated industrial estate including a SEZ. In Prague the main airport has been at Ruzyne on the western edge of the city, but British and North American companies are involved in building and operating a new terminal on the former Soviet airfield of Milovice, 30 km northeast of the city, with Maglev (magnetic levitation) links to the centre. Meanwhile, Bratislava has tended to integrate with Schwechat, Vienna and there is now a prospect of an integrated Bratislava–Vienna complex including tariff-free manufacturing zones close to the frontier. Such an arrangement would transfer a half share in Bratislava's Ivanka airport to Flughafen Wien who would then carry out modernisation and staff training.

Turning to the SEECs, major extensions have been made at Bucharest (Otopeni) and Sofia (Vrajdebna) with additional runways and new passenger/cargo terminals. Overloaded air-traffic-control systems are being overhauled and extended under EIB loans in both Bulgaria and Romania. Otopeni now has its dedicated bus link into the city centre, provided after sustained opposition from the taxi cooperative. The new capitals emerging in former Yugoslavia are now upgrading their airports, notably in the cases of Ljubljana and Zagreb. Macedonia has had a difficult job building up its airport at Skopje after it was stripped of much of its equipment when the Yugoslav Army left in 1992. There are fewer trained personnel and little traffic since much of the activity was formerly of a military nature. Traffic levels are still quite modest at Rinas in Tirana although a credit from Germany has enabled Siemens to modernise the runway, taxi lanes, lighting/navigation equipment and airport buildings in 1996. Rebuilding was undertaken by NATO in connection with the Kosovo War.

THE PROVINCES

Throughout the region provincial airports are becoming more prominent and some are being upgraded to handle international traffic. Kraków can now celebrate the operation of John Paul II International Airport while a new international airport to serve the Saxon urban triangle of Dresden–Leipzig–Chemnitz is planned. International flights are operating out of Constanţa and Timişoara in Romania (with Arad, Cluj-Napoca, Iaşi and Satu Mare to follow). However, several airports in former Yugoslavia found themselves in a difficult financial position even before the outbreak of ethnic violence: Ohrid airport closed in 1989 because of lack of funds to operate an unsafe runway. However, Kukës airstrip in Albania was improved to international standards by NATO in connection with its Kosovo operation.

Transition: telecommunications

Telecommunications are being most thoroughly overhauled since good intercity and international links are essential for foreign investment while a poor infrastructure constitutes a barrier to economic development. Under communism there was an increasing technological backlog with systems several decades behind the international average by 1989. Would-be subscribers had to wait up to 20 years for a connection, and while 30 per cent of Prague residents had telephones at the onset of transition, the proportion outside was only 5 per cent. Many villages – in the SEECs especially – had no telephone links at all and where connections did exist the service was frequently operator-controlled from the post office. Transition has seen the overloading of unmodernised fixed-line systems, given the huge scale of the pent-up demand with waiting times of up to five years in the Czech Republic. Considerable capital can be raised domestically through new subscriptions while useful administrative changes have been made by separating telephone companies from post office managements: the East German telephone service has been merged with the West German Telekom while Deutsche Post has been absorbed by Bundespost. There is also a trend towards privatisation, with a block of shares available for foreign investors: 30 per cent of shares in Matav (the Hungarian telephone company) have been taken by a consortium led by Deutsche Telekom. But there has also been international assistance – from the EBRD in the case of Albania – to finance transition to full digital exchanges with fibre-optic cables along main routes to which local rings are connected. Now waiting lists have largely gone and Internet-based telecoms are widely available – with quality improvements in the run-up to liberalisation and competition expected in Hungary in 2002.

The lack of previous investment in the south may prove an advantage because installations can now be designed from scratch, taking advantage of modern technology. There is World Bank finance for Bulgaria's telephones including long-distance optical-fibre cables to serve new users in Sofia. However, thanks to the country's role in electronics under Comecon, Bulgaria has enjoyed a high telephone density with 14 lines per 100 people in 1980 and 34 in 1993. But equipment is obsolete and hence the need to modernise. Bulgaria's first cellular network was established in 1994 through 'Mobikom', a joint venture between the state-owned Radio Electronic Systems and two private companies: Cable & Wireless and Bulgarian Telecom. It provides direct connections for Bulgaria's largest cities with more than 70 countries. In parallel with the situation in Poland, this has been followed by two analogue systems: one a wholly Bulgarian venture and a joint venture between 'Intracom' of Greece and Bulgarian Telecom. But there is also to be a digital overlay network (DON) involving 1,700 km of optic-fibre lines. In this case a joint venture between Telecom Netherlands and the Bulgarian State Telecommunications Company will benefit corporate users in eight city regions.

Other major companies are involved because Siemens have installed a new international digital exchange, along with trunk and local digital exchanges in the north to complement the work of Alcatel Cable (France) in the east, Ericsson Telefon (Sweden) in the south and Northern Telecom (Canada) in the west. Meanwhile, Satellite Transmissions Systems of USA have constructed an Intelsat ground station. Development of telecoms makes for increasing Internet penetration – 21 per cent expected for Hungary in 2005 – given the tendency for Internet service providers (ISPs) to offer free access: just now they find themselves with declining net capital expenditure while waves of new customers are coming forward.

Pressure on fixed-line telephones has increased interest in mobile phones, quite apart from their special usefulness for executives and businesspeople on the move (Table 4.7). In Poland where a fourfold increase in the number of subscribers was anticipated during the 1990s the telephone company (Telekomunikacja Polska) found demand increasing faster than its capacity to install new lines. Hence the growth of mobiles which began in Warsaw in 1992 with 82,000 subscribers countrywide after four years for the state monopoly's joint venture with France Telecom. However, the pace quickened as the original analogue systems were supplemented by the more technologically advanced GSM digital systems, two of which were licensed in 1996 with initial coverage in Gdańsk, Katowice, Kraków, Poznań and Warsaw subsequently extended through a dense network of transmitters enabling the network to expand to 160 towns and linking routes. The business is highly competitive with 0.6 m. users in 1997 and 3.0 m. expected in 2000. By the end of 2005 Hungary is expected to have 6.6 mobile subscribers ptp while the Czech Republic – with almost 8.0 – will be virtually on a level with Germany. Again, private capital is largely responsible for development as state monopolies break down: Albanian Mobile Communications has been bought by Greek-Norwegian consortium Cosmote-Telenor.

Table 4.7 Telecommunications, computing and the Internet 1994–9[1]

Country	Main phone lines		Mobile subscriptions		Internet hosts		Computers	
	1994	*1999*	*1994*	*1999*	*1994*	*1999*	*1994*	*1999*
Albania	1.3	3.6[2]	0.0	0.2[1]	na	na	na	na
Bulgaria	24.8	34.4	0.1	4.3	–	0.2	1.8	2.7
Croatia	29.6	35.0[2]	0.5	3.9[1]	na	0.9[1]	na	na
Czech Republic	15.7	37.0	0.3	18.9	0.1	1.2	4.4	10.7
Hungary	9.6	40.7	1.4	16.1	0.1	1.2	3.4	8.4
Macedonia	17.1	22.9[2]	na	na	na	na	na	na
Poland	8.7	26.0	0.1	10.2	–	0.4	2.2	6.2
Romania	10.2	16.6	–	6.2	–	0.2	0.4	2.7
Slovakia	13.4	30.7	0.1	17.0	–	0.5	2.8	7.4
Slovenia	21.1	38.1[2]	0.8	8.2	0.1	1.2	7.5	25.3

Sources: Eurostat 2000: 184–7; Eurostat 2001: 512–16

Notes
[1] Calculated per hundred of the population
[2] 1998

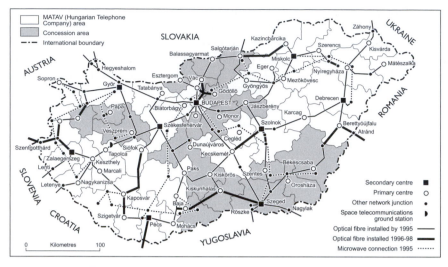

Figure 4.3 Optical-fibre networks in Hungary 1998
Source: Nagy (2001: 95)

However, while there is no doubt that enormous improvements have occurred, the spatial pattern is not one of uniformly high capacity. In the first place there are still many limitations in the availability of the international system of digital networks (ISDNs), involving optical cables with large data transmission capacity, without which it is impossible to access the Internet. This is a major drawback in terms of the development of SMEs which may depend on Internet contact with potential customers in export markets, where the greatest potential exists. Furthermore, even where modern systems are available, levels of penetration vary according to the social-occupational structure of the population. In Hungary, despite the widespread extent of ISDN (Figure 4.3), the contrast between Budapest and the provinces remains because penetration correlates with income and social status. And within individual provincial regions differences across the settlement hierarchy have increased. The county seats are in an advantageous position, with a relatively large number of well-educated people earning high wages and hence a high level of information activity. But, with poorer provision and penetration lower down the hierarchy, 'it is to be feared that, in the short run, the majority of the settlements and a large proportion of the population will be unable to adapt to the novelties accompanying an emerging information society' (Nagy 2001: 108). Perhaps new policies are needed. Albania is offering licences for local private operators who will install digital phone services for communes around cities provided all the villages in each area are covered.

Energy

The communist period

Great increases in the production of electricity reflected the growth of industry and the use of relatively heavy machinery consuming large amounts of power by

Western standards. However, the supply-driven logic of the communist system and the subsidised prices discouraged serious thought of conservation, although winter shortages impacted severely on domestic consumers. National grid systems, which did not exist outside Germany before the Second World War, assured a power supply for domestic and industrial consumers almost everywhere and this was a major factor in the changing distribution of industry. International connections organised through Comecon's 'Mir' grid allowed surpluses and deficits to be balanced out, but each country tried to be self-sufficient in electricity and the result was the development of large power stations on the principal domestic fuel bases such as the lignite fields: more than 5,000 MW of capacity was installed in areas like Cottbus in East Germany and Maritsa Iztok in Bulgaria; with more than 2,000 MW at Bełchatów/Konin in Poland and Craiova/Rovinari in Romania. However, there remained a substantial primary energy gap, which meant increasing dependence on the Soviet Union.

Initially this was not a problem because Comecon states enjoyed concessionary prices for Soviet oil and gas under the so-called 'Bucharest Principle' whereby the price was set at the average world market price over the previous five years. But this system – particularly attractive during periods of rapid world price increases – was revised in 1986 and world market prices have prevailed since 1991. Moreover, the Soviet Union reduced oil exports in the early 1980s and ECE's dependence on its own brown coal resources increased. In East Germany mining was set to reach 450 m.t eventually (West German figures say 660 m.), since 'brown coal had to be mined for survival without attention to environmental impact' (Böhmer-Christiansen 1998: 93), partly because fuels derived from brown coal (briquettes and coal powder for pulverised bed combustion) had to be exported to the West in growing amounts to earn hard currency and buy more political stability. Energy remained 'dirty', with East Germany one of the worst transfrontier polluters at a time when West Germany was cleaning up in response to 'Waldsterben' or forest death. While ecological disaster might be exaggerated, there was considerable environmental damage which 'stemmed from the inability of the socialist system to extract sufficient wealth from a society to which it had promised western living standards to invest in technical progress' (ibid.: 90).

The huge open-pit exploitations in Czechoslovakia, East Germany and Poland were mirrored on a smaller scale in the SEECs. In Romania, the Anina project envisaged a 900 MW power station to burn bituminous schist extracted from a huge quarry and then enriched by natural gas: the first 300 MW generating set was started in 1989 only for a generator breakdown and insoluble problems in coping with ash to force the abandonment of the whole scheme soon after the revolution. Other ways of boosting power output involved the use of hydro resources. The most attractive sites in the SEECs, like the Iron Gates, were developed within a few decades, though some systems performed badly on account of dry seasons and silt accumulation in reservoirs. But the potential in the north was more limited although the GDR installed pumped storage hydropower capacity in the Harz Mountains (Rübeland) and also at Saalfeld, following an earlier project in the Dresden area in the 1930s. Nuclear power offered a further solution that was used increasingly during the 1970s and 1980s as Soviet technology became available and the engineering industry in the ECECs gained the necessary expertise. Nuclear cooperation between the Soviet Union and ECE began in 1960, but a standing committee for nuclear energy was created only in 1972 to allocate research responsibilities while Interatomenergo carried out the planning on the basis of standardised Soviet reactors. The first stations in Bulgaria (Kozloduy), Czechoslovakia (Jaslovské Bohunice) and Hungary (Paks) came on stream in 1974, 1978 and 1983 respectively. Commissioning delays were considerable but, more seriously, the Chernobyl disaster swung public opinion against further nuclear projects even when non-Soviet technology was involved.

Transition: integration and deregulation

Demand for electricity is falling due to factory closures and the pressure of realistic energy prices. However, the energy gap remains. Increased flexibility has been achieved through the coordination of electricity grids: in 1991 the Visegrád states agreed to coordinate efforts to bring their electricity transmission systems into the Union for the Coordination of Production and Transmission of Energy. This would allow emergency imports of power from the West but would also facilitate regular exports where surplus capacity was available, e.g. from Poland to Germany which became a possibility in 1996. Interconnections are now widespread and Hungary – which is very strategically placed – is linked with the grids of virtually all its neighbours. Since overall energy self-sufficiency remains highly desirable in order to reduce the balance of payments deficits, exports of electricity (nuclear power where public opinion is supportive) or coal/hydrocarbon fuels take place wherever possible. Power engineering firms have considerable export potential especially if Western links are forged, like Škoda Plzeň in the Czech Republic and Zamech of Elbląg in Poland. The expertise of General Electric (GE) in gas turbines has been linked with Škoda's skills in steam turbines and its low-cost manufacturing base. So GE will supply gas turbines and engineering expertise for Škoda-designed power plants installed in the ECECs as well as the CIS and various Asian countries.

There has been considerable privatisation especially in East Germany where the three big West German utilities – RWE Energie (Essen), PreussenElektra (Hannover) and Bayernwerk (Munich) – took over most of the power stations from Treuhand in 1991 (while some were taken over by municipalities) and set up VEAG (Vereinte Energiewerke AG) to manage the high-voltage grid. As Böhmer-Christiansen (1998: 92) points out, the aim was to buy the East German assets cheaply and make high profits to finance the environmental clean-up and encourage energy saving. About half the Verbundnetz gas monopoly has been sold to the two main West German corporations (BEB Erdgas and Ruhrgas) with the remainder made available to other German and foreign interests. Opencast lignite mining complexes (such as the Mitteldeutsche Braunkohle interests at Leipzig, linked with the 1,600 MW Lippendorf power station) have been taken over; while the sale of the Minöl oil distribution business to Elf Acquitaine of France has led to the modernisation of some 1,000 filling stations. The project is linked with the German steel and engineering company Thyssen through the latter's acquisition of the 10.0 m.t refinery at Leuna (raising the possibility of a pipeline from Rostock) and the partnership has also taken over the petrochemical unit at Zeitz.

Progress has not been so rapid elsewhere since there is little money within the region while the poor condition of many installations discourages foreign interest, not to mention some strategic barriers to majority stakes as in the case of the Hungarian oil company MOL. But privatisation is gathering momentum and former national power companies are caught up in global processes of oil sector consolidation. Thus MOL has acquired an interest in the Slovakian oil giant Slovnaft and other national oil companies will need to forge partnerships given that there is now competition within each state. Arguably they are 'doomed' to becoming subsidiaries of TNCs. The MOL strategy makes the company big enough to survive, especially in view of its enhanced interest in petrochemicals following the stake acquired in the Hungarian petrochemical company TVK. Austria's ÖMV is well situated to develop as a pan-regional conglomerate with more petrol stations in the region than any other Central European company, while Hellenic Petroleum of Greece could well penetrate the region more exten-sively from the south (Greek companies are now prominent in Durrës). However, Croatia declined to sell their state oil and gas company INA to MOL because the price was too low and ÖMV may offer a better deal. It remains to be seen if any agreement will include the Janaf (Adria) oil pipeline system based on Omišalj which serves Bosnia & Hercegovina, Hungary and Yugoslavia as well as Croatia. Some states are still reluctant to sell what may be seen as national assets despite pressure from global institutions. The Poles want their national oil company PKN to be a regional leader and not the victim of a carve up. The Romanians take a similarly defensive attitude towards Petrom, fearful over the long-term future of oil refining in the country.

Power stations are being privatised like Poland's Konin-Adamów complex and the scope for competition in electricity generation may make it attractive for large consumers to build their own power stations. Those which already have their own capacities, like the Resita iron and steel works in Romania which controls the hydropower of the Bârzava river, may achieve overall viability as a result. Of course an industrial company may not always be able to raise capital to build or extend power stations. For the Dunaújváros steelworks has arranged with Tenneco Energy (USA) the sale of its Dunaferr gas-fired station so that capacity can be increased from 60 to 200 MW. But if industrial users do opt to build their own plant this could create problems for other generators and distributors. Emasz, serving the northeast of Hungary (one of four such companies across the country), is threatened by the interest of BorsodChem, the petrochemical company TVK and Diosgyör Steel Works who could save considerably through operating their own plant. Further reform to create competitive energy markets could thus be dangerous. Domestic consumers, currently subsidised by industry, could face steep price increases, as new distrib-utors capture corporate business with lower prices or large industrial consumers make their own generating arrangements. But there is also a risk that competition to supply power will drive down prices charged by generators to the point where further investment in capacity is brought to a halt. There is no guarantee that privatisation and deregulation will maintain prices at levels sufficient to maintain interest in electricity shares while encouraging greater fuel efficiency and reduced energy intensity.

There are various ways in which Western companies are becoming involved in the energy sector of ECECs without actually owning mining or power companies. Apart from the links with power engineering companies already referred to, several foreign oil companies are setting up filling stations – likewise some domestic oil-refining companies like Gdańsk – to challenge the state distribution companies which are being forced to modernise their outlets in order to remain competitive. There has been a remarkable transformation in the outlets of the Romanian national oil company Petrom which are now largely equal to the competition. There is scope for upstream joint ventures like Pol-Tex Methen which is to exploit coal-bed methane at Rybnik through underground gasification: 920 m.m³/yr with total reserves of 600 bn.m³. Foreign help is being engaged in the renovation and operation of the Albanian oilfields and refineries, which experienced a sharp fall in production during the early 1990s, and also in prospecting for oil: offshore in Albania and Bulgaria – but not Romania which runs its own Black Sea operation – and on land in Poland and Romania. Downstream collaboration relates to oil refining and distribution. British Gas is using its 'swagelining' technique (inserting plastic pipes into steel pipes) to repair pipelines for the West Bohemian Gas Company which has built three environment-friendly boiler houses in Prague. Desulphurisation plant at Poland's Jaworzno power station is linked with the German company Knauff who will supply a factory and the necessary technology to use the dehydrated gypsum annually for construction materials.

The transition: the fuel mix

A considerable reduction in energy consumption has occurred in the ECE candidate countries covered by Table 4.8: 292.1 m.t oil equivalent in 1990 to 246.9 in 1998, with an improvement in energy intensity at the same time. However, solids are playing an increasing role in domestic production (up from 70.0 per cent in 1980 to 72.8 per cent in 1998, given the sharp decline in oil and gas from 55.9 to 25.8 m.oe – a reduction of 35.1 m.t of which Romania alone accounts for 25.9). There is no immediate prospect of recovering this loss which has kept the hydrocarbon imports of the seven countries stubbornly in the 80–90 m.t oe bracket for the last two decades. Import dependence is therefore increasing relatively if not absolutely: 22.5 per cent of the region's consumption in 1980; 27.1 per cent in 1990 and 27.6 per cent in 1998. Trends vary for individual countries. In Bulgaria the proportion of 50.5 per cent in 1998 is down from 73.1 per cent in 1980, thanks to the development of nuclear power while maintaining output of solid fuels. Hungary has compensated for reduced solid fuel output with nuclear capacity (dependence up slightly: 49.1 to 52.8 per

Table 4.8 Energy balance for EU candidate countries 1980–98 (m.t. oil equivalent)

		Total		Solids		Oil-Gas		Primary Electricity			Other Imports	
		A	B	A	B	A	B	Aa	Ab	B	A	Dep.
Bulgaria	1998	9.8	10.0[2]	5.0	2.2[2]	0.0	7.8[2]	4.4	0.1	–	0.3	50.5
	1990	9.9	17.0[2]	5.4	3.4[2]	0.1	13.3[2]	3.8	0.2	0.3[2]	0.4	63.1
	1980	7.7	21.0[2]	5.2	4.3[2]	0.4	16.4[2]	1.6	0.3	0.3[2]	0.2	73.1
Czech Rep.	1998	31.3	10.5[2]	26.6	4.4[3]	0.8	14.9[2]	3.3	0.2	–	0.4	25.1
	1990	39.5	5.4[2]	35.2	7.9[3]	0.4	13.4[2]	3.3	0.1	0.1[3]	0.5	12.0
	1980	43.5	4.5[2]	42.2	8.7[3]	0.5	13.3[2]	–	0.2	0.1[3]	0.6	9.3
Hungary	1998	12.3	13.8[2]	3.9	1.1[2]	4.6	12.5[2]	3.6	–	0.2[2]	0.2	52.8
	1990	14.2	14.2[2]	4.1	1.6[2]	6.1	11.6[2]	3.6	–	1.0[2]	0.4	50.0
	1980	14.8	14.3[2]	6.3	2.2[2]	7.6	11.5[2]	–	–	0.6[2]	0.9	49.1
Poland	1998	91.8	6.8[2]	82.6	19.1[3]	3.6	26.1[2]	–	0.2	0.2[3]	5.4	6.8
	1990	101.6	2.1[2]	94.5	18.9[3]	2.6	21.1[2]	–	0.1	0.1[3]	4.4	2.0
	1980	124.0	2.6[2]	115.9	20.5[3]	4.8	23.1[2]	–	0.2	–	3.1	2.0
Romania	1998	26.1	10.8[2]	4.8	1.9[2]	16.6	8.9[2]	1.4	1.4	–	1.9	29.2
	1990	39.7	21.6[2]	7.6	4.3[2]	30.6	16.5[2]	–	0.9	0.8[2]	0.6	35.2
	1980	52.7	12.0[2]	8.1	4.0[2]	42.5	8.0[2]	–	1.1	–	1.0	18.5
Slovakia	1998	4.4	12.6[2]	1.2	3.4[2]	0.2	9.1[2]	2.6	0.3	0.1[2]	0.1	74.1
	1990	5.3	16.4[2]	1.4	5.7[2]	0.4	10.3[2]	3.1	0.2	0.4[2]	0.2	75.5
	1980	3.4	17.4[2]	1.7	6.0[2]	0.1	11.1[2]	1.2	0.2	0.3[2]	0.2	83.6
Slovenia	1998	2.9	3.8[2]	1.0	0.2[2]	–	3.7[2]	1.4	0.3	0.1[3]	0.2	56.7
	1990	2.7	2.5[2]	1.2	0.1[2]	–	2.5[2]	1.2	0.3	0.1[3]	–	48.0
	1980	na	na	na	na	na	na	na	na	na	na	na
Total	1998	178.6	68.3[2]	125.1	14.7[3]	25.8	83.0[2]	16.7	2.5	na	8.5	27.6
	1990	212.9	79.2[2]	149.4	11.7[3]	40.2	88.7[2]	15.0	1.8	2.2[2]	6.5	27.1
	1980[1]	246.1	71.8[2]	179.4	12.7[3]	55.9	83.4[2]	2.8	2.0	1.1[2]	6.0	22.5

Source: European Commission Annual Energy Review 2000

Notes
[1] Excluding Slovenia
A Domestic production
Aa Nuclear
Ab Hydro and Wind
B Net Trade ([2] denotes import; [3] export)
Dep. Overall import dependence (percent total energy consumption)

cent) while nuclear has helped to reduce the level in Slovakia from 83.6 to 74.1 per cent. Meanwhile, reduced coal mining in the Czech Republic and Poland has meant an increase in dependence: the Czech Republic from 9.3 to 25.1 per cent in 1980–98 (despite nuclear power) and Poland from 2.0 to 6.8 per cent. Despite a drastic reduction in consumption, Romania's import dependence has increased (from 18.5 to 29.2 per cent) due to the catastrophic reduction in hydrocarbon output – 42.5 m.t oe to 16.6 – and a reduced coal output as well.

Demand for coal is falling with the decline of heavy industry and the switch to gas for electricity generation and domestic heating. This requires economic restructuring in the coalfields through light industry and small businesses to provide alternative employment since job losses are disproportionately high due to higher productivity. Reform of energy markets again poses a threat to coal through an accelerated switch to gas at a time when more nuclear capacity is coming on-stream in some countries. However, there is a widespread view that the use of lignite should be reduced even though better filtering of emissions could reduce pollution and that any surviving coal industry should ultimately be profitable. It should also find acceptance among local communities. Thus in his work on Most in northern Bohemia, Pavlinek (1998a: 230) explains how the new coal companies have had to build new relationships with communities and work within new ecological limits. However, in the short term the communities see the same people running the business and inevitably associate them with the 'megalomania of coal mining' when communist authority could not be challenged (ibid.). There is still some apprehension over threats posed to villages; just as in Slovakia's Nitra Valley the expansion of mining – with resettlement of people in the Prievidza area from Koš to Kanianka – has continued since 1989. However, the more significant trend appears to be union resistance to contraction, very evident at Petroşani in Romania's Jiu Valley coalfield where miners have been connected with a series of violent interventions in Bucharest, especially in the early transition years. However, this is just one of many cases where the working of low-grade minerals can no longer be justified in a global economic context.

Coal in Poland

Coal production fell from 180 m.t in the late 1980s to 133 in 1997, but home demand is now below 100 while export to Western Europe has recently been unprofitable due to falling world prices and is maintained only for social reasons. Various agreements have been made with the unions, most significantly the 'Katowice Contract' (1995) for downsizing over a ten-year period. Closing some 20 mines will lower capacity by 25 m.t while employment will be reduced to 135,000 from 400,000 in the late 1980s and 240,000 in 1997. The contraction will be concentrated overwhelmingly in Upper Silesia (and especially along the 'saddle' in the heart of the conurbation) but some closures will also occur in the area of more recent development around Jastrzębia and Rybnik where reserves, quality and productivity offer no prospect of viability. The loss-making coal companies into which the state-owned industry has been split are also interested in selling off individual pits. The agreement includes severance payments and diversification measures including the Katowice SEZ set up in 1996 with sites in Gliwice and Dąbrowa Górnicza on the saddle and at Jastrzębia and Tychy to the south. New ventures include the Opel car factory in Gliwice where about a fifth of the 1,200 workers are ex-miners. Indeed, coal is now a burden for Silesia which has been developing IT since the 1970s and spawned its first private computer company by 1984.

The changing economic profile can also be seen in the landscape with the planting of spoil heaps, desalinisation of water discharged from the pits, stone crushing and sludge treatment. The hope is that a stable and profitable industry will ultimately emerge but the companies have to keep borrowing to pay wages and service debts while the difficult geological conditions and low technical level (perpetuated through low investment) suggest further decline especially if the end of price controls in 1999 gives Polish power stations the incentive to import coal. However, limited port capacity means that Polish power generators will remain dependent on high-cost indigenous coal supplies in the short term. Poland's energy guidelines provide for diversification in favour of gas and liquid fuels, but 60 per cent of all electricity generation remains coal-based and mines are tied to particular power stations with production levels pegged close to market needs. But over the longer term Poland may increase gas consumption from some 10.0 bn.m^3 at present to 22–7 bn. by 2010, with a proportion of this (estimated at 4.0 bn.) coming from offshore fields in the Baltic to supply the Gdańsk–Gdynia–Sopot Tricity and perhaps the Zarnowiec power station as well. Overall it is envisaged that coal consumption will fall from 98.8 m.t oe in 1988 to 71.5 in 2010, while oil and gas will rise from 27.4 to 45.4 and other energy sources from 2.4 to 7.8.

Hydrocarbon fuels

Although it is desirable to reduce coal consumption in power stations for environmental benefits this can mean greater foreign dependence for hydrocarbons, even with the fullest exploitation of domestic sources. Western technology has been used to restore oil/gas production levels. Offshore prospecting for oil has been successful in Poland and Romania. Petrobaltic landed its first oil from the Baltic at Port Północny in 1992 and daily output has been reported at 1,300 t. Methane gas is also being tapped offshore at Cape Rozewie, while further prospecting is taking place on the edge of the Carpathians and also in Romania. Oil prospecting on land has been undertaken in Poland and Romania but without success. The last discovery on land in Romania dates back to 1969, but it was hoped that a concession competitive in a global context might attract complex technology and risk capital with a prospect of success. It now seems that a better strategy will be to increase production in existing fields through secondary and tertiary recovery. Albania – an oil producer since 1918 – is also rehabilitating oilfields to increase output from 323,000 t in 1999 to 2.5 m. in 2016. Foreign companies are drilling offshore but they have not yet found commercial quantities. So it seems that the region will remain heavily in deficit over oil supplies with the need to import 60 m.t out of an estimated demand of 75 m.t. In 1991 Albania had a small oil surplus (0.1 m.t), which continues – and will be an even greater asset when Ballsh refinery is recommissioned. Further production is signalled by the Premier Oil deal at Patos Marinza which involves central processing and an export pipeline. However, Albania has an overall energy deficit because of net electricity imports since drought has reduced hydropower output. Hungary, Romania and former Yugoslavia together had a substantial oil production of 15.8 m.t to set against a consumption of 38.7, but the other ECECs managed only 0.5 m.t production against a consumption of 59.3.

It is not clear how much of the hydrocarbon deficit will be met by Russia, given the need to pay in hard currency, the poor state of the pipelines and difficulties in increasing production, for oil deliveries in the early 1990s actually fell. Gas could be supplied from Norway, Algeria and the Middle East by tanker to terminals on the Baltic and Black Sea coasts, but Russia is in a strong position with a distribution system already existing through the ECECs to Western Europe and new deals have been negotiated. The resources of the Yamal Peninsula were plugged into the system in the late 1990s while distribution amongst the SEECs is being facilitated by extensions to the Bulgarian pipeline into the Western Balkans and a new pipeline across Romania through Iaşi and Satu Mare to Hungary. Oil pipelines are split between links connecting Russia with ECE and others supplying West European refineries from ports like Genoa, Marseilles, Rotterdam, Trieste and Wilhelmshaven. ECE does have ports which can handle crude oil. Constanţa (Romania), Rijeka/Omišalj (Croatia) and Varna (Bulgaria) all have pipelines to transport oil inland while Rostock has a pipeline link with the system originating in Russia. Storages are also available in Bulgaria (Burgas), Poland (Port Północny, Gdańsk) and in former Yugoslavia at Bar (Montenegro), Koper (Slovenia) and Zadar (Croatia). Thus Western oil companies could get involved through the extension of West European pipelines or by tankers supplying direct to terminals in the ECECs. Competition is very desirable from the customer's point of view, given the high transit charges Bulgaria has to pay to import Russian gas across Romanian territory and obligations to pay 75 per cent of the price of any part of the quota not taken up.

Poland could receive oil and gas at Gdańsk to reduce dependence on Russian supplies and could build a pipeline from Gdańsk to the Płock refining and petrochemical complex; alternatively delivery could be taken through Rostock, where the port could be improved by dredging to accommodate larger tankers. Gas could come in liquid form from Africa or the Middle East in order to avoid over-dependence on Russia, but it is now decided that Poland will receive 3–4 bn. m^3/yr of gas from Norway by a Baltic pipeline which could also be used for deliveries to Lithuania and Slovakia. Again, the pipelines based on Rotterdam and Wilhelmshaven could be extended to join the former Soviet Friendship system. Meanwhile, Hungary could join the Janaf system based on Rijeka in either Croatia or Vojvodina; and Slovakia could get alternative supplies by extending the Schwechat pipeline (based on Trieste) to Bratislava. This leaves the Czech Republic (served by the Friendship system as far as Záluží in western

Bohemia) with the choice between pipeline extensions from Gelsenkirchen or Cologne (based on Rotterdam), Strasbourg (Marseilles) and Ingolstadt (Genoa and Trieste). The third option was taken up and supplies from Ingolstadt now reach the Czech system at Kralupy nad Vltavou, following the proposal from AGIP, Conoco and Total to construct the pipeline and invest in modernisation of Kralupy and Litvinov refineries. Meanwhile, Macedonia is supplied with Russian gas piped via Bulgaria, but it remains to be seen if Serbia will be happy to continue using the Janaf oil pipeline now that it is under Croatian control. Caspian hydrocarbon supplies are another option with pipelines from Burgas to Vlorë and Constanţa to Trieste featuring among the possibilities. However, switching to alternative fuels may be difficult: where power stations cannot convert to imported fuel and domestic coal there will be closer business contacts between coal producers and power generators.

Given the need to import energy it is all the more desirable to modernise industrial plant to limit energy consumption and reduce pressure to bring new capacity on stream – all the more so given extremely high energy intensity levels (Table 4.9) and the challenge to achieve a better quality of life commensurate with 'fair shares' of global energy consumption. Figures quoted by Kramer (1999) show that Hungary had a per capita energy consumption of 69.4 per cent of the EU for a per capita GDP output of 38.7 per cent in 1995. In other words GDP per unit of energy was only 55.6 per cent of the EU level; though this was better than all the other states examined: the Czech Republic 53.1 per cent; Poland 49.3 per cent; Romania 44.5 per cent; Bulgaria 42.6 per cent and Slovakia 38.9 per cent. 'East European nations cannot afford to keep pouring energy down the drain' (Kats 1991: 865). Overall, the region is relatively more sustainable than Western Europe but may lose this advantage – given the growth in consumption expected over the next few years – if the region cannot leapfrog towards a better model of dematerialised sustainable development. Progress is slow due to technical and financial difficulties: many industries cannot afford energy-efficient measures, 'but without them they will find it increasingly difficult to survive and compete in the European market particularly as energy costs increase' (Green 1998: 99). Government support is needed to meet higher national standards of energy efficiency; but the banks also need to make energy-efficient equipment a priority. The Czech Republic's Kyjov district heating system has benefited from a Global Environment Facility (GEF) grant of $5.8 m. to reduce emissions of greenhouse gases and increase efficiency in the power/heat supply to Vetropak Moravia glass factory. Lignite will still be used, but with a reduced share of the total fuel mix. Over the long term, Poland would like to transfer from heavy, energy-intensive industries to agriculture and high-tech processing but it needs international cooperation over credits and technology transfer. Greater efficiency is being encouraged by the Alliance to Save Energy and by organisations such as REC and USAID working in liaison with the Polish Network of Energy Cities (covering around 40 small towns) and the Romanian Energy Policy Association. Slovakia has produced a practical guide to 'Energy and Buildings in Slovakia' distributed among the construction sector.

Table 4.9 Energy intensity 1980–98

	,000t oil equivalent per € GDP				
	1980	*1985*	*1990*	*1995*	*1998*
Bulgaria	2.35	2.04	1.69	1.79	1.58
Czech Republic	2.51	2.48	2.16	1.95	2.00
Hungary	1.19	1.13	1.01	1.01	0.94
Poland	2.71	2.73	2.20	1.92	1.60
Romania	2.16	1.85	2.03	1.69	1.52
Slovakia	1.98	1.90	1.76	1.66	1.36
Total (ECE)	2.21	2.09	1.85	1.67	1.57
European Union	0.29	0.27	0.25	0.24	na
World	0.54	0.52	0.50	0.43	na

Source: European Commission Annual Energy Review 2000

New sources of electricity

Whatever happens to total demand, new power stations will have to be built in the interest of modernisation, rationalisation and reduced import dependence. In Poland the coal industry is being helped by public unease over nuclear power, and a new coal-fired capacity at Bełchatów (800 MW) has been linked with a lignite mine at Szczerców. Indeed a further 1,000 MW may be needed to meet peak demand and provide a surplus of power for export. But it seems inevitable that gas will emerge as the best alternative to nuclear power and will account for some of Poland's extra capacity, including the Zarnowiec project which was selected as the site for Poland's first nuclear power station. Meanwhile, Macedonia plans to import high-sulphur lignite from Greece, where environmental standards will not permit its use, in return for electricity from Bitola power station which is not subject to the same constraints. On the other hand, a referendum in Slovenia in 1999 decided against the new 200 MW Trbovlje power station, proposed by Termoelektrana to maintain support of the Hrastnik and Trbovlje mines when the present plant is closed by 2004. Alternative energy options will require the economic and social restructuring of the Zasavje region. Croatia wants to avoid dependence on imports of electricity from neighbouring states by building three regional power stations for north/central Dalmatia, Vukovar-Sirmium county and Zagreb but they will function on 'ecologically safe fuel' (not coal or nuclear energy).

Renewable energy

There are wind farms in East Germany where the Ministry for Economics and Labour for Saxony is keen on small biomass methane generators up to 100 kW; solar to 200 kW; hydro to 300 kW. Messerschmidt Blohm und Bolkov (MBB) plan a 4.0 MW photovoltaic pilot plant at Bad Langensalza in Thuringia (the site of a plant for the building and marketing of solar cells). In 1999 a wind turbine was set up in Hungary on a 30 m tower adjacent to the Bakony Power Plant's Inota thermal station at a cost of HUF60 m. It should pay for itself in

20 years, but there will be an evaluation after two years to see if further investment is justified. Meanwhile, in the Czech Republic, the Děčín geothermal
energy heating project will improve the environmental health of the town by
reducing consumption of non-renewable energy. Geothermal energy has been
in use at Galanta in Slovakia since 1996 where the Podhajska station (8 MW)
supplies the local community using thermal waters of Oravice. There are similar
plans for Bojnice and Bánovce nad Bebravou, while biomass is used to incin-
erate wood and agricultural waste, including excrement from cattle and pig
farms. Hydro sources remain a possibility although high development costs
including the loss of agricultural land and social disruption constitute a disincentive. It is possible that small hydro schemes may be cost effective in yielding
significant amounts of power while contributing to water storage and flood
control without the costs of massive relocation of settlement and extensive
flooding of farmland. However, Albania envisages eight small hydro stations for
the Vjosë valley in the south of the country: a high demand area which is inadequately served due to problems with the grid-carrying power from the principal
hydro stations in the north. There is scope for a similar approach in Romania.

Since 1990 there has been rapid growth of small private power stations in
Poland based on renewable energy – 283 hydropower, 11 biogas and seven
using windpower – and the sale of power direct to the country's 33 distribution companies. This is encouraged by five-year tax exemptions and stimulative
prices from the distribution companies (set by parliament at 85 per cent of the
average for industrial, municipal and domestic consumers) and easy access to
preferential loans. The trend is also helped by developments in technology globally during the 1990s with regard to turbines and control equipment, not to
mention family traditions in running small power plants. Only 1.5 per cent of
Poland's electricity is from renewable sources (5.5 per cent for the EU) and
government is committed to increasing this proportion. Under Polish law, utilities must obtain 2.4 per cent of all the energy that they sell to consumers from
green sources; rising to 7.5 per cent by 2010. The Netherlands has offered
E55 m. to Poland for the construction of wind turbines, 30 of which are to be
installed in the northwest by 2003. Businesses are also playing a part in this
trend: the Dutch confectionery company Vam Melle has installed a 250 Kw
wind turbine at its Warsaw factory which helps to reduce sulphur dioxide.
Reference may also be made to geothermal sources which are being used for
heating in several areas: Mszczonow – dubbed the 'ecological town of the 21st
century' – and the Zakopane area of the Tatra Mountains.

Hydropower has always been regarded as an attractive option for the SEECs
in view of their mountainous terrain and limited high-quality coal endowments.
Virtually all the major potentials in the Romanian Carpathians have now been
exploited, although some projects that were halted by the revolution (like the
Buzău-Bâsca catchment) will now be completed with foreign help (Figure 4.4).
However, the initial investment is high and the environment effects are being seen
as increasingly unacceptable. There is much concern over Croatia's plant on the
Drava where a 3.5 km diversion canal could seriously endanger the flora and
fauna; causing Hungary to decline any involvement. In the face of strong ENGO

pressure Hungary has also withdrawn from a joint project with Slovakia on the Danube which is examined in a case study below. Meanwhile, severe drought in the Balkans during 2000–1 has led to greatly reduced output from hydro stations. Autumn 2000 saw the Drin at its lowest for 30 years, while Macedonia felt unable to increase the flow from Lake Ohrid for fear of environmental damage; so with 1,668 MW of hydro capacity (eight large stations and 83 small ones, mostly in a poor technical state and due for privatisation) and only 224 MW of thermal power, mainly at Fier, it was not possible to produce more than 10 m.kwh daily to meet a winter demand of 16–18 m. Hence Albanian consumers faced prolonged daily power cuts and rationing throughout the winter. It will take time for greater security to emerge from new power station reconstruction (at Korçë) and a 400 Kv link between Podgorica and Elbasan – to be implemented under the Stability Pact with a grant from the German Bank for Reconstruction and Development – to increase the power that can be imported: at present the limit is 6.0 m.kwh daily, given the obsolete grid system, with Montenegro and Slovakia the main suppliers. After building no new power plants for 20 years, a Chinese construction company is to construct the 80 MW Bushat plant in the north of Albania, while Italy will invest in the 100 MW Kalivac plant on the Vjosë. The rehabilitation of the Bistrica stations is also sought.

Hydro-electricity and the Gabčikovo–Nagymaros controversy

Meanwhile, large hydropower projects have not been consigned to history. Given the difficulties over water supply and alternative sources of energy, Bulgaria remains interested in large-scale projects and the Gorna Arda hydropower plan was agreed jointly with Turkey in 1999. This will certainly continue to attract opposition, although the greatest controversy arose over the Gabčikovo–Nagymaros project on the Danube, involving two dams in the section where the river marks the frontier between Hungary and Slovakia and short sections on either side where the river lies wholly within the one country or the other. It was given serious thought in the 1950s under Comecon auspices, but was shelved in favour of an integrated electricity grid. Interest was rekindled on a bilateral basis in the 1970s with particular enthusiasm from Czechoslovakia, firmly recentralised after the Prague Spring of 1968 yet still dedicated to regional development projects for the benefit of Slovakia. The new plan was organised on a joint basis with Hungary in 1977 and work started in Czechoslovakia in 1978 and Hungary in 1980. The Soviet Union was broadly supportive, through the prospect of an upgraded continental waterway (with the section from Bratislava to the Black Sea effectively freed from the historic problems of low water) and a perceived political advantage in linking Hungary more closely with her more reliable northern neighbour. However, a clear philosophical difference arose between Hungary, responsive to the environmental lobby's plea for sustainable development, and Czechoslovakia where 'real conflict between civil society and the central leadership could not evolve' (Galombos 1993: 218) prior to the Velvet Revolution; also after 1993 independent Slovakia was highly committed to the project through its continuing drive for industrial growth and modernisation (ibid.: 221).

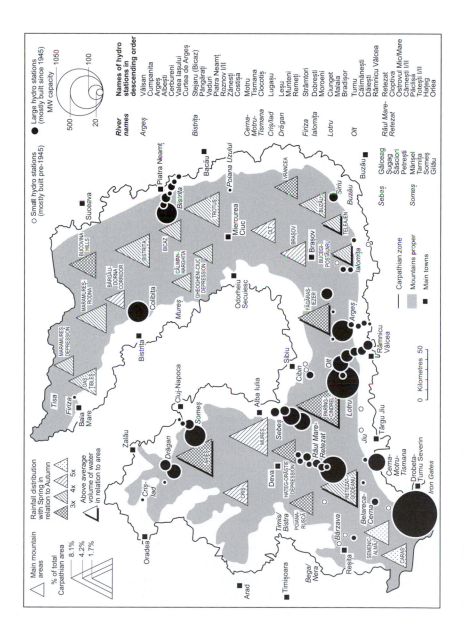

Figure 4.4 Hydropower in the Romanian Carpathians

Source: Based on the Romanian National Atlas

Opposition on environmental grounds surfaced in Hungary in the early 1980s and despite renewed commitment to the project in 1985 (when financial problems were overcome) Hungary later suspended its contribution and subsequently the Antall government decided to withdraw from the project altogether. At a time of political change, public opinion considered that farmland and wetland would be lost, while earthquake damage might precipitate a major flood disaster. However, Czechoslovakia (and Slovakia after 1993) claimed that the Gabčikovo dam (along with an artificial 25 km canal 300–650 m wide at a level 11–18 m above the natural river) would protect the Zitney Ostrov (Rye Island) zone from flooding and would allow water to be released into the natural drainage system to protect the wetlands. This would also safeguard Hungary's drinking water which is obtained by seepage from the Danube by means of a complex process of filtration into the water table. Navigation would benefit and the electricity generated (eight 90 MW turbines) would mean less dependence on coal-burning power stations and so reduce gaseous and solid emissions. There was also a wider issue because of the deepening incision of the river on its natural course, attributed to dam construction in Austria in the 1950s, holding back gravel that was formerly carried downstream to the delta. Carrying less material, the river eroded more effectively. The impact was exacerbated by dredging carried out by the Slovaks at Bratislava (to avoid flooding and improve the port) in the 1970s and 1980s; also through work to straighten the channel and increase flood capacity. Thus the river was progressively cut off from the wetland. But arguably river regulation would continue to have a knock-on effect, with increasing difficulties for navigation and ecology, and so make Gabčikovo essential. In the same way, the Nagymaros dam would become unavoidable as erosion made navigation increasingly difficult, while a further dam would eventually be needed to the south of Budapest.

Abrogation of the treaty in 1989 resulted in a stoppage at Nagymaros near Visegrád (where a 160 MW power station was planned) where the Danube lay wholly within Hungarian territory. But at the upper end of the section affected it was possible for Slovakia to proceed unilaterally in 1991 with a variant of the Gabčikovo dam project which diverted water into an artificial channel at Cunovo – instead of Dunakiliti further downstream – where that state controlled both banks of the river. This part of the scheme was completed although Hungary still considered it an environmental hazard while the diversion of the navigation channel away from the frontier and into Slovak territory was a blow to national pride, through a perceived 'attack on the country's borders' as well as the local interests along the section of river affected. Hungary is also sensitive to the fact that it was the Hungarian minority population in Slovakia that bore the brunt of the inconvenience (while nationalists branded opponents within the country as Hungarian agents). The Gabčikovo scheme is now complete and the environmental situation appears to bear out Czechoslovak and Slovak assessments, while accessibility has improved for the 'island' between the two channels. Some were apprehensive about 'disaster' but chose not to move and now appreciate the potential for rural tourism.

Hungary challenged Slovakia's right to proceed unilaterally through the International Court of Justice during 1994–7 during which time the work in

Slovakia was completed. Although the court ruled that Slovakia should not have proceeded unilaterally, it also upheld the legality of the original agreement and ruled that Hungary should proceed with the Nagymaros section. The former Horn government was prepared to negotiate on this basis, but there was opposition in Budapest from the junior coalition party ('Fidesz') and Horn had to delay signing an agreement and call for more impact studies. In the meantime the election has brought Fidesz to power and subsequent governments do not accept any obligation and the matter remains in dispute. It is certainly true that the project is a survival from the communist past and the Hungarians are captive to a hasty commitment made in 1977 'based on the environmental paradigm of the 1950s and energy supply projections of the mid-1970s' (Ostry 1988: 22). On the other hand there is a school of thought still arguing that since the regulation of the Danube has started in German and Austrian territory, the adjustments to the bed make it essential for similar works to continue on downstream.

Transition: the nuclear dilemma

It has already been noted that several ECECs went in for nuclear power before 1989, with stations on-stream in that year in Bulgaria, Czechoslovakia, East Germany, Hungary and former Yugoslavia, though only the latter used non-Soviet (CANDU) technology (Table 4.10). Countries were driven by limited fuel endowments and access to technology in a way that would allow their engineering industries to participate in the production of equipment. Poland and Romania had plans, based on the Soviet and CANDU systems respectively, leaving Albania as the only state limited to conventional power. Nuclear power continues to be widely regarded by many as an essential part of the region's energy profile, but the issue is now more vigorously contested. Arising out of Chernobyl, much opposition is down to fears about safety, since Soviet-designed reactors in the ECECs were not placed in containment vessels as a precaution against 'worst case' scenarios such as melt-down. There is also a conviction that nuclear capacities will become unnecessary if adequate conservation measures are taken with greater emphasis placed on renewable energy. These arguments have helped to neutralise the strong pro-nuclear lobby led by people working for the nuclear industry which now faces a sluggish market in Western countries; as well as the 'true believers' who see nuclear power as the least of the evils in the present energy supply and demand picture.

East Germany is exceptional in having seen all its nuclear stations closed. The small Rheinsberg station was shut down even before 1989 while Griefswald was closed in 1992 on the grounds of safety and high operating costs. But in other countries closure will be hugely burdensome, given the loss of power (sometimes exceeding half the national electricity output) and the costs of decommissioning. On the other hand, safety has been improved, for although in-built weaknesses and deficiencies in operational safety standards were revealed by the International Atomic Energy Agency in 1991, the feasibility of modernising to Soviet-designed reactors attracted Western financial assistance and generated competition among the nuclear power companies wishing to carry out the work.

Table 4.10 Nuclear power 1980–95

| | Capacity (Giga.Watt -10^9) | | | | | | Generation (TeraWatt.hour -10^{12}) | | | | | |
| | 1980 | | | 1995 | | | 1980 | | | 1995 | | |
	A	B	C	A	B	C	A	B	C	A	B	C
Total	70.7	1.7	2.4	93.7	8.8	9.4	323.3	10.7	3.3	361.7	55.0	15.2
Bulgaria	8.2	0.9	11.0	12.1	3.5	28.9	34.8	6.2	17.8	40.7	17.3	42.5
Czech Rep.	na	na	0.0	13.9	1.8	12.9	52.7	0.0	0.0	60.6	12.2	20.1
Hungary	4.8	0.0	0.0	7.0	1.8	25.7	23.9	0.0	0.0	34.0	14.0	41.8
Poland	24.7	0.0	0.0	29.5	0.0	0.0	120.8	0.0	0.0	137.0	0.0	0.0
Romania	16.1	0.0	0.0	22.3	0.0	0.0	67.5	0.0	0.0	59.3	0.0	0.0
Slovakia	na	na	na	7.1	1.6	22.5	20.0	4.5	22.5	25.6	11.4	44.5
Slovenia	na	na	na	2.5	0.6	24.0	8.0	0.0	0.0	12.6	4.8	38.1

Source: European Commission Annual Energy Review 1998

Notes
A Electricity
B Nuclear
C Nuclear share (%)

There was also assistance from the World Association of Nuclear Operators which extended into twinning arrangements linking East European power stations with utilities in Western Europe. However, there is also a firm belief that the older VVER 440 MW reactors will remain a risk and the EU enlargement project is now a potent force for eventual closure. Thus a condition for negotiations with Bulgaria over EU membership is the announcement of a firm closure date for the four oldest reactors at Kozloduy: this has now been set at 2010, though the two oldest reactors could go four years before this.

At the same time, the Bulgarian government is legally obliged to conduct an in-depth impact assessment report at the plant every five years and in 2001 a consortium of European firms – consisting of Germany's Siemens, France's Framatome and Russia's Atomenergoexport – signed a $235.5 m. contract to modernise the two 1,000 MW reactors at Kozloduy, with further upgrading by the US company Westinghouse Electric. Meanwhile, Slovakia – which has started negotiations with the EU relatively recently – operates the old reactor at Jaslovské Bohunice near Trnava and wishes to extend the closure date from 2000 to 2005 even though the new Mochovce nuclear plant came on-stream in 1998 and there is acute concern in Austria about the danger that lurks very close to its borders. Former Yugoslavia is a special case because Western technology was used for the one 600 MW station which opened in 1983 at Krsko, Slovenia, close to the border with Croatia. Built by Westinghouse, the safety of the CANDU plant has not been a major issue, although in the aftermath of Chernobyl the green movement in Slovenia advocated closure and there was concern over security in the fighting that accompanied Slovenian independence in 1991. However, the main problem recently has been between Slovenia and Croatia over the joint financing and operation of the plant. It should also be noted that the post-Chernobyl furore was sufficient to stop plans for further nuclear stations in former Yugoslavia.

New capacity has in some cases been deferred. In East Germany, preparatory work at Stendal was stopped leaving the land available for industrial purposes. In 1990 Siemens were contemplating two 'Convoy' reactors on the Baltic coast at Greifswald but the project failed partly because the utilities declined full responsibility for clean-up and safety improvements. Moreover, there was no 'Energiekonsens' given opposition to nuclear power from the Social Democratic Party. Meanwhile, Poland abandoned its Zarnowiec (Gdańsk) nuclear project and the site is now being used as an industrial estate and gas-fired power station. Dubbed 'Zarnobyl' by anxious Poles, fearing a possible repeat of the Chernobyl disaster which affected them greatly, there was particular opposition in the local area where the villagers of Kortoszyna (obliged to move to Odargowo) felt they would be sitting on an atomic bomb. Completion was originally expected in 1984 but the project was rescheduled for 1991. However, the final abandonment may have been due as much to economic pressures as environmental concerns and it is possible that Poland may reconsider the position in future. Meanwhile, Bulgaria placed its 1986 plan for a 2,000 MW nuclear power station at Belene on hold in 1990, but the planned closure of the four 440 MW reactors at Kozloduy, plus reduced coal-burning capacity, is likely to rekindle interest in the project. In 1993 resumption was declared, on the grounds that the country had no alternative, but protesters forced a halt in 1995. Although the site is not seismically unstable, opponents of the project are concerned about the poor safety record of Kozloduy and the cost of fuel including its import and reprocessing. Hungary may embark on a further nuclear power station in collaboration with French companies, but no specific proposal has emerged.

FORMER CZECHOSLOVAKIA

This reduces the active nuclear projects based on Soviet technology in the FCS to those at Mochovce near Levice (Slovakia) for four 440 MW reactors and Temelín north of České Budějovice (Czech Republic) for two 1,000 MW reactors. It was intended that as nuclear power increased its share of total production from 14.6 per cent in 1980 to 62.1 in 2000, the share for thermal power would decrease from 67.1 to 23.2 per cent: thus avoiding increasing dependency on imported hydrocarbons while reducing pollution from coal-burning stations, along with disruption linked with lignite mines. The projects incorporate Western safety systems which will be paid for by deliveries of electricity for which suitable transmission links are now available. Slovakia however deliberated over the preference between Western and Russian options until funds for Western assistance were obtained from a consortium of banks in 1996 and the first reactor entered service in 1998. Despite Austrian reservations Slovakia insists that the plant is safe. However, it is doubtful if work at Mochovce will proceed beyond completion of the first two reactors. Meanwhile, work originally began at Temelín in 1980 with a view to installing four 1,000 MW reactors. But the first non-communist government in Prague was against completion because the design was considered flawed and the Klaus government of 1992–6 only agreed to two reactors, under pressure from Westinghouse (the contractors for the modernisation of the project), who arranged for the Czechs to receive substantial loan guarantees from the American Export–Import Bank. A total of 78 design changes posed a challenge for Russian–US cooperation at the site from 1993, given uncertainty as to how well the modifications would harmonise with the Russian design. At the time attempts were made to neutralise criticism from ENGOs by cutting off their funding.

Domestic tensions within the Czech Republic have been mainly concerned with the economics of nuclear power and the need for the extra capacity in the context of energy conservation. After investigations, Zeman's Social Democratic government allowed the project to continue though it went over-budget and fell behind schedule. A Czech study claimed that not putting the Temelín nuclear power plant on-line would cost taxpayers almost $3.0 bn. for the building costs and disposal of fuel. But the project attracted much hostility from Austria which is against Czech accession to the EU without proper evaluation of the environmental impact. The province of Upper Austria even proposed in 1998 that Austria and the EU should pay for three gas-fired power stations so that Temelín could be abandoned. Upper Austria also set aside

$8.0 m. to fight Temelín's completion by gathering safety and economic infor-
mation. The final run-up to commissioning in the autumn of 2000 was marked
by demonstrations by Austrian nuclear activists on all the country's frontier posts
with the Czech Republic and a European Parliament resolution claiming insufficient environmental impact study. There was also an Austrian government
resolution that accession negotiations with the Czech Republic should not be
regarded as complete until Temelín is shown to meet EU standards and a reve-
lation from the German government on potential problems with the plant's
safety mechanism in an emergency, although the Czech government insisted the
station was safe, given the US control system in use. A new German law reportedly provides for a ban on imports of electricity from any ECE nuclear plants
suspected of being unsafe. Meanwhile, the Czech government has accepted an
international EIA which will ensure that the views of Austrian and German
experts are heard.

ROMANIA

Romania is also developing nuclear power although it has not created great
controversy because the technology is generally considered to be acceptable.
The Canadian CANDU system was selected before the revolution, and the reactors at Cernavodă were to be supported by a heavy water plant operating at
Halinga near Drobeta-Turnu Severin and an existing natural uranium-processing
facility at Feldioara near Braşov, drawing ore from several domestic sources. A
nuclear strategy was first envisaged in the 1960s when Soviet technology was
to sustain 1,000 MW of capacity by 1970 and 2,400 by 1980. However, it was
only in 1970 that agreement was reached over a single 440 MW reactor for
Cernavodă and even this limited project was abandoned in 1978 after cracks
occurred in the foundations as a result of the 1977 earthquake. A switch was
then made to Canadian technology: the CANDU system (Canadian Deuterium
Uranium) was considered an extremely safe design for an earthquake zone while
requiring natural uranium and heavy water that could be produced within the
country. The first power station was to be built during 1980–5 with the possibility of obtaining some uranium from Zambia through a joint project which
also covered coal and nickel. Braşov and Piatra Neamţ were mentioned as sites
for further nuclear power stations in connection with total national nuclear
capacity, which was set at 4,500 MW for 1990, rising to 9,600 in 1995!

Financial difficulties proved insuperable and the Zambia deal collapsed, but
a start was made at Cernavodă and the supporting installations went ahead,
though the project was still being considered in 1985 with only supporting
capacity installed. After the revolution the project was revived as the best way
of boosting generating capacity to avoid winter shortages. The Canadians took
over managerial responsibility and financial backing was secured from Atomic
Energy of Canada and Italy (Ansoldo) with a complex of five 700 MW reactors as the ultimate objective. After further delays the first unit opened in 1996
and has been operating successfully since. The discussion has now moved to
the future of the other four units and especially the second which is partially

constructed. Public opinion is positive while the scope for electricity exports is appreciated and there is a desire to retain the skills and manufacturing capacities acquired at huge expense. However, the government cannot find the money and foreign investment – perhaps in the form of a 'build–operate–transfer' scheme – has not yet materialised. Only in 2001 was agreement reached between the Romanian operator Nuclearelectrica, Atomic Energy Agency of Canada and Ansoldo for the construction of the second station over four-and-a-half years at an estimated cost of $689 m. Meanwhile, the third and fourth units are still contemplated, but without any firm commitment.

Fuel reprocessing and nuclear waste

A considerable part of the nuclear argument relates to the handling of nuclear fuel and waste. The import of nuclear fuel from Russia (necessary for countries using former Soviet technology) is a matter of concern for security reasons, but waste is a much more sensitive issue. Arrangements were initially negotiated with the FSU (bringing some benefit to the country through the plutonium yield from the processing of the spent fuel rods) but they will not be satisfactory for all time since Russia will now require increasingly large payments – in hard currency – and in 1998 Minatom contemplated costs as high as $1,000/kg since fuel reprocessing generates a lot of liquid waste. Also, Moscow passed a law in 1992 against disposing of nuclear waste from foreign countries on its territory. Slovakia's arrangements to send fuel rods to Mayak have recently been put on hold. Countries are also sensitive about allowing nuclear waste to pass in transit. Bulgaria despatched its spent fuel to the FSU until 1990 but then used temporary storage at Kozloduy until the link with Russia resumed in 1993. This established a pattern whereby Bulgaria's spent fuel was handled in Russia (at costs 25–30 per cent below quotations from Western Europe) through barter deals involving Bulgarian food and medicines. But Moldova's demand for higher transit fees then prompted consideration of the Romanian–Ukrainian route avoiding Moldova altogether. But in 2001 Russian ENGOs petitioned the Ukrainian government to stop a rail consignment of Bulgarian spent fuel due to the risk of accident or terrorist attack.

These considerations have provided a stimulus for the ECECs to provide their own storage and reprocessing facilities. While there was no significant opposition to Hungary's nuclear power station at Paks, NIMBY attitudes towards the storage of nuclear waste surfaced when the inhabitants of Ófalu (Baranya County) maintained an effective opposition to waste being dumped in their area and considerable research has focused on the issue of storage in a loess-covered hill environment. In the Czech Republic ten local authorities have been offered substantial inducements to accept nuclear waste on a temporary basis. However, the public in the SEECs is somewhat less sensitised on the issue. Bulgaria is confident in finding suitable sites in loess formations close to Kozloduy and a new unit has started up at Kozloduy for reprocessing and conditioning low- and intermediate-level solid radioactive waste generated by the plant's six reactors. Croatia plans a low/medium radioactive waste site in the Moslavačka and

Trgovska Mountains in Sisak-Moslavina County, while Yugoslavia has considered a nuclear dump drawing waste from all over Europe in the former Gabrovnica uranium mine at Kalna on the Bulgarian frontier, even though it is situated in the Stara Planina national park and has attracted some opposition from the ENGO Zeleni Sto (Green Table). Poland, of course, has been spared controversy because although a storage site was identified near the Warta–Noteć confluence (a former German Second World War military installation intended to halt the Red Army's advance on Berlin) no nuclear development ever took place.

Tourism

Early development

Tourism and recreation opportunities were, of course, in place well before the Second World War. It was in the eighteenth century that visitors to the Šumava started to use the trails opened up by woodcutters (catering for a growing glass industry) to appreciate the alpine Black and Devil lakes. Momentum was maintained by nineteenth-century romantic writings on 'Bohmerwald' by Adalbert Stifter (1805–86) and Karel Klostermann (1848–1923) to stimulate small resorts like Eisenstein (Železná Ruda) which enjoyed rail links to Plzeň and Prague in 1877: supporting four hotels and a population of 1,500 by 1880. Alpine clubs were developing at this time and the Czech Tourist Club (1888) built hostels for student parties and marked out 'Šumava hiking trails'. Meanwhile, in Croatia, modern tourism originated with the steamboat services from Trieste to Kotor which started in 1837, though the first tourist association (at Hvar) did not emerge until 1868. The railway to Rijeka created conditions for the growth of the seaside resort of Opatija, where the first hotel (Kvarner) was built in 1884, followed by Crikvenica with its Therapia Hotel in 1893.

Carpathian resorts were also developing at this time in the wake of the railway, e.g. in Transylvania Höhe Rinne (Izvoru de Sus – now known as Păltiniş) was developed at 1,452 m from 1892 by Siebenbürgische Karpatenverein. In Russian Poland, convalescence in the Włocławek area was popular through the unique climate of the Ciechocinek salt springs, 'discovered' in the 1820s and endowed with their famous 'teznias' (tall timber structures, first constructed in 1824–8) which drip salt water for speedy evaporation for the benefit of people with respiratory problems who can readily inhale the chloric, bromic, iodic and ferrous mineral salts. Of course these are just a few examples of what was cumulatively a considerable growth in capacity. By the end of the nineteenth century the region was endowed with inland climatic stations and spas (taking advantage of the excellent endowment of mineral waters) and coastal resorts. The business continued to flourish in the inter-war years as the provision of more basic accommodation for ramblers of modest means grew up alongside the more traditional and elitist forms. Some pre-1945 country houses were built as second homes for recreational use.

The communist period

By 1945 most hotels were in a run-down state and many of those that were not converted for other uses under the central planning process were maintained largely for domestic use. State companies were set up to arrange for travel and accommodation: for example, Balkantourist (Bulgaria's State Enterprise for Travel and Tourism) dates to 1948. During the 1960s it became common for the SOEs to provide recreation centres for the use of their employees in the coastal areas or mountain regions and from the 1970s many families were able to secure simple second homes on the edge of large cities and in major recreation areas like Poland's Tatra Mountains and lake country. Purpose-built 'dachas' were built in tightly-packed groups on isolated sites, thanks to a planning system with insufficient powers of control where low-grade agricultural land was concerned. Meanwhile the larger towns provided facilities for sports and recreation in their immediate vicinities on the basis of surveys indicating needs and potentials.

Tourism did attract some major investments in order to build up the international side of the business, for there was potential for rapid growth (from a low base level) to provide summer holidays for people from other communist countries and by the 1960s there was a substantial movement from the northern countries to the warmer and sunnier SEECs. Slightly later, investments were made in facilities and marketing to attract Western visitors and thereby obtain some of the hard currency needed to import machinery and technology and also to repay loans. Tourism helped to boost ECE's drab image and also offered benefits through regional income and employment; while investments in infrastructure improved the quality of the environment for the population as a whole. The greatest growth occurred in Fmr Yugoslavia where hotels were built along the Dalmatian coast, opened up by the new coast road of 1964–5. The number of foreign tourists increased from 2.6 m. in 1965 to 8.4 in 1985 (with particularly rapid growth in the late 1960s and early 1970s); and foreign exchange from tourism grew from $81.1 m. to just over 1.0 bn. during the same period. Yugoslavia accounted for 4 per cent of Europe's tourist arrivals but only 2 per cent of foreign exchange earnings in 1985. Elsewhere in the Balkans there was development in the coastal areas, although it was relatively modest in scale by the standards of Yugoslavia (and Greece). There was also a rapid growth of international tourism in Hungary during the 1960s and 1970s after the post-1956 'ice age' to become 'one of Europe's numerically most tourist-saturated countries' (Borocz 1990: 21). Business with Austria was particularly brisk and a 'Mini City' opened on the Austrian border at Hegyeshalom in 1988 to attract further spending.

Momentum flagged in the 1980s in the absence of long-term strategies to ensure 'improvements in the environment for tourists including freedom of movement and exchange problems and upgrading the quality of tourist products' (Buckley and Witt 1990: 17–18). Bulgaria tried to remain competitive by abandoning compulsory currency exchange (in force between 1974 and 1977) and providing premium exchange rates. But the bulk of the foreign visitors were always from socialist countries. The total income from tourism was high in some cases (6.5 per cent of total exports in Hungary and 18.6 in Yugoslavia) but this should be seen against much higher figures for Austria (20.1 per cent), Greece (20.8) and Spain (25.8), leading to the conclusion that the potential, arising through climate and scenery along with cultural attributes and traditional hospitality, was not being fully exploited. There was also a lack of excellence in marketing and training as well as problems of coordination among the ECECs which complicated joint development of holiday packages. Even in the major complexes there were problems of seasonality and the recruitment of skilled manpower. Also, poor and unreliable equipment affected the quality of catering and black market activity was liable to take off when home-produced goods were inferior to the stocks in the foreign currency shops. But in both Bulgaria and Romania there were many customer complaints (with the limited availability of evening entertainment a particular issue in the latter case), although the winter sports' resorts fared better. Some progress was made through joint ventures with Western equity participation, particularly in Hungary, Poland and Yugoslavia. Hotels could be run by Western chains and new accommodation was provided, especially in Budapest. But quality control remained difficult with a shortage of trained staff and visa complications for personnel selected for training in the West.

Arguably it was always difficult to cope with Western tourism under the socialist system with its controls on small private businesses which usually provide the best standards of service in the industry. In Fmr Yugoslavia, support for small businesses should have allowed for a greater spread of tourist activity away from the coast, but although some investment did come from private funds (generated by migrants who spent time working in Germany), most capital came from within the industry and other sectors of the Yugoslav economy. With security a factor discouraging dispersal, segregation was most marked in Albania in order to 'preserve ideological discipline and to minimise the demonstration effect of the presence of numbers of (usually richer) foreigners' (Allcock and Przeclawski 1990: 5). Albania's Stalinist government was always concerned about the subversive effect of the tourist industry and business was restricted to 6,000 visitors a year. Only organised tour groups were accepted, while foreigners had to meet the regime's sartorial standards and were banned from using public transport. In addition the citizens of four countries were excluded altogether: Israel, South Africa, USA and the USSR. Under such circumstances Hall (1990: 51) believed that Stalinist tourism in Albania could only be a transient phenomenon and conservative forces would eventually break down. Nevertheless, while the international arrivals from outside the communist bloc were insignificant in Albania and reached only 7 per cent in Czechoslovakia, the share was 20–5 per cent in Bulgaria, Hungary, Poland and Romania and over 80 per cent in former Yugoslavia (and over 90 per cent in Croatia and Slovenia).

There were also problems of concentration. Large complexes on the Adriatic accounted for much of Yugoslavia's tourist bednights during the decade 1978–88: one-fifth falling to the Dubrovnik and Makarska areas and a tenth to the Šibenik and Zadar areas. In such places large organisations ran much of the industry, with Union Dalmacija in Split employing 4,000 people. Concentration arose because of the need for a good infrastructure covering hotels, roads, airports and ferry links and tourist objectives, including museums and monuments. The larger centres were also the places where further growth could most easily be accommodated; so tourism made a significant contribution to urban development. But ecological conflicts became more acute. Tourists visiting ecologically endangered areas helped to 'make the problem of environmental degradation a nationwide issue and, with respect to foreign tourism, an international concern' (Kruczala 1990: 77). The 'big tourist complexes' of Bulgaria – Slunchev Briag (Sunny Beach) with 60,000 beds (25,000 in hotels) along with Sveti Constantin and Zlatni Piassatsi (Golden Sands) – were all built from the early 1960s and were partially in need of renovation and upgrading by the late 1980s, all the more so since the rooms tended to be rather cramped. There were smaller 'holiday centres' with 2,000–3,000 beds spread across groups of hotels and villas: Roussalka, Beli Briag, Eleni, Duni, Primorsko, Kiten, Lozenets. However, pressure on Bulgaria's Black Sea coast was eased by various conservational projects, including nature reserves and conservation areas like Nessebur.

Tourism in transition

As with other businesses state control has now been relaxed. Tourism ministries are tending to change into broad-based quangos forging partnerships between governments, professional associations and NGOs. Segregation is breaking down and two-tier price systems, with higher prices for foreign tourists, were becoming relatively rare by the middle of the decade, though some restaurants may still discriminate. Privatisation of travel agencies and hotels has occurred, with much capital invested from abroad. The Hungarian travel company Ibusz has been privatised and the US Marriott Corporation has won control of the Duna Intercontinental. Meanwhile, the Danubius chain – with a leading position in medical tourism – was the first company to be privatised through public offering to Hungarian investors, and both the Pannonia group (acquired by Accor of France) and Hungar Hotels were also placed on the market.

Many hotels remain undercapitalised by Western standards but the backlog of refurbishment is gradually being overcome. In Poland, Orbis run a network of 55 hotels and are investing heavily in modernisation. At the same time, new hotels in the cities are catering for both tourists and business people. In the early 1990s Warsaw could provide only 6,000 rooms for visitors: less than half of what was needed. However, high land values and local authority interest in prestigious projects have discouraged accommodation in the medium-quality range. Thus the 368-bed partly German-owned Grand Hotel Corvinus Kempinski, which opened in Budapest in 1992, joined the list of four- or five-star hotels which already included the Atrium Hyatt (which has just opened a casino), Duna Intercontinental, Forum and Hilton. There is also scope for more provincial hotels convenient for visiting cultural and sporting events as well as historic monuments, while more building is also needed along motorways, complementing simple but well-located hotels under local authority management (including Poland's Gromada Tourist Cooperative) and small quality hotels owned by Polish companies and joint ventures. Poland has great scope for the provision of rural hotels in major historic buildings available for privatisation through the breakup of state farms. The case of Krobielowice Palace near Wrocław is outstanding and another example is the Zamkowy Hotel – with concert and conference facilities – completed in 1999 beside an angling lake close to the old Leszczynski family seat of Baranów Sandomierski: the castle, which dates back to the 1590s, was renovated in 1966 after wartime military use and became a store for the nearby state farm.

While domestic tourism was depressed by transition, due to reduced spending power and a scaling down of the subsidised vacations provided by SOEs and trade unions under communism, international business took over in several countries. There were Western perceptions of a region 'open' for travel for the first time for half a century: an exaggerated view – for communist bureaucracy was generally user-friendly where hard-currency spenders were concerned – encouraged by the removal of most visa requirements and currency restrictions so that the process of entry and exit was greatly simplified, while standards were improved by successful privatisation and greater Western involvement in the business. Although the SEECs seemed to have the better long-term prospects through climatic and scenic advantages, an immediate response to the changed situation was seen very clearly in the north where the number of foreign visitors increased sharply in 1990 and 1991, reflecting natural and cultural assets, a relatively good infrastructure (transport and telecommunications), investment by a resurgent private sector which gained the confidence of Western tour operators, plus World Bank and IMF validation of the tourist potential on the basis of currency convertibility which made vacations in ECE eminently affordable (Table 4.11).

The north versus the southeast

There was considerable potential for attracting visitors from Western Europe, given the initially low levels of travel generation. The main cities did best – given the hotel improvements – as did the tourist facilities provided along the principal routes. Capital from West Germany helped in the revamping of tourism on East Germany's Baltic coast through facilities such as the IEA hotel complex Fereinpark Rügen and Hapag-Lloyd's cruises by *Saxonia* between the Rügen ports and Griefwald, Stralsund and Zingst on the mainland. Prague soon became regarded as 'the place to be' and the capital city was backed up by the spas of western Bohemia while Budapest in Hungary was complemented by Lake Balaton. Foreign visitors to Poland in 1993 contributed a third of all foreign currency earnings and a significant proportion of this arises through visits by members of Polish communities abroad. Germans and Scandinavians appreciated the Baltic coast: Berliners are holidaying around the Oder estuary at the Inskie lake area and on the coast at Międzyzdroje and Świnoujście where seaside properties are much sought after. There was a revival at Sopot (part of the Gdańsk–Gdynia–Sopot 'Tricity') where the old bath house has been converted into a

Table 4.11 Tourism arrivals and overnights 1990–8

	Annual tourist arrivals (,000)					Indexes		
	1990–1	1992–3	1994–5	1996–7	1998	A	B	C
Albania	na	na	322.0	203.5	184.0	na	32.0	0.06
Bulgaria	4250.0	3788.5	5284.0[1]	4923.0[2]	3266.0	3.1	32.4	0.40
Croatia	na	na	17270.6	21372.5	25499.0	na	na	5.66
Czech Rep.	na	na	98060.5[1]	108644.3	102844.0	na	32.9	9.99
Hungary	21185.0	21496.0	39538.0	38574.0	33624.0	0.6	42.7	3.32
Macedonia	na	na	2908.5	2117.0	1848.0	2.2	na	0.92
Poland	na	na	78248.5	87428.0	88592.0	na	41.0	2.29
Romania	3004.5	3339.5	4120.5	5177.3	4831.0	na	39.5	0.21
Slovakia	728.5	609.5	24584.5	32427.5	32735.0	3.5	32.3	6.07
Slovenia	na	na	3261.5	3711.0	3297.0	3.2	36.0	1.66
Yugoslavia	na	na	227.5[1]	300.0	na	3.5	na	[3]

Source: World Tourism Organisation and Eurostat 2000: 156–63

Notes
A Overnights per visit 1996–7
B Average net rate of utilisation of hotel beds 1998
C Visitors ptp 1998 (Yugoslavia 1996–7)
[1] 1995 only
[2] 1996 only
[3] less than 0.01

hotel–service–catering complex. Poland also has attractive lake and forest country in the northeast with scope for further growth; Sudetes (Karpacz and Sklarska Poręba) and the Carpathians (Zakopane) are slated for massive expansion with local geothermal energy – hot mineral water, already used in the experimental Banska Nizna plant – and the possibility of tunnels through the mountains: both long-distance, e.g. Poroniec–Suchá–Zab–Nowe Bystre, and also locally at Guba-łowka Hill to give villages on the other side of the hill a chance of development.

The SEECs lost out in the medium term through lack of capital and expe-rience, but also because most Czechs and East Germans began to exercise their new-found freedom to visit tourist areas previously closed to them. Accessibility was generally poorer and political instability became evident in Albania and former Yugoslavia, especially the latter in 1991–2 when the largest tour oper-ator (Yugotours) was forced to abandon its programme. Bulgaria's resorts picked up in 1994 thanks to the return of the Russians, although this business was again damaged by the subsequent collapse of the currency. Meanwhile, Albanian tourism witnessed the greatest changes when Stalinist restrictions were with-drawn. New hotels were provided and a link with Corfu was provided by the Sarandë ferry inaugurated in 1989. However, the scale of activity was still very modest when the events of 1997 forced the fledgling business into recession. Romania has found it difficult to rebuild its image as an attractive tourist desti-nation after the problems of the later Ceauşescu years were marked by austerity and oppressive secret police activity. However, there has been a gradual improve-ment in performance, given political stability, despite a fragile economy.

Croatia however has witnessed a striking revival after initially subsiding into civil war. While Istria remained aloof from the fighting that plagued other parts of the country, Serb shelling of Dubrovnik – continuing until 1994 – appeared to be a calculated bid to scare tourists away, while accommodation of refugees has contributed to the rundown condition of many hotels. The Dalmatian coast is now doing better, while Croatia also has attractive inland areas including the famous waterfalls in the national park of Plitvice (on the World Heritage List since 1979) and Medvednica which is a hilly area of beech and oak forests on the edge of Zagreb. The highest point of Sljeme (1,000 m high) has cablecar access and both mountaineering and winter sports are available. Like other Yugoslav suc-cessor states (and Slovakia), Croatia needs a readily-identifiable tourism image. Moreover, the issue of rehabilitation is pressing. However, a boom in 2001 should generate some funding for refurbishment and make assets more attractive for pri-vatisation. The aim is to attract quality operators to the Dubrovnik, Hvar and Makarska areas while Croatians are to be encouraged by concessionary bank loans to open small hotels. One way forward for the state is to link the big resorts with smaller tourist settlements in order to combine the marketing experience of the former with the superior recreation activities of the latter. Holiday centres may be leased out to Western companies like Roussalka – the oldest such com-plex in Bulgaria (dating back to the 1970s) – which is a typical 'village de vacance' leased to Club Méditerranée. Some centres are in remote places, with transport problems and limited entertainment opportunities, and may have difficulty sur-viving now that subsidised holidays are no longer provided on a large scale.

Tourism and regional development

Tourism tends to be seen as a panacea. It may generate large injections of money to help promote economic stability, with the foreign visitors acting as potential carriers of the region's new 'body language': prioritising Western links and signalling a tacit retreat from the communist values which confronted visitors before 1990. It is important to identify areas of comparative advantage and invest to encourage higher per capita spending and maximise the industry's contribution to regional development. As a labour-intensive business tourism can generate significant employment, with opportunities for the unemployed after limited retraining. But not all entrants are successful. 'Small privatisation' in the Czech Republic and Slovakia was initially dominated by people with liquid capital and collateral, but with little previous experience in the tourist industry: hence there was much short-term and illegal activity with 'wild east' capitalism very evident, while new owners with greater experience and commitment have emerged much more gradually. Tourism is a catalyst for change in less-industrialised regions where an infrastructure of shops and restaurants can serve the community through the year; while there are clear benefits in the context of pluriactivity among small farm owners. On the other hand the business is seasonal and labour can be diverted from traditional occupations while social tensions can be generated by disparities in wealth when the cash injection from tourism benefits only a few. There can also be overdependence on foreign investment and patronage from specific countries which may not prove reliable in the long run. According to Balaz (1994), state investment in the infrastructure is a precondition not yet satisfied in Slovakia, although much could be done by strong 'autonomous regions' networked with local authorities in countries experienced in tourism like Austria and Switzerland. Long-term ecological stability is also a consideration. Risk of environmental damage has been highlighted in the Tatra Mountains (a candidate for the Winter Olympics).

Indeed, the region is now replete with ecological controversies which tend to be most exhaustively discussed in the northern countries, whereas in the SEECs many of the warnings come from foreign experts: the Albanian coast, which has attracted the notice of international finance, and the Danube delta where even modest growth could conflict with the conservation programme. However, in Bulgaria where three new ski runs and facilities in the heart of Pirin National Park will violate a protected area by damaging forests – cutting century-old pine and spruce forest – and cause erosion, opposition is coming from the local ENGO Za Zemiata! ('For the Earth'). On the other hand, Albania is hoping to develop sustainable tourism with World Bank assistance on the coast at Butrint national park, south of Sarandë, in the vicinity of Buthrotum ruins with Roman and Byzantine development phases, while the attractive border areas of Prespa and Shkodër Lake–Buna River offer potential for cooperation with Macedonia and Montenegro respectively. In the northern part of the region, reference may be made to the Harz Mountains where there was much opposition to the reopening of the branch of the Harz narrow-gauge railway from Schierke to Brocken mountain at a cost of DM20 m. However, despite the danger of tourist pressure on delicate peat bogs supporting dwarf willow and fir, the railway was accepted as a sustainable option for handling visitors who are channelled along walkways to viewpoints on the summit of the mountain. Attempts are now being made to promote public transport in other parts of the Harz to reduce pressure on Brocken and give other communities access to tourist spending. Finally, there has been a resurgence of public interest in the Šumava following the communist period when alpine clubs were suppressed and ramblers discouraged from entering protected core areas like Boubínský Prales: a primeval mountain of mixed forest which was declared a nature reserve back in 1858. Since 1989 there has been a surge of foreign interest which has focused on both the wild areas and the town of České Krumlov, designated a UNESCO World Heritage site in 1992. By the mid-1990s up to a fifth of the population of the bioregion were amenity migrants: roughly half of them Czechs living permanently in the area and the other half Austrians and Germans established in second homes. Šumava is now a key area for debate on sustainable tourism, with considerable capacity now available in České Krumlov.

RURAL TOURISM

New forms of 'niche' tourism are being marketed in the region at a time when conventional holidays at mountain or seaside resorts are being increasingly complemented with vacations geared to a specific recreational activity. Cultural tourism is of particular relevance to the region, since the demise of communism means the removal of inhibitions over the presentation of societies – especially in rural areas – that value a range of traditional practices concerned with agriculture, handicrafts and social activities. Tourism based on such resources is also facilitated by the restoration of private enterprise which allows individuals and communities to formulate their own investment plans. Thus, in the Slovak Carpathians there is much historic and cultural interest, including a range of towns (and rural areas like Spiš and the Ore Mountains) with historic buildings and specific cultural or ethnic associations. The cultural factor is again significant in the Polish–Slovak frontier region through a religious angle due to the importance of pilgrimages at Kalwaria Zebrzydowska. There is also an interesting mix of accommodations comprising agrotouristical farms, modern hotels and motels, as well as villas built earlier in the century in the spas and towns; also a network of local railways and roads in addition to the principal routes.

Some areas may lack potential because of language barriers, poor sanitation and medical services and a lack of tourist office back-up. The potential cannot be exploited immediately by people who may have no capital and virtually no experience of private enterprise after half a century of central planning. Financial help from domestic or external sources is another precondition (in the form of grants, loans and/or tax concessions). And a modern legal framework should be put in place, preferably in harmony with European Union (EU) legislation, so that the Ministries of Tourism can regulate the industry and exercise control through subordinate organisations. Rural cultural tourism also needs promotion as a distinct product to attract the interest of tour companies as well as individual travellers. In Poland, it has been shown how rural tourism can contribute to sustainable development in an old industrial area on the edge of the Sudetes. However, while effective promotional work has been carried out across Lower Silesia by the Agricultural Consultative Centres, greater efforts are needed to increase interest in rural tourism in Poland and attract more visitors from abroad. Promotion could be linked with tourist axes, such as the 120 km route through the karst region of caves, canyons, lakes and waterfalls in Bulgaria's central Pre-Balkan region. This could help protect unique natural sites since 'the best guarantee of their future preservation is their rational use as tourist objects' (Petrov 1996b: 174) (Figure 4.5). The small towns at either end have a well-developed tourist economy and local industries concerned with food, wine and tobacco, and the best sites – Negovanka Canyon near Emen and Devetaki Cave – are very close to one centre or the other.

Community leaders and the mass media need to get involved in satisfying prerequisites for rural tourism, including local services such as small shops and other retail outlets, handicraft workshops, entertainments and other local events. Coordination is also needed to ensure the availability of good-quality souvenirs. Older people should be motivated to pass on their skills so as to safeguard the quality of production and diversify into new craft products with the help of links with archaeological and folk museums to ensure authentic handicrafts of good quality. More thought will have to be given to an ecological agriculture appropriate to local physical conditions and the demand for high-quality organic food in the spas and other leading resorts. Moreover, sensitive social integration is vital where it is difficult to 'identify' rural communities and where there

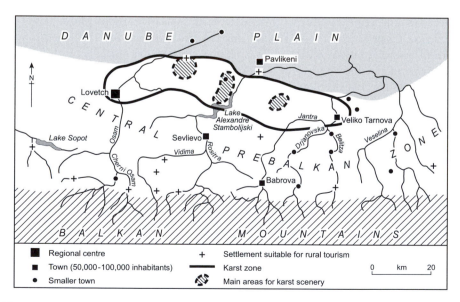

Figure 4.5 The concept of the tourist axis applied to Bulgaria's Central pre-Balkan zone
Source: Petrov (1996a: 159).

is a conspicuous absence of local consensus. The resurgence of traditional social systems and the empowerment of local actors in cases where organised crime has taken root pose major problems for rural diversification in parts of Albania. Sustainability must also be considered with reference to the host culture in the event of spontaneous, uncontrolled development; also the environment, for 'population growth and the uncontrolled expansion of the buildings reduce the tourist attractiveness of some areas' (Kurek 1996: 196).

5 Urban and rural settlement

Urban development under socialism

In 1945 the region was still predominantly rural with far too many people on the land to allow the productivity needed for comfortable living, but the post-war period saw a massive redistribution of population in favour of the towns. Leaving aside East Germany, the urban population increased almost threefold – from 27,400 to 70,700, yet this growth came largely from natural increase so that the rural population declined absolutely by less than 9,000 (Table 5.1). The greatest changes occurred in Bulgaria and former Yugoslavia where the urban share increased by more than 30 per cent while Albania saw the smallest change with only 15 per cent. In 1990, apart from Albania with 35 per cent, all the communist ECECs had an urban majority of 50–70 per cent, with 78 per cent in East Germany. Although the capital cities varied greatly in size, related to total population, they all grew with particular speed, while there were progressively lower rates of growth further down the hierarchy: regional administrative centres; other towns with substantial resources for industry; and small towns, which in some cases actually declined. Despite these increases in the population of the towns, the region remained 'under-urbanised' because housing shortages induced high levels of daily and weekly commuting. Figure 5.1 shows the main commuting routes in the Romanian Carpathians, and the sharp decline in the rural population in the western areas. There is, however, no simple correlation between accessibility to large towns and rural population trends because of variations in natural increase (referred to in Chapter 2) and in the availability of local employment, e.g. in wood processing. In some parts of ECE, bureaucratic controls on migration into the larger cities typically led to large-scale growth in suburban areas just beyond the administrative boundaries. Hungary's 'National Settlement Plan' of 1971 tried to slow down the growth of Budapest by strengthening the five largest provincial cities (Debrecen, Györ, Miskolc, Pécs and Szeged) and five others of intermediate size (Békéscsaba, Kecskemét, Székesfehérvár, Szolnok and Szombathely).

Reorganising capital cities and conurbations

The capital cities were substantially modernised. Budapest's position was then boosted by the NEM through enterprise self-financing from profits, as in the

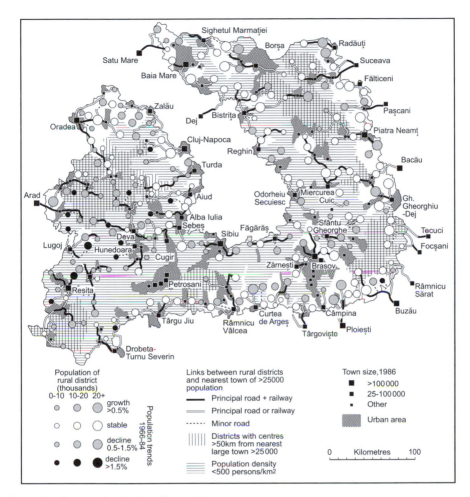

Figure 5.1 Commuting in the Carpathians

Source: Census of Romania 1966 and 1984 estimate

case of R&D spending by successful Budapest companies like Videoton and also government spending on infrastructure, plus the political 'clout' of the capital city, Budapest, and the whole northern industrial region. However, there was some change within Budapest through displacement of industry from the inner city, associated with slum clearance, and the growth of peripheral housing estates extending the rail/river-based complexes of Csepel and Óbuda. There was also some relocation in the wider region affecting places such as Gödöllo, Jászberény, and Törökszentmiklós. Meanwhile, the leading provincial centres were less radically transformed. Poland's Upper Silesia industrial region, specialising in coalmining and metallurgy, was subjected to an elaborate decongestion programme drawn up in 1950–3 to transfer industry and population from the inner 'Zone A' (area 700 km^2 and population 1.85 m. in 1950) to the outer 'Zone B' (1,700 km^2 and 0.31 m.), partly through new towns like Tarnowskie

Table 5.1 Urban population 1950–2030

| | *Population (,000) in* | | | | | | | |
	1950[1]	*1970[1]*	*1990[1]*	*1995[1]*	*2000[1]*	*2010[2]*	*2030[2]*
Albania	250	679	1176	1183	1218	1485	2244
Bulgaria	1856	4397	5796	5802	5763	5710	5408
Fmr. Czechoslovakia	4687	6944	9661	9901	10079	10535	10967
Czech Rep.	na	na	na	6754	6787	6938	7028
Slovakia	na	na	na	3147	3292	3597	3939
Hungary	3667	5016	6432	6602	6719	6863	6745
Poland	12673	16958	23570	24577	25414	27231	29690
Romania	4160	8472	12442	12669	12997	13563	13887
Fmr. Yugoslavia	3154	7422	11608	12086	12966	14399	16334
Bosnia&H.	na	na	na	1403	1714	2085	2538
Croatia	na	na	na	2506	2582	2734	2911
Macedonia	na	na	na	1174	1256	1421	1701
Slovenia	na	na	na	1021	1045	1100	1168
Yugoslavia	na	na	na	5982	6369	7059	8016
Total	27381	49948	70685	72910	75156	79776	85275

	Urban population as a percentage of the total							Annual change (%)			
	1950	1970	1990	1995	2000	2010	2030	1950–1990	1990–2000	2000–2030	1950–2030
Albania	20.3	31.8	35.8	37.2	39.1	44.4	56.7	+9.26	+0.36	+2.81	+9.97
Bulgaria	25.6	51.8	66.5	68.3	70.1	73.6	80.0	+5.31	−0.06	−0.21	+2.39
Fmr. Czechos'kia	37.8	48.6	62.1	63.1	64.5	67.9	75.3	+2.65	+0.43	+0.29	+1.67
Czech Rep.	na	na	na	65.4	66.3	68.9	76.2	na	+0.10[3]	+0.12	na
Slovakia	na	na	na	58.8	61.1	65.7	74.0	na	+0.92[3]	+0.65	na
Hungary	39.3	48.5	62.1	64.6	66.9	71.3	78.2	+1.88	+0.45	+0.01	+1.05
Poland	38.7	52.3	61.8	63.7	65.6	69.9	76.8	+2.15	+0.78	+0.56	+1.68
Romania	25.5	41.8	53.6	55.9	58.2	63.0	71.8	+4.98	+0.45	+0.23	+2.92
Fmr. Yugoslavia	19.3	37.7	50.9	53.9	56.1	61.0	70.3	+6.70	+1.17	+0.87	+5.22
Bosnia&H.	na	na	na	41.1	43.2	48.2	59.7	na	+4.43[3]	+1.60	na
Croatia	na	na	na	55.8	57.7	62.1	71.0	na	+0.61[3]	+0.42	na
Macedonia	na	na	na	59.8	62.1	66.3	74.4	na	+1.40[3]	+1.18	na
Slovenia	na	na	na	51.3	52.6	56.4	66.4	na	+0.47[3]	+0.39	na
Yugoslavia	na	na	na	56.6	59.9	65.6	74.0	na	+1.29[3]	+0.86	na
Total	31.2	46.4	57.9	60.1	62.0	66.2	71.4	+3.95	+0.63	+0.45	+2.64

Source: FAO database

Notes
[1] Estimate
[2] Forecast
[3] 1993–2000

Góry/Miasteczko Śląski to the north and Mikołów/Tychy to the south. This was only partly successful because the urban population of Zone A actually increased by 48 per cent (1.29 m. to 1.91) during 1950–75 while Zone B grew four times (0.09 m. to 0.36) and the Rybnik coalfield six times (0.06 m. to 0.36).

The Rybnik area of Upper Silesia continued to grow rapidly in the 1970s and 1980s, with in-migration by young people from all parts of Poland in the context of the 50 per cent rise in the region's coal production to 150 m.t anticipated in 1980. New towns subsequently emerged at Jastrzębie-Zdrój and Żory, where the radicalism of a large, young migrant population became evident through strong support for Solidarity in the Rybnik coalfield in the 1980s and also in the support for strikes called by new labour unions in response to austerity and the contraction of mining in the early transition years. Industry was also decentralised: lead-zinc working shifted from Bytom to Olkusz and Siewierz-Zawiercie while steel production started in Kraków ('Nowa Huta') and Dąbrowa Górnicza ('Huta Katowice') and the chemical industry moved further down the Vistula valley from Oswięcim/Auschwitz to Alwernia. However, there was no comprehensive modernisation of the steel industry and old obsolete plant remained in the Katowice area where the percentage of new industrial plants (built after 1945) was much lower than in Poland as a whole. The planners also failed to modernise the infrastructure of the core and coordinate the growth of the periphery (including Olkusz and Rybnik) with plans for the adjacent industrial regions of Bielsko-Biała, Częstochowa and Kraków so as to ensure an adequate protective band of forest and farm land. A motorway and express railway for the conurbation remained a priority to complement the electrified railway and dual carriageway links with Warsaw.

Variations in urban growth

Growth rates among the larger towns varied tremendously. Those that were initially quite small but were recognised as regional administrative centres achieved rapid growth, thanks to tertiary sector employment and the chance to benefit from a disproportionately high share of the investment allocated for regional development. But leading provincial cities might gain relatively little owing to the initial collapse of the capitalist economy following on from some cases of severe war damage. Jobs tended to increase faster than population in Czechoslovakia's 71 national and district capitals which had 41.4 per cent of the population in 1970 and 43.9 per cent in 1980, but 52.0 and 55.0 per cent respectively of the jobs. It was in the 176 smaller towns where the share of population (17.3 and 18.4 per cent respectively) grew faster than the share of jobs: 19.9 per cent in 1970 and 20.4 in 1980 (Hampl 1999a: 41).

An example of chequered fortunes is provided by Chemnitz in East Germany where the residential areas were ravaged by the allied bombing at the end of the war, leaving the Soviets to dismantle equipment from the relatively intact industrial zone. However, Chemnitz derived some benefit from being renamed Karl Marx Stadt (despite the lack of any historical association with this famous socialist philosopher) and recognised as both the capital of one of the new administrative regions ('Bezirken') and the headquarters of the huge uranium mining company 'Wismut' which exploited the Erzgebirge to provide the Soviets with 'by far the most important source of raw materials from their entire sphere of influence' (Bochmann 1995: 543). Indeed, had the value of this resource not been greatly reduced by the 1980s, German reunification might have been further delayed. The mining industry meant there was considerable purchasing power in the city and its vicinity; while the massive ore shipments protected the railway to Dresden from being dismantled: it retained its two tracks and was eventually electrified. The city was rebuilt as a socialist showcase, but there was a great loss of population (businesspeople, entrepreneurs, the engineering/technical elite and intellectuals) 'due to politically motivated flight' compounded in the run-up to unification by massive migration to the West (ibid.: 546). The airport disappeared under a new housing estate, typically endowed with a very limited infrastructure while historic buildings were neglected as part of a 'traumatization of cultural history' evident throughout East Germany when unification occurred (ibid.).

Most smaller towns were – relatively-speaking – neglected in favour of the larger towns with agglomeration economies and a mix of large enterprises which continued to draw in migrants when they might more logically have sought higher productivity to avoid labour shortage. Some planners wanted to expand smaller towns where land was more readily available and costs were lower. According to Perger (1989: 102), 'the most favoured type of settlement was the medium-sized town consisting of an industrial plant and a housing estate. This was the cheapest way of providing infrastructure facilities.' Moreover, country people perceived a move to a medium-size town as offering greater probability of advancement than a move to the city. Hence the argument advanced in Poland that the regions ('voivodeships') should try and stem the decline of district ('powiat') towns by strengthening their sub-regional service role and reducing competition among the higher-order economic and cultural centres (ibid.: 233). However, in Hungary small towns on the Great Plain – strongly supported under communism – grew too quickly because socialist industry was not backed by adequate services and a 'small town society' (Csatari 1993). There were even references to the 'ruralisation' of the towns because of the large numbers of rural migrants and the limited infrastructure which did not always include the full range of elements needed for effective town–country integration: secondary school, hospital, department store, law court (prosecutor's office) and a police station. While the rural dwellers had some compensation for their isolation through small-scale farming, the labourers in the country towns might find few opportunities in the second economy and Karpaty (1986: 132) remarked how 'a lack of additional income sources delays the rise from social marginality in the case of people still coping with serious housing and income difficulties'.

The nature of socialist cities

During the communist era Western geographers gave considerable thought to 'socialist cities' that might be fundamentally different from their capitalist counterparts, although the restrained development of new towns and the limited scope for remodelling what had previously been capitalist cities suggested 'socialised cities' rather than urban settlements that reflected the new ideology more comprehensively (French and Hamilton 1979). Great importance was attached to the centre containing monumental buildings with an ideological inspiration. Warsaw's Palace of Culture was built during 1952–5 on a specially-levelled site of 22 ha as a 'gift' from the Soviet Union. It was 234 m high and to maintain its dominance other buildings in the vicinity were not to exceed 32 m until the 'Eastern Wall' (built along an adjacent shopping street between 1962 and 1970) brought about a 'counter-revolution in space' (Grime and Weclawowicz 1981). Certainly the GDR's 'Sixteen Principles of Urban Development' (1950) included the role of the centre in projecting the victory of socialism using tall buildings (competing with the cathedrals of pre-industrial cities), boulevards and central squares (Haussermann 1996: 216). Yet all totalitarian regimes seek expression through buildings while scarce resources must limit the scale of the ostentation. And while socialist planners were not too constrained by narrow economic considerations they had their own ideas of what was desirable. High-rise building in the centre of Budapest was restrained in order to maintain the urban vista enjoyed from the Buda Hills: high-rise development was restricted to the outer boulevard to close the vista rather than disturb it (Szelenyi 1996: 301–2). On the other hand, in the late 1970s/early 1980s Romania's President Ceauşescu was able to indulge his passion for 'building socialism' by demolishing substantial properties in the centre of Bucharest in favour of a 'Victory of Socialism Boulevard' (now B-dul. Unirii) leading to the monumental 'Casa' which is one of the largest buildings in the world with ample space for the Romanian parliament.

Suburban development occurred on greenfield sites with near-total emphasis on large apartment complexes linked with community facilities and industry. The Soviet leader N. Khrushchev commended the modern, industrial housebuilding which took root in the GDR in 1955. A system of segregation by age tended to occur since young couples received priority in the allocation of new apartments in the high-rise blocks. But while the policy was quite rigid it is argued that this approach through prefabricated concrete housing, uncomfortably hot in summer and cold in winter 'was increasingly unacceptable to inhabitants and to dissenting urban planners' (Haussermann 1996: 219). It was also counterproductive because central planning could not guide social development in a normative fashion: 'living at close quarters did not engender collectivism but rather social tension and neurosis' (Enyedi 1996: 104). Such development has been linked with the 'microrayon' model, but it is difficult to make a meaningful distinction from the Western 'neighbourhood unit'. A close association of residence and work was anticipated, but this quickly broke down in favour of cross-city commuter flows. Communal living was never seriously attempted, while green spaces were limited to save on infrastructure and agricultural land. However, it could be claimed more convincingly that socialist cities placed strong emphasis on collective functions (parks and parade grounds) rather than increased living space for individual apartments, and also on public transport rather than private car ownership, which was explicitly banned in Albania.

More fundamentally, socialism afforded only limited evidence of a Western-style CBD, since residential functions were retained in the central area through lack of any land value gradient (and the very limited development of services above the level of the neighbourhood unit). The inner city remained function-ally mixed (a legacy of nineteenth-century railway-related growth) with redevelopment featuring opportunity sites, especially the main boulevards where some façades were transformed. Although East Berlin was favoured with rela-tively high shopping and residential standards in the 1980s, new housing was still restricted to vast suburban complexes like 'Neubaugebiet' Hellersdorf, much larger than their counterparts of the 1950s and 1960s. But in Hungary if the new estates lost some of their attractiveness through the introduction of more working-class elements, ethnics and Roma, there was scope for the formation of cooperatives in Budapest finding 'niche' sites in the hills or 'a rebourgeoisifi-cation of traditional middle-class neighbourhoods and even the first signs of gentrification of inner-city areas' (Szelenyi 1996: 307). Thus the reforms of the 1980s favoured some private lower-density housing and an element of gentrifi-cation. Meanwhile, more traditional rural-type housing might intrude as the cities expanded to take in formerly-separate villages, sometimes as a politically-motivated gambit to boost the apparent rate of urban population growth. Such housing in the suburbs (often without running water) provided rural migrants with a foothold, but it took a long time for facilities to improve.

Much inner-city housing remained in private hands 'although the right to profit from such ownership was generally abolished and other matters, notably rents, were strictly controlled' (Harloe 1996: 15). Leaving such areas well alone avoided the problems of redevelopment and rehousing, but with an ageing popu-lation and lack of maintenance, these old residential areas were marginalised as dumping grounds for nonconformists, dissidents, 'trouble makers' and unskilled workers in the less important industrial branches (Haussermann 1996: 220) and accelerated the transformation to slum conditions. Moreover, in East Germany strict controls on rents meant that private owners lacked the resources to maintain their traditional town houses which were frequently 'gifted' to the local authorities. Although boarded-up, because of the preference for low-cost development of greenfield sites, these properties do at least offer the potential for revival of historic town centres once the repairs have been undertaken.

Shopping also had its distinct characteristics. Provision was planned in the GDR on the principle that everyone should be within 15 minutes of convenience goods and one hour in the case of higher-order requirements. Shops were over-whelmingly owned by the state or by cooperatives and the few private businesses did not, however, have the power to decide on prices or the goods to be sold. The GDR had separate state organisations for small shops and department stores (with a small minority of shops belonging to ministries like the Ministry of Tourism), but daily needs were often met from shops in factory and office canteens which reduced the need for provision on housing estates. Despite some concentration in city centres – a trend reinforced by the consumerism of the Gierek era in Poland – most shops had shabby exteriors and unimaginative indoor displays, with a perennial shortage of goods; a situation somewhat relieved by (a) dollar shops geared mainly to tourists like the Pewex chain in Poland; (b) sales from private farms and (c) weekly private markets for both food and non-food items which charged higher prices for greater choice and quality. There was also a big gap between East and West in terms of food manufacture and distribution. Whereas Western firms were obliged to become more consumer-oriented, ECE enterprises remained production-centred. 'The absence of competition in food production, processing, wholesaling or retailing impeded the marketing revolution which was taking place in the Western world', although the emphasis on small-scale production, involving two-thirds of households in Hungary, has helped overcome food shortages and improved access to quality food at reasonable prices (Mueller 1993: 4).

East–West comparisons and contrasts

Both socialist and Western cities were characterised by tertiary employment (despite industry's special role in the ECECs); functional separation of urban land uses; suburban development; an urban hierarchy which encouraged some growth in smaller towns (less noticeable however in the SEECs than further north); and rural–urban migration. Regulation involved different mechanisms – land values in the West and models and commands in the East – but there was the same efficiency logic and 'consequently the locational map of an ECE city did not differ substantially from a Western one of the same size, importance and functional type, even though governments had more formal power to shape the urban environment in the East than in the West' (Enyedi 1996: 104). The same author remarks that when ECE tried to close the gap with the West after 1945, it was necessary to follow Western approaches to technological development in cities based on economic prosperity and so – despite Marxist principles of egalitarianism – urbanisation led to increased social differentiation. Moreover, the urban dwellers in the ECECs had a goal of family well-being achieved through choices which 'expressed a certain perception of urban space which is a part of a shared European culture' (ibid.: 105). Although there was an ideological emphasis on decentralisation to diminish regional inequalities, 'centralisation and concentration were [in reality] highly valued in the socialist political and decision-making system' (ibid.: 113). Some changes in urban areas during the later years of the socialist era so closely resembled Western experience as to suggest a gradual process of convergence between the once separate socioeconomic systems. Notwithstanding the communist housing model of state ownership and subsidised accommodation with continuing shortages despite the prominence of high-rise apartment blocks, some of the region's housing was managed in the context of a mixed economy which included a measure of privatisation. Inequalities in housing and services were frequently noted, especially in Hungary and Poland and the capital cities in particular.

Reconstruction of war-damaged central areas

New urban landscapes were balanced by substantial efforts of reconstruction after the Second World War. The GDR had enormous responsibilities through control of the centres of the old Prussian and Saxon capitals (Berlin and Dresden respectively). Berlin will be referred to in the next chapter, but it may be noted that Dresden's Zwinger Palace – an eighteenth-century royal playground for dances and tournaments – was largely reconstructed, including the art gallery with its collection of paintings by seventeenth–eighteenth-century Saxon kings returned by the Soviets through persuasion by the GDR leader W. Ulbricht. The Semperoper was reopened in 1985 and there was also some sensitive building in the Hauptstrasse and concern for the early twentieth-century garden city of Hellerau, though the famous Frauenkirche was left as a ruin and is only now being resurrected in all its Gothic splendour following German reunification. In Hungary the Buda Castle project was outstanding while Poland provided for the rebuilding of Warsaw's old town square after wartime destruction. Wrocław was reduced to a state of near-total devastation through the German strategy of 'Fortress Breslau' but much of the old fabric was restored including the City Hall and the famous Market Hall which regained its familiar exterior despite internal reconstruction involving massive concrete beams. Elsewhere in the city redevelopment involved appropriately sensitive architectural styles while some redundant public buildings like the Świebodzki railway station (superseded by the main station Wrocław Głowny) were put to new uses. Another Polish project concerns Elbląg where the centre was devastated during the Second World War and grassed over. Archaeological work led on to a reconstruction plan in 1975 for the 15 ha site, involving 300 historic tenement houses, faithfully reconstructed in many cases.

Cities in transition: business, housing and shopping

During the transition the growth in the urban population has slowed down and is estimated at over 1 per cent per annum only in former Yugoslavia (specifically B&H, Macedonia and Serbia). In the future (to 2030) the rate of growth is expected to be even lower, though the same countries should remain well above the average – along with Albania where expansion should be very rapid if recent exceptional levels of out-migration are contained. However, urban growth will be limited by the dire demographic predictions for parts of the region so that in Bulgaria, although the urban population should continue to rise, there is likely to be a fall in absolute terms – faster than the slight decline estimated during the 1990s. There are considerable variations in the urban profile between the ECECs, apart from the level of urbanisation. The smaller towns (below 100,000) generally account for more inhabitants than all the larger centres – and by a very large margin in Albania, B&H, Croatia, East Germany, Macedonia, Slovakia, Slovenia and Yugoslavia – while the Czech Republic and Romania see the larger towns in the ascendancy (Table 5.2). However, Romania also has a disproportionately large capital city 6.6 times the size of the largest provincial city while the disparity is much smaller in the Czech Republic. And while Hungary has a considerable urban population in cities of over 100,000 this is concentrated in one metropolitan region of over four million, with a city of over two million which is almost 22 times larger than the second city (Debrecen) (Table 5.3).

Table 5.2 Urban population size groups 1996[1]

	Number of towns and share of total population (%)					
	A	B	C	D	E	F
Albania	–	–	–	1 (7.3)	2 (6.6)	na (23.4)
Bosnia & H.	1 (13.2)	–	1 (9.5)	–	1 (3.3)	na (36.2)
Bulgaria	–	1 (15.0)	2 (7.9)	–	7 (10.2)	na (39.4)
Croatia	1 –	–	–	–	3 (9.5)	na (39.9)
Czech Republic	1 (11.8)	–	2 (6.9)	3 (7.1)	9 (13.7)	na (25.9)
East Germany	1 (7.9)	–	2 (6.0)	5 (7.8)	7 (4.5)	na (51.8)
Hungary	1 (18.9)	–	–	1 (2.1)	7 (9.5)	na (34.2)
Macedonia	–	–	1 (22.7)	–	–	na (37.2)
Poland	1 (3.8)	4 (7.2)	5 (5.1)	10 (6.0)	22 (7.4)	na (35.2)
Romania	1 (9.1)	–	7 (10.1)	4 (4.0)	13 (8.1)	na (24.1)
Slovakia	–	–	1 (8.3)	1 (4.4)	–	na (46.1)
Slovenia	–	–	–	1 (13.5)	1 (5.3)	na (44.7)
Yugoslavia	1 (11.1)	–	–	–	5 (6.8)	na 38.6
Total	7 (7.4)	5 (2.9)	21 (5.4)	27 (4.5)	75 (7.4)	na (35.4)

Source: Statistical Yearbooks (individual countries)

Notes
A Over 1.0 m.
B 0.5–0.9 m.
C 300–499,000
D 200–299,000
E 100–199,000
F Below 100,000
[1] East Germany 1989; Bosnia & Hercegovina 1991; Romania 1994

Table 5.3 Capital cities 1950–99

Country	City	Population (,000)			Percentage of National Total			Multiple of Pop'n of second city		
		1950	1980	1999[1]	1950	1980	1999[1]	1950	1980	1999[1]
Albania	Tirana	40	198	600[2]	3.3	7.6	11.0	1.4	3.2	4.8
Bosnia&H.	Sarajevo	na	na	416	na	na	9.5	na	na	2.9
Bulgaria	Sofia	437	1057	1384[2]	6.1	11.9	15.8	3.4	3.0	4.0
Croatia	Zagreb	na	na	1047[2]	na	na	15.0	na	na	5.5
Fmr. Czechosl.	Prague	933	1193	1225	7.5	7.8	7.8	3.3	3.2	2.7
Czech Rep.	Prague	na	na	1225[2]	na	na	11.9	na	na	3.2
Hungary	Budapest	1600	2061	4487[2]	17.2	19.2	43.9	12.5	9.8	21.9
Macedonia	Skopje	na	na	444	na	na	22.9	na	na	5.8
Poland	Warsaw	804	1617	2413[2]	3.2	4.5	6.2	1.3	1.9	3.0
Romania	Bucharest	1042	1861	2286[2]	6.6	8.4	9.2	8.8	6.1	6.6
Slovakia	Bratislava	na	na	448	na	na	8.3	na	na	1.9
Slovenia	Ljubljana	na	na	330	na	na	13.5	na	na	2.5
Fmr.Yugoslavia	Belgrade	389	911	1594	2.5	4.1	4.9	1.3	1.4	1.5
Yugoslavia	Belgrade	na	na	1594	na	na	13.3	na	na	6.0

Source: United Nations Yearbooks

Notes
[1] B&H 1995; Bulgaria 1997; Macedonia 1994; Yugoslavia 1998
[2] Population of district 1995 (Bucharest 1999)
Second city calculations 1991 for Banja Luka and Split

Nevertheless, the revolutions in the ECECs have impacted very heavily on the towns in all cases, in view of the concentration of business, especially in large urban areas. The political changes have given power to local government and new forms of citizenship are developing. Meanwhile, the workplace has been depoliticised and this has eroded the influence and privileges which some workers (e.g. union representatives) enjoyed under the old system. At the same time there has been massive disruption through the downsizing of industries and consequent unemployment and poverty, linked with the breakdown of Comecon, while the curtailment of public works initiated under communism left many apartment blocks unfinished until restructured building companies were able to take them over. Of course, there is significant socialist legacy, given the emphasis on multi-storey apartment blocks and limited shopping provision through non-competitive state and cooperative outlets, which points to an element of hybridisation. But new development is taking place in a wider global context where investment arises from the decisions of individual transnational companies and also through the EU project of political and economic integration and the financial resources of PHARE and other pre-accession funds plus the EIB and EBRD. Pressure is particularly great in the new state capitals which must now accommodate parliaments, presidential residences and embassies. Some countries, such as Croatia, will have concerns over their urban systems in general and the need for greater national cohesion.

But everywhere the rights of private property owners will break down the inheritance of mass uniformity in favour of a relatively chaotic 'mosaic of projects' expressed architecturally through a 'postmodernist hotch-potch'. There has been much resentment over the collapse of communist enterprises with the risk of social conflict enhanced by alcoholism in some parts of the region, with criminality nourished to some extent by the legitimacy problem facing the law-enforcement agencies. Yet the very essence of the new system is conflict and inequality as interest groups compete for influence. However, in the early stages the new regulation system was weakened by a legislative vacuum and some large commercial buildings have lacked authorisation, like the ultra-modern Columna Bank headquarters which appeared in a select residential area of Bucharest. Urban development is also complicated by restitution and disputes over property ownership, combined with the availability of property vacated by the demise of parts of the communist bureaucracy.

Forces shaping post-socialist cities

In East Germany where restitution covers Nazi seizure of Jewish-owned property as well as communist expropriations after the war, some 2.5 m. separate properties and one-third of the population have been involved and the issue was a protracted one throughout the 1990s. Until claims were settled property could not be modernised or sold, while the interim owners had no interest in carrying out maintenance work for tenants left struggling for improvements, with the longer-term prospect of rent increases – if not displacement in favour of higher-income families. Much of the property was looked after by the East Berlin Community Housing

Association (with 16,000 such properties on its hands in 1996) and corresponding organisations in other cities. However, after limited opportunities during the 1990s, the construction industry is starting to benefit from a surge of modernisation work where a majority of owners and tenants agree to refurbishment of a tenement block. However, in other countries settlements have been made quite quickly. Thus for example, following the return of property to the churches in Nitra, Slovakia, some university departments found sanctuary in office space made available by the demise of the communist party and the state security apparatus. Some former party/trade union properties are being turned over for cultural use.

What is clear however is that the new behavioural environment is producing new geographies and these will be contested. In communist Leipzig comprehensive redevelopment threatened the inner city, but after 1989 the question 'can Leipzig be saved?' was debated by local neighbourhood committees (some of which originated through 'Die Wende' in 1989). Meanwhile, the town planners, who resented challenges to their professional judgement as they began to react to market forces, found themselves having to accommodate social needs. Certainly it is in the towns where most things are happening and it would appear inevitable that the towns will grow in terms of business. Whether they grow in terms of population will depend on whether their 'absorptive capacity' arising from superior economic performance sustains a migration surge without a complementary outflow of people wishing to commute from the comfort of the urban–rural fringe. In the SEECs the urban share will probably increase, but in the north the opposite may occur as urban growth impinges on the metropolitan ring with its lower land and housing costs. The situation could be affected by vacant land within the city boundary and the effect of this on the land value gradient. To some extent it will come down to the entrepreneurialism of local authorities. Having separate authorities for individual districts of cities could lead to a polycentric structure. But a compact city with development concentrated in the core requires a stock of land and a disposition to sell where much of this is publicly owned. On the other hand, given the planning vacuum of the early transition years, it was easy – especially with the stimulus of tax incentives – to find large parcels of land for shopping malls and car dealerships accessible to urban dwellers with their own means of transport.

Where the rural fringes of a large city are particularly accessible, as they are in Berlin through the railway and autobahn rings, councils can accommodate industrial and business parks, as pro-development authorities in Brandenburg have been doing for some time. Indeed, the option has been made all the more attractive in East Germany by restitution and property ownership complications in city centres. The transfer to the suburbs is well illustrated by the case of Leipzig which has experienced a sharp fall in population from 545,000 to 440,000 between 1988 and 1998. Warsaw illustrates the choice between using vacant land in the centre to create a more compact city and development on the edge of the city which will accentuate the sprawl and require more investment in infrastructure. However, to achieve substantial savings it may be necessary to sell publicly-owned land, promote high-density development and redevelop underused areas. In a bid to woo some of the wealthy middle-class families out of their city apartments, up-market housing has been provided at Myśiadlo (15 km from the centre of Warsaw) with around 140 properties on a five hectare site where each home has its own character as a result of designs by Canadian and Polish architects. Thus the urban–rural fringe is changing. Not only are the service industries more prominent – garages, motels and modern distribution complexes linked with the emerging motorway systems – but the military units from the communist period are complemented by more recreation facilities including stables and golf courses which – in Hungary – conflict with the subsistence rationale of the surviving 'tanyas'. Transport in these areas has improved with the expansion of housing. In Hungary the need for a unified treatment of the urban–rural fringe has been raised in a situation where the state rather than the town council takes responsibility beyond the administration boundary.

The growth of business

The 'socialist city' image of the region's cities is rapidly being set aside through the privatisation of business and commerce, as well as the growth of the private sector in housing. Business is taking over from residential use in urban centres where land now acquires a relatively high value which may only be sustained through conversion into shops, offices and hotels. But the clearance of industrial blight and further plant relocation may provide the best opportunities. Some relocations effected before 1989 have provided immediate opportunities: for example, a former wood-processing factory in Nitra, strategically situated between the CBD and the bus and railway stations, has become the Octagon Market. And as further businesses are closed and the private railway sidings disappear there are more substantial opportunities. The process has been gradual (especially in small towns where land value gradients and development pressures are relatively moderate) unless pollution problems or market collapse have led to the closure of the business. In Kraków a former soda factory site is now occupied by a shopping centre while in Leipzig the greening of the city centre has followed from the demolition of industrial premises and railway sidings at Plagwitz, and elsewhere in the city a spinning complex has been converted into a largehotel, restaurant complex and shopping centre. The demise of textile industries in towns like Leinefelde in Thuringia has provided a good opportunity for the tertiary sector. But there may also be a compromise whereby industry migrates vertically to allow new users ground-floor access for small shops and offices, specialist wholesalers or small craft units. But in the early transition years there was neither a 'mature commercial property market' nor a 'stable and widespread leasing market' (Redding and Ghambari Parsa 1995: 314). It took time for companies with plenty of space to realise that it was good business to downsize and move to more modest accommodation on the edge of the city in order to lease the old premises to a foreign firm (where suitable) or sell the land for redevelopment.

Hotel and office accommodation: the case of Warsaw

There has been a big growth in hotel and office accommodation. In the communist period functionaries would make do with cheap lodgings, often available through enterprises. But Western companies require higher standards for accommodation and work environments. Capitals have attracted the leading hotel chains through privatisation or new projects. Even Tirana has been affected through the four-star Europa Hotel opened in 1995 by an Austrian group – with the Inter-Continental Hotel Corporation following – while the 'Skanderbeg' trade complex has emerged in the Mezez district. Provincial cities are also acquiring new hotels like the Ibis Hotel in Łódz which includes offices and entertainment, thereby helping the city to attract lucrative conference business. The need for offices created a similar spurt of activity especially in the capital cities. The stock of accommodation was quite insufficient and very little – 5 per cent in Warsaw in 1993 – was acceptable to international tenants. After flats and hotel rooms were converted on a temporary basis, 1.68 m.m² of new space has been opened in Warsaw over the past decade, following an initial wave in 1993 to meet the needs of industry, computer, insurance and telecoms companies, and pension funds. For example, in 1992 the Skanska International Building of Malmö and the Warsaw district of Wola agreed on a plan for a new Atrium business centre which completed its first phase in 1995, bringing together the best of Polish and Swedish architecture. The total complex involves 80,000 m² of office and retail space with secure underground carparking. There is a cluster of developments in Centrum borough's Sródmiescie district, mainly situated on Al. Jana Pawla II, where the rise in land values was rapid during the late 1990s (Figure 5.2). Through the Dutch developer ING Real Estate, Zlota Centre (125,000 m²) is emerging near Centralna Station with shops and offices as well as a hotel, cinema and entertainment facilities, while the Palace of Culture is itself being promoted as a 'sky high business centre'. This extremely high tower, which is out of proportion with the rest of the central area, is to be balanced by an inner ring of tall buildings for offices, hotels and residential use (while retaining much of the parkland), with a further ring of new buildings of conventional height along the main thoroughfares.

Rents were reaching exorbitant levels in 1997 but there was some reduction in 1998 as more space became available (around 15 per cent of space was unrented in 2000) though costs still exceed the European average. Rents for new accommodation are $28–30/m³ in the best locations ($22–5 for modern space outside the centre) while the average European rent is $22–5 outside London and Paris. Indeed, taking living costs in general, Warsaw scores 92.6 per cent of the New York figure, ahead of Paris (91.8), Berlin (83.0) and Prague (74.4). Lower rental costs apply in B Class buildings (renovated industrial buildings often less-conveniently located) and with longer leases of seven years replacing those of three to five years. There have been some difficulties, notably in the case of Daewoo's miscalculation over the Warsaw Trade Tower where 34,000 m² of Class A accommodation in the working-class Wola district have not proved attractive, partly due to the company's wider problems and competition from the Warsaw Financial Centre. The pace of development is still rapid with much international finance in Polish real estate, while completion may be advanced by contracts with clients prior to construction and use of designs with legally-valid construction permits. Far more companies intend to open offices

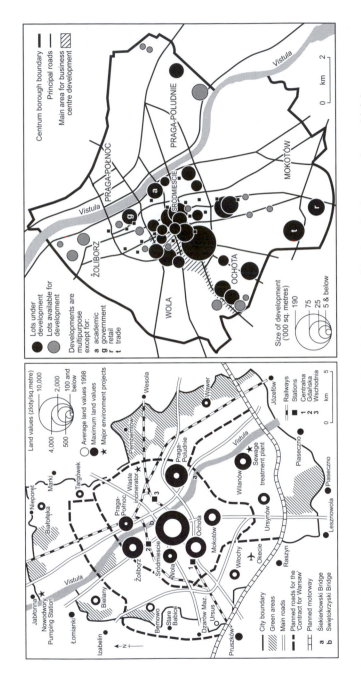

Figure 5.2 Warsaw, showing land values across the city (left) and the development of business in the central area (right)

Source: *Warsaw Voice*

or branch locations in Warsaw than in other European cities because Poland is perceived as Europe's third most attractive market for manufacturers.

There is now a tendency for office development to move further from the centre, where relatively little space is now available: only small, scattered lots, apart from Defilad Square near the Palace of Culture which is to be surrounded and 'camouflaged' by new high-rise buildings to create Warsaw's equivalent of Berlin's Potsdamer Platz. Much B Class accommodation has been provided on Ulita Kasprzaka in Wola – several kilometres from the centre – which is being transformed from an industrial to a banking area through conversion of former radio and pharmaceutical industry premises. Rents can come down to around $10/m^2 as has also happened in the Jerozolimskie, Ochota, and Wisniowy Business Parks. Other developments are being drawn south to Mokotów district where Mokotów Business Park occupies a former factory site and the European Investment Centre has absorbed a building formerly used by the communist trading company Pewex. These districts are convenient for Okęcie airport where Warsaw Distribution Centre and Warsaw Industrial Centre are located within the Warsaw International Airport SEZ.

Budapest and Prague

Again in Budapest, new investment has gone overwhelmingly into the historic inner city. Japan International Development Organisation and two other companies have completed a joint venture for an 8,000 m^2 Central Business Centre in Buda. This provides a mix of offices and serviced building sites as well as retail and residential areas. There is also a Korean–Hungarian joint venture for 6,300 m^2 of floorspace close to Andrassy Street in Pest with a shopping basement and parking space in addition to five floors of offices. However, there are also some high-technology parks and shopping centres on the edge of the city. In Prague pressure on the centre is again intense with ratios of 10:1 for land values between centre and periphery, given the scarcity of property available in 1991 when deregulation of rents was first allowed. Industry here has been pushed out, with the relocation of the Tatra tram factory from Smíchov to Zličín being a case in point. The demand for space raises the danger of inappropriate tall buildings in the centre to stabilise office prices. It also absorbs much open space, as it is doing in the central areas of large towns throughout the region, thus restricting recreational areas much to the displeasure of residents. One solution would be for former socialist enterprises like the state railways to release more land. But there are also pressures on the outskirts where large suburban complexes were typically endowed with few services. In Prague the open space adjacent to the blocks has been privatised and a rash of small developments (filling stations and cafés) threatens the environment of each neighbourhood. Landscaped open space is transformed into a straggle of small buildings with increasing hazards for road users.

Housing

There has been a change from high state housing subsidies under communism (often amounting to 3–5 per cent of GNP and the most important consumer subsidy after food) to decreasing subsidies and a substantial withdrawal of the state from the housing market. A change in the political culture, with the eclipse of corporatist welfare approaches by neo-liberal forces, has emphasised restitution and 'right to buy' as a way of gaining distance from the former regime. In Prague there has been a growth in the private rental sector because of property restitution to former owners rather than 'give-away privatisation' to tenants at a fraction of market value. This latter policy has been used most intensively in Budapest, resulting in higher levels of satisfaction with the privatisation procedure than in other capital cities. Massive privatisation of housing constitutes the greatest wealth transfer in the whole of Europe (about a quarter of GDP in the countries involved) and since the benefits seemed to be enjoyed most by those favoured by the old regime, the market mechanism seemed to increase social inequality and offered little 'emancipatory ability' (Bodnar 1996: 634). Those who lack the means to take advantage of privatisation are now pressurised by rising rents (following cuts in housing subsidies) and may eventually have to move out under much less favourable circumstances. 'The most radical change has been in East Germany where the Western system of property law was imposed, leading to selective gentrification and displacement of low-income populations' (Harloe 1996: 22). Households still in the state sector are generally younger with lower incomes and educational qualifications: this shrinking sector (with no significant renovation programme) is increasingly a shelter for marginal households.

Increased housing differentiation is very obvious in Budapest where flats in the attractive wooded fringe of the Pest Hills command three times as much rent as similar accommodation in the industrial districts of Csepel and Újpest whereas in small towns the contrasts in house prices between different neighbourhoods are quite moderate. In Berlin it is assumed that segregation will increase so that some areas (like Spandauer Vorstadt which is close to the centre) may become high-income domains – commanding rents beyond the means of most East Berliners, while low-income families (including Turkish 'Gastarbeiter') may move into the less popular East Berlin estates. In Leipzig, while Gohlis has regained its reputation as a residential area for the wealthy middle class, Volkmarsdorf and Kleinzschocher are in a poor state with many houses either empty or needing repair. In Brno the east and southeast is a refuge for the socially-deprived population. Meanwhile, inequality in service provision between rich and poor neighbourhoods is increased (especially in the State capitals) by the 'persistent tendency towards the formation of functional units within the cities that are more homogenous than the Metropolitan population at large' (Surazska 1996: 367).

It seems that middle-class families (as well as the elite) are benefiting from privatisation, having exercised the right to buy after the revolution under very favourable terms. Meanwhile, in the early 1990s, uncertainty made house building too risky a business except where there were very high potential profits and even now new building tends to be for the wealthier families, as exemplified by a high-class residential development for some 30,000 inhabitants on the edge of Warsaw at West Wilanów. As already noted, poorer families unable to buy are stranded in a spiral of rent increases with the withdrawal of subsidies. They are displaced from attractive inner-city neighbourhoods and trapped in deteriorating properties, fearing renovation because of the rent rise that they could not afford. Although economic difficulties make it necessary for the state to reduce its commitment, the ideological appeal of the communist period in terms of providing for all sections of the population will not be forgotten and in the future it will be necessary to return to the question of housing as a form of welfare. This may well occur through support for political parties which have evolved from the communist parties of the past and retain elements of their social policy. For there remains the poorer stock – which people cannot and will not buy – to be refurbished by local government using income from better properties.

The idea of a prohibition list to ensure that part of the better housing stock remains in local authority ownership has not been favoured, but other systems could emerge through management of housing by non-profit-making companies, cooperatives or tenant organisations – although these options have not been adopted to any extent. Apartment blocks inherited from the communist period are a particular problem. In the Czech Republic alone an estimated 3.2 m. people are living in such buildings – often poorly built and inadequately maintained – which 'can lack even minimal aesthetic values, community-creating institutions and suitable transport connections' even if they guarantee a minimum standard of living under satisfactory environmental conditions on the edge of towns (Schmeidler 1998: 71). Regeneration is needed and partnerships are suggested, bringing together the residents with local authorities and business communities. Problems arise with apartment blocks of the communist era through high energy costs arising from ill-fitting windows, corroded pipes and poorly insulated walls; with the departure of the better-off reinforcing impressions of low quality. But although refurbishment is cheaper than redevelopment the cost in East Germany exceeds DM40,000 per apartment. In Poland where money might come from EBRD and PHARE, the Warsaw authorities are keen to launch a pilot project involving community participation that would extend to the communal areas (lift and staircase, reroofing and insulation) as well as play areas and parking arrangements.

Retailing

East German town-centre shopping lost out to West German facilities immediately after unification, but standards are now higher throughout the region and the competition comes from supermarkets and hypermarkets on the edge of urban areas which can offer easy parking. There has been a rapid growth of fast-food outlets, hair-dressing and a wide range of professional services. Following their initial successes in Hungary, McDonalds are now developing chains of fast-food stores throughout the region, though they are much less prominent in the SEECs than in the north. Pizza Hut and Kentucky Fried Chicken are also active. A variety of public house environments are now available: UK's Allied Lyons opened its first British-style pub in Prague, selling its own brand of beer. There has been a decline of traditional laundry facilities but new modern shops are opening. Backed by British capital, Luxomat in Warsaw offers washing and dry cleaning. It provides a fast collect/deliver service and attracts large orders from institutions.

There have been radical changes in the way that trade is organised. Western goods, competing against domestic products on quality and price, became widespread when the liberalisation of international trade – enabling most goods to be imported without licence – allowed small importers into a business previously reserved for the large state trading organisations. New retailers quickly appreciated the need to stimulate customers, and the higher quality private shops (with improved fittings and Western brand names) have driven up standards. Keen young assistants displaced older staff associated with state shops in the communist era. Western-style media advertising quickly caught on, though it could be condescending and counterproductive at first when a straightforward 'educational' treatment linked to familiar scenes seemed more effective. The change was especially rapid in East Germany with unification and the surge of consumer-led growth through enhanced spending power in the light of exchange of Ostmarks for DM.

State stores went in for leasing and some withdrew from retailing altogether, while retailing chains were broken up through sale as individual units. Restitution also created a mechanism whereby property could be purchased and refurbished at fairly low cost. And where ownership remained uncertain, shops went on a tenancy basis to the highest bidder (with minimal prices, varying for the type of shop and the zone of the city). A high birth and death rate for shops in the early years of transition has been replaced by a more stable situation. The 'rent gap' between use under socialism and the potential 'best' use under market conditions provided for a big increase in the shops available and their profile, except where the existing occupiers were protected. Typically a loss has occurred in lower-order (food) shops and a gain in higher-order shops selling clothes and luxury items – such as jewellery which was only rarely available before 1989 – in elegant shopping streets like Warsaw's Nowy Świat, improved by renovation and pedestrianisation.

Many small private businesses are springing up through purchases of shops by auction, while the limited finance for major schemes of urban redevelopment (bank loans are expensive) means that many new businesses are forced to overspill from the 'High Street' into courtyards and residential properties. Many small kiosks and booths have appeared (sometimes unregistered to avoid taxation): 'boutiques' constructed of metal sheets in standard variants have become one of the minor features of transition cities, along with graffiti, advertising and new tastes in street names and monuments. The more successful ventures can generate resources sufficient to launch successful companies. There has been a mushrooming of 'pubs' in working-class residential districts: just small boxes with the sole purpose of selling alcohol. They lack sanitation and may attract criminal elements which prey on local residents. There is also much market trade along with other forms of 'proto retailing' which involves considerable ethnic diversity in Budapest, with 'Hungarian peasants from Transylvania, Polish black-marketeers, Vietnamese and Chinese smugglers, all offering their goods to tourists', not to mention the hosts of suitcase traders from former Soviet republics (Szelenyi 1996: 312). In Bulgaria, there has been great success for Oriflame: a cosmetics operation originating in Sweden which has taken on 15,000 distributors after the first generation of highly-motivated sellers encountered great interest among the public.

Western penetration and the arrival of the supermarket

While Western manufacturers found a general acceptance for their products on the grounds of quality, wholesaling tended to be haphazard at first because most new private operators had only limited capacities (perhaps a single lorry), could not guarantee delivery and forced shopkeepers to buy in bulk. The time needed to build stable relationships based on retailer loyalty helped to accelerate the establishment of new shopping. The contribution of émigrés can be seen in the case of T. Bata who returned to Czechoslovakia in 1989 to rebuild his empire in retailing and become the leading footwear retailer. He took over shops and factory space in Zlín to create a design centre and quick-response factory geared to both the home market and export. However, most new businesses were initiated by Western retailers with no previous ECE connections who often experienced problems getting title to properties due to bureaucratic delays and linguistic complications. The best immediate prospects seemed to lie in Czechoslovakia, Hungary and Poland where the Western invasion set up bidding wars that drove up prices and triggered a great demand for qualified staff including managers and buyers. It also created tensions with small shops given the lower prices charged by the foreign-owned chains thanks to their superior ability to extract discounts. Meanwhile, chains of shops welded together by tightly centralised purchasing and logistics can provide competitive prices for proximity outlets which don't involve travelling to supermarkets and this can be attractive in the context of low car ownership and habits of buying locally each day.

In Hungary the state-owned wholesaler Duna was purchased by a Belgian group for use as the distribution network for its own Profi supermarket chain. Meanwhile, the UK supermarket chain Tesco acquired a stake in the Global chain. The German retailer Tengelmann also built up a chain in Hungary, including a stake in Skala, the country's biggest retail chain, while Julius Meinl (Austria) took over a hundred state-owned Csemege shops in 1991 and became Hungary's major foodstore chain with some 50 outlets in Budapest alone. Rapid privatisation in Hungary meant that foreign investors could buy up networks cheaply and test the market at only modest risk, leading to expansion after several years' experience with new purchases and new stores plus a growth in efficiency through procurement associations between Meinl, Profi and others. Meanwhile, Metro (Germany) have built up a chain in Poland to anchor large shopping centres including other Metro chains like Adler (clothing), Praktiker (home improvements) and Roller (furniture). Rewe (Germany) are also active through boosting the Billa supermarket chain into the hypermarket class with 50 stores by 2000 (also 'MiniMal' hypermarkets for cities smaller than 100,000). Ahold & Allkauf are prominent in Poland (combining the Dutch retailer Ahold and German hypermarket operator Allkauf) along with several French companies: Auchan, Casino, Carrefour and Leclerc. Marc Pol set up supermarkets in Poland while the Dutch wholesale chain Makro opened its first Cash & Carry in Warsaw and went on to build up a network across Poland, while Ahold of the Netherlands (already mentioned) has developed its Mana supermarket chain in the Czech Republic and Rewe expanded aggressively through the small towns of Hungary with its Penny Market stores. Ikea of Sweden has furniture stores in FCS, Hungary and Poland. When retail sales are considered, Hungary's top 15 rankings are dominated by the food chains – including Metro, Csemege–Julius Meinl, Spar and Tesco Global. But also prominent are petrol stations and convenience stores operated by the Hungarian oil company MOL and a number of larger international companies led by Shell, ÖMV (Austria), Aral, Agip and Conoco.

Further south foreign access to city centres has been complicated by hostility from indigenous business networks and the continued presence of former state shops paying low 'historic'/concessionary rents that have not been updated (not to mention low spending power and car parking problems, aggravated by delay in introducing parking charges). However, supermarkets are now appearing in the SEECs, with the pacesetter in Romania being the Turkish company Fiba which plans a chain of 30 'Gima' superstores, starting with its outlets in the Bucureşti Mall (1999) – indeed Fiba built the whole of this complex as a joint venture with Banca Turco Română – and the Prisma shopping centre on the edge of the capital, plus the first provincial venture in Iaşi. An initial emphasis on Turkish products is now balanced by the display of more Romanian goods. Romania's provincial cities are now experiencing a supermarket boom: none more so than Timişoara where Cora (France), Metro (Germany) and Tesco (UK) are setting up supermarkets to compete with the indigenous companies Bega, Kappa and Terra. The wider spread of large modern stores will erode the cross-border dimension of supermarket shopping formerly relevant in Hungary where the country is well placed to draw customers from Romania, Ukraine and former Yugoslavia (Croatia and Serbia). Yugoslavs close to the borders in Vojvodina go to Szeged where several foreign-owned supermarkets are established: Metro, Tengelmann, Spar (Austria) and Tesco. The attraction has been all the greater because of difficulties for foreign manufacturers in selling to shops in Yugoslavia because of the absence of large foreign-owned cash-and-carry chains now so common in the north. Distributors employed by companies like Nestlé, Procter & Gamble and Unilever have to try and sell to hundreds of wholesalers and thousands of boutiques and kiosks as well as the retail chains. Of course foreign custom is not a major consideration for supermarket development in Hungary as it spreads down the central place hierarchy from Budapest – with Tesco stores in Kaposvár near Croatia, Debrecen near Romania, Györ near Slovakia, with the latter also blessed by Spar and the Makro (Metro-owned)

'Önkiszolgáló Nagykereskedelmi Áruházlánc' (self-service wholesale department chain). Business will in any case be complicated by more complex border controls after Hungary joins the EU.

Department stores and shopping malls

Department stores are another facet of the boom in shopping, with the outdated 1960s' inheritance in the northern countries now refurbished after privatisation. Poland's Centrum department stores are now in the hands of Bank Handłowy and a group of other domestic and foreign investors, while East Germany's legacy is now in the hands of Karstadt. Meanwhile, the American company Supershop found FCS a stable environment and negotiated for 13 stores – including properties in Bratislava, Brno, Nitra, Plzeň and Prague – with a total employment of 6,000. Trading at K-Mart, the company set about modernising the old 'Prior' stores and American-style retailing was introduced with simplified management hierarchies and a customer-friendly approach (which paradoxically resulted in a less substantial service than before). Tesco have now acquired these outlets. Where department stores were less prominent under communism a rapid growth of shopping malls is evident, especially in Budapest since Duna Plaza opened in 1996 (Figure 5.3) and was quickly followed by the Polus mega-mall: a 56,000 m² development with 400 shops and 2,500 parking spaces near the motorway in Rákospalota using an abandoned Soviet military base. It is the work of TriGranit, a joint venture between Hungary's Polus Investments and the real estate giant TrizecHahn (Canada), whose chairman was born in Hungary. Capital has been subscribed by diaspora Hungarians and good results have been obtained by anchor merchants like Cineplex Odeon (cinemas) and Tesco; so Polus Two is now under way. Another project for a mall with office accommodation beside the Polus Centre arises from the 'Alliance of Chinese People Living in Hungary', for it seems that Chinese shop-owners – some 3,000 of whom have small retail outlets in Józsefváros – feel threatened by EU regulations over open-air markets.

Development seems likely to slow down as the large complexes, which have recently taken so many customers from small shops, compete increasingly with each other to the point where price wars become inevitable (with the added complication of the virtual mall already established in Poland where many stores can now be accessed through <www.vendi.pl>). Purchasing power is limited and the smaller malls, without entertainment and large anchor attractions, are in some difficulty, with the risk of downward spiral if tenants move out. Many tenants in Csepel have failed. Moreover, additional control was introduced in 1999 through a Budapest zoning plan, whereas previously the prospect of local taxation income has encouraged the individual boroughs to compete. Meanwhile, the front runners include Campona Centre, Duna Plaza, Europark, Mammut, Polus Centre and Westend City Centre: all large and well-designed. Campona Centre has its own railway station to help attract residents from outlying towns and it also attracts people with its cinema and 'Tropicarium'. Duna Plaza is located at a major metro station and has extensive office space; while Westend

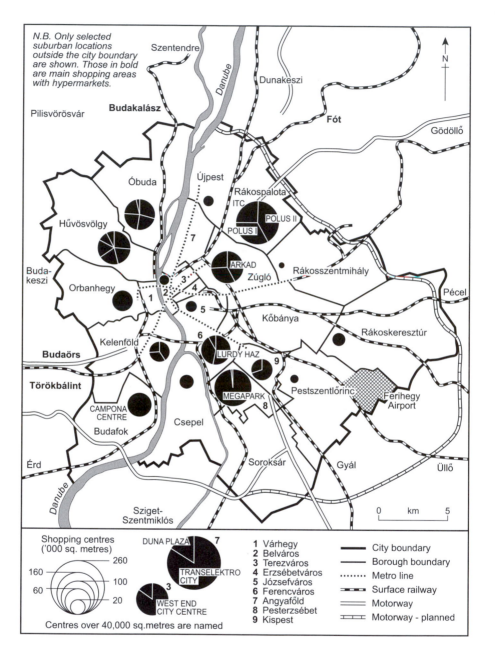

N.B. Only selected suburban locations outside the city boundary are shown. Those in bold are main shopping areas with hypermarkets.

Szentendre

Dunakeszi

Pilisvörösvár

Budakalász

Fót

Gödöllő

Óbuda

Újpest

Rákospalota

ITC

POLUS II

Hűvösvölgy

POLUS I

ARKAD

Zúgló

Rákosszentmihály

Buda-keszi

Orbanhegy

Rákoskeresztúr

Kőbánya

Pécel

Kelenföld

LURDY HAZ

Budaörs

Törökbálint

MEGAPARK

Pestszentlőrinc

Ferihegy Airport

CAMPONA CENTRE

Csepel

Budafok

Érd

Soroksár

Gyál

Üllő

Danube

Sziget-Szentmiklós

0   km   5

Shopping centres ('000 sq. metres)

DUNA PLAZA   7

260

160   100

TRANSELEKTRO CITY   3

60   20

WEST END CITY CENTRE

Centres over 40,000 sq.metres are named

1 Várhegy
2 Belváros
3 Terezváros
4 Erzsébetváros
5 Józsefváros
6 Ferencváros
7 Angyafőld
8 Pesterzsébet
9 Kispest

City boundary
Borough boundary
Metro line
Surface railway
Motorway
Motorway - planned

Figure 5.3 Shopping malls in Budapest

Source: Economist Intelligence Unit

– close to the main Nyugati (West) station like Duna Plaza – has office space, water attractions and a Hilton Hotel. It seems that the malls need catering and cinema/leisure facilities to strengthen their role as destinations for family outings. This can also be seen in GTC's $100 m. 'Galeria Mokotów' mall in Warsaw which has a muliplex cinema, bowling alley and Atomic entertainment centre and day-care centre, as well as involvement in cultural life through concerts, fashion shows and schoolchildren's art contests. However, while these developments certainly put life into city centres, they can also give rise to environmental criticisms on the grounds of increases in traffic, truancy and crime as well as unsympathetic design, suggesting the need for more thorough impact assessment. Large developments in Budapest will in future be restricted to undeveloped areas, although bribery remains rife as a means of getting planning permission. Meanwhile, the mall is moving down the hierarchy as provincial cities benefit: the 'Csaba' complex is under construction in Békéscsaba and across the border in Romania, the Dutch company Plaza Centers Europe are building Center Plaza in Timişoara.

Out-of-town shopping

Despite the encirclement of the central area by shopping malls, Budapest has also witnessed a hypermarket boom on the city edge, reflecting the accessibility of suburban locations given rising car ownership and the developing orbital expressway MO. Also land is cheaper and suburban municipalities tend to offer tax advantages which also attract tenants to the industrial estates or 'Uzleti Park' (business tax in Budaörs is only one-third the level in the city). However, Budapest is not yet at the stage of boasting mammoth retail parks with central parking surrounded by half-a-dozen warehouse stores, though existing complexes combine clothing, DIY, food, furniture and office supplies. Popular locations lie on the northern side: both east of the Danube at Fót and west of the river at Budakalász (Bricostore-Cora in both cases), and in the south: east of the Danube at Soroksár (Auchan) and on the west side at Budaörs (Auchan and Baumax-Ikea, with Tesco-Media Markt in prospect) and Törökbálint (Atlanta Centre-Bricostore-Cora). Just within the city boundary at Budafok is the large Interfruct wholesale fruit market. Super- and hypermarket development contributes to lack of shopping on low-prestige housing estates. Bank headquarters are also pushing outwards but historic central districts should do better as local authorities sell off rental rights to shops: in 2003 legal changes will force Budapest to equalise rents for all shops and many existing businesses will have to close. However, change in city centre shopping is already occurring as many shops become highly specialised and foreign-owned to cater for the new elite, a trend which is not satisfactory for low-income shoppers isolated from edge-of-the-city hypermarkets by lack of car ownership.

East Germany best exemplifies the Western fashion for out-of-town shopping with the ring of 'Einkaufszentren' around Berlin. Leipzig has a new 40 ha shopping centre (Sachsenpark) and the Quelle mail order complex on the northern edge convenient for the autobahn, but the largest developments lie beyond the city limits with the 150 ha Saalepark between Leipzig and Merseburg and the 250 ha 'Güterverkehrszentrum Wahren' near Schkeuditz airport along with other industrial and business estates. In mid-1995 only 223,400 m² of retail space out of a total of 655,400 was situated in the city centre because of uncertain land ownership and inadequate planning in the early days, which gave rise to a 'Wild East' phenomenon. A similar trend is evident in Chemnitz with Chemnitz Park situated near the autobahn junction Chemnitz Nord. Meanwhile, the company involved with the Polus project in Budapest now aims at 25 malls in the region including one in Bratislava which will provide Slovakia's first multi-use shopping complex. Austria's Spa International has linked with Slovenia's Mercator to open Ljubljana's largest shopping centre Interspar with 48,000 m² of retail area plus parking. American-led Prague Investments are renovating and developing a 3,700 m² plot at Na Prikope for retail and office space. And a group of American, European and Asian developers are creating a Warsaw Distribution Centre, combining manufacturing, warehousing and distribution facilities. Near Okęcie airport 55,000 m² of space will become available in ten separate buildings.

Cities in transition: infrastructure and conservation

Transport

Tramways are being retained but with better interior design (seating, illumination, insulation, heating, ventilation and passenger information), low floors and secondary suspension for greater comfort and reduced damage to track. Germany's 1992 'Verkehrsfinanzierungs-Gesetz' provided 90 per cent federal grants towards the cost of new tramcars for East Germany. ECE tram builders like Tatra need foreign partners to help with the transition to modern low-floor designs which are necessary to compete now that orders can be placed in the West and second-hand trams (from the Netherlands and West Germany especially) are readily obtainable. Hungarian Ikarus buses which used to predominate in most socialist towns are being replaced by more modern vehicles from Neoplan (Germany), Volvo (Sweden) and other Western builders. Meanwhile, the expansion of metro systems is occurring in most capital cities: Berlin, Budapest, Bucharest and Prague, while Belgrade, Bratislava, Sofia and Warsaw have opened their first lines since the revolution, leaving Tirana and the other capitals of former Yugoslavia as the only exceptions. The older systems are being modernised (new metro cars for Prague will be provided by a Siemens-led consortium) while the Berlin S-Bahn is still adjusting to reunification with the impending completion of the inner circle following the reopening of sections closed through the erection of the Berlin Wall. Provincial cities are also building rapid transit rail systems. A new 38 km S-Bahn has been provided for Halle–Leipzig and a metro for Łódz is planned. Main-line railway stations are being reorganised in the light of metro links.

Such developments naturally dominate urban politics over long periods of time. The Warsaw metro was first proposed in 1925 (north–south between Mokotów and Muranów complemented by an east–west Praga–Wola axis) but momentum was halted by the depression of the 1930s and then the Second World War. Mounting pressure on surface transport forced serious reconsideration in the 1970s with a breakthrough when the FSU offered assistance in the aftermath of martial law (given a perceived dual role as a civilian bomb shelter as well as a public transport facility). Once again a north–south axis has emerged with the first line advancing from Kabety in the south through Mokotów to Gdańska railway station and Młociny (Lucchini steelworks) in the north, with complementary east–west links for Bradno–Jelonki and from Gocław to Dwórzec Zachodni bus station. All this is linked with population trends: plans for more construction east of the Vistula have been revised due to the relatively slow rate of population growth. At the same time housing and office development crystallise around planned metro stations: house prices in areas close to the metro have risen up to a fifth faster than the market average, given the strong desire of Poles to own their own flats and avoid dependence on a poorly-developed rental market. However, metro lines are very costly to build and objectors argue that with more land for surface transport such investments could be transferred to other regional development projects.

Surface railways are also evolving, notably in Berlin with the new north–south main line to link with the east–west axis – and the S-Bahn – at a new underground Lehrter station and with the U-Bahn at Potsdamer Platz (referred to in more detail in Chapter 6). Kraków's Transportation Centre includes a new railway station, bus station, flight terminal – and eventually a metro – as well as hotels, shops and a business centre. A tunnel connecting the east and west parts of the city will cater for motor vehicles and trams while a new Vistula bridge will provide a new north–south tram route. Meanwhile, relocation of the main railway station is contemplated at Brno where the central area could be enlarged and a closer link could be made with the bus station at Zvonarka which is one of the largest in the region. Chemnitz and Dresden are other possibilities, while the approaches to Leipzig could be simplified by a tunnel linked with underground services to the airport and trade fair as well as the main station.

Research has shown that where metro systems are well established – as in Budapest and Prague – all groups appreciate subsidised public transport. Whereas in Kraków and Warsaw people in high-status households rarely used public transport (at least not until the recent arrival of the metro in the case of Warsaw) and subsidy of public transport is more contentious. Nevertheless, the use of cars for urban transport has increased sharply in all cities. In Budapest the public transport/private car split was 84:16 in 1975 but 80:20 in 1990 and 64:36 in 1993, exceeding estimates made in 1990 for the 2010–20 period (Nelson 1997: 47)! In part, the figures reflect decreased use of public transport caused by financial pressure faced by operators and customers, with too little investment in fleet renewal and provision of bus lanes. But this is placing city centres under very heavy pressure and to prevent further deterioration a Budapest Transport Association will operate a unified tariff system and offer concessions. Efficient integration of bus, tram and metro has been achieved in Prague – with ticketing which allows transfers between modes – while new terminals for regional bus lines plus high-speed rail tracks and better airport connections are envisaged to retain public support. Meanwhile, car parking is strictly controlled with increasing use of underground car parks, though this does not solve the problem of clutter in the immediate vicinity of apartment blocks. In Warsaw, where parking payments have been long delayed, parking meters – with charges rising exponentially per unit of time – will now generate funds to improve public transport and provide eight underground car parks. Some department stores are providing multi-storey parking and community garages are to be provided underneath schools. Provincial towns are taking firm action, with Pécs in Hungary using a European Commission programme to help fund environmentally friendly transport projects by introducing a clean zone that will limit car access to the city centre.

Ring roads are gradually being provided to avoid pressure by lorries in city centres and Warsaw has actually banned heavy trucks except for the evening period (1900–2300). Motorway links are also part of the strategy of reducing through traffic, an issue largely ignored under communism when transit vehicles were relegated to waymarked routes through back streets. Finally, there is a growth of interest in cycling. This is very evident in Berlin and Brandenburg where an extensive network of paths is being promoted. In Warsaw, environmental protection activists began work in the 1980s to gather momentum for the present Warsaw 'Bike Path System' project which should eventually exceed 700 km. It is being piloted by a system connecting student dormitories with the main university campus. Gydnia is developing bicycle routes as a means of limiting motor traffic in the city centre.

Transport and town planning

There are close links between transport projects and the zoning of land for future recreational and institutional use. Industrial/warehouse developments are being sought near to airports and motorway intersections. Industrial sites to the south of Berlin are close to the motorway ring and are also convenient for Schönefeld and Tegel airports. At Wrocław the districts of Długołęka and Kobierzyce will benefit in the context of the developing motorway system and ring road. In Budapest, the orbital expressway (MO) influences hypermarket locations on the edge of the city – as already noted – and also had a bearing on the abortive plan for the Expo (World Exposition) in 1996 to coincide with celebrations marking 1,100 years of Magyar settlement in the Carpathian Basin: this focused on a 44 ha site accessible from the MO and the Csepel Island suburban railway. An extended university campus will now use this site. A new west–east metro line is envisaged from Kelenföld – linked with a 40 ha commercial development for District XI – to a central intersection at Astoria with the prospect of an intermodal complex and motorway focus. Already Budapest Technical University is present plus several companies (including IBM, Matav and Mercedes). Other projects for a highly dynamic city include an Information Park (a miniature 'Silicon Glen' with a number of buildings already in place), a town centre programme for Albertfalva requiring foreign investment for housing, offices and a trade centre, the 107 ha residential development for

Madarhegy, and provision of water tourism on the underutilised swamps of Lágymányos gulf.

Warsaw's railway stations are development points, with Gdańska – on the modernised Berlin–Moscow line – attracting development by Carrefour for shopping, with a hotel, theatre and offices. Also Warsaw demonstrates the importance of new bridges over the Vistula for the improvement of the east bank. Although there are exceptions, Praga is generally perceived as an area of poverty, crime and filth with low rents and concerns over personal security. But after neglect under communism, in favour of the city centre, there has been change with the renovation of Rozyckiego Bazaar in the form of a three-storey shopping centre with 400 stalls on the ground floor, 160 shops on the two upper floors – plus a parking lot. Change is also evident around the Wschodnia (East) railway station and in Targówa Street with late nineteenth/early twentieth-century buildings transformed from residential use to office space as residents are rehoused and buildings in danger of collapse are renovated. By the river at Praski Port, 0.70 m. m^2 of office, retail and residential building will occupy a 47 ha site. It is an opportunity for original ideas by imaginative people, taking into account the entrance to the port and the need to protect green spaces. Praga is close to central Warsaw and yet may cost only a tenth of land in Srodmiescie. The new Swiętokrzyski Bridge, designed by a Polish–Finnish consortium, is the key to the transformation, but the Siekierkowski Expressway to the south – catering for heavy vehicles – is also important, along with the corresponding project on the northern side of the city between Białołęka and Bielany boroughs. Swiętokrzyski Bridge will also give access to a modern campus including an ultra-modern library for Warsaw University which will also contribute to renewal in Praga. Finally, planning for recreation facilities includes a 21 ha area of natural landscape at Czerniakowski Point on the river – close to Lazienkowska Bridge – which will receive the first large-scale recreation and sports investment in the city.

Infrastructure in Wrocław

Before 1990, water and sewage services suffered from lack of funding but a new management model adopted during the mid-1990s is leading to better quality services with local government responsibility with finance through a grant from the National Fund for Environmental Protection. Wrocław has never had a modern sewage-treatment plant, but a new station became fully operational in Pracze Odrzańskie in 2001, served by a network of main sewers to intercept the local systems already in place which generally discharged into rivers without treatment. The project was started in 1976 but only made rapid progress after 1991 when the local authority took over. The new treatment works has a daily capacity of 90,000 m^3 and meets both Polish and EU standards (Figure 5.4). Odra and Sleza trunk sewers will serve new factories built in the west, and on lands stretching from Wezel Bielanski in the southwest of the city past the airport to the west of the city. The whole collection system extends to 1,050 km and covers 90 per cent of the populated area of the city. For water, the old Na Grobli system, dating to the end of the nineteenth century, serves the city core with water from the Nysa Kłodzka and Oława rivers and also from ground water from over 1,000 ha of catchment. The newer parts of the city get water – all taken directly from the two named rivers – from Mokry Dwór. Leśnica serves a local area with water from a deep well. The two main treatment plants have a combined production of 0.18–0.22 m. m³/day and the distribution system of 1,660 km reaches 99 per cent of the population. The medium-term investment programme 1997–2003 will secure better water quality through modern treatment technology, and reduced consumption through more economical use, including the renovation and renewal of pipes. The entire water and sewage programme was enlarged after the 1997 floods: pumping stations have been upgraded to work in extreme situations, with submergible equipment installed at Bierdzany and Czechnica.

District heating systems are a distinctive feature of former socialist cities. In Wrocław there are some small units linked with factories (Hydral and Polar) while the main heating plants serving one integrated network are Czechnica and Wrocław. The system began as a 3.6 km heating conduit but now comprises a 440 km network – including a central ring system – and 4,500 substations heating 60 per cent of Wrocław's buildings. The urban heat energy enterprise (MPEC) – founded in 1961 – became a limited liability company in 1993 (wholly owned by the local authority) but was privatised in 1999 and then launched a share issue the following year to raise capital to expand the network and replace 200 km of pipes by 2005 using pre-insulated pipes that are almost 100 per cent secure and guaranteed for more than 30 years. The company has successfully increased its role through the liquidation of some 40 local boilers in the centre of the city, where alternative supplies can most easily be provided, with state-of-the-art technology to regulate the system. A pilot programme is operating on the Huby estate whereby the company takes responsibility for internal installations as part of the building operation: new buildings have the system installed while old buildings have to be modernised at high cost. Such has been MPEC's success that it has now taken over Poznań's heating company. The heat itself comes from a separate company: the Wrocław Group of Combined Heat & Power Plants: now named Kogeneracja to highlight the efficiency of combining electricity and hot water production, which has achieved its own environmental and efficiency gains. As a large company (now half owned by a foreign group headed by Electricité de France) it can afford costly dust-extraction and desulphurising equipment: hence the pollution is negligible compared with the emissions from small boiler chimneys.

There is also World Bank and Dutch government assistance for a waste incinerator to be run by the local garbage company WPO. Attempts to find a location for an incinerator during 1990–4 failed and a site near the sewage plant was secured only in 1996. Large new modern vehicles are being introduced for garbage collection (involving bins, containers up to 1,100 l and plastic bags), but in parts of the city where they cannot be used smaller Polish-made trucks have been acquired and there are special vehicles to handle containers. In total, 550–600 t are handled daily. Mention should be made of the green area

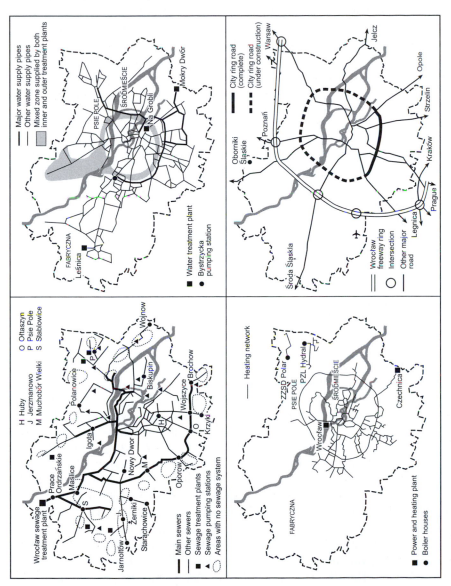

Figure 5.4 The infrastructure of Wrocław, showing sewage treatment (top left); the heating network (bottom left) and the expressway system (bottom right)

Source: *Warsaw Voice.*

management programme for renovating recreation facilities in old parks (also the Japanese gardens in Szczytnicki Park). New green areas are planned along streets, along with 50 km^2 of municipal forests. Finally, Figure 5.4 shows Wrocław's integration with the Polish road system. The motorway intersection on the southwestern edge of the city will launch an urban freeway which will continue northeastwards to Łódź and Warsaw. This will require a 900 m viaduct over the railway at Gádow, but it is very necessary to relieve pressure on the city's historic bridges because of heavy traffic in a situation where there are no other crossings of the Oder River for 30 km in each direction. There is also a proposal for a complete 30 km ring road and soundproofing against noise along main streets. As described elsewhere, the railway link with Warsaw will benefit in the same way through a connection with the Central Trunk.

Urban conservation

The townscape is changing in a host of ways: the renaming of streets and replacement of monuments raised in honour of communist leaders, to recognise others connected with other epic events in history. Graffiti is an art form which expresses itself prominently on the walls that protect building sites and in public transport environments, especially railway rolling stock, while advertising is often more obtrusive than in Western cities, claiming a contribution to the ECE urban style for the turn of the millennium. Modern architecture is much in evidence and while the style of many new buildings is generally in keeping with the existing stock, some do not conform because they are massive structures of great height which symbolise international capitalism erected in the absence of adequate planning control. 'Gold teeth in a decaying jaw' is a phrase used to describe some luxury buildings, brought to Warsaw from Western design studios, which ignore their drab, dirty and lacklustre surroundings. However, there are interesting projects on the outskirts where predominantly residential functions encourage more sophisticated designs to satisfy both future tenants and existing styles (since fewer buildings were destroyed here during the war). Also more of the buildings are designed by Warsaw architects aware of local traditions. Former socialist cities are good repositories for nineteenth-century – and earlier – architecture; so whatever consensus develops over the blending of old and new there will be a higher priority for environmental quality and the preservation of historic buildings.

City plans are seeking to provide an adequate context to guide the development of the townscape. The Warsaw plan covers not only the central area commercial development and housing zones, with industrial and warehousing requirements, but also green areas and corridors reserved for environmentally friendly activities. Budapest's long-term plan of 1989 has been revised to make greater use of the Danube, generating a green belt round the city and preserving architectural character. The latter will be difficult given the development pressures but Budapest could be a pace-setter and its experience will be critical for what happens elsewhere. Competitions have played a role in the planning concepts for sensitive places like the areas below the castle in Bratislava and around the parliament building in Bucharest. Major buildings from the commu-

nist period may be accepted pragmatically as part of the picture, but with attention to the surroundings to modify their dominance. The dominance of Warsaw's
Palace of Culture has long been an issue but only in 1991 was a competition
staged to find the best development plan to integrate the palace with surrounding
structures and minimise its symbolic weight in the area. The winning design
will ensure that in future the palace will be seen only through narrow streets
and passages intersecting with a circular boulevard surrounding the redesigned
district.

However, there are many neglected buildings that require refurbishment – both
medieval structures and nineteenth-century buildings linked with independence
and the rise of capitalism – not to mention recent cases of war damage in
Dubrovnik and Sarajevo. Inherited conservation enterprises are playing important roles in the new situation: the Polish organisation for conserving buildings
(PKZ), formerly financed by the Ministry of Culture, is now operating profitably
with a number of domestic projects and contracts abroad (in Europe and Asia).
Money has to be found, sometimes through special funds, like the Prague Heritage
Fund set up in 1991 by Václav Havel and the Prince of Wales. The capital cities
are seeing the greatest efforts to protect urban heritage, like the Budapest Opera
House of 1893, finished in the run-up to the millenary festival of 1896 celebrat-
ing 1,000 years of Magyar settlement. And some preservation is possible in cooperation with business: McDonalds' interest in the restaurant has helped in the
restoration of the Njugati (West) railway station in Budapest, while the Slovak
architectural group HUMA is refurbishing historic buildings in Bratislava for
office and executive residential use. However, as well as individual buildings there
is a case for rescuing entire architectural groups which in a Western situation
would probably have been lost beyond recall under the comprehensive redevelopment ethos of the 1960s and 1970s. In Prague the sale of historic buildings by
local authorities who lack the means to look after them is being shelved until an
adequate system of conservation control is in place.

There is much conservation in provincial towns. Gdańsk old town is now a candidate for the UNESCO World Heritage List – established in the 1970s to identify sites of primary importance for cultural heritage – with Mariacki Church the
largest brick church in the world, complemented by the late medieval structures of
Wyspa Spichrzów (Granary Island) between the Old and New Motlawa 'water
streets' comprising a sixteenth-century moat protecting the grain stores from fire
and requiring connection with the rest of the city by drawbridges. Although much
was lost during the war, the essence of Gdańsk's historic fabric – recalling the city's
role as a pre-modern exporter of grain surpluses floated down the Vistula – justi-
fies housing revitalisation combined with the sensitive filling of gap sites and some
reconstruction of entire frontages. A Polish–American company has built a luxury
hotel, architecturally in keeping with the granaries which are being partially reconstructed for a maritime museum, and the Gdańsk Philharmonic will be accommodated in the old Olowiany Dwor power station. Meanwhile, at Toruń –
included in the World Heritage List – a local initiative was launched in 1989 to
conserve some 1,500 historic structures (including three Gothic churches with fine
interiors and Europe's largest Gothic city hall built of brick) and highlight the

Copernicus connection through international academic and cultural events. In 1997, the council in the nearby town of Płock adopted a 30-year plan for the revitalisation of the old town, including historic buildings, fountains and other structures, with help from sponsors including Płock's partner city of Darmstadt.

In Łódz there is the work of the 'Łódzermensch': cosmopolitan entrepreneurs in a city where Polish, Russian, German and Jewish cultures have merged throughout the modern period. The result has not only been a factory system, but also townhouses, villas and palaces, and churches, shops and banks. The wealthiest entrepreneurs had private 'duchies' that included suburban working-class areas and parks like Ksiezy Mlyn (Priest's Mill) belonging to the cotton king Karol Scheibler, which is now an open air museum of nineteenth-century industrial architecture, with many of the buildings still serving their original function. The complex – completed in 1883 – includes the first spinning mill in the city (1870–3 – now Uniontex cotton works), workers' housing (1875) with a hospital, school, fire station and shop. Another candidate based on German-Jewish entrepreneurialism is the Poznański complex in Ogrodowa Street embracing a factory – with tied housing, hospital, Catholic church and concert hall – along with a historically important palace of 1878 which houses the Łódz History Museum. Elsewhere, the Auschwitz project is also being taken forward in a broader context which includes the old town of Oswięcim and a buffer zone from which inappropriate development will be excluded. Fortifications are regarded more ambivalently. The army is established in Warsaw citadel but Poznań's fortress was destroyed in post-war reconstruction, while Dęblin is a fruit and vegetable warehouse, Kraków's St Benedict Fort is a centre for modern art and others have been converted into apartments. The future for the remainder may well depend on finding utilitarian functions.

Town centres are being refurbished throughout the region, even in relatively small provincial towns like Banská Bystrica. But finance is a serious problem in the SEECs, though the 'Historical Centre' Agency for Bucharest is restoring some buildings to their original shape with help from the EBRD and income will then be gained from their commercial use as shops or banks. In Albania there are many inappropriate uses for historic buildings. A seventeenth-century 'hammam' (bath house) in Elbasan, restored under communism (along with its fountain), is now a disco and bar fitted out with plywood and plastic. Unfortunately, the restitution law of 1992 did not make allowance for historical monuments and the state cannot buy back the historic buildings because there is no money. Car repairs are being done in a valuable nineteenth-century town house in Shkodër while the homeless are being accommodated in the amphitheatre in Durrës. This is unfortunate in terms of future tourism.

The rural areas

Communist restructuring

Under communism the expansion of the rural population came to an end. Calculations by Eberhardt (1994), which cover the entire region apart from Albania, see the rural population rising from 49.81 m. in 1900 to 62.21 m. in

1941 although the share of the total fell from 73.3 per cent to 66.5. Thus the region remained essentially rural in 1945 and any post-war regime would have encouraged urbanisation. However, despite a massive turnaround with 68.8 per cent of the population rural in 1950 and 57.9 per cent urban in 1990, the rural population only declined absolutely by 0.37 per cent per annum. Indeed, Albania's rural population increased by 2.89 per cent annually (admittedly due to the special circumstances of strict internal migration controls) while Poland's rural population continued to grow during the first half of the communist period (Table 5.4). There was not so much a catastrophic fall in absolute figures as a diversion of the natural increase to the towns. The changes were not surprising, but there was political significance in that the 'new socialist power in the region was urban based through centralisation and aimed to control the cities and to govern the country from them' (Harloe 1996: 14). There was also a social revolution which affected rural areas through the elimination of former landowners and relatively prosperous peasants. Traditional peasant life disappeared with the formation of cooperatives and commuting to towns which provided money for new houses and consumer goods.

Socialist farms eliminated the differences between rich and poor peasants, although in poorer countries like Romania with a relatively low level of mechanisation, continued dependence on carts driven by the experienced 'conductori' of Romanian villages – distinguishable from the working peasants or 'maistri' – gave the remnants of the 'kulaks' a grip on the cooperatives' affairs. But instead of sharing in the modernisation launched in the towns, the villages tended to be marginalised and the conflict between equity and efficiency was all too often resolved in the latter's favour by large cities and industrial projects. The rural areas remained underdeveloped with higher energy costs and poorer infrastructure by urban standards. In the GDR an average telephone provision of 10 phones per 100 of the population (compared with 50 in West Germany) worked out at one phone per two apartments in East Berlin (but only 1 in 10 in Dresden and Rostock), while at the other extreme 2,000 villages had no connection with the telephone network at all. The 'urbanisation' of the countryside was far from complete. Given the lack of material progress and limited choice of employment, out-migration became a universal feature of rural change, except in Albania where internal migration controls were maintained.

There were some cases of radical changes, most notably through resettlement in Poland's 'Recovered Territories' where Germans were expelled to make way for Poles displaced further east. There was coercion in the way that Germans were expelled from the Sudeten territories of Czechoslovakia while Turks were removed from Bulgaria's southern frontier and Ruthenians had to vacate the Bieszczady district of southeastern Poland. And when Tito was expelled from the Cominform in 1952, fear of pressure on Romania's western frontier led the authorities to clear the border of unreliable elements who were transported – temporarily in this case – to the Bărăgan steppe in 1952. There was also some immediate effort to rebuild after war damage, as in the case of Lucimia in the Gura Puławska area of Poland and Lidice near Kladno in Czechoslovakia. And gradually houses were replaced by new structures in more modern styles. Some

Table 5.4 Rural population 1950–2030

	Population (,000) in						
	1950[1]	1970[1]	1990[1]	1995[1]	2000[1]	2010[2]	2030[2]
Albania	980	1459	2113	1934	1895	1862	1713
Bulgaria	5395	4093	2922	2697	2462	2043	1358
Fmr. Czechosl.	7702	7372	5901	5779	5552	4997	3588
Czech Rep.	na	na	na	3571	3457	3128	2201
Slovakia	na	na	na	2208	2095	1869	1387
Hungary	5671	5322	3933	3625	3317	2764	1882
Poland	15217	15508	14549	14033	13351	11959	8990
Romania	12151	11781	10765	10032	9330	7962	5448
Fmr. Yugoslavia	13191	12261	11200	10342	10129	9189	6893
Bosnia&H.	na	na	na	2012	2258	2245	1712
Croatia	na	na	na	1987	1891	1669	1187
Macedonia	na	na	na	789	768	721	585
Slovenia	na	na	na	969	941	851	592
Yugoslavia	na	na	na	4585	4271	3703	2817
Total	60307	57796	51383	48442	46036	40776	29872

	Rural population as a percentage of the total							Annual change (%)			
	1950	1970	1990	1995	2000	2010	2030	1950–90	1990–2000	2000–30	1950–2030
Albania	79.7	68.2	64.2	62.8	60.9	55.6	43.3	+2.89	−1.03	−0.32	+0.93
Bulgaria	74.4	48.2	33.5	31.7	29.9	26.4	20.0	−1.15	−1.57	−1.49	−0.94
Fmr. Czechosl.	62.2	51.4	37.9	36.9	35.5	32.1	24.7	−0.58	−0.59	−1.18	−0.67
Czech Rep.	na	na	na	34.6	33.7	31.1	23.8	na	−0.64[3]	−1.21	na
Slovakia	na	na	na	41.2	38.9	34.3	26.0	na	−1.02[3]	−1.13	na
Hungary	60.7	51.5	37.9	35.4	33.1	28.7	21.8	−0.77	−1.57	−1.44	−0.84
Poland	61.3	47.7	38.2	36.3	34.4	30.5	23.2	−0.11	−0.82	−1.09	−0.51
Romania	74.5	58.2	46.4	44.1	41.8	37.0	28.2	−0.29	−1.33	−1.89	−0.69
Fmr. Yugoslavia	80.7	62.3	49.1	46.1	43.9	39.0	29.7	−0.38	−0.96	−1.06	−0.60
Bosnia&H.	na	na	na	58.9	56.8	51.8	40.3	na	+2.45[3]	−0.08	na
Croatia	na	na	na	44.2	42.3	37.9	29.0	na	−0.97[3]	−1.24	na
Macedonia	na	na	na	40.2	37.9	33.7	25.6	na	−0.59[3]	−0.79	na
Slovenia	na	na	na	48.7	47.4	43.6	33.6	na	−0.58[3]	−1.21	na
Yugoslavia	na	na	na	43.4	40.1	34.4	26.0	na	−1.37[3]	−1.13	na
Total	68.8	53.6	42.1	39.9	38.0	33.8	28.6	−0.37	−1.04	−1.17	−0.63

Source: FAO database

Notes
[1] Estimate [2] Forecast [3] 1993–2000

small apartment blocks were provided for state employees, but most rural housing remained owner-occupied with the emphasis on single-storey cottages. There
were some new functions, for investment arose not only from the development
programmes of cooperative and state farms, but through the plans of silvicultural
and woodcutting enterprises. There were also mining and manufacturing companies, particularly those with interests in light engineering, food processing and
textiles. Rural industries multiplied in the 1970s at a time of labour shortage in
the towns. It was a spontaneous process and many units were badly equipped
with obsolete machinery, though some rural growth areas emerged, and improved
rural living conditions often resulted. Rural development also occurred through
tourism in the form of 'resort villages' created by refurbishing accommodation
provided for building workers engaged on road and hydropower projects, while
some camp sites and motels were provided by cooperatives, and private households might provide bed and breakfast accommodation.

Services and central places

Services improved as the budgets of education and health ministries impinged
on rural services but rural education was often deficient. Public transport was
particularly sensitive to the needs of the commuter because falling employment
in agriculture gained some compensation from commuting to work in the cities,
particularly the regional centres where light industry was promoted. Small
commercial developments provided additional shopping and catering facilities.
These various strands of development tended to affect the larger, best-placed
villages and introduced increasing differentiation between the 'key villages' and
the outlying settlements with relatively few facilities. Electrification failed to reach
the smaller places, like the remoter individual farmsteads ('tanyas') and hamlets
on the Hungarian Plain where the dispersed patterns of settlement appropriate
under a system of small largely self-sufficient peasant farms were not so satisfactory in the context of large commercial farms with labour deployed from
central points. Variations in the employment structure also related to population size, and levels of educational attainment were similarly structured with
respect to settlement groups. A population of some 2,000 marked the threshold
for key village characteristics. State farms and agricultural schools played an
important role in the expansion of villages, but the key villages were also favoured
by investment in light manufacturing (already referred to), services and an inten-
sified agriculture, although rural industry was initially discouraged through
nationalisation and the concentration of capacity in large urban-based units.
Small-scale enterprises persisted in the larger villages while wood processing
made diversification common throughout the mountain regions.

Before 1948 there was relative stability in the villages but the formation of
cooperatives led to some consolidation because as their average size increased
selective migration favoured the villages appointed as administrative centres for
the larger farms. This led first to Stage 1 key villages (or 'temporary main rural
settlements') and then to a smaller number of Stage 2 key villages ('settlements
of local importance') with many cases of a change in employment (linked to the

expansion of cooperatives into non-agricultural functions) as well as a switch in residence. This process has been documented for the Nitra region of Slovakia (Drgona *et al.* 1998) (Figure 5.5). Strangely, the key villages did not always harmonise with the cooperative centres. Although most of the Stage 2 key villages in 1989 were also centres for cooperative farms, Zbehy (for example) was not; and equally Ivanka was one of a few villages acting as a cooperative farm centre that was only a Stage 1 key village. Special circumstances affected the Nitra and Zlaté Moravce areas where the agriculture was organised by the university and high school respectively (both with a specialised profile). Since 1989, however, villages have not been tied to the established hierarchy and many have opted for administrative independence, especially where people believe there is considerable growth potential. This includes some of the old cooperative farm centres and other villages which have set up their own democratic organisations.

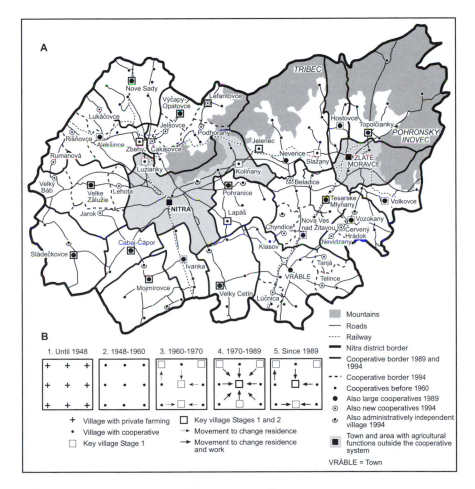

Figure 5.5 Cooperatives and key villages in the Nitra region of Slovakia

Source: Drgona *et al.* (1998: 270)

Some efforts were made by the planners to bring the town closer to the remoter rural areas. In some cases large-scale industrial development led to the emergence of a completely new industrial centre. This occurred where new mineral resources were found, e.g. Dimitrovgrad in Bulgaria and Gh. Gheorghiu-Dej (now Oneşti) in Romania. They were also found at new assembly points for manufacturing like Eisenhüttenstadt in East Germany and Leninváros (now Tiszajváros) in Hungary, and again at decongestion points on the edge of conurbations, as at Tychy in Poland (already noted). However, gradual urbanisation based on well-situated villages seemed to offer the best way forward. On this basis some radical programmes evolved to consolidate rural settlement: notably in Hungary in the 1950s (when there was a plan to eliminate 'tanyas' and concentrate population in 'agrogorods' on the Soviet model), and Romania in the 1980s (with Ceauşescu's plan for agro-industrial towns and the destruction of 'non-viable' villages). The Romanian idea seemed likely to succeed, but the plan was overtaken by the revolution in 1989, while the Hungarian project was not pressed too strongly and consolidation was often frustrated when people from small settlements chose to migrate directly to the towns. In fact, most cases of village abandonment have resulted from the extension of lignite quarries in Czechoslovakia and East Germany and soil contamination in the Glogów Copper Basin of Poland. In Bohemia 65 villages were destroyed to make way for the expansion of huge lignite quarries: eventually there were five separate complexes each covering around 25 km²; and these required resettlement of 180,000 people. This was a far greater upheaval than the much-publicised programme of 'sistematizare' in Romania in the late 1980s which saw the elimination of only a handful of villages before the revolution.

Transition in the countryside

Out-migration has been a persistent feature of rural areas over the last 50 years, but decline has been accelerated by the demographic transition working its way through the region during the twentieth century. Areas of rural increase are now restricted to the urban-rural fringes and parts of Albania and former Yugoslavia – where relatively high rates of natural increase persist among the Muslim population – in contrast to peripheral areas of demographic disaster. However, the strict control of migration under communism has been followed by a massive outpouring of population that has seen Tirana grow to a reported 0.60 m., from 0.20 m.: surely the most outstanding case of population redistribution apart from during war-related upheavals in former Yugoslavia. During the transition rural decline has accelerated – only B&H has grown in 1995–2000 due to the end of the war – and the rate of decline is expected to accelerate during 2000–30, though relatively slowly in B&H, Albania, Macedonia to suggest a distinct West Balkan scenario, in contrast to Bulgaria where forecasts suggest that only a fifth of the population will be rural in 2030, compared with 40.3 per cent in B&H and 43.3 per cent in Albania (Table 5.4). This contrast between Bulgaria and Albania is remarkable considering that in 1950 Bulgaria had a rural share of 74.4 per cent with Albania only slightly higher at 79.7. Given the anticipated sharp reduction

in Bulgaria's total population this will mean only 1.36 m. in the countryside compared with 2.46 m. now. There will be major implications for rural services.
However, through prolonged out-migration, some border areas are already
endangered demographically, as in the case of Slovenia's Alpine and karstic
regions, where the tradition of polycentric development is being reversed. But
more stable areas are rarely sustained by adequate job creation and hence there
is heavy dependence on small farms with coping strategies that may well extend
into illegal trade. However, while the Czech Republic has problems over its 'inner
borderland' – manifest through a declining and ageing population, as well as a
dilapidated housing stock with only limited opportunities for recreational use –
people in western Bohemia can now take advantage of work opportunities in
Bavaria with daily commuting by bus.

Rural infrastructure

All rural areas show a superficial transformation from state-ownership, cooperative organisation (linked to stock-rearing in individual courtyards), public
transport and communist propaganda to private agriculture and business
(including shops and kiosks which offer a variety of beers and cheap country
delicacies), along with more advertising (including political party manifestations),
satellite television, second homes, second-hand car ownership and church-based
cultural activity. But more fundamentally there has been a demise of services
formerly maintained by cooperatives whose enterprise managers 'were in practice the most powerful local authority' (Rey and Bachvarov 1998: 347) to local
administrative autonomy on the basis of small territorial units which are not
viable for economic planning. There is particular need to regain cohesion where
large agricultural enterprises have collapsed and destroyed microregional cooperation, thus undermining local independence. Reference should also be made
to problems of ownership, especially for East German villages: 'without detailed
ownership information rural towns and villages cannot engage in such redevelopment programmes as housing repairs, new housing construction or general
improvements of the residential environment' (Kobayashi 1996: 390). A more
problematic situation in many areas is highlighted by fewer industrial units
(usually operating independently of cooperative management); an ageing housing
stock in need of renovation; less shopping (with reliance on savings from the
communist period); simpler food with reduced resources for medical and social
services; and reduced public transport with limited car ownership (with a predom-
inance of older and second-hand vehicles).

Rural infrastructure is slowly improving, aided by considerable foreign assistance including the EU PHARE programme. Water supplies reached 47,000
additional houses each year in Poland during the 1980s, rising to 116,000 during
1989–94. An even still faster pace of change has occurred in the installation of
gas-pipe networks, although the use of exchangeable metal cylinders is still very
popular (Gorz and Kurek 1998: 185). Retailing has also developed in Poland:
the number of shops and kiosks – selling a variety of goods – increased from
70,000 in the late 1980s to some 92,000 in 1993; a change generated almost

exclusively by the private sector because state services are now being closed down quite rapidly. However, the quality of services is extremely poor in parts of the SEECs. Some Albanian villages are still only accessible by foot or mule and farmers may travel, with difficulty, up to 3 km to their fields and 20 km to the nearest market. Water is often in short supply. Some Albanian farmers are able to water their crops from wells, but ground water is often saline, notably along the coastal plain. Improvements in East Germany have been financed by the application of the West German 'Dorferneuerung' programme. Participating villages receive subsidies to cover 80 per cent of costs of community works (like replanning a village centre) and up to half the cost of private works like housing renovation. Improvements in village housing, infrastructure, environment and social and cultural amenities are important to maintain the present population, particularly the younger people, and to stimulate a measure of counterurbanisation.

The rural employment problem

Since 1989, most rural areas have experienced some degree of impoverishment which has undermined the relatively comfortable lifestyles of even the village-based elites. The changes in agriculture have created new tensions. Much land is now owned by townspeople and even if they lease it to country people the rents take a third of production, thus adding to the traditional rural–urban resentments which have been marked by an 'electoral chasm' as rural areas vote for welfare rather than reform. Some government reorganisation appears misguided. Much of Poland's 'setaside' land falls into areas of high unemployment, caused by the laying-off of the former state farm workers. They receive benefits which cost almost twice as much money as was previously paid in budget allocations to the state farms. Thus the social costs of untilled land are very high and the economic rationale overall is debatable in the short term. Meanwhile, unemployment undermines the work ethic and aggravates problems of crime, alcoholism and drug addiction although the problem may be 'hidden' on small private farms which seek – albeit inefficiently – to absorb the additional labour. Fear of pauperism leads to higher involvement in small-scale farming to ensure self-sufficiency although, at the same time, it leads to higher consumption of animal fats which are not good for public health. And people may also be forced further towards the margins of legality, e.g. by smuggling cigarettes across Poland's borders with Belarus and Lithuania.

The most likely scenario is that far fewer people will be needed on the land. Agricultural employment across the region rapidly declined from about one-fifth in 1990 to a tenth in 1995 and an anticipated 5 per cent at the turn of the century because a lot of manual work can easily be replaced by machinery. The decline was very steep in East Germany: 884,000 at the end of the communist era but 208,000 when Bergmann (1992) saw the need for a further reduction by as much as one-half. In Poland the decline in the farming population has been estimated at 0.4–0.5 m. people between 1988 and 1993 and even so many more are considered to be surplus: to maintain structural change and farm consolidation, eliminating 0.75 m. very small farms of between one and three hectares, 0.15 m. new jobs must be created in Poland every year (Gorz and Kurek 1998: 197). However, if there is no alternative employment then small farms will continue to be valued as a means of subsistence. Farm units in Croatia have declined more slowly than production, reflecting increased diversification and feminisation. Land is valued on the grounds of sentiment and security.

Reduced employment in farming could well trigger renewed rural–urban migration, as in Albania where strict controls were in force during the communist period. But urban unemployment could bring about movement in the opposite direction, following the rationale of Western-style counterurbanisation, especially if there is diversification in rural areas. In the Czech Republic there are now more opportunities in the countryside for non-agricultural activity and a section of the rural population can pursue business at home in areas that were effectively closed in the past: not only small farms but also workshops, retail outlets, public houses and tourist accommodation: a trend reinforced by the loss of many urban jobs and the transfer of several thousand town dwellers to new homes and occupations in the countryside where houses have been purchased or obtained through restitution (Bicik and Gotz 1998: 115). There are also new jobs connected with wholesale trade and cottage industry, environmental protection, landscape and water management and agrotourism. A case has been made for resettlement in Serbia's border villages which have experienced steady depopulation since the Second World War. With houses available for refurbishment and arable land cheap, resettlement of refugee farmers could make an immediate impact on food production. Over the longer term there could be sustained economic growth in the context of a newly-constituted Euroregion for the Danube–Criş–Mureş–Tisza area.

In rural Hungary, family-based non-agricultural businesses using the family house as an office or workshop can have an important demonstration effect and help regain cohesion where 'early euphoria about local independence' has been eclipsed by the loss of cooperative farms (Meszaros 1996: 408). Small-scale enterprise is widely discussed in Slovakia, even though innovators may have to face ill-will and envy from passive co-inhabitants. At Stará Bystrica near Čadca, new businesses are being accommodated in buildings leased from the cooperative including a small cheese factory while Rejdová near Rožňava is experiencing a revitalisation of traditional crafts: carpet weaving, embroidery, shingle-splitting and wood carving. In the case of Romania, some resourceful peasants from the mountains around Sibiu have responded to the decline of the sheep business by moving into surrounding lowlands (sometimes taking over the houses of departing Germans) where they have invested their savings – gained from the Caritas pyramid investment scheme – into garages and other service industries. But although the secondary and tertiary sectors are growing – and some factories are appearing in old cooperative farm buildings – the rate of change is very slow. Nagy (1993) suggests a 'vitality index' to help identify small settlements which are entrepreneurial about their future development, using such criteria as flexible leadership and fund-raising schemes aimed at private and business sources to support local infrastructure.

The rural settlement system

Conditions vary considerably according to the situation of settlements in the hierarchy, for the best chances of growth occur on the edge of large towns, now that a turning point in urbanisation has occurred. Toth (1992) sees evidence of counterurbanisation with some large Hungarian towns registering a decline in population, beginning in the 1980s, in response to the attractions of village life for gardening and recreation. The trend is assisted by surfaced roads and bus services, but it is largely restricted to the suburban zone and there is little long-distance migration from the towns. Depopulation slowed down in the 1990s in parts of rural Bohemia through the building of urban-type detached houses on cheap land for middle-class commuters, around České Budějovice and Tabor for example. Urban–rural migration has been an option for the under- or unskilled unemployed escaping from high living costs and moving to areas where they can at least take up subsistence agriculture, while those with modest reserves of capital may start small businesses. Although bus services have been reduced, car ownership makes the acquisition of a housing plot in a village a more viable option. Szelenyi (1996: 312) even refers to 'rurbanised' villages as working-class suburbs functionally linked with industrial centres, given the quality of the housing stock and the potential of the privatised countryside for small businesses. East Germany is experiencing considerable non-agricultural development on the urban fringes of Berlin, Magdeburg, Halle, Leipzig, Dresden and Erfurt as well as rural areas bordering key transport corridors (Wilson 1998: 138–40). A roughly parallel situation can be seen in the Czech Republic with the fastest growth occurring in and around Prague, where there is a low unemployment rate of below 1 per cent and an intensive rhythm of construction activity. Many second homes are being converted into permanent homes in the city's immediate hinterland.

Away from the cities, contrasts between rural areas occur in terms of history and morphology, but more significantly through regional economic conditions which help to determine the employment and income possibilities within commuting range. Centre–periphery distinctions at the national level have been expressed in one case through the Poland A/Poland B phenomenon with the latter an unofficial designation for an area of high unemployment offering little hope of career development for young people and therefore carrying an element of social risk. But in either case there is a valid distinction to be made between small villages, with very limited infrastructure and poor living conditions – typically experiencing depopulation – and medium-sized villages with a range of industrial and service functions, especially where there is a history of small town status. Evidence from the Czech Republic reveals clear contrasts in employment structure and housing quality according to village population size (Table 5.5). Only a small proportion of the rural population are getting richer: 'local leaders, entrepreneurs, the highly qualified and those who own land or shops' (Kiss 2000: 232); and since these people tend to live in the larger settlements there is a relationship between village size and living standards.

Table 5.5 Rural issues in the Czech Republic 1993 (a) Employment structure in rural areas (b) Households and dwellings by settlement size

(a) Village size	Percentage of agriculture and forestry	Employment in industry	Construction	Other services
0–199	40.0	29.4	6.3	24.3
200–499	34.9	31.2	6.4	27.5
500–999	28.1	34.6	6.9	30.7
1000–2000	22.2	38.2	6.9	32.7
All rural areas	28.4	34.7	6.7	30.2

(b) Sett. size	A	B	C	D	E	F	G	H	I	J
0–199	18.5	53.2	78.6	30.8	45.7	9.6	48.7	1.5	34.4	66.4
200–499	17.7	59.0	80.8	31.0	47.6	10.6	47.2	1.7	25.7	51.4
500–999	17.2	62.7	81.9	31.5	48.3	12.1	45.1	1.8	20.7	38.7
1000–1999	16.9	65.5	82.9	32.9	49.2	14.0	43.0	2.5	17.1	36.1
2000–4999	16.4	69.1	86.1	37.3	51.4	16.5	42.1	4.2	12.8	28.8
Total	16.7	74.9	91.7	46.1	58.9	29.4	45.7	12.2	10.1	34.4

Source: Trnkova 1994

Notes
A	Square metres of living space per person	F	Telephone
B	Percentage of dwellings with central heating	G	Own car
C	Bathroom/shower	H	Weekend house
D	Washing machine	I	Per cent dwellings with no permanent resident
E	Colour television	J	Percentage of I used for recreation

SMALL TOWNS, VILLAGE CENTRES AND THE ORGANISATION OF THEIR HINTERLANDS

Much local development is likely to gravitate towards small towns which are the logical cores for rural associations. Small towns are numerous in the northern part of the region and examples of their dynamism are provided by recent research on Jemnice in Moravia and on groups of towns in the border regions of Poland. However, substantial areas of the SEECs remain remote from market towns. A more even urban network could achieve a better regional balance, along with continuing progress in education and growth in the tertiary sector, for this encourages economic activity to be 'more evenly distributed across the country, than would otherwise be dictated by manufacturing and farming activities which rely on natural, local resources and therefore lead to regional variations' (Nemes Nagy 1994: 367). Fortunately, key villages act as surrogate urban centres to some extent, as has been shown in the case of Pătârlagele in the Buzău Subcarpathians of Romania which provides services for a population of some 20,000 and has experienced a considerable growth in small private shops – including many roadside kiosks – since 1989 (Figure 5.6). The strengthening of these centres, leading to formal urban status, would discourage long-wave migration and make for equality of living conditions and incomes.

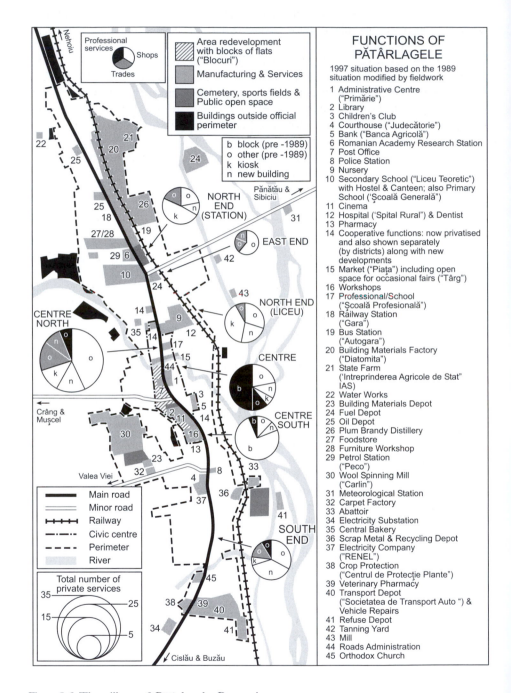

Figure 5.6 The village of Pătârlagele, Romania

Source: Pătârlagele Town Hall

Dobraca (1999) sees the daily markets of some 450 villages as an indicator of significant rural central place status. There are also some 750 villages with periodic markets and fairs.

Away from the small towns, the relatively poor services in rural areas will be a discouragement for counterurbanisation in the foreseeable future. There is a modest flow of capital into villages as pensioners return to the countryside and build or repair houses, while younger people may improve property inherited from their parents to have the option of returning permanently in the future. Villagers are investing the profits derived from farming in home improvements and there is a growth of food shops, bars and small manufacturing units. But unemployed people who live on small family farms cannot contribute greatly to output. A quarter of Poland's workers are active in rural areas but they produce only 7 per cent of GDP. Productivity is hit by low educational standards, poor services and the inefficient use of farm machinery. Services are very limited, because of polarisation on district centres, and this even applies to sheltered accommodation which means that old people must invariably leave their home village when closed care is required.

The weakness of rural districts has been related in part to communism, on account of strong urbanising–centralising forces and the abolition, in Romania, of the traditional 'plase' (a cluster of communes) and also (in 1968) the 'raioane' which had replaced them. Out-migration was strongly encouraged, after villages had previously developed their own facilities. Crisan (1999: 202) therefore calls for a rediscovery of identity reflecting the physical resources, 'ethnographic values' and local economic conditions. There is still some tendency for fragmentation as villages seek independent commune status, especially among localities absorbed into larger units through the reorganisation of 1968 which now see autonomy essential for their future development. In some cases the motive arises through ethnicity, where a village feels alienated from the rest of the commune, but generally there is a desire to have a budget which is not centralised in the key village of a larger commune.

However, there could be benefits through village associations to combine settlements polarising around local centres, discussed in Chapter 6. This tendency clashes with the desire of small communities for administrative independence and even when stronger groupings do emerge the limited tax-raising powers constrain investment in development projects because of the priority to maintain the administration, improve local education and infrastructure and cope with the housing needs of young married couples. Few of them engage directly in economic activity. However, local authorities are keen to improve roads and telecommunications as well as gas, water and sewage services: seen as essential prerequisites for attracting foreign investment. Local authorities also influence growth prospects through the general management plan which includes an overall development conception and appropriate zoning of industrial, residential, agricultural and woodland areas. Such decisions can certainly influence economic activity and the prospects for gaining inward investment. And in the context of strong rural units meshing into development axes they could increase their integration into the global economy (Vrabete and Popse 1999: 263).

And transportation can be problematic, for although no village in Hungary is more than a half-hour journey away from the nearest town there are variations in the quality of transport and services within each small town hinterland (which may involve 20,000–30,000 people, though usually no more than 15,000 for the newer small towns). The removal of buses from routes between central rural towns and outlying villages makes it difficult now for people to shop in the central settlements when they lack a car. The rising cost and reduced provision of public transport also makes it difficult for people to retain jobs in nearby towns. Hungary has particular accessibility problems in demographically-weak areas containing small farms where central places at the local level are not easily identifiable, and border areas which are a legacy of the imposition of Trianon frontiers after the First World War. There are similar problems for the hamlets of Romania's Western Carpathians and the 'kopanitse' zones of Slovakia.

Meanwhile, information flows into small towns have benefited from an increase in the number of local and district newspapers and radio stations, while cable-television systems now exist in the towns and larger villages. These improvements are important for the flow of information and for social cohesion and identity in each district or locality.

> Civil society is being more effectively mobilised so that each small town, along with the villages in its neighbourhood, can find a basis for cooperation with regard to both economic growth and cultural life. Simultaneously new connections are being established with other parts of Hungary and with foreign countries as the actors in each local economy seek out partners and try to enlarge their market areas.
>
> (Nagy and Turnock 1998: 20–1)

Larger firms can use the new information technologies to extend their networks at the local or district level, while flexible specialisation through subcontracting and supply networks provides opportunity for SMEs in backward areas to cooperate with business elsewhere. This enhances the central place functions of small towns and strengthens their economic image so that they may attract larger industrial plants and other investments. They can thereby develop their business services and improve living conditions through the widening range of non-agricultural employment.

Coping on the edge: pluriactivity

Low levels of non-agricultural employment make for a greater interest in restitution holdings for subsistence, especially in the SEECs. Labour-intensive cultivation (including rabbit breeding, fruit-growing and the cultivation of mushrooms and vegetables) may be advocated to maximise the potential for agricultural work. It is an important question how far the region will identify with the Swiss-Bavarian model of relatively small farms combined with a significant rural industry. Families involved in pluriactivity (where farm occupancy is supported by a range of activities with the flexibility of family labour as the critical issue) have been seen as 'social anomalies' in a world dominated by capital. The economic factors of uneven development and constrained choice may be complemented by a web of social relations supporting an ideological commitment to family-based farming. South Moravia shows the tendency towards a dispersal of employment in manufacturing, reinforced by workshops on the former state and cooperative farms. Population was stable during the 1980s even in areas that were not the most favourable for agriculture. Smith (2000) therefore raises the interesting question that while pluriactivity is certainly driven by stress it may well be a more durable phenomenon reflecting satisfaction with a small farm lifestyle.

Rural tourism

This is widely discussed, for all ECECs enjoy a wealth of natural and cultural heritage, which provides opportunities for the development of small-scale tourism carefully integrated into the local economy with respect to organic farming (vegetables, fruit and dairy produce), quality brandy and wine production and sustainable fishing. In Poland, local agrotourism associations developed after 1990 and there are now some 38 groups with information and booking systems which represent over 4,000 farmers (and maybe four times this number if unregistered rooms are considered). PHARE funding for training has been provided through the Polish Federation of Country Tourism and there are facilities for preferential loans and tax exemptions. Further institutional support comes from ECEAT Poland (European Centre for Ecological Agriculture and Tourism) which produces publicity on 'Holidays on Organic Farms in Poland' in several languages. As young urban entrepreneurs buy old declining farms and start business, one agrotouristic farm will stimulate others and whole villages may revive. Referring to Lucim 23 km northeast of Bydgoszcz, Halasiewicz (1995: 97) sees rural tourism as a means of boosting income and also of 'eliminating conventional methods of intensive agriculture', protecting traditional landscapes and restoring traditional village architecture. In the Polish Carpathians, where the private sector is relatively well established, much progress is being made in combining small-scale agriculture with woodcutting, handicrafts and tourism. Rural tourism has been particularly successful in a number of villages close to the Tatra and Babia Góra (Kurek 1996). Large modern houses have been built on the strength of increased incomes and the demand for tourist accommodation while summer visitors generate demand for locally-produced food which in some cases is ecological produce grown by members of the Ekoland Association established in Toruń in 1989.

Other countries have tried to develop this particular tourism model: notably Romania where the Commission (later Agency) for Mountainous Regions – which existed between 1990 and 1996 – generated considerable momentum which is now being maintained by the National Association for Rural and Ecological Tourism, with considerable external support from Western Europe through 'Opération Villages Roumains'. The potential for agrotourism linked with careful management of the mountains to prevent erosion by both overgrazing and excessive visitor pressure is demonstrated in Romania's Părâng Mountains where national park designation may provide the necessary protection in an area first opened up to visitors by the expansion of wood processing and the development of hydropower. The business has taken off most strongly around Bran near Braşov and it is here where the possibility of connecting higher tourism earnings with reduced levels of sheep stocking and consequent enhancement of large carnivore conservation has been proposed. There is also a strong cultural component through Bran Castle, popularly supposed to have been Vlad the Impaler's fortress and the model for Dracula'a castle in the Bram Stoker novel, while the Bran–Rucăr corridor has potential for winter sports at Fundata and Şirnea, Peştera and Bran-Poarta. Other regions are now following suit, notably Maramureş which has outstanding cultural resources (including remarkable wood-carving skills) and scope for cross-border cooperation with Ukraine across the Tisa Sighet.

Hungary demonstrates the way in which rural tourism can connect with specific local activities: for example, viticulture in Gonc village and others in the famous Tokaj wine region. New vineyards like Neszmély near Tata – supplying affordable dry white wines – will also have this opportunity. Hungary also has thermal springs, one of which is being promoted at Siófok to encourage more year-round tourism. Sustainable rural tourism could be significant in Albania's efforts to replace communism's prescriptive approach and the coast certainly has considerable potential for both small developments in secluded settings (where high tariffs could be charged) and larger 'village' projects in the areas of Vlorë-Fier, Lushnje-Kavaje and Shëngin, making use of former Greek villages, now heavily depopulated. In the mountains, tourism could integrate with farm modernisation, given the potential of the national parks and other protected areas. Bulgaria offers small family farms in the clean pastoral environment of the Balkan Mountains, with a benign climate and a diverse landscape including luxuriant forests. The villages in the Lovetch district of the Central Prebalkan Zone have made progress with the help of specialists from Sofia over tourist information and maps, but the lack of institutions is a common problem in the SEECs.

Rural policy

The ECECs are being restructured at a rapid rate and agriculture is experiencing a major shake-out of labour which will continue into the foreseeable future. The question is whether there will be an intensification of urban expansion along Third World lines, despite limited employment opportunities and housing shortages – including a poorly-developed housing market – or a stabilising of population on the basis of farm diversification and pluriactivity. The evidence from Albania, where there has been an outpouring of population from the northern highlands, suggests the Third World scenario, yet there are also counterurbanising trends which suggest a period of change in the functions of rural settlements. There is a case for taking an integrated approach to rural issues through farming systems. Agriculture may not merit support on a permanent basis, but there is certainly a need for a rural social policy to ensure welfare for the farming population by resolving conflicts arising from land reform, monitoring the new social situation, coping with poverty and unemployment and helping with regeneration and diversification. There may also be a case for intervention in the interest of structural change and the smooth operation of a land market, with legislation to facilitate land leasing and farm consolidation. And resources to stimulate SMEs through training and credit are very desirable in line with Poland's 'Opportunities for Rural Areas and Agriculture' programme of 1992. Meanwhile, Pak and Brecko (1999) consider that the modernisation of agriculture, with more market production and food processing, requires the villages to improve the range of functions and provide a better environment for the non-agricultural population which will become increasingly important for its future development. However, this raises questions about the extent to which services can be dispersed among the smaller villages. The Polish government would like to help rural areas by consolidating village schools so they can provide specialist teaching, e.g. languages and computing. But the smaller villages have fought hard to retain their schools which they can run themselves with the help of state grants. And small villages will find it more difficult to gain access to digital telecommunications with implications for Internet access.

More jobs could be created in agricultural services, building on the favourable natural conditions and the established profile over farm production and food processing. Maintaining rural transport in the interests of commuting and providing more collective water-supply systems require attention, especially in the SEECs where all aspects of farm infrastructure are inadequate. In the area of telecommunications, local tele-information centres ('telecottages') have a role in the development of tourism and dissemination of information for potential visitors by fax and email. Finally, to strengthen county centres, Njegac and Toskic (1999) argue that decentralisation from Zagreb should be accelerated in the interests of rural diversification and the retention of population in the countryside. The county centres, as regulators and stabilisers of spatial processes, are critical for the reduction of contrasts in regional development. However, policy will need to consider the remote rural areas distinguished by a fallowing of plots and a growth of afforestation, combined with inadequate opportunities in manufacturing and tourism which leads to depopulation. These stand out against the more favoured areas with optimal conditions for agricultural development, significant flows of capital investment and an increasing average farm size, albeit with signs of soil erosion and pollution by agrochemicals and soil erosion. Rural poverty in the SEECs may be linked with outlying rural communities where people do not possess their own land, as in the case of Turkish communities in Bulgaria. In such areas low spending power and outright poverty limit the scope for business. Regional policy needs to develop a welfare agenda to maintain services and employment, sometimes through cross-border cooperation in frontier regions.

Finally, better local leadership is being sought in backward areas like Hungary's Cserehat, and emergency 'self-support' programmes are being introduced, especially among the older people in the poorer regions of Slovakia. Such measures however depend on organisation and institutions that can project local opinion and stimulate individuals to take a leading role. Cohesion is well established among family groups and the churches have traditionally provided some community leadership, but now there is a greater need for organisation along Western Europe lines through groups which are well endowed financially, are environmentally aware and are often founded on a stable intellectual nucleus which covers both long-term inhabitants and incomers. Where agricultural cooperatives have been retained there is already an infrastructure in place, but where the former communist collectives have dissolved (and equally in areas where they never existed) rural communities may lack effective leadership and, in the case of ethnic minorities, fragmentation and isolation may amount to exclusion.

RURAL POLICY: THE ROMANIAN CASE

After initially providing special assistance for mountain regions, the initiative was restricted in 1996 to an NGO ('Federaţia Română pentru Dezvoltare Montană') concerned with rural tourism, marketing and the transfer of relevant foreign experience on local development in mountain regions, but resources have been greatly reduced. The change in policy occurred because of a review of government expenditure and a decision to direct limited welfare spending to the areas with greatest poverty. This reflected the rapid run-down in factory employment in lowland areas of the northeast and southwest, reinforcing extremely high levels of dependence on agriculture. At the same time increasing involvement with the EU as a candidate country made for the formulation of an overall rural planning concept. As already noted in Chapter 3, Romania has produced a 'National Programme of Rural Development' aiming at a better quality of life in rural areas, with more entrepreneurship and cooperation between sectors. The key elements of the plan are: an integrated and dynamic rural economy, based on the development of agriculture and agricultural services, the rationalisation of holdings and pluriactivity; improved living conditions and infrastructure; conservation of culture and patrimony through appropriate modernisation of villages; demographic reinvigoration with skills training and stimulation of business; environmental protection through reafforestation, water management and more sustainable agriculture. The 'Agenţia Naţională pentru Consultaţii Agricole' (ANCA) will run 500 advice centres which will provide information on harvesting prospects and help with entrepreneurial skills and offer limited credit for seeds, chemicals and machines. Funds will also be directed into improvements in rural households, with a stimulus to marketing and food processing, while the Planning Ministry has limited funding for rural infrastructure. Rural diversification has been assisted in the past by EU cohesion funding through PHARE and should benefit in the immediate future from the SAPARD programme.

Comparing Romania with other countries of the region, Surd and Zotic (1999: 227) highlight the need not only for better infrastructure and more investment in the economy, but stronger and more transparent local government which may depend on larger professional elements in rural communities. Local development associations are beginning to emerge – for example in the Arieş valley of the Western Carpathians where 'Izvoarele Arieşului' brings together five communes which polarise around the key village of Albac, with particular emphasis on the development of a rural tourism network. This is an area of extremely dispersed settlement where many outlying villages have very poor communications and depend heavily on farming and wood processing (Figure 5.7). But through their local organisations stakeholders can identify themselves, forge partnerships and compete for resources to implement local projects. A tourist network is now being built which promotes excursions to the remoter places, as well as the services of the main village centres along the main road. At the same time, more substantial funding will be needed to support such 'less favoured areas' (LFAs) where 'it will be impossible to break the poverty circle in these areas if agriculture is not supported to recover' (Ramboll Group 1997: 5). One approach is based on EU Objective 5b (which seeks to reduce a high level of dependence on agriculture along with low agricultural incomes and a tendency to depopulation) and recognises 'poor' regions where agriculture exceeded 60 per cent of all employment. Research has revealed areas with similar patterns to the zones of 'high poverty' previously identified by Puwak

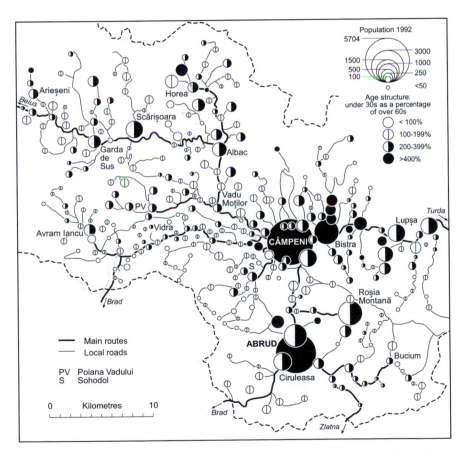

Figure 5.7 Dispersed settlement in the Arieş Valley of Romania's Apuseni Mountains

Source: Census of Romania 1992 and Albar County Council

(1992: 39–40). Under the National Poverty Alleviation Strategy of 1998, regional gender empowerment projects in rural areas were carried in two deprived areas: the Buzău Subcarpathians and Vaslui in Moldavia. In the case of Pătârlagele (a village in the former area, referred to earlier in the chapter), potential projects for meat processing, fruit and vegetables, tailoring and knitwear workshop were considered. A problem remains to select a methodology which may be used to identify 'poor' or 'under-developed' areas in which a special regime of support may apply over the longer term. For the only LFAs currently enjoying fiscal concessions are those recognised by the government on the basis of high unemployment associated with restructuring in the mining industry. Finally it is arguable how far a purely rural programme can succeed apart from a parallel initiative to stimulate more enterprise in the market centres, for it is growth in these places that is needed to stimulate commuting and a greater demand for the products and services the countryside can provide.

6   Regions of East Central Europe

The changing status of regions: communism and the transition

This section looks at the significance of regions and the prospects for development in the different states across ECE, bearing in mind that potentials do not just vary between countries but also between smaller units characterised by unique combinations of human and physical resources. Under socialism, regional development existed only at the level of ideological proclamations. 'Urban and regional planning – like other state socialist policies – was the preserve of politicians, bureaucrats and experts, involving dialogues from which the general public were excluded' (Harloe 1996: 14). According to Enyedi (1990) regional development was essentially the haphazard outcome of various sectoral decisions taken by ministries. Capital allocation was obviously based on surveys of resources which planners exploited as they saw them, but the regions were largely passive and had little opportunity to initiate programmes on an autonomous basis. Dostal and Hampl (1994: 204) refer to 'the policy of nivelization of interregional disparities in living standards based on the industrialization campaign'. This 'implied an extraordinary suppression of any important selective tendencies at the interregional level' (ibid.).

The regions could muster considerable influence and the national leaders struggled to keep each regional party apparatus active in controlling waste and achieving mobilisation, while preventing them from 'building up their own regional empires from the resources they were supervising' (Surazska *et al.* 1997: 442). Conservative leaders made regional party 'barons' the pillars of their regimes, while reformers like Gierek in Poland and Kadar in Hungary might have to fight the regional apparatus through territorial reorganisation. This could be damaging because Poland's 1975 administrative divisions often meant splitting the districts or 'powiaty' which were vital for intelligence gathering: a major setback because the strikes of 1980 occurred in the areas where most change had taken place. However, the vested interests of the regions could hardly be counted a progressive force since they were fundamentally undemocratic in being based on the party and not on state structures (ibid.: 439).

Urban growth potential was linked closely with administrative functions since projects tended to cluster in the regional centres for ease of supervision and also to take advantage of a relatively good infrastructure. Regional centres tended to grow rapidly under communism because they benefited from a disproportionate share of the capital invested. However, there were contrasts between relatively advanced urban regions (with large cities as regional centres and with the urban population in a majority overall) and more backward rural regions with a low level of urbanisation and net out-migration to stronger regions: for example, in Poland there was a steady transfer of population to Warsaw from the surrounding voivodeships of Łomża, Płock, Siedlce and Radom, and also to Poznań from Konin, Leszno and Piła. The larger regions now in force generally reflect these traditional functional relationships. In the case of Slovenia, the centres of the country's 73 administrative units (of which only 58 are true towns) increased their share of GDP from 76.0 to 80.3 per cent between 1966 and 1990. Towns over 20,000 increased from 38.1 per cent to 48.9 per cent while the smaller towns declined from 37.7 to 31.4 per cent and rural areas from 24.2 to 19.8 per cent.

Given these variations in potential, communist governments were ambivalent towards the issue of regional disparities: while the 'equity' principle delivered some support to rural regions, 'political clout' and economies of scale in export-orientated industry tended to limit decentralisation. Despite migration controls (reflecting perceived congestion costs and severe housing shortages in some large cities) and some relocation of production – especially for polluting and labour-intensive industries – the leading regional centres continued to grow. In Czechoslovakia a Regional Planning Decree in 1977 prescribed a 'balanced and even development of territory' and attention being given to specific problems including housing and infrastructure in Prague, professional services in Košice, and environment in North Bohemia and Ostrava. But this was not a recipe for checking uneven growth and at the end of the communist era there were great variations regarding demographic and employment structures between stronger and weaker regions across ECE (Figure 6.1; Table 6.1). However, Hungary was unusual in setting up some regional development machinery. In 1985 a parliamentary decree defined long-term tasks to develop backward areas with particular emphasis on two counties in the northeast: Borsod-Abraúj-Zemplén and Szabolcs-Szatmar-Bereg. However, the funding was limited, institutions were inadequate and sectoral priorities continued to take precedence.

Regions in transition

Contrasts have widened as a consequence of the transition according to calculations for per capita GDP related to the European average (Figure 6.2). The low values clearly demonstrate that after enlargement the bulk of the cohesion funding will be going eastwards to the detriment of the present periphery. While only one of the regions shown (Prague) is above the EU average and most do not reach 75 per cent of the average – some in the SEECs have barely a fifth – the capitals always do relatively well (Table 6.2). Hampl (1999a: 38) has produced figures which show the Prague metropolitan area with 13.3 per cent of the Czech Republic's population in 1998 (down 0.1 on 1991) but 16.9 per

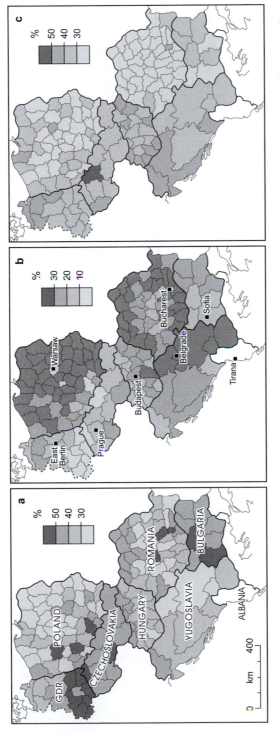

Figure 6.1 Regional variations in employment at the end of the communist era: (a) industry and construction, (b) agriculture, forestry and fishing, and (c) services. Shadings indicate the share of total employment falling to each sector

Source: Bachtler (1992b)

Table 6.1 Regional variations on the eve of transition 1988 (%) (a) Population age structure and (b) Employment structures

(a)

	Child population				Aged population				Working-age population			
	A	B	C	D	A	B	C	D	A	B	C	D
Bulgaria	21.0	22.8	18.1	4.7	21.9	30.2	16.4	13.8	57.1	61.7	49.5	11.8
Fmr. Czechosl.	23.0	27.3	19.3	8.0	19.4	21.7	15.9	5.8	57.6	61.5	56.8	4.7
East Germany	19.4	22.3	17.6	4.7	16.2	19.6	12.7	6.9	64.4	67.1	62.3	4.8
Hungary	na	na	na	na	na	na	na	na	57.3	59.0	55.0	4.0
Poland	30.7	34,4	23.5	10.9	12.8	16.2	9.1	7.1	56.6	61.0	53.1	7.9
Romania	23.7	30.0	15.1	14.1	13.0	18.9	9.0	9.9	63.3	69.5	56.8	12.7
Fmr. Yugoslavia	25.4	29.1	21.0	8.1	10.2	13.1	7.2	5.9	64.2	65.7	62.7	3.0
Total[1]	18.4	28.3	11.7	16.6	14.2	20.1	7.6	12.5	67.2	71.2	60.7	10.5

Note
[1] Excludes Hungary for child and aged populations

(b)

	Industry				Agriculture				Services			
	A	B	C	D	A	B	C	D	A	B	C	D
Bulgaria	46.3	51.8	41.6	10.2	19.3	26.2	1.8	24.4	34.9	53.0	28.2	24.8
Fmr. Czechosl.	36.1	55.0	35.3	19.7	13.7	21.1	2.0	19.1	40.2	62.1	35.3	17.1
East Germany	47.0	58.5	30.1	28.4	10.8	26.8	1.1	25.7	42.3	63.9	35.3	28.6
Hungary	38.6	47.8	28.2	19.6	16.0	32.3	0.7	31.6	45.5	83.6	37.8	45.8
Poland	36.4	60.9	12.0	48.9	28.9	61.3	6.0	55.3	34.2	46.6	22.7	23.9
Romania	40.2	61.4	25.8	35.6	27.9	48.1	3.8	44.3	27.0	43.1	16.8	20.1
Fmr. Yugoslavia	33.1	43.1	28.8	14.3	30.7	38.4	14.6	23.8	35.1	49.1	31.2	17.9
Total	41.1	61.4	12.0	49.4	21.0	61.3	0.7	60.6	37.0	83.6	16.8	66.8

Source: Bachtler 1992b: 41, 52–3

Notes
A National Average
B Regional Maximum
C Regional Minimum
D Regional Difference

cent of 'labour opportunities' (up 2.0), 21.5 per cent for 'economic aggregate', taking wage levels into account (up 5.7) and 36.6 per cent for the financial sector, including banking and insurance (up 3.8). Eleven other metropolitan areas have 43.6 per cent of the population (up 0.2 on 1991) but only 33.5 for economic aggregate (down 1.7) – a poor showing which reflects the problems of metropolitan coalmining areas which contrast with the dynamism of České Budějovice, Hradec Králové and Plzeň – while the rest of the country has 43.1 per cent of the population (down 0.1) and 45.0 for economic opportunity (down 4.0). Taking four countries (the Czech Republic, Hungary, Poland and Slovakia) there are also widening disparities between the highest and lowest regions regarding wages during 1989–96 especially in the Czech Republic (1.16 to 1.34) and Slovakia (1.14 and 1.41) compared with Hungary (1.47 and 1.50) and Poland (1.56 to 1.64) where disparities were relatively high at the outset: each value shows the highest regional rating as a multiple of the poorest. Disparities for unemployment are much greater and Poland again shows the greatest variations in both years (4.43 and 7.57). Slovakia has seen a big widening of the gap (1.47 to 4.34), whereas the change in the Czech Republic is very small (3.33 to 3.44) and in Hungary the disparity has narrowed (3.91 to 3.57) (Tomes and Hampl 1999: 140).

However, democracy means that all the regions are now able to take development initiatives (Enyedi 1996). Small grass-roots organisations disseminating 'underground' literature provide some continuity with the NGOs of today through 'path dependence', as the forces dimly evident through the urban activism of the late 1980s have now come to the fore. As Plut (1997: 44) explains, there has been a switch towards the sustainable European model of regional development 'based on administrative decentralisation and a greater role for endogenous regional sources'. The process is encouraged by the EU which seeks interaction not just with national government but with the regions as well. Nevertheless FDI is seen as the financial mechanism that really makes a difference. Regions are in competition and may be evaluated in terms of investment attractiveness. However, not all the regions shown in Figure 6.2 have elected councils and those which do find themselves facing great disparities in income and expertise when it comes to making promotional efforts. There are also problems of 'identity' in the minds of investors, for the weaker regions are not well known enough to foreign business interests and struggle to make an impact.

Regional development prospects will inevitably be grounded to some extent in the present industrial structure. Thus, Bachtler (1992b: 136–7) highlights 'diversified industrial regions with a relatively good material and technical base and experienced personnel, good infrastructure and international links'. Also attractive are capital cities and major regional cities in general, as well as areas close to EU borders and areas of tourism potential with an unspoiled environment, including highland regions where the successes of Western 'mountainology' could be repeated. Agricultural work must be safeguarded through greater efficiency and competitiveness in export markets. V. Rey (1994) has raised the importance of reasserting comparative advantage, highlighting the richer areas of FCS such as Czech Silesia where small peasant farms persisted

GDP per capita as a percentage of EU average 1996
- 60
- 50
- 40
- 30
- 20

States not in the EU Pre-Accession Phase

● Capital cities in non-EU candidate countries

— European transport corridors

● Largest provincial towns (over 75,000 population)

ALBANIA
1 Durrës
2 Elbasan
3 Shkodër

BOSNIA & HERC.
1 Banja Luka
2 Doboj
3 Mostar
4 Prijedor

BULGARIA
1 Burgas
2 Plovdiv
3 Russe
4 Stara Zagora
5 Varna

CROATIA
1 Osijek
2 Rijeka
3 Split
4 Zadar

CZECH REP.
1 Brno
2 Česke Budějovice
3 Olomouc
4 Ostrava
5 Plzeň

EAST GERMANY
1 Chemnitz
2 Dresden
3 Halle
4 Leipzig
5 Magdeburg

HUNGARY
1 Debrecen
2 Györ
3 Miskolc
4 Pécs
5 Szeged

MACEDONIA
1 Bitola

POLAND
1 Gdańsk
2 Kraków
3 Łódz
4 Poznań
5 Wrocław

ROMANIA
1 Cluj-Napoca
2 Constanța
3 Galați
4 Iași
5 Timişoara

SLOVAKIA
1 Banská Bystrica
2 Košice
3 Nitra
4 Prešov
5 Žilina

YUGOSLAVIA
1 Kragujevac
2 Niš
3 Novi Sad
4 Podgorica
5 Priština

0   Kilometres   300

━ State boundary
— Regional boundary (regions including capital cities are shown schematically)

Figure 6.2 The regions according to per capita GDP 1996

Source: Eurostat 1999

Table 6.2 Regional variations in EU candidate countries 1996

	Regions distributed according to percentage of EU GDP pc									
	+100	90–100	80–90	70–80	60–70	50–60	40–50	30–40	20–30	Total
Bulgaria	–	–	–	–	–	–	–	2	7	9
Czech Rep.	1	–	–	–	2	4	1	–	–	8
Hungary	–	–	–	1	–	1	1	4	–	7
Poland[1]	–	–	–	–	1	–	6	18	24	49
Romania	–	–	–	–	–	–	1	6	1	8
Slovakia	–	1	–	–	–	–	–	3	–	4
Slovenia[2]	–	–	–	–	1	–	–	–	–	1
Total	1	1	–	1	4	5	9	33	32	86

Source: Eurostat 1999

Notes on National Units for Territorial Statistics (NUTS)
[1] Calculations were made on a NUTS III basis using the former voivodeships
[2] Slovenia is not subdivided: the NUTS I level operates

under communism, partly because the farms were too small for the peasants to be branded as kulaks. Almost certainly, the surviving SOEs will have an important role in the foreseeable future, for despite the many cases of closure there may be 'insular areas that would represent the frameworks for stability of employment in individual areas and in the country as a whole' (Bucek 1992: 2). These 'islets of development' may well be developed by joint ventures linking former SOEs with foreign companies supplying investment capital, new technology and a marketing system; and state governments (especially in the SEECs) may be persuaded to assist with low-interest credits and tax exemptions.

Meanwhile, there is very little money available to support the weaker regions and agencies set up to provide assistance, like the Regional Development Agencies which were first set up in Poland in 1991, have not been able to make a major contribution despite their concern for high unemployment areas like Suwałki in the northeast. Industrial regions like Łódz and Wałbrzych suffered high unemployment in the early days due to the contraction of some large SOEs, although they have generally found favour with foreign investors because of the infrastructure and adaptable labour available. Rural regions tend to face the greatest difficulties, especially where there is a very heavy dependence on agriculture and subsidies supporting large socialist farms have dwindled. Meanwhile, the eastern border districts have been marginalised by European investors because they are relatively remote from the main markets of the continent. So sharp are the distinctions between the favoured and less-favoured regions that Poles refer to 'Poland A' and 'Poland B' with the latter group offering only limited career prospects for young people coming on to the labour market. These areas retain traditional roles including some manufacturing based on local raw materials, often for local customers, and small-scale commerce for a population with little spending power. There is much evidence of such regional systems in the SEECs where trade is more usually based on cheap manufactures drawn from warehouses in Istanbul and Thessaloniki. A 'bazaar economy' based on short-term speculative business is seen as a barrier to regional development since stable institutions are lacking, but former cooperative farm buildings could be transformed into 'embryonic business parks' (Swain 1994: 9), with some success where émigrés support their home villages and influential local officials can make use of Western contacts to promote the secondary and tertiary sectors in rural areas. An alternative way forward, as noted in Chapter 5, is to adopt the 'Bavarian model' of pluriactivity, combining agriculture with a range of ancillary activities.

Administrative units

The issue is complicated by the fact that regional systems are themselves changing through reorganisations that are to some extent contested between state and regional interests. Some countries have not made any alterations to the units inherited from communism but two trends have been evident. First, a strong grass-roots desire for small municipalities with many individual settlements frequently opting out of the lowest level units established in the communist period. They have often been motivated by a desire to restore historical arrangements revised under communism (especially where ethnicity is a factor) and also by a desire to have an independent budget. In Hungary there are now 3,200 basic units, reflecting a strong decentralising trend encouraged by a policy of enhanced power for municipalities 'substantively shaped and carried endogenously by a constitutional consensus which, as a result of Hungary's negotiated transition, was borne both by the opposition forces as well as by the reformist communists' (Surazska *et al.* 1997: 477) (Table 6.3). In the Czech Republic 11,500 municipalities in 1948 were reduced to 4,100 in 1989, but some 2,000 have now been restored, despite populations of less than 200 people in some cases. Naturally, the capacity of many of these small villages can be questioned because they lack the resources to discharge all their functions and promote investment (Blazek 1997: 41).

Second, some governments have considered larger regions for the coordination of planning and development. In Bulgaria in the early years of transition and in Poland more recently, regions have been introduced which replace the smaller top-tier units of the communist period (Koulov 1992) (Figure 6.3). But Bulgaria has already had 'second thoughts' and there are now 29 regions, while in the Czech Republic and Slovakia an additional tier has been added with the region ('kraj') on top of the district ('okres') and the basic municipalities. In Slovakia, the regions were first introduced by the Mečiar government for central coordination of planning and it was only in 2001 that elected councils were provided (Figure 6.4). Furthermore, in the Slovak case the districts are now much smaller (there are 79 instead of 39), but the result is three tiers of administration which can create some confusion for the population in terms of appreciating which services are discharged at which level. The issue has been further complicated by EU cohesion funding which requires large regions at the NUTS II level (the layer immediately below the national level of NUTS I). But in most cases these regions have a coordinating role only, because regions built

Table 6.3 Units of regional development and local government

| | EU NUTS II | | Tiers of local government | | | | | |
| | Regions | | Upper | | Middle | | Lower | |
	A	B	A	B	A	B	A	B
Albania[1]	–	–	–	–	36	86.5	915[1]	3.4
Bosnia & Hercegovina[2]	–	–	10[2]	397	na	na	145[3]	27.4
Bulgaria	9	914	29	284	–	–	259	31.8
Croatia	–	–	21	213	–	–	481	9.3
Czech Republic	8	1280	14	732	72	142.5	6196	1.7
Hungary	7	1434	20	502	–	–	3135	3.2
Macedonia	–	–	–	–	–	–	123	16.5
Poland	16	2423	16	2423	373	103.9	2459	15.8
Romania	8	2791	42	532	–	–	2951	7.6
Slovakia	4	1347	8	673	79	68.2	2825	1.9
Slovenia	1	1986	–	–	–	–	148	13.4
Yugoslavia	–	–	29	367	210	50.7	7634	1.4

Source: Europa Yearbook 2001

Notes
A Number of units
B Average population (,000) 2000
[1] 66 towns, 306 town boroughs and 537 village unions
[2] 10 cantons for FB&H
[3] Municipal councils mainly in Republika Srpska

up around the great provincial capitals are not satisfactory for representative government. Only in Poland are the top-tier administrative units acceptable for EU purposes, and although the NUTS II regions are constructed elsewhere by grouping smaller regions together, they do not have elected councils and play a coordinating role only. The NUTS II regions are used in Figure 6.2 (already referred to) although in Poland the mapping exercise was based on the old voivodeships because the new regions were only just being introduced at the time. Finally Macedonia, Slovenia (and the Republica Srpska in B&H) do not have regions but only national governments and municipalities.

Of course, delimiting the new regions is a highly contested business. Poland's centre-right government of 1997–2001 wanted only 12 regions (average population 3.22 m.) but there were strong campaigns in support of Bydgoszcz, Kielce, Opole and Zielona Góra as regional centres, backed by the parliamentary opposition, which made the difference; so the average population for the 16 regions was brought down to 2.41 m. Slovakia's regional system has also generated much debate among the political parties between the extremes of the communist period and the 'zúp' system of small city regions based on cultural–historic links and adopted by the Hungarian system of the late nineteenth century. The dispersed Hungarian minority population campaigned strongly against large city regions – though their functions were limited – because they could never hope to form a majority in any of them, but retention of the 'okres' is helpful because Hungarians can control some local governments at this level. However, there is

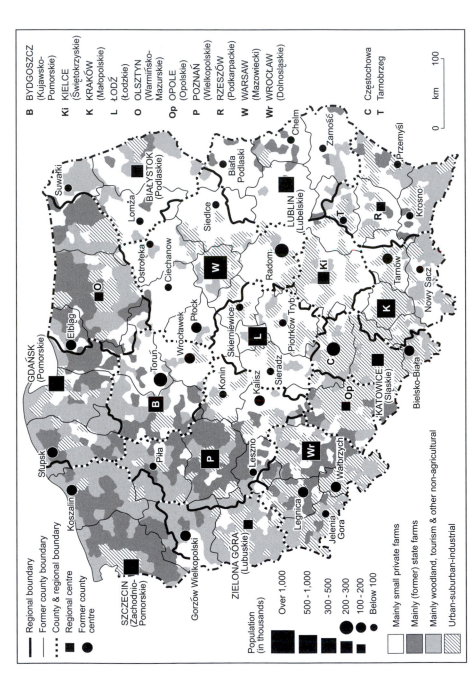

Figure 6.3 Poland's administrative regions in the 1990s

Source: Polish administrative maps

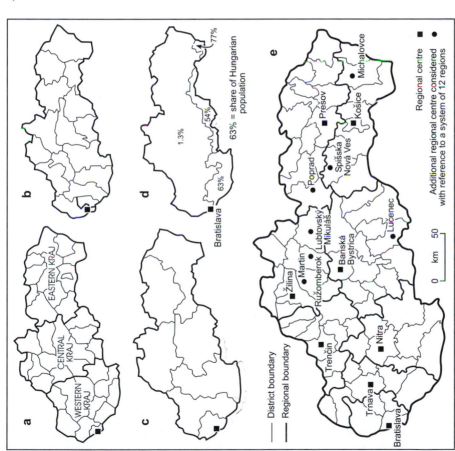

Figure 6.4 Slovakia's regional reform: (a) The two-tier system inherited from communism, (b) The 'Zúp' decentralisation proposal for counties based on cultural and historical links, (c) The Mečiar government proposal for a centralising strategy of seven functional regions for government coordination; based on Banská Bystrica, Bratislava, Košice, Nitra, Prešov, Trenčín and Žilina, (d) The Kormarno proposal recognising three Hungarian majority areas, (e) The present system which is based on variant (c) but with an eighth region (for Trnava) and provision (starting in 2001) for regional councils. There are also elected councils at the district (Okres) and municipality levels. Note that the districts for Bratislava and Košice each comprise four units which are not individually shown.
Source: Slovak administrative maps

an on-going tension and when the elected councils were introduced in 2001 (by the successor government to Mečiar's centralising administration) there was considerable support for a modified 'zúp' system of 12 regions including one for the predominantly Hungarian area of Lučenec with a population of only 267,000. But the socialists advocated the other extreme of returning to the communist system of three large regions plus Bratislava. Parliament therefore voted for the status quo, with regions with an average population of 671,000 instead of 447,000 under the 'zúp' system. It is evident that a system of 12 regions would have given rise to some disagreement as in the northern region where Martin, Liptovský Mikuláš and Ružomberok all laid claim to a higher administrative status. Likewise further east where Poprad and Spišska Nová Ves were in competition.

While democracy tends to favour smaller regions, there are instances where large provinces might seek autonomy, as in Croatia where a system of 20 counties ('županije') plus the city of Zagreb was introduced in 1991. This system has some advantages for Istria where a regional party (the Istrian Democratic Alliance) represents an ethnically mixed population including Albanians, Italians, Serbs and Slovenes as well as Croats. Despite a political culture of tolerance, the party seeks a federal system which would allow more taxation revenue to accrue to the regions and, in the process, would allow Istria to keep much more of its tourist revenue. But there has been opposition to the new administrative arrangements from Dalmatia over the high level of centralisation of government in Zagreb and the division of Dalmatia into four counties: Dubrovnik-Neretva; Šibenik; Split-Dalmatia and Zadar-Knin. The region retains some cohesion through the Association of Communes and a regional Chamber of Commerce. The political party Dalmatia Action feels that region is treated by Zagreb in the same insensitive manner that characterised previous relations with Belgrade: hence the slogan 'ZG=BG'. The region enjoyed a de facto autonomy between 1990 and 1995 through the isolation of the 'Dalmatian Island' after the Krajina Serbs blocked the Zagreb–Split road in 1990 at the height of the tourist season and subsequently destroyed the Maslenica Bridge in 1991. For a time even the air links were severed and the only safe road route to Zagreb was via the Pag bridge and ferry crossings.

Central government and regional policy

The national level remains of decisive importance, despite some 'hollowing out' through a shift towards more complex regulation involving the supranational, national and subnational scales. Regional policy is contested between advocates of efficiency on the one hand and protection on the other. G. Barta (1992: 378) argues that 'the major aim of public policy should be to strengthen the competitiveness of national economies and major urban centres'. Growth poles should be strengthened and population mobility should be encouraged. Similarly, J. Bachtler (1992a) stresses that regional policy should identify potential growth areas which can lead the national restructuring process, with potentials most likely to be found in diversified industrial regions with a relatively good technical and material base. But other opinion favours an emphasis on welfare. So P. Pavlinek

(1992) calls for effective regional social policy to deliver education and welfare, including job training and incentives (capital grants, tax exemptions and reloca-tion allowances) to protect the weaker regions. The protection of rural settlement through development of services on a key village basis has also been mentioned.

In the early days of transition 'countries such as Poland, Hungary and Czechoslovakia perceive it to be premature to discuss important regional economic development strategies. The initial priority is clearly to develop strate-gies at the national level' (Bachtler 1992b: 135). Resources were very limited, the neo-liberal philosophy prevailed and regional fortunes could quickly change in some cases: Mladá Boleslav in the Czech Republic was initially embarrassed by its one-sided industrial structure biased towards engineering and vehicles, yet a joint venture with a foreign company (Volkswagen) brought about a trans-formation and showed that the area could be attractive for investment. Thus the efficiency approach predominated in the short term while monitoring by multi-disciplinary regional research institutes such as those in Hungary has been a welcome trend. So it is evident that, with more intervention from govern-ment, a compromise position may offer limited help to the weaker regions while allowing the most attractive areas to advance. Small businesses are needed to support the large enterprises and develop the service sector and rural areas might support clusters of SMEs grouped into 'small enterprise spatial systems'. In this way the 'top down' process of breaking down the SOEs can be comple-mented by a 'bottom up' programme to integrate the transformed state enterprises with existing handicraftsmen and family-managed workshops.

Aid in East Germany was made available through 'Gemeinschaftsaufgabe' (GA) and there was additional help through 'firefighting' in areas particularly affected by structural change. Rural infrastructure was improved through the West German 'Dorferneuerung' programme which was extended to the East after unification, while low-interest loans were available to improve the housing stock. Innovation Centres aimed at technology update but were 'the only element of a regional technical infrastructure which had to be built "from scratch" after reunification' (Sternberg 1995: 94). Elsewhere there are regional dimensions in efforts to reduce unemployment and environmental problems, e.g. regional employment agencies in Poland which date to 1992. This initiative developed into the system of special economic zones which enjoyed fiscal concessions to make good use of assets and infrastructure in areas not coping well with the transition, such as cities dominated by one troubled industry where restructuring would require massive lay-offs. A 'Northern Economic Zone' was established in response to problems in the ports and at the abandoned nuclear power station site of Zarnowiec. Land in Gdynia, Tczew and Zarnowiec attracted interna-tional interest in production of electronic and telecommunications equipment, plastics/household goods and food in view of links with the Gdańsk–Gdynia–Sopot Tricity and proximity to Kaliningrad. It has already been noted how a SEZ has been created for Mazovia, based on the former military airfield of Modlin 40 km northwest of Warsaw. This area will focus on the food industry, electronics and transportation services and boost the economy of the former Ciechanów and Płock voivodeships.

Former Czechoslovakia

Under communism, a Regional Planning Decree in 1977 aimed at 'balanced and even development of territory', but a more elaborate programme emerging during 1989–92 brought almost everywhere into at least one category of assisted area and there was an excessive dispersal of funds, although total expenditure was modest and low unemployment prevented any sense of major crisis. It was also significant that the swing to local government decentralisation – with the emphasis on very small municipalities – meant that there was very little interest in a truly regional scale of planning and development. In 1992 the right-wing coalition introduced a new Regional Policy Act to support the problem regions looked after by the Ministry for Economic Policy and Development. SMEs were encouraged in high-unemployment areas in the expectation of job creation and export growth, with additional help provided to develop consultancy agencies and establish scientific/technological parks in the most serious cases: mainly in Slovakia. From 1993 (in the context of the Czech Republic only), economically weak regions were identified, mainly in Moravia, while other areas were considered vulnerable to structural change while some border zones (mainly in northern Bohemia) suffered from unfavourable environmental conditions. In assisted areas funds were made available for SMEs and for the financial restructuring of large companies; also for environmental improvements in the polluted regions, for public transport in rural areas, for farming in environmentally difficult zones and for job creation in high-unemployment areas.

The Czech system has been criticised for its institutional fragmentation – with different ministries responsible for the various elements – including a Ministry of Regional Development set up in 1996, along with Regional Development Agencies. For example, while it is hoped that Northwest Bohemia will qualify for an EU 'Pilot Regional Development Programme', it is already assisted by PHARE CBC (as a frontier region), by the Ministry for Regional Development (MRD) – separately for 'economically weak' and 'structurally affected' regions, and by MRD and the Ministry of Trade and Industry (MTI) together for revitalisation of former military training areas. Meanwhile, MRD and MTI have additional programmes specifically for this coal mining region. Hence a more straightforward regional strategy is recommended, that will be easier to integrate with the EU. There can be no question of the need for regional assistance however. While most industrial regions have managed to restructure and diversify, the lignite mining region of North Bohemia has found the going difficult, given the decline of mining and traditional manufacturing sectors (chemicals, steel and textiles) while most foreign investment has gone to Prague and West Bohemia. Infrastructure is being overhauled – with German government help for the upgrading of sewers in Most and Teplice – and interest by German and Japanese firms in the new Chomutov industrial park may complement the expanding computer components factory established in Most by the Irish firm M. Ward Manufacturing. Chomutov is one of 18 locations where industrial zones were established in 1999 as part of a programme – launched in 1998 – to stimulate foreign investment. The main incentive is corporate tax relief for ten years but duty is waived on imported machinery and training grants are available for large projects, along with low-cost building land and infrastructure.

Slovakia It inherited the programme for problem regions and assistance was eventually made available in 1994 for five areas with unemployment exceeding 20 per cent: Rimavská Sobota, Rožňava, Spišska Nová Ves, Svidník and Trebišov. Four more areas were added in 1996 (Lučenec, Michalovce, Velký Krtíš and Vranov nad Toplou). These were mainly depressed rural areas where a substantial reduction in agricultural and industrial production had taken place. Pilot projects were launched while PHARE supported a development agency for Upper Považie (Čadca, Považská Bystrica and Žilina). The programme was revised in 1997 to encourage new entrepreneurs in industry (including handicrafts), construction, tourism and transport, and to improve technology in existing firms. Slovakia also set up regional economic monitoring systems and SME advisory centres – some with PHARE funding – growing into a network of 'Regional Advisory and Information Centres' to stimulate business. The RAICs were established in 12 towns and cooperated with Business Innovation Centres in Banská Bystrica, Bratislava, Košice, Prievidza and Spišska Nová Ves (see Figure 3.1). But these were widely seen as state institutions, not necessarily responsive to local interests, although the districts were able to establish economic and social councils which generated promotional literature. The problem here was that the Mečiar government withdrew support from district offices adjudged to be 'non-sympathetic' and in such cases the leading state enterprises did not cooperate.

But in addition to the arrangements for new business in the weaker regions, there is a perceived need to stimulate the growth of the central Slovakian metropolis so that Bratislava and Košice will be complemented by Banská Bystrica-Zvolen to make for enhanced 'internal compactness' (Ivanicka 1996: 63). There is a strong argument for large local government areas which can sustain an expanding regional city 'capable of larger investment absorption and concentration' (ibid.: 64). Slovakia does not want to copy Austria where one city dominates and rural areas are depopulated, so development in central Slovakia is being encouraged, especially Banská Bystrica-Zvolen which is seen as the core area providing the high-level services for the local communities of central Slovakia distributed among the Hron valley (Pohronie), Váh valley (Považie), the volcanic depressions and even the upper Nitra. The region lies strategically on the axes Bratislava–Košice and Upper Silesia/Ostrava–Budapest (both of which could be extended to Vienna–Moscow and Baltics–Balkans respectively). Government support for the region is evident through the university faculties and the development of banking, commerce and telecommunications.

Hungary

Hungary reformulated its region policy after the initial launch under communism in 1985. Following the creation of a separate Ministry of Environment and Regional Policy (ME&RP) in 1990, a Regional Development Fund was provided in 1991 to finance regional development. But no specific regional strategy was formulated and the position of the ME&RP within the government was relatively weak (despite stronger parliamentary supervision of government and an increased role for NGOs). Then in 1993 four categories of assisted areas were identified: socio-economically backward settlements; settlements not themselves backward but located in backward regions; settlements with an unemployment rate exceeding 150 per cent of the national average; and settlements in need of modernisation (combining the three elements) (Csefalvay 1994). Altogether 1,325 settlements were listed with a total population of 1.79 m. (17.4 per cent of the total), but Borsod–Abaúj–Zemplén and Szabolcs–Szatmár–Bereg still took the lion's share – 70 per cent of the Regional Development Fund during 1991–4 – while some counties had less than 1 per cent of their settlements gaining benefit. There were limited resources and the highly centralised programme allocated 70 per cent of the funding to long-term infrastructure projects, e.g. gas and telephones (though with insufficient resources to close the gap with the more advanced counties) rather than job creation.

A new stage was launched in 1996 when the Law on Regional Development and Physical Planning provided for a more decentralised and transparent regional programme, extended to 1,740 settlements with a 3.41 m. population by 1998 (Table 6.4). By now the counties were being grouped into regions conforming to the EU's recognition of socio-economic development disparities at the 'NUTS II' level. There is also a clearer regional concept based on the four categories of developing cores referred to above. The National Regional Development Concept seeks balanced/sustainable development with modification of Budapest's dominance by promoting the spatial spread of innovation through improved infrastructure and exploitation of regional potentials. A specific aim is decentralisation of R&D activity to settlements with appropriate intellectual and production profiles – and better relations between these centres and their regions.

There is also a clearer regional concept in which four categories of developing cores are recognised: I Budapest; II a 'national semi-peripheral belt' extending to a constellation of towns up to 100 km from the capital: Veszprém, Székesfehérvár, Dunaújváros, Kecskemét, Szolnok and Salgótarján; III the 'semi-peripheral enclaves' of Pécs, Szeged, Debrecen and Miskolc which are territorially isolated; and IV 'commercial gates' on national borders: Györ-Sopron and Záhony which may become alienated from their respective regions. Regional development is to reflect these realities through the development of incubation centres, industrial parks, innovation parks, and technology transfer centres. Regional advice centres and capital risk associations are expected to facilitate the rational spread of these innovations with attention to synergy between entrepreneurs, managers, workers and consumers.

As a candidate for early accession to the EU it is an important question whether the present cohesion policy will apply fully to new members: it is estimated that Hungary will need E1.5–2.0 bn. in order to enhance endogenous potential in the regions, and foster local developments (SMEs and cross-border cooperation in frontier areas) in keeping with the principles of decentralisation,

Table 6.4 Hungarian regional development policy 1985–96

Policy elements	Periodisation		
	STAGE ONE 1985–90	STAGE TWO 1991–95	STAGE THREE 1996–
	Bureaucratic	Transitory	Decentralised
Aim	Equalisation	Equalisation	Restructuring
Object	Underdevelopment	Underdevelopment	Negative market effects
Target Area	U'rdeveloped region	Ditto settlement	Problem region
Tool	Reg. Dev. Org. Fund	Reg. Dev. Fund	Earmarked R&D funding
Financing	Centralised	Centralised	Decentralised
Direct Finance	0.05% GDP	0.20% GDP	0.30–0.50% GDP
Incentives	Automatic	Discretionary	Discretionary
Authority	County Council	Local Government	Reg'l Development C'l
Effect	Isolated	Isolated	Integrative
Key Sector	Industry	Gas and Telephones	Manufacturing
Population	4%	17%	28%

Source: Horvath 1998

partnership and subsidiarity. A successful outcome in terms of equalisation at regional level (guided by Regional Development Agencies) will inevitably involve some concentration of investment within which regions will tend to heighten tension between marginalised rural settlements – lacking their own regional development objectives and resources – and the regional centres. But there is provision for small area associations (combining urban and rural areas) which can then be assisted by the appropriate county development council. Eastern Hungary is attracting more foreign investment because of incentives at a time when labour is getting scarce and expensive in the West. Calsonic (Japan) and Delphi Automotive Systems (USA) are producing compressors for car air conditioning at Balassagyarmet, while US electronics firm Jabil Circuit has invested $80 m. at Tiszaújvaros and Singapore-based Flextronics has opened a $75 m. project to produce printed circuit boards at Nyíregyháza (which will be within 50 km of a motorway to Budapest by the end of 2002). But all these decisions have gone in favour of the larger urban centres.

Small towns in Hungary's settlement system

More development should gravitate towards small towns which are the logical cores for rural associations at the 'kistersegi' or micro-regional (NUTS IV) level. It is the role of small towns to disperse urban characteristics from the cities to the rural areas. The towns stimulate the growth of the tertiary sector, depending on a skilled educated workforce, and contribute to equality in development, whereas agriculture and industries dependent on natural resources produce regional variations (Nemes Nagy 1994: 367). Small entrepreneurs may also be more evenly distributed than large businesses. So a strengthening of district centres would place individual farmers and small rural communities in better contact with the market and with investment opportunities. During communism small towns developed non-agricultural functions under the aegis of the cooperative farms, especially in the context of mergers in the 1970s when some farmers became commuters by cultivating land in other villages. Communist centralisation policy was based on new district centres where 34 types of public institution were to be located, while isolated farms were to be physically eliminated and their population resettled. Not all new centres were successful because some complexes of public buildings failed to attract settlement, but others – like Martfű near Szolnok – became towns. Services for agribusiness are now shifting to small towns and shopping is improving as the 'Penny Market' supermarket network (belonging to the German Rewe group) expands rapidly and aggressively in order to secure a presence in every Hungarian town: already there are approximately 125 outlets. However, on the other side of the coin, the number of small shops has been declining quickly since the mid-1990s.

Small towns are numerous in the northern part of the region and examples of their dynamism are provided by recent research on Jemnice in Moravia and on groups of towns in border regions of Poland. However, substantial areas in the SEECs remain remote from market towns. A more even urban network could achieve a better regional balance, along with continuing progress in education and growth in the tertiary sector, for this encourages economic activity to be 'more evenly distributed across the country. Fortunately, key villages act as surrogate urban centres. Weekly markets are held in hundreds of Romanian villages, which were seen as having potential for urban status under Ceauşescu's 'sistematizare' programme. Strong rural centres have also emerged in border regions of Hungary where the imposition of new frontiers after the First World War separated many towns from much of their hinterland. Hungarian planners now believe that cohesion should rest on urban networks including small and medium towns as well as large ones with their international links. So new towns are being established gradually in all micro-regional centres and the development of good relations between urban centres and their rural hinterlands is a priority. The number of towns in Hungary has increased from 125 in 1988 to 184 in 1993 and 218 in 1998 (Beluszky 1999: 37).

Additional growth poles on Hungary's Great Plain are needed to extend urban influence and enhance the adaptability and creativity of local society, thereby diversifying employment structures and relieving unemployment in rural areas (Csatari 2001: 63). Figure 6.5 portrays a range of centres of micro-regional importance (Hódmezóvásárhely, Kalocsa, Kiskunfélegyháza, Kiskunhalas, Makó, Orosháza and Szentes) which are to be endowed with industrial parks and financial services. Small town locations are not always appreciated by foreign investors. They score well in terms of corporate loyalty but not for the import of outside talent. A single large employer in a small town (e.g. Electrolux in Jászberény, Hungary) can 'fire up' community spirit, especially when people pull together in a crisis, and some small towns close to major transport routes are being noticed. Population resources could be enhanced by a process whereby rising living costs in large centres stimulate a shift down the central place hierarchy: to occupy empty houses, restore natural population growth and overcome the problem inherited from communism when parents encouraged the young to leave the villages while intellectuals (school teachers and cooperative farm agronomists) commuted in daily and left rural communities short of leadership. A cost-effective banking network is important since customers want a personal style, despite the arrival of electronic banking. Hungarian banking is traditionally concentrated in Budapest, but regionalism requires money markets and institutions to finance regional policies, including regional branches of the Hungarian Regional Development Bank.

Thus, some small towns still lack adequate services, especially those that expanded under socialism without a proper 'small town society' and local governments are often reluctant to spend local taxes on improving the infrastructure: hence government funding for gas and sewerage. There is a need for better banking networks and computer technology. Favourable tax regimes can help, while small town networking could offer a way forward. 'As the larger cities try to organise the space around them, groups of towns with similar and complementary functions could cooperate to increase their chances in the competition for development resources' (Nagy and Turnock 1998: 19). Examples could include the Nagykunsag area east of Szolnok and the small town trio of Battonya, Mezőhegyes and Mezőkovácsháza in the south of Békés county. Local area

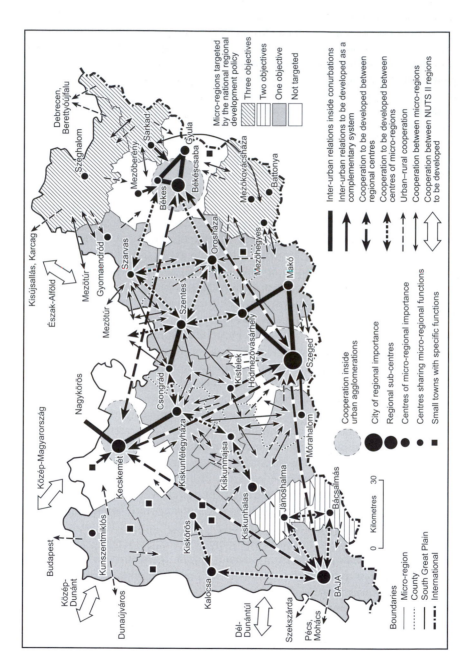

Figure 6.5 Hungary Southern Great Plain Region, one of Hungary's regions at the 'NUTS II' Level, taking the three counties of Bacs–Kiskun, Bekes and Csongrad. The map shows the official town–county associations of municipalities along with the relations between these areas and the extent to which specific regional development objectives apply.

networks might focus on a technologically-strong nucleus such as a town offering higher education facilities. The plan for the South Great Plain Region envisages cooperation between microregions with some provision of enterprise zones, industrial parks and financial services. Enterprise zones projected specifically for Makó, Sarkad and Szeghalom arise from the preconditions laid down: backward areas in a border region. Nevertheless, without stronger regional policy there could still be social and economic disruption of the eastern regions of Hungary (Nemes Nagy 1994: 368). It is important that national regional policy should be complemented by local institutions which can also take their own actions to relieve high unemployment.

COHESION AMONG RURAL COMMUNITIES IN THE GREAT PLAIN

This issue has to be seen in the context of fragmentation under Hungary's Local Government Act of 1990 which allowed settlements of more than 300 people to become administratively independent. Over 700 villages can now make their own decisions (including environmental regulations), though much depends on the level of organisation and the effectiveness of local actors. The option was generally discounted where new residents in suburbanised villages were keen to remain part of a wider urban community with the resources to develop the infrastructure. On the other hand, separation was motivated by several considerations: the prospect of income through Corporation Tax from local business (as at Algyő near Szeged); lower social expenditure obligations (Bocskaikert near Hajdúhadház in Hajdú-Bihar county); resources for local infrastructure and escape from over-emphasis on town-centre refurbishment at the expense of the suburbs (Csabaszabadi near Békécscaba); and an escape from inappropriate functional groupings (as in the case of villages associated with the small town of Edeleny in the northwest). However, where local leaders were able to forge alliances, village associations started to form spontaneously in order to apply for EU and central government funding of road improvements and natural gas supply. And the 1996 Regional Development and Physical Planning Act went on to require village associations – officially drawn up as a nation-wide network of statistical areas based on small towns – to make 'grass roots' contributions to county and regional plans in line with EU principles on decentralisation, subsidiarity, partnership and transparency. Indeed without such organisations – and payment of dues to the county councils – the relevant areas may be excluded from county development plans. Horvath (1998: 50–1) reported that many rural areas were not yet members of associations, or else they were operating within pre-1996 associations covering only very small areas which did not combine town and country. This is still the case, for some settlements may participate in several spontaneous associations covering essentially rural areas, and the political parties work to their own agenda. But the emphasis is clearly shifting towards the official regions which are considered as partners by the central government, although this does not prevent the continuance of the more informal arrangements which help communities to make the most of their geographical situation and influence decision-making. However, the growth of

institutions should not be limited to local government but should include partnerships with NGOs concerned with local development and environmental
management.

When matters were discussed at a regional planning congress in Hungary in
1997 involving the National Regional Development Centre and Development
Committees at regional and county levels – at a time when institutional and
financial arrangements relating to the 1996 act were still unclear – it was evident
that these alliances had no precedents and that much would depend on the
cooperation of local leaders in implementing the 'bottom up' process. Where
local plans could not be reconciled with each other, it was clear that coordination between micro-regions would be needed in order to build up to county
and regional plans, as well as the national development plan which will eventually crystallise as a complement to the existing National Regional Development
Concept. It was also appreciated that locally-defined goals might be difficult to
achieve without decentralisation of finance. Tensions would appear to be
inevitable since the primary aim of EU policy is equalisation at regional level
which will be based largely on concentration in key centres. This will 'fuel
antipathy among the rural settlements: resistance against the territorial centres
(towns) intensifies and chances for regional cooperation decline' (Horvath 1998:
72). Tensions are all the more likely because the change from a county to a
regional basis for the development councils has brought reduced county and
micro-region representation (and exclusion of CCIs) and more central government input. And this is occurring at a time when more EU cohesion funding
is being allocated.

When political problems are solved it is evident that transportation can be
problematic, for although no village in Hungary is more than a half-hour journey
away from the nearest town there are variations in the quality of transport and
services within each small town hinterland (which may involve 20,000–30,000
people, though usually no more than 15,000 for the newer small towns). Hungary
has particular accessibility problems in demographically-weak areas of small
farms ('tanyas'), where central places at the local level are not easily identifiable, and border areas which are a legacy of the imposition of Trianon frontiers
after the First World War where the old road patterns have been disrupted.
Meanwhile, information flows into small towns have benefited from an increase
in the number of local and district newspapers and radio stations; while cable-
television systems now exist in the towns and larger villages. These improvements
are important for the flow of information and for social cohesion and identity
in each district or locality. 'Civil society is being more effectively mobilised so
that each small town, along with the villages in its neighbourhood, can find a
basis for cooperation with regard to both economic growth and cultural life.
Simultaneously new connections are being established with other parts of
Hungary and with foreign countries, as the actors in each local economy seek
out partners and try to enlarge their market areas' (Nagy and Turnock 1998:
20–1). Larger firms can use the new information technologies to extend their
networks at the local or district level, while flexible specialisation through subcontracting and supply networks provides opportunities for SMEs in backward areas

to cooperate with business elsewhere. This enhances the central place functions of small towns and strengthens their economic image so that they may attract larger industrial plants and other investments. They can thereby develop their business services and improve living conditions through the widening range of non-agricultural employment.

The role of local government

While central government can evolve policies best calculated to stimulate growth, there is scope for specific responses to opportunities from individual regions where local government now has a critical role in regional development, including the freedom to establish direct contacts with foreign business. The stakes are high because the successful regions will become partially 'disembedded' through the development of strong inter-regional linkages and growing dependence on external institutions. But at the same time, new practices relevant to the market economy should be internalised so that there is local integration through the old second economy transforming itself into a self-sustaining sector able to foster economic growth. On the other hand, the more backward regions may become 'overembedded' and circumscribed by the rationality of the old business environment in which institutions try to insulate themselves from an increasingly pressing reform ethos. Adjustment has been difficult for some of Poland's newer industrial regions, especially those with a heavy dependence on Soviet orders: textiles and vehicles in the case of Kielce, Łomża and Starachowice (Czyz 1993: 494). In a situation where local political leaders tend to think too narrowly in terms of infrastructure, local actions are needed, not only to attract capital but to bring existing firms together and create stronger industrial branches. But local resources are variable in terms of institutions (both local government and NGOs) and finance. It is suggested that the poorly resourced units are those which are 'no longer rural . . . but not quite urban either, with incomplete urbanisation as a barrier which exerts a negative impact on the level of local self-organisation' (Surazska *et al.* 1997: 453).

Toth (1993) sees a logical evolution towards greater decentralisation and regional autonomy, as local government becomes more involved with business and competes for inward investment. Diversification will be all the greater because ethnicity will exert an influence on the evolution of distinct economic and social profiles. Thus the tremendous opportunities of the post-Fordist era make the quality of regional leadership all the more critical. However, capacity rests on powers and finances and the former may increase faster than the latter, though some local authorities have been able to borrow money through municipal bonds, and partnership schemes are being considered. At the same time, public expectations are high when it comes to environmental protection and public safety followed by improvement of utilities and social services, although such priorities detract from direct engagement in economic activity and in any case there is a general lack of managerial skills (familiarity with business law and foreign languages) among council leaders. There is certainly a need for more enterprise specialists on councils. One way of gaining expertise is collaboration with local authorities in the West, exemplified by the links between Košice in Slovakia and Halifax (Nova Scotia) in Canada. The process is being assisted through an official 'Local Government Partnership Programme' (LGPP) with grants to strengthen 'gmina' (parish) capacity in financial management, sustainable economic development and public relations by involving more individuals, businesses and NGOs in providing services to local government: the best results will be publicised as examples of innovation and good practice. Through the Stefan Batory Foundation the Soros Foundation is helping to reduce rural isolation with support for a democratic and open society.

Some communes have taken substantial initiatives, for example Siedlec near Zielona Góra, with a 12,000 population spread over 20 villages concerned with cereal farming and meat processing. When the state farm collapsed small businesses appeared – including small stores, service outlets and video shops – and a brochure was published outlining a development plan. Although the people were generally receptive to innovation, progress arose from the flair of the commune administrator Stanislaw Piosik: a former school principal who pressed successfully for a new school building and became commune chairman in 1972. He carried on in 1989 as 'administrator', securing a gas-distribution system and a centre for elderly people. In 1996 the decision was taken by referendum to supplement funding of the health service with an additional local tax. As a result the commune's three health centres were upgraded with new diagnostic equipment and new doctors were hired. A new indoor sports arena has also been provided for young people with help from the Office of Physical Education and Tourism. Money for new roads has been secured from the European Fund for the Development of Rural Poland. Evidently, although cash is limited it is possible for go-ahead local authorities to make progress.

When foreign businesspeople deal with local authorities it may be that rapid settlement of land deals is crucially important for consumer industries – like producers of detergents, household chemicals, food and tobacco – that wish to build up market share. Hence considerable growth can occur where there are attractive sites and efficient local administration, as at Kobierzyce (which lies close to the motorway intersection on the edge of Wrocław) where labour costs, local infrastructure and environmental conditions are also satisfactory. The Swedish Ikea company has built a shopping centre here (including a furniture store and supermarkets) and there is cooperation with Wrocław for an agricultural wholesale market. Cadbury's have a distribution depot and the American food processor Cargill has opened a starch and gluten factory yielding various foods, pharmaceuticals and cosmetics. Taxation income to the commune will finance a new health centre, church and school (drawing students from seven villages) as well as improved water, gas, telephones and sewage. A local government industrial zone has emerged at Kleszczów in the southern part of Łódz region where 200 ha of land is available with full technical infrastructure (plus another 150 ha zone at Bogumiłow for recycling plants). Land and electricity are cheap, with pole positioning in relation to the Polish rail and road networks. In addition to Polish plants there are investors from Austria (food industry packaging), Finland (polyethylene pipes), France (Renault distributor), Germany

(Knauf: producer of gypsum-based building materials), Italy (household goods), Netherlands (Kertsen steel constructions) and Portugal (Colep cosmetics). The commune offers ten years' exemption on local taxes, plus legal and administrative assistance and subsidies for staff training.

Gateway cities

It has already been noted that the large city regions are likely to attract the greatest growth, but fertile agricultural areas should benefit from a readjustment of priorities between sectors, while 'trickling down' to smaller towns may be evident along growth axes where 'psychological distances' may be reduced. Thus 'network-led integration' is needed to tie rural peripheries with city region cores that score highly for both accessibility and receptivity to innovative ideas. The process could be accelerated by technologically strong nuclei like the university centres in Poland which provide a good regional spread. A well-integrated settlement hierarchy is very desirable, with small towns and cohesive rural districts to facilitate a dispersal of investments. Such objectives may be facilitated by investment in village infrastructure, as in the case of the Czech Republic's 'Programme for Renewal of Rural Municipalities'. But according to the synergetic evaluation of Chalupa (1993), integration depends not only on good infrastructure, but also on human resources and buoyant labour markets, supported by relevant institutions such as incubation centres, innovation parks, technology transfer centres, regional advice centres, and R&D. For these innovations to spread there is a need for good 'communications' between entrepreneurs, managers, workers and/or consumers. Environmental quality is also a relevant consideration and culture may have a bearing: valorising what Hardy and Rainnie (1996: 180) refer to as 'cultural specificity' and microcorporatism. Where large centres are lacking, as they are in most parts of Hungary's Great Plain region, it is difficult to provide the further education needed to produce the managers and experts who can make business profitable in a situation where most active wage earners are not globally competitive.

On the whole, 'an emancipational process of nationalities and regions can be expected' through adaptability and motivation as well as cost (Vaishar 1992: 397). Yet there are differences in approach, reflected in levels of ambivalence towards the reform process and foreign economic penetration in particular. Better-educated Poles – especially those who have done well during the transition – are positive about reform while those with a rural background are most distrustful. Significantly, negative attitudes are much less apparent in Wrocław in the west than they are in Lublin in the east, where German business is associated with a modern 'Drang nach Osten'. For although foreign penetration is linked with modernisation (new machinery and new technology), a new work ethos of increased productivity and a revitalised economy, it is also associated (rightly or wrongly) with unemployment and the decline of local businesses. But there may be problems of morale and adaptability in areas of high unemployment like eastern and northern Hungary (where social problems may be compounded by a poor environment and infrastructure) and ethnic tensions may also discourage investment. The underdeveloped areas of the communist period are now even further disadvantaged because 'unemployment in these regions may mean the last step in the disintegration of the local societies which is a real danger now' (Dovenyi 1994: 396). Border areas (especially large towns in the northern and western borderland of the Czech Republic) are likely to attract an ethnically mixed population including Asians and refugees trying to get into Western Europe. This creates instability in the area from which the Germans were earlier expelled, for 'a greater part of the population has not identified themselves with their social milieu' (Vaishar 1993: 171). There are relatively low levels of adherence to religious cults 'which indicates lower social control' (ibid.: 171). There are high levels of crime, divorce, suicide and illegitimacy in the north borderland; and high levels of abortions in the southwest. People are now finding work in Bavaria and this enhances local purchasing power, but the frontier zone with Saxony remains difficult.

Cross-border cooperation

In 1991 the European Commission saw border areas disadvantaged at the extremities of transport systems planned on a national basis. Artificially separated from natural hinterlands and handicapped by distorted patterns of commerce and local authority services, the mobility of local inhabitants was further constrained by different languages, tax and welfare systems as well as employment practices. The situation has been particularly difficult in areas adjacent to the closed borders of communist ECE. But now there is a strong 'cohesion effect' resulting from the transition, especially along the main international transport axes guided by the EU's 'European Spatial Development Perspective'. While the first INTERREG programme in 1990–3 covered only internal borders – though including the former Inner-German frontier – INTERREG II for 1994–9 covered the German, Italian and Greek frontiers with the ECECs, while INTERREG III (2000–6) will extend to the frontiers of candidate countries. The key objective is economic and social cohesion, with sustainable development through balanced urban systems and parity of access to infrastructure and information and prudent management of the natural and cultural heritage. A vision of cooperation on spatial policy involves seven large overlapping regions (and four smaller ones). These have been identified to include a 'Central Adriatic Danubian and South Eastern European Space' (CADSES). CoE has also become involved through a 1997 'European Conference of Ministers Responsible for Spatial Policy' to establish guidelines for future spatial development.

The logic of closer cooperation across frontiers can be seen in a variety of ways. With the impending eastern enlargement of the EU to include Poland and the Baltic States, Russia's Kaliningrad territory will create complications noted by the European Commission in 2001. This survivor of the Cold War has been perceived as a threat on the grounds of a potential nuclear confrontation, not to mention the city's reputation as a centre of crime, drugs, prostitution and pollution. These fears are somewhat exaggerated, but, with Moscow's cooperation, progress can be made in such fields as fisheries, migration, trade and transport. Mutual trade benefits are of particular importance throughout the region. Thus Petrakos (1996: 18) sees an expanding regional market in the Balkans because Albania's agricultural bias (with a population dynamism not found elsewhere in Europe) complements Bulgaria's manufacturing and Greece's strong tertiary sector. Greek investment across the border is attracted by low-cost labour; with the additional stimulus of a Greek population in the Albanian border region. Thus ethnic minorities may be seen as 'assets that increase cultural diversity and bring nations closer to a truly international society' (ibid.: 261). However, increased trade will not necessarily have a significant impact on border regions (rather it may circulate between large cities and core regions) and while Slovenia has made progress in developing 'transnational functional regions' through early collaboration with Friuli-Venezia Giulia, foreign companies do not generally choose to invest near the country's borders.

However, to exploit the 'wages precipice' on the EU frontiers, substantial investments are being made in the western border areas of the Czech Republic and Poland. Free trade zones have been established in Cheb and Ostrava, while centres of innovation (Economic and Technology Parks) have been established in České Valenice and Cheb. Regional universities (or departments) have been founded in České Budějovice, Cheb, Frýdek Místek, Jindřichův Hradec and Ústí nad Labem. Other border areas of the Czech Republic have been boosted by the opening of the frontier with Austria and Germany. Tax holidays and other financial incentives are available for selected private firms and there are also grants to improve technical infrastructure. German investment in Poland is also being steered to various locations where there is state-owned land and an adequate infrastructure, discussed further below.

The potential for tourism is considerable in areas with scenic and cultural attractions. The opening of a new frontier crossing in 1994 on the route from Nowy Sącz in Poland to Bardejov in Slovakia (Baran 1995: 101–4) brought pilgrimage tourism to Uście Gorlickie, a centre of Poland's Ruthenian 'Lemko' population noted for its Orthodox churches and folk events. Slee (1999) has discussed the potential for sustainable tourism in the Carpathians which was examined by a CoE mission during 1997–9. This is an area dependent on small farms, following the decline of communist industry, with a distinctive cultural landscape and a good environment (low fertiliser/pesticide application and lack of Chernobyl-related damage). Slee comments that 'the furtherance of trans-frontier tourism makes both political sense in binding countries to a shared vision of sustainable tourism and economic sense in that the combined product of the different countries provides an additional attraction to visitors' (ibid.: 3).

Hungary's special interest

Hungary is heavily committed to cross-border cooperation in the interest of more permeable frontiers for its minorities abroad, given the imposition of Trianon frontiers after the First World War which caused widespread disruption (Figure 6.6). Despite the strains of sanctions against Yugoslavia, Croatian and Slovak hydropower schemes on border rivers, and drugs and refugee smuggling from Romania and Ukraine, Hungary has been unswervingly supportive. Hungary cooperates with Slovakia over the karst region and it is also possible to control pollution in a situation where there are many polluted rivers flowing into the country like the Bodrog, Hernád and Sajó, all linked with industries in Slovakia. There are also highway developments that may be expedited, such as the Budapest–Osijek–Sarajevo–Ploče road which is underpinned by the Danube–Drava–Sava Euroregion. In addition, 12 Hungarian municipalities are involved in twinning with Romania, 11 with Yugoslavia and three with both. At the same time some of its cities such as Szeged are attracting migrants from the neighbouring countries, sometimes to avoid military service and despite the formation of a Cultural Association of Ethnic Hungarians in Ruthenia, Hungarians from Ukraine are settling over the border in Hungary due to economic hardships. Following the Carpathian Euroregion, referred to below, 1996 saw a breakthrough in the east through cross-border cooperation leading to the present Area Development Project for the Hungarian–Romanian Border Zone.

Business centres are developing in eastern Hungary (e.g. at Bihar and Záhony) in order to stimulate cross-border trade while Romanians bring agricultural produce to Békéscsaba market and buy medium-quality manufactures to sell at a premium back home. However, there is much apprehension over the future of the road programme, very necessary to enhance confidence, and the post-accession frontier regime which carries the fear of a new 'iron curtain' induced by the Schengen Agreement. Probably the best record of progress applied to Austria, with agreement after 1990 to tackle problems of crime, drugs and prostitution. The Austro-Hungarian Cross-Border Regional Council founded in 1992 (combining Austria's Burgenland province with the Hungarian counties of Győr–Moson–Sopron and Vas) has harmonised public utility maps to facilitate joint planning of infrastructure including sites for solid waste disposal. Then a 'West Gateway' Area Development Association was founded in 1999 on the Austrian–Hungarian–Slovakian border, as well as the West Pannonia Euroregion. However, business opportunities may be limited by differences in technology. There is a positive attitude to visitors on both sides and Austrians crossing for shopping and pleasure are complemented by Hungarians travelling for business. But there is little FDI from Austria's Burgenland province in Hungary because of the lower industrial quality standards.

Environment

This has become a very important aspect of cross-border cooperation. In 1990, A. Langer wrote on 'Ecological Bricks for our Common House in Europe' and produced a list of frontier areas where action was merited (Figure 6.7). No integrated programme has emerged from this initiative but several of the areas are now being more adequately managed, with encouragement from WWF and financial support from PHARE and the World Bank. Poland is involved in cooperation with neighbouring countries as part of the 'Green Lungs of Europe' project to coordinate activities in networks of national parks and other protected areas in the northeastern lakeland, extending into neighbouring states: for example, Belarus in the case of Białowieza forest. Poland also cooperates with the Czech Republic in respect of the environmental problems of Karkonosze National Park and, most extensively, with Slovakia involving several Carpathian districts, including the Tatra. Meanwhile, the Czech Republic has transboundary programmes with Austria and Germany in respect of national parks and other protected areas, while Slovakia collaborates with Hungary over nature heritage and conservation in the Danube–Ipoly area (including Naszály mountain near Vác with a wide range of habitats). The Stability Pact, which brings together the countries of Southeastern Europe, has given rise to a Regional Environmental Reconstruction Programme which has been taking shape since the Kosovo War. It includes a Swiss-funded programme for five biodiversity-rich transfrontier areas requiring cooperative management. The shortlisted areas are the Belsitza Mountains (Bulgaria, Greece and Macedonia), Shkodër Lake (Albania and Montenegro), the Neretva River (Bosnia & Hercegovina and Croatia), Šar Planina (Macedonia and Kosovo) and Stara Planina (Bulgaria and Serbia).

Of particular interest is the support of the World Bank's Global Environmental Facility (GEF) for cooperation between Poland, Slovakia and Ukraine in an Eastern Carpathian International Biosphere Reserve of 164,000 ha established in 1991 which includes Europe's largest stand of beech forest as well as distinctive mountain meadow country called 'poloniny'. The Polish section includes the Bieszczady National Landscape Park and the Dolny San Landscape Park, while the Slovak part includes the East Carpathian Region of Protected Landscape and the Ukrainian section covers the Stuzhitsa Nature Reserve. The concept of a joint reserve was first discussed in the 1960s by a group of scientists and foresters in the context of a small area of some 2,760 ha. This initiative was not successful but now the signatories have agreed to comply with international standards for biosphere reserves over a far larger area. The prime focus is on forest management to ensure the long-term presence of viable populations of flora and fauna most sensitive to habitat loss. But sustainable development is also being applied to the sparsely-populated cultural landscape of wooden Orthodox churches and wayside shrines traditional to the Boyko and Lemko peoples. Conflicts have been overcome through negotiation with local people and there are now good prospects – discussed in the next chapter – for extending the network of protected areas throughout the Carpathians interconnected by ecological corridors, appropriate for the conservation of large carnivores, which could form part of the further Europa 2000 system. Experience gained in conservation management across a series of fragmented territories will provide valuable experience for the wider project. Meanwhile, the Lower Morava is seen as a model area for cross-border cooperation between Slovakia, Austria and the Czech Republic, with potential for tourism, agriculture and small industries and agrotourism. The area was marginalised on the frontier of the communist world but the Danube River Protection Convention (DRPC), which came into force in 1998, now offers a framework for international cooperation and more integrated river basin management which was previously lacking.

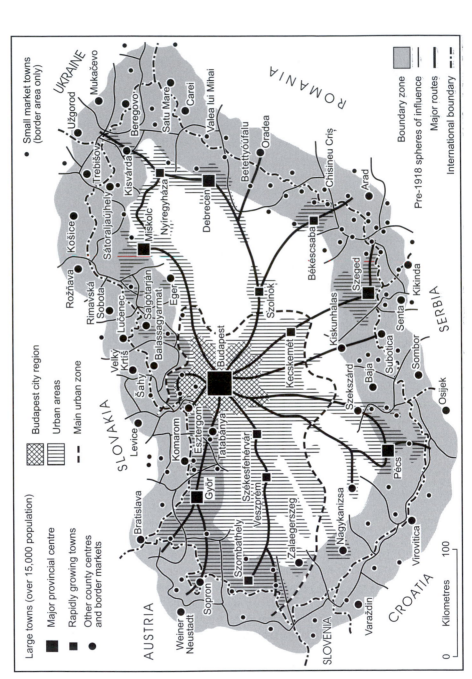

Large towns (over 15,000 population)
■ Major provincial centre
■ Rapidly growing towns
● Other county centres and border markets

⊠ Budapest city region
▥ Urban areas
—·— Main urban zone

● Small market towns (border area only)

▨ Boundary zone
═══ Pre-1918 spheres of influence
——— Major routes
—·—·— International boundary

Figure 6.6 Hungary's border regions

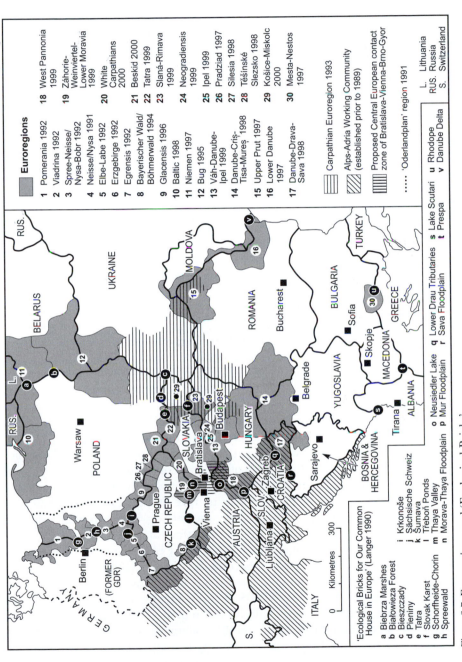

Euroregions

1	Pomerania 1992
2	Viadrina 1992
3	Spree-Neisse/ Nysa-Bobr 1992
4	Neisse/Nysa 1991
5	Elbe-Labe 1992
6	Erzgebirge 1992
7	Egrensis 1992
8	Bayerischer Wald/ Böhmerwald 1994
9	Glacensis 1996
10	Baltic 1998
11	Niemen 1997
12	Bug 1995
13	Váh-Danube-Ipel 1999
14	Danube-Criş-Tisa-Mureş 1998
15	Upper Prut 1997
16	Lower Danube 1997
17	Danube-Drava-Sava 1998
18	West Pannonia 1999
19	Záhorie-Weinviertel-Lower Moravia 1999
20	White Carpathians 2000
21	Beskid 2000
22	Tatra 1999
23	Slaná-Rimava 1999
24	Neogradiensis 1999
25	Ipel 1999
26	Pradziad 1997
27	Silesia 1998
28	Těšinské Slezsko 1998
29	Košice-Miskolc 2000
30	Mesta-Nestos 1997

Carpathian Euroregion 1993

Alps-Adria Working Community (established prior to 1989)

Proposed Central European contact zone of Bratislava-Vienna-Brno-Gyor

'Oderlandplan' region 1991

L. Lithuania
RUS. Russia
S. Switzerland

'Ecological Bricks for Our Common House in Europe' (Langer 1990)

a	Biebrza Marshes	i	Krkonoše
b	Białowieza Forest	j	Sächsische Schweiz
c	Bieszczady	k	Šumava
d	Pieniny	l	Třeboň Ponds
e	Tatra	m	Thaya Valley
g	Slovak Karst	n	Morava-Thaya Floodplain
g	Schorfheide-Chorin	o	Neusiedler Lake
h	Spreewald	p	Mur Floodplain

q Lower Drau Tributaries
r Sava Floodplain
s Lake Scutari
t Prespa
u Rhodope
v Danube Delta

Figure 6.7 Euroregions and 'Ecological Bricks'

Source: European Commission and Langer 1990

Euroregions

These became the conventional vehicle for close coordination in border regions over the past decade. The concept evolved in Western Europe when such trans-frontier regions were first recognised in the 1960s with the support of the Regional Development Fund and the CoE. Experiences of cross-border programmes were shared through Linkage Assistance and Cooperation for the European Border Regions (LACE), an EU project administered by the Association of European Border Regions, created in 1971 by the Rhine regions. During the 1990s Euroregions have proliferated, especially in the northern half of the region. Close cooperation is not, of course, dependent on the creation of Euroregions. But funding is available through the EU's INTERREG programmes and there is a growing consensus that Euroregions are a force for stability and proof of good relations with neighbouring countries, which in turn contributes to the progressive image sought by EU candidate countries. Moreover, with the threat of tighter adminis-tration on the borders of the EU, Euroregions offer concessions with regard to visa formalities and help to maintain intimate links with former Soviet states, especially where there are strong cultural reasons for doing so, as between Moldova and Romania. Institutions dedicated to permeable frontiers and the removal of bureaucratic blockages against socio-economic progress may now be seen as a sig-nificant part of a global world with enhanced mobility and integration.

The development of Euroregions has not been without problems. Resources are limited and improvements in infrastructure can only take place slowly. The Carpathian Euroregion Association suffers from an acute lack of funding, comprising as it does the poorest regions of the countries in question. Cohesion may be compromised by a history of antagonism arising from a contested fron-tier in the Spiš region, now divided between Poland and Slovakia. Some governments 'are afraid of an economic or cultural annexation of their own territories' (Baran 1995: 34). Weclawowicz (1996: 167) sees regional problems for Poland in terms of a western border where 'the great wealth and economic disparity between Germany . . . and the relatively underdeveloped former Regained Territories raises the threat of losing economic and political control to Germany'. More fundamentally there is concern over the Euroregion concept because local organisations may appear to take initiatives independently of central governments. Of course because sovereignty is in no way affected, supporters of cross-border cooperation feel that alarmist views flatter the Euroregion concept with a coherence that goes way beyond reality. Nevertheless sensitivity arises when such regions become formalised within the EU NUTS hierarchy through the combining of basic units into transfrontier regions which may then be recognised at the NUTS II level. Finally, Euroregions are really prominent only in the northern half of the region. This is partly due to the marginalisation of some SEECs from the EU enlargement process which reduces the likelihood of West European institutions being adopted. Also, given the recent history of separation psychological barriers remain strong.

Cross-border cooperation exists along the Greek border (Petrakos 1996), but few formal Euroregions have been established south of the Banat where the

Danube–Criş–Mureş–Tisza Euroregion brings together Hungarian, Romanian and Serbian territories. It has been suggested that 'meso-regions' may emerge in the future, perhaps in the context of a Danubian zone based on Bratislava (Slovakia), Brno (Czech Republic), Györ (Hungary) and Vienna (Austria) while the 'Alps–Adria' concept might also be expressed in terms of a large region based on the Adriatic ports. However, the prime importance of national cohesion makes such ideas academic for the moment. Economic specialisation has led N. Lutzky (quoted by Maggi and Nijkamp 1992: 41–3) to divide Europe into seven regions with the ECECs involved marginally in 'Baltic Hanse' (the Baltic ports and the lines of communication running inland along the Elbe, Oder and Vistula valleys) with opportunities for shipping, shipbuilding and oil refining. Much more substantial opportunities arise in the region of 'Middle European Capitals' where the Czech Republic, Hungary, Slovakia and Slovenia would have opportunities for trade, heavy industry and administrative activity along with research and development in the social sciences. The picture is completed by a 'Balkan Takeoff' region, with a focus on the ports of the Adriatic, the Aegean and the Black Sea and routes along the Danube, Morava-Vardar; as well as Trans-Dinaric routes which might compete in food production and household goods. Globalisation may well generate clearer patterns of specialisation in future but for the moment there are vested interests in retaining the more diverse industrial structures established under communism.

The Neisse–Nisa Euroregion

This links the Zittau area of Germany with Liberec in the Czech Republic and Bogatynia in Poland. After contacts were first made at Zittau in 1991 new border crossings were opened and EU support was provided for ecological studies in the 'Black Triangle' with woodland degraded by pollution. It is recognised that the local economy must be rebuilt using clean technology, alternative sources of energy, tourism, agriculture and forestry to provide the base for sustainable development. The Liberec area has been isolated by closed frontiers and modernisation of the industrial structure was constrained by the policy of expanding the manufacturing base in Slovakia. But the reopening of the frontier with Saxony and access to the German market (especially in Berlin) offers a better future for light industry. And Liberec will have the advantage of adjusting to the West European market without the burden of the communist legacy of heavy industry dominance. As Liberec (including Jablonec) gains momentum as a dynamic centre, attention will focus on the sub-regional scale and the possibility of divergent tendencies for the fringe areas: positive perhaps in Rumburk-Varnsdorf close to Germany but less favourable in Frydlant and Česka Lípa with the burden of former uranium mining. Meanwhile, the Poles have the opportunity of developing land belonging to the Agricultural Property Agency at Porajow where a few hundred metres of Polish territory intervene between the borders of Germany and the Czech Republic on the edge of Zittau. Given the declining Wałbrzych coalfield, dispersal of industry will include a growth of private business concerned with (organic) farming and tourism, especially in the vicinity of the resorts of

Karpacz and Sklarska Poręba which attract many Germans as well as Poles in contrast to the heavy concentrations of population in the main urban zones of Jelenia Góra Kłodzko and Wałbrzych. A great improvement in housing and services is also needed. Polish lands closest to Germany will also benefit from international tourism, including 'shopping tourism' at new markets close to the frontier and transit through Szczecin between Germany and Scandinavia.

The environmental programme – for an area rather larger than the Euroregion – follows from a three-country agreement signed in 1991, involving a Black Triangle Commission with a secretariat in Ústí nad Labem. The problem is grounded in advanced soil acidification, retarded nutrient cycling and a major decline in biodiversity linked with atmospheric pollution and spruce monocultures. Forest damage has also occurred through pollution aggravated by insects and the spread of parasitic fungi. With cooperation among the national parks of the Sudetes, regeneration should aim at broadleaved or mixed forest stands to enhance the stability of forest ecosystems. Meanwhile, serious health problems in the Turów area of Poland have been eased by coordination of electricity production in 1993 with some reduction in capacity in Germany and modernisation of the Turów's own power station. Benefits also arise from wider use of gas in Lusatia and greater energy efficiency through the supply of Turów's surplus steam to Görlitz, Liberec and Zittau. More sustainable tourism is needed, but many short visits are made by people who do not appreciate the environmental problems and so damage continues. Suitable control measures are needed, including investment in other areas to divert some of the pressure. Nevertheless, there has been massive improvement over the decade 1989–99 with reductions in emissions of sulphur dioxide, nitrogen oxides and particulates in excess of 80 per cent

The Carpathian Euroregion (CER)

With generous logistical and financial support from the United States, through private funds and the work of the Institute of East–West Studies (IEWS), efforts have been made to enhance identity and highlight the advantages of an ethnically diverse multi-language low-wage workforce, with traditions of cooperation, supported by a general commitment to regional development on the part of both NGOs and local government. Cooperation has proceeded on many fronts. Regional trade fairs began in 1993 and have taken place in such places as Jasło (Poland), Lviv (Ukraine) and Nyíregyháza (Hungary); overall there has been a growth in small-scale commerce, assisted by the fact that all the CER states except Ukraine are now members of the CEFTA. The Carpathian Border Region Economic Development Association (EBA) was set up in 1994 through links between Hungary's Zemplen Local Enterprise Development Foundation in Sátoraljaújhely (where the main office is situated), and both the Carpathian Society of Hungarian Intellectuals at Užhorod (Ukraine) and the Velké Kapušany Regional Enterprise Development Centre (Slovakia) where small offices are maintained. Ultimately a network of centres of small business should be created including Dámóc, Sárospatak, Sátoraljaújhely and Záhony (Hungary); Královsky Chlmec, Michalovce, Trebišov and Velké Kapušany (Slovakia); and Beregovo,

Mukachevo and Užhorod (Ukraine). Meanwhile, an Association of Carpathian Region Universities (ACRU) was established at Lublin in 1994 to facilitate mobility among the university community, cooperate with local authorities and NGOs and provide a communication channel to debate regional issues. And Užhorod Enterprise Development Agency has initiated a 'Movement for Central/Eastern European Rural Hospitality'. Festivals of traditional folk games and sports are also being arranged as well as theatre cooperation and artistic events.

Case study: Berlin and its region

No major city in Eastern Europe has experienced such a transformation since 1989 as Berlin. Unification has reversed the post-war experience of division, first into occupation zones and then formalised (in 1961) by the Berlin Wall which separated the Soviet Zone (East Berlin) from the combined American, British and French zones (West Berlin). The wall remained in force until the virtual collapse of the GDR in 1989. However, in the meantime various agreements were made to reduce tension: the Four Power Agreement 1971; the Transit Agreement 1972; cooperation over water pollution and garbage disposal; also access to the Steinstucken exclave near Potsdam. Berlin's population growth ceased: 1946 3.13 million (4.9 per cent of the population of Germany) with 1.17 m. in the East and 1.97 in the West; 1984 3.04 m. (3.9 per cent): 1.19 m. in the East and 1.85 in the West. West Berlin experienced heavy natural decrease, balanced by some net in-migration by Germans from West Germany (benefiting from incentives), refugees from the east and, especially, large numbers of Turkish 'Gastarbeiter'. These movements transformed the demographic fortunes of West Berlin during the 1980s, bringing the population back almost to the 1950 level by 1989.

The post-war division followed hard on the heels of monumental-scale planning under the Nazis. Since nineteenth-century Berlin architecture was not felt to be sufficiently distinguished, Hitler (through his competent architect Albert Speer) was able to start with a clean sheet. After building the new Chancellery (1938), Speer created a detailed master plan of Berlin for 10 m. inhabitants during 1937–40. Whole streets were immediately purchased for demolition and plan implementation. But while Hitler viewed the plan as an instrument of political propaganda, Speer worked in the 'Beaux Arts' tradition to envisage a major north–south axis (from the southern station to Hitler's huge hall) with a series of circular and square plazas, each with a façade separately designed to evoke the new Rome. Then after the war came the era of separate development. In the West: the Congress Hall (1957), the restored Chancellery, the Philharmonic Concert Hall (1963) and the new National Gallery of Modern Art. There was administration in Charlottenburg and Schöneberg with commercial foci in Tiergarten (Kurfürstendamm/Breitschiedplatz) and Steglitz (Schloss-Strasse/Forum Steglitz). The wall prevented any decentralisation and forced new housing development into small vacant areas like Gropiusstadt in Neukölln and Falkenhagener Feld in Spandau. Meanwhile in the East, the Alexanderplatz became a focal point with buildings like the television tower (1969) and the housing of Karl Marx Allee (originally Stalinallee); followed by some belated restoration of historic buildings along Unter den Linden as part of a general refurbishment of the war-damaged Mitte (the old city centre, which was allocated to East Berlin). Suburban growth in the East was tied strongly to the railway lines which has made Berlin a star-shaped city with a compact centre. Planning favoured high-density projects on the eastern edge of the city (where Hellersdorf, Hohenshönhausen and Marzahn accommodated 450,000 people). Typically for socialist cities, residential functions remained strong in the heart of East Berlin with 80,000 people still living in the Mitte when unification took place. There was little suburban shopping in either part of the city: just five district centres in the West to complement the main focus in the 'Kudamm' or Kurfürstendamm.

The differences in management can be seen very clearly in the way that the historic monuments were treated. In the case of the war-damaged cathedral (Dom) which lay in the Mitte, the shattered cupola was sealed in 1952–3 and then abandoned for many years until the authorities accepted outside help for restoration (with some toning down of the ornamentation): much of the money (DM45 m.) came from churches in West Germany which in effect provided a hard currency boost to the East German economy through payment for building materials and labour. The authorities could claim to be acting responsibly towards German heritage while signalling the superiority of socialism in the juxtaposition of modern buildings. The exterior work was finished in 1984 but the interior was not completed until 1993 when the cathedral was rededicated. Meanwhile, work on the ruined Gedächtniskirche in West Berlin began very quickly. It offered a highly visible presence in the centre of West Berlin's most significant intersection. There was an attempt to transfer the ruin away from Breitscheidplatz in order to alleviate traffic problems. But the church has a privileged position in German society and public opinion insisted that the site should not be changed. The amalgam of a newly-designed church retaining the old tower enabled the ruined tower to be perceived as a moral message: a symbol of victimisation. The new church was dedicated in 1961 and came to symbolise the revival of the city.

Unification

Unification has brought East and West Berlin together as one of the new federal units. Berlin is again the capital of a united Germany and the parliament returned to start the new millennium. Taxation incentives to stimulate investment in property since 1991 have helped Berlin develop as a global city. Compared with nominal rates of DM600/m^2 in 1989–90, values for prime locations soared to DM10,000–15,000/m^2 in 1991 and DM25,000 in 1992, after a property scramble of 'gold rush' proportions (Berry and McGreal 1995: 388). The transformation is best exemplified by the rise of the futuristic Sony Centre at Potsdamer Platz, replacing a ramshackle caravan park and trading area in the shadow of the wall. Momentum has been maintained by the transfer of government and the development of official buildings, including the Chancellery and Bundestag/Bundesrat offices, around the refurbished Reichstag, but with some impact on the east around Alexanderplatz and Unter den Linden where the historic Hohenzollern Palace (opposite the cathedral) is destined to be followed into oblivion by its successor – the East German 'Palast der Republik' – which will make way for the new Foreign Ministry, though the security implications could impact negatively on this tourist zone. Such has been the transformation of the north–south central belt previously blighted by the wall that it is difficult to reconstruct the landscape of the divided city and a red line has been suggested to help visitors fix the exact alignment of the wall, of which only a few examples have been preserved, as at 'Checkpoint Charlie', which is now a museum.

In 1991 Deutsches Architectur Museum invited ideas for 'architectural reunification', raising questions about the optimum layout in central Berlin: should there be a park to replace the wall; should Kurfürstendamm be extended as a sort of New York Fifth Avenue; or should there be islands of monuments and buildings to reflect the suburbanisation of the city? International architects have prepared plans for the area of Brandenburg Gate–Alexanderplatz and Lustgarten–Mehringplatz. For example, Sir Norman Foster (responsible for the new glass dome on the Reichstag building) has proposed a revitalised Friedrichstrasse intersecting with Unter den Linden where the wall would be preserved as a public park (though the city authorities would prefer to build over it). The result is a new phase of planning after the totalitarianism of the Nazi and communist periods through a strong desire to reassert the vernacular and assimilate unsympathetic architecture in the East where linear blocks clash with traditional tenements. 'Critical reconstruction', with dialogue between traditionalists and modernists, would commend a 25 m height limit and modest plot sizes for individual buildings. New hotels on Unter den Linden involve modernisation in the linear tradition. But some reaction against unadulterated modernism can now be seen in the Potsdamer Platz and Leipziger Platz with smaller units of building to offset gigantic projects like the Sony Centre.

The 'Flachennutzungsplan' (F-Plan or integrated land use plan) recognises the importance of the inner city (within the S-Bahn ring), with opportunities on the ring itself, along the Spree and in the open space along the line of the former wall. Both the former centres (Alexanderplatz and Tiergarten) will be developed with commercial and service sector activity, with government functions near the Reichstag and in the historic core around Kronprinzpalais, and Spree Island. There will be a diversified development for the Mitte. The residential function of the inner city will remain (by designating adequate space for offices) and pressure for gentrification is to be resisted, given segregation between Tiergarten and the peripheral slums, with middle-class housing in between. While Spandauer Vorstadt, close to the centre of East Berlin, is becoming attractive to high-income groups, low-income households could finish up on the unpopular East German estates. In Weddingstrasse and Brunnenstrasse in Kreuzberg there is rehabilitation of tenements: maintaining the façades but thinning out the buildings behind in order to reduce densities. This approach is now being extended to the East with some innovative courtyard concepts, though water and sewage infrastructure is not always adequate for suitable upgrading of the housing stock. Also in East

Berlin, where there is too much servicing in the centre rather than in the housing estates, the centre's dominance is being loosened by service centres at various points within the Stadtbahn ring. There are still separate housing markets in East and West Berlin. The West has socio-economic segregation while the East has demographic segregation, given the estates built for young couples in the 1970s and 1980s. Also the West has high levels of foreigners while immigration in the East was insignificant until small numbers of labour migrants arrived in the 1980s, but increasing after unification due to asylum applications.

Deindustrialisation is taking place. Wiegandt (2000: 10) cites the case of the old industrial district of Oberschöneweide in the southern part of East Berlin with 30,000 workers before unification but now only some 3,000, mainly in the service sector which has taken over the old factories. Manufacturing and service industries in West Berlin are benefiting from market links with East Germany since they have greater efficiency than their Eastern counterparts (and there is often spare capacity too). In the East however there are problems in retaining links with East European customers who must now pay in DM. Treuhand has handed over many businesses to former owners and has found entrepreneurs to take over the rest (though Western businesses were initially slow to arrive in sufficient numbers). It seems that there are few management buy-outs because 'often the old management elite has been either politically discredited or intimidated by culture shock experienced after the wall fell' (Ellger 1992: 45). So a rundown of industry seems likely, especially in view of the lead already gained by cities like Munich and Stuttgart for high-tech industry. Decline in industry could be balanced by a growth of office accommodation (though this will be complicated by ownership disputes in the East where a US/French consortium has plans for Friedrichstrasse). A likely site for further office development is Potsdamer Platz (between the rival city centres) as part of a possible relocation of German company headquarters following the transfer of the parliament.

Transport

The transport infrastructure will need radical overhaul and the rail network is experiencing radical change (Figure 6.8). The railway system was badly broken up by the wall although some West Berlin suburban trains were able to make short transits through the East. Routes crossing from West Berlin were greatly reduced while the GDR completed the 'Aussenring' round the city (1961) which kept its trains firmly within its own territory. With reunification the former links are being restored. This is a complex business because advantage is being taken of the option of establishing a single major interchange at Lehrter where north–south and east–west lines will cross. The latter axis is already in existence through Zoologische Garten and Ostbahnhof (though the alignment is being improved to eliminate some sharp curves as part of a refurbishment of the Stadtbahn). But the north–south route is being built underground and will bisect the S-Bahn ring roughly between Gesundbrunnen and Papestrasse, with a

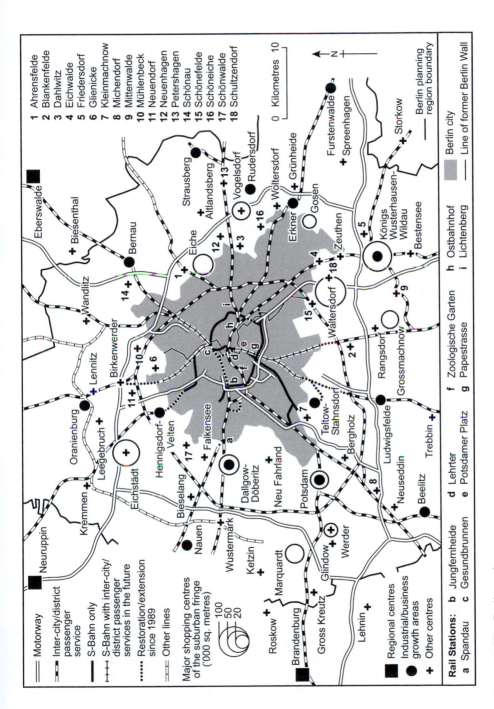

Figure 6.8 The Berlin region
Source: Universität Potsdam

3.25 km tunnel under the Tiergarten. This project adopts a mushroom concept ('Pilzkonzept') because of the shape of the interconnections. Trains which for many years have been diverted round the ring into Ostbahnhof will in future be able to follow their historic routes towards the city centre. Meanwhile, trains from the west will use a high-speed line from Hannover through Stendal to Spandau and Zoologische Garten. This route into Berlin was available for electrified trains to Hamburg via Falkensee but elaborate works have been needed to connect Spandau with the centre. District trains from the north and west will terminate on the S-Bahn ring at Gesundbrunnen and Jungfernheide respectively. Since the use of the 'Aussenring' must continue for passenger trains until the new underground route is open, it was inevitable that interim main stations for the reunified city would be in the East at the reconstructed Ostbahnhof and Lichtenberg. Finally the revamping of the S-Bahn includes a possible extension to Teltow. There has also been a debate as to whether the S-Bahn should be extended further into Brandenburg or whether the outer suburbs would be better served by regular half-hourly services on the DR network to Brandenburg, Cottbus, Eberswalde, Frankfurt/Oder, Juterbog and Neuruppin.

The urban fringe

The reunified Berlin has a fine regional setting of forests and lakes with scope for recreational facilities and suburban housing. The need now arises for a wider spread of housing after decades of high-density development and many more facilities – including shopping and recreation – on the urban fringe. Prior to unification Berlin was highly contained. The wall prevented any decentralisation in the West and forced new housing developments into small vacant areas like Gropiusstadt and Falkenhagener Feld as already noted above. In the East planning favoured high-density projects on the eastern edge of the city, as already noted. Movement is taking place from the inner city ('Wilhelminian Belt') – Kreuzberg-Neukölln and Wedding-Reinickendorf – in response to commercial and gentrification pressures. Some large housing developments are taking place in Brandenburg, e.g. the small town of Werder is to double its population. Shopping centres on the edge of the city are also springing up, e.g. Werder Park on the former refuse dump. There is a danger of too much capacity and some businesses, car showrooms for example, have failed. The new peripheral centres inevitably exert pressure on central shopping (a project to develop shopping in central Werder was abandoned), but there is a new market in Potsdam and the Stern Centre has been built nearby in Babelsberg. Some Brandenburg local authorities keen to encourage development have already given planning permission for 'Gewerbeparks' (business parks) close to the Berlin Ring. With the reorganisation of local government there are larger districts ('Kreise') while the local units ('Gemeinden') have been grouped into 'Amter', each with a range of services. The Amter provide a logical structure for central places and pro-development authorities have been taking initiatives on the fringes of Berlin in areas like Mahlow.

Growth offers some compensation for the decline of agriculture around Berlin. Fruit growing is traditional, but specialisation was reinforced by FGDR government in 1973 when four major fruit/vegetable enterprises were created, each with up to 2,000 ha of land. Holdings were apparently grouped in certain areas and other farms (growing potatoes and cereals) had to give up land for fruit trees. Processing industries were started, but the system has largely collapsed and the area under fruit and vegetables has already declined by 80 per cent due to competition for imported fruit handled by new wholesaling complexes (one of which is located within the old fruit-growing region at Gross Kreuz). Many of the local fruit trees are old and relatively unproductive, while irriga-tion is too expensive: the water company in Glindow tried to diversify into waste disposal but is now bankrupt. Glasshouse complexes are not profitable either, not to mention the additional problem of pollution through burning lignite. Former Junker families have got land back but they too lack the funds to modernise. However, there is hope of modest recovery to exploit the advantage of proximity to Berlin and some new orchards are being planted.

A wider region?

The growth taking place on the edge of Berlin raises questions about a wider Berlin region extending into Brandenburg. There will be migration into the rural areas and coordination of planning between Berlin and Brandenburg will be necessary. The Berlin planning region covers the main commuter region, including the adjacent 'Kreisen' of Bernau, Fürstenwalde, Königs Wusterhausen, Nauen, Oranienburg, Potsdam (both Landkreis and Stadtkreis), Strausberg and Zossen. This makes for an urban region of just over four million inhabitants, with the proportion of one-fifth resident in the suburban zone – stable throughout the post-war period to date but now set to rise significantly. Half the building scheduled for industrial purposes lies outside the city and there are also housing developments south of Berlin in several pro-development administrative areas. Such growth, however, could compromise a green belt around Berlin, with devel-opment axes on the main transport routes up to 60 min travel time from the city centre. Business centres arranged in such a way would better stimulate development in surrounding rural areas.

The merging of Berlin and Brandenburg has been much discussed as a logical development in planning coordination terms and a trendsetter for larger and more financially viable provinces (Berry and McGreal 1995: 377). This will probably happen, although Brandenburg is reluctant to take on Berlin's finan-cial problems. It may contribute to a unified planning strategy, which obviously makes sense in the context of counterurbanisation and recreational pressures. But there will still have to be a compromise between Brandenburg's concern over job creation to relieve unemployment in the Cottbus coalfield area in the south of the province – where major projects like the Lausitz motor racing circuit have been located – and Berlin's interest in the motorway ring which a number of Brandenburg local authorities support. Unification may also settle the uncertainty over the choice of Schönefeld as Berlin's main airport. It could absorb the traffic presently handled at Tegel and Tempelhof, but the issue is complicated by Brandenburg's desire to use a former military airport further to the south which would draw further growth towards the declining lignite mining area of Cottbus. However, rejection of the union also comes from East Berlin and is seen as a strong local desire to preserve identity and sub-culture.

Cross-border cooperation

However, there is potential for a much larger urban region to be constructed around the city since Berlin is likely to be the dominant service centre in the northern part of ECE with influence extending through Brandenburg into the eastern parts of Sachsen-Anhalt. The influence of Berlin also extends eastwards across the Polish frontier. Formal planning arrangements have been proposed by Brandenburg through the 'Oderland' plan in 1991 for 34 German districts ('Kreisen') and five (former) Polish voivodeships. Complementary economic interests feature in German industry's consolidation: with a growth of high-tech manufacturing in the south and penetration into Poland, combining with Frankfurt a.d. Oder's important functions as a transport node and economic–cultural meeting point. In addition, there is potential for high-quality light manufacturing and a growth of tourism and national parks on infertile land. Finally, there is an emphasis on agriculture and light industry in Poland. Transport axes were identified first as Berlin–Poznań–Warsaw, second, Dresden–Wrocław–Katowice–Kraków and third the two routes from Szczecin southwards to Prague via Berlin–Dresden and Poznań–Katowice, with growth likely through motorway extensions, easier frontier formalities, EU enlargement and the growth of business in the east with the greater economic stability in Ukraine.

Despite reluctance in Poland to be drawn into such a structure, a tier of Euroregions has been established with EU financial support. They do not fully reflect the area of German influence as defined by the Oderplan, for local authorities made their own decisions over membership. Institutional development has proceeded through town–commune partnerships, e.g. Frankfurt a.d. Oder/Słubice (exemplified by Viadrina University with Collegium Polonicum under construction in Slubice), while the Gorzów-based German–Polish Society for Economic Support – a joint stock company combining Berlin, Brandenburg, Mecklenburg-Vorpommern and Saxony with Szczecin, Gorzów, Zielona Góra and Jelenia Góra – has contributed to joint ventures, fairs and training conferences. The German side stresses cooperation in technical infrastructure and production, while business-supporting institutions and organisations for regional development are relatively weak in Poland where too many communes are looking to tourism. However, public opinion is generally in favour of cross-border cooperation, despite a lack of knowledge and fear of neo-colonisation.

The growth of the bazaar economy in Poland has been impressive. In 1995–6 shopping tourism from the 16 largest bazaars on the Polish border constituted 16.4 per cent of official Polish exports to Germany: the seven largest were Cedynia, Gubin, Kostrzyn, Łęknica, Słubice, Świnoujście and Zgorzelec – averaging almost Zl 200 m./yr in 1995–6 – reducing local unemployment and stimulating German companies to build cheap hypermarkets in Poland to compensate for lost trade. At Kostrzyn alone 1,200 jobs in the town's municipal bazaar have been created due to a large German clientele (including many from Berlin) patronising the petrol stations and catering establishments as well as the local market. There is also much German investment in Szczecin (the former Stettin) related to imports of building materials, chemicals and timber and also the sailing potential of the Oder estuary (for Świnoujście is closer to Berlin than Rostock). However, the difference in wage levels means that while German companies may move their head offices to Berlin they are unlikely to develop labour-intensive manufacturing capacities. Instead, an appropriate conception would be an industrial grouping combining Cottbus, Eisenhüttenstadt and Gorlitz in East Germany with Polish centres extending to Legnica and Jelenia Góra where a growth of low-wage manufacturing jobs might be anticipated.

Industrial estates around Zielona Góra provide attractive economic and financial conditions for investment. The list includes Brody on the Neisse, close to the Olszyna border crossing, and other places further away where state-owned land has a favourable infrastructure: Babimost close to Zielona Góra airport, Torzym at a major road junction in an environmentally attractive area, Zbąszynek which is a railway junction, and the old Soviet airfields of Szprotawa and Żagań, where buildings and services are already available. There is the danger of a 'maquiladora' syndrome (typified by the US–Mexican frontier) with high-quality production in USA and low-pay manufacturing jobs in Mexico. But with proper attention to skills through coordinated training projects, 'cheap manufacturing in the border area can only be an initial phase on the way of Polish enterprises towards long-term growth of their technological competence and an improvement in the quality of cooperation' (Stryjakiewicz and Kaczmarek 2000: 51). There is a prospect of technical progress through the plan for an industrial complex to develop on both sides of the German–Polish frontier and link the EKO steelworks of Eisenhüttenstadt with metal industries in Cybinka on the Polish side. Diversification is occurring in the oil-refining town of Schwedt through a growth of light industry which could be integrated into an east–west trade corridor over the longer term. However, while there are also increasing numbers of joint ventures, the stimulating free flows of capital, technology, information and people to enhance integration, is still largely a matter for the future. There is also a need for coordination in environment matters since the FGDR installed power stations and major industrial enterprises (like the Eisenhüttenstadt metallurgical complex and the Schwedt oil refinery and several power stations) without considering their impact on Poland.

Cross-border issues extend southwards into the triborder zone where the Neisse Euroregion links the Zittau area of Germany with Liberec in the Czech Republic and Bogatynia in Poland. German investment is occurring here, for like western Poland, northwest Bohemia lies within ten hours' travel of a market area of over 80 m. people. Such growth is advantageous for the Liberec area where modernisation of the industrial structure was constrained by the FCS policy of expanding the manufacturing base in Slovakia. Access to the German market (especially in Berlin) offers a better future for light industry, including glass manufacture and textiles. Growth can also occur in Poland where the Dresden–Zittau–Liberec road runs through the Porajow district of Bogatynia, and land belonging to the Agricultural Property Agency is being set aside for development as noted above.

Regional development in Romania: the potential of the West Region

An important initiative was taken by the Romanian government in 1998 with the setting up of organisations to coordinate regional development in eight large regions which now serve as the building blocks for EU cohesion policy. Although the constituent counties remain the top-tier units of local government, Romanian social scientists have identified larger functionally-coherent areas polarising around major provincial cities and roughly similar in population (in excess of two million) as the basis for durable economic and social development and international (cross-border) cooperation. These constitute the Romanian version of the 'NUTS II' regions recognised by the EU and referred to earlier in this chapter. Regional development councils will determine policy, while the corresponding regional development agencies will formulate and implement the plans, backed by a national council and a national agency to approve EU structural funding and allocate resources from the national regional development fund. While there are historic contrasts between regions, the planners have been particularly careful to emphasise the variations that exist within regions, demonstrating contrasts between stronger and weaker counties in every region except Bucharest. Hence the new programme is commended in part as a means of limiting the danger of polarised sub-regions through action to combine stronger and weaker counties. 'Growth poles located at the border between centre and peripheral sub-regions could play an important role in solving regional problems' (Ramboll Group 1997: 5). This means addressing unemployment in depressed industrial regions but also assisting agriculture where this is the key to the poverty problem. The West Region is used to illustrate the opportunities and problems that exist. The region as a whole is doing relatively well and is poised to develop its international relations, but there are also serious problems of unemployment and the internal contrasts are very evident when Caras-Severin and Hunedoara are compared with Arad and Timiş (Figure 6.9; Table 6.5).

A regional profile

The modernisation of the region is based on the Habsburg colonisation of Banat involving diverse ethnic groups which greatly extended a cultural profile based on earlier contacts between Romanians and Hungarians. Although not the region with the largest non-Romanian population, Romanian Banat must now be an area with the largest number of different non-Romanian ethnic groups. And it is apparent that the experience has been predominantly one of tolerance and mutual respect to the point where all contributions are valued as parts of the region's economic and cultural identity. Nationalism has not significantly detracted from the positive contribution of ethnicity for the human resources of the region. However, the emphasis on agricultural colonisation on the plain was complemented by priority on extractive industries and metallurgy in the mountains. Reşiţa (in Caras-Severin) has a history of ironworking dating back to 1771 which makes the West one of the longest-established industrial

Table 6.5 Romania's West Region: a profile

CRITERION (1995 except where otherwise stated)	A	B	C	D	E	F
Per Capita GDP – ,000 ECU purchasing power std.	5.7	4.7	6.0	7.0	6.0	5.6
Ditto: % of the EU average	23	27	35	41	35	32
Ditto: ECE candidate countries' average	78	74	94	110	94	87
Index of global development (a)	55	48	58	63	na	na
Human capital index (b)	106	16	57	157	na	na
Index of labour use (c)	125	15	48	174	na	na
Share of the aged population – 65 years and over	14	12	10	12	12	12
Share of urban population	52	57	76	62	63	55
Urban population in cities 100,000+ inhabitants	75	0	0	77	40	57
Participation rate – salaries per 1,000 inhabitants	293	260	355	304	307	272
Dynamic of participation 1995 as % of 1990	81	66	81	74	76	76
Percentage employed in industry	23	28	44	26	30	29
Percentage of public roads modernised	24	37	32	21	28	24
Unemployment rate	5	10	12	4	7	10
Infant mortality (d)	17	21	26	22	22	23
Out-migration (e)	7	9	10	4	7	9
Private cars/,000 inhabitants 1994	99	95	81	136	106	89
Telephone subscriptions/,000 inhabitants	146	101	104	137	124	129
Culture: Consumption/,000 inhabitants (f)	5.1	4.4	5.4	6.2	5.4	4.6
Media: Radio & TV licences/ ,000 inhabitants (g)	212	175	197	243	212	190
Hospital beds/,000 inhabitants	8	8	10	11	9	8
Water use: m³ per capita	50	51	63	58	56	54
Tourist overnight/,000 inhabitants	3	5	2	2	3	2
Average criminality rate 1990–5 (h)	253	279	365	343	311	327
Foreign Investment $ per capita 1990–7	52	33	30	180	85	99

Source: Ramboll Group 1997 and Romanian statistical yearbooks

Notes
A Arad county; B Caras-Severin county; C Hunedoara county; D Timiş county;
E West Region; F Romania
(a) 17 indicators relating to economy, infrastructure, demography and living standards
(b) Based on education, infant mortality and communication environment. All positive values are above the national average
(c) Unemployment rate, participation rate and out-migration
(d) Deaths under one year/,000 live births 1993–5
(e) Temporary migrants to other counties/,000 inhabitants
(f) Sum of sportsmen (10%), ,000 cinema attendance (multiplied by two) and ,000 library books lent divided by two
(g) Average of the two sets of figures
(h) Convictions per 100,000 inhabitants

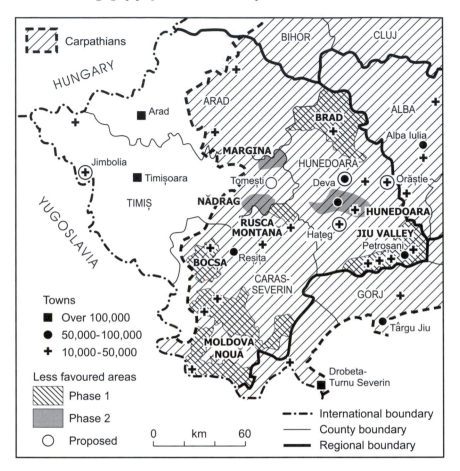

Figure 6.9 Romania's West Region
Source: Romanian administrative maps

zones in the southeast of Europe: a situation reflected in urban expansion, early electrification, the growth of a relatively dense railway network and the emergence of food-processing industries based mainly on the rich agriculture of the Banat Plain. The capacity of the Resita metallurgical and engineering complex to act as a mother factory, stimulating a series of transfers of production to new locations, also indicates the high level of experience and skill.

The historic strength of the extractive industries is extended by consideration of Hunedoara's mineral wealth in coal, in the Jiu Valley at the southern end of the county, complementing the long-established working of non-ferrous ores in the Brad area of the Apuseni Mountains to the north. The autarkic stance of the communist regime was conducive to heavy investment in high-cost deep-mining ventures which have now been stranded as unviable in the more global market-oriented climate of the last decade. The problems of restructuring are plainly complex and progress can only be sustained if a willingness to reprofile production and boost efficiency is complemented by an inflow of funds and know-how from international financial institutions and foreign investors. The West Region has done relatively well with a range of new international links as well as intensification of contacts established in the context of the planned economy. The development organisations established in Arad (ADAR), Caras-Severin (ADECS) and Timiş (ADETIM) with help from Germany's Nordrhein-Westfalen province helped to provide expertise before the wider West Region came into existence.

Per capita GDP sees the region above the national average with Timiş almost 50 per cent greater than Caras-Severin: although Timiş only scores 40 per cent of the EU average, it is above the average for candidate countries. The region has a participation rate well above the national average, but the very high figure for Hunedoara – with a dynamic of participation higher than both the regional and national average – applies to 1995 when mining work was still heavily protected. However, since 1996, restructuring has been undertaken and the weakness of Hunedoara's position – with 44 per cent employed in a poorly-diversified manufacturing sector, against regional and national averages 30 per cent and 29 per cent respectively – has become apparent through rising unemployment. At present therefore Caras-Severin and Hunedoara are relatively depressed through contraction of the extractive and metallurgical industries, though there is potential in wood processing, and the situation is not helped by significant pollution problems in mountain depressions. On the other hand, Arad and Timiş have diverse processing industries, with food processing particularly strong. Moreover, 'the lowland countryside has been revitalised by reinstatement of private property [helping] farmers increase their revenues, develop local services and intensify the village/town relationship' (Ramboll Group 1997: 31). The most impressive indicator however is FDI where Timiş has nearly doubled the national average, more than double the regional average and six times the level of Hunedoara.

Services are relatively good throughout the region but Timiş is outstanding, being well above the regional average for culture, media and hospital beds, though Hunedoara is a major consumer of water in view of its industrial profile. Arad and Timiş have advantages in accessibility and large dynamic cities. On the other hand, medium-sized towns of 20,000–100,000 are unevenly spread, for eight of the region's 11 such towns are in Hunedoara. Arad's urban profile is particularly unbalanced for below Arad City, with 184,000 inhabitants, there are only two towns which (just) clear the 10,000 threshold, while the remaining five towns in the county are all below it. Across the region one-child families make for negative natural increase, inducing an influx of young workers from the north (Maramureş) and the southwest. In the past, Hunedoara has improved its natural demographic indicators in this way, but in view of redundancy many are returning home and 'this drain of young labour may create serious long-term development problems' (Ramboll Group 1997: 32). Hunedoara scores high for criminality and for infant mortality. Since 1989 there has been much emigration from the region and there has been a sharp decline in the German community, so there has been rural depopulation despite relatively high living standards in the countryside. There is a net import of population from all other regions of Romania, especially from the North East and North West, but most go to Arad and Timiş.

A strategy for growth

Regional plans will inevitably capitalise on the leading assets. Timişoara, the obvious centre of Romania's West Region, can sustain self-reliant internationally-competitive development. With a central location in Timiş County, Timişoara offers qualified labour, education/training facilities and research institutes (with the university seeking the status of 'centre of excellence'), and environmentally friendly industry (at an advanced stage of restructuring) with a stock market, business centre and supply networks. It has a network of SMEs to serve large enterprises, good banking and local government support, fiscal incentives and low-cost sites and premises. Pollution is not a serious problem in view of the bias towards engineering and light industry, while services, including waste processing, are relatively modern. It offers institutionalised international relations with Hungary and Yugoslavia, an international airport, a potential shipping lane (Bega Canal) and plans have been drawn up to greatly improve the traffic circulation, including the routing of the Bucharest motorway (Figure 6.10). The settlement pattern includes strong suburban communities which contribute to the economy through warehousing and industrial sites as well as tourist and recreational facilities. In the early years of transition there were investments in the city from Alcatel, Coca-Cola and Procter & Gamble, with a second wave at the turn of the millennium which has included the American enterprise Solectron and such German companies as Continental, Draxlmeier and Siemens.

A new study of Timişoara promotes the city as a growth pole in the context of an expanding Europe and a growing CEFTA. The city could act as a pilot for the adoption of the EU 'acquis' and the process of Romanian integration into the EU generally. An exhibition centre and commercial park are anticipated as part of a satellite town which will help to gear up Timişoara for the creation of 15,000 new jobs during 2000–7 and hopefully a reduction in unemployment from 8.6 to 3.5 per cent, plus a boost in average income by 20 per cent. However, improvements in local infrastructure and international communications are needed (see below), while the city is constrained by population decline and the emigration of qualified people. The city would like more autonomy in education, with higher pay for teachers to retain them for the profession. But there is a fear of renewed centralisation and a 'closed' economy; also an awareness that because rapid growth in Timişoara will not be welcome in depressed areas, there will be conflicts of interest both locally in Timiş County and more widely across the region. Instability in the SEECs will also have a negative effect, as demonstrated by the Yugoslav War of the early 1990s.

The West Region seeks a restructured agriculture and rural diversification under the guidance of a regional information centre for agricultural problems and a rural development agency. County plans are supporting such arrangements in respect of small towns and communes with central place functions as

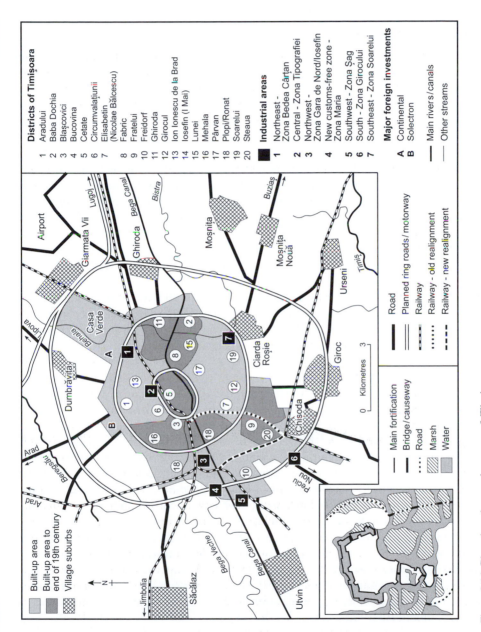

Districts of Timişoara

1 Aradului
2 Baba Dochia
3 Blașcovici
4 Bucovina
5 Cetate
6 Circumvalațiunii
7 Elisabetin (Nicolae Bălcescu)
8 Fabric
9 Fratelui
10 Freidorf
11 Ghiroda
12 Girocul
13 Ion Ionescu de la Brad
14 Iosefin (I Mai)
15 Lunei
16 Mehala
17 Pârvan
18 Plopi/Ronat
19 Soarelui
20 Steaua

Industrial areas

1 Northeast - Zona Bedea Cârtan
2 Central - Zona Tipografiei
3 Northwest - Zona Gara de Nord/Iosefin
4 New customs-free zone - Zona Maria
5 Southwest - Zona Șag
6 South - Zona Girocului
7 Southeast - Zona Soarelui

Major foreign investments

A Continental
B Solectron

— Main rivers/canals
— Other streams

Built-up area
Built-up area to end of 19th century
Village suburbs

— Main fortification
— Bridge/causeway
··· Road
Marsh
Water

Road
Planned ring roads/motorway
Railway
Railway - old realignment
Railway - new realignment

0   3
Kilometres

Figure 6.10 Planning the road system of Timişoara

Source: West University of Timişoara

centres for non-agricultural functions in predominantly rural areas. Integrated local programmes, implemented by SMEs in agriculture, food processing, marketing and tourism (reflecting in part the diversity of ethnic minorities), may be encouraged. The small town of Deta is seen as suitable location for a specialised livestock market and for fruit processing, including production of a range of brandies, while Făget could be a tourist centre for the Poiana Rusca and a suitable location for a small business incubator (generally linked with the larger towns). Each town should also have an organisation for businesspeople: 'Club ale Oamenilor de Afaceri'. Stronger rural centres could also emerge at Bethausen, Ciacova, Fârdea, Gataia and Recaş, although deficiencies in the infrastructure (medical services, natural gas, roads, telecommunications, water and sewerage) are also highlighted. Local halls could be refurbished as multi-functional centres for educational and social–cultural functions, but at Izvin, where the high cost of transport into Timişoara provides a basis for independent development, the hall has been destroyed and the bricks stolen. The reality is a long way from the ideal prospect of cinema, culture house, general store with repair workshops, swimming pool, children's park and new school, and the community feels largely powerless.

Less-favoured areas (LFAs)

However, regional development is being linked with 'firefighting' in areas of high social risk, especially in areas of mining rationalisation. According to arrangements made during 1998–9, the government is recognising LFAs ('regiuni defavorizate') which benefit from fiscal concessions for 3–10 years, to stimulate enterprise: where the unemployment rate is 25 per cent higher than the national percentage; where a single industrial branch accounts for over half the salaried population working in industry; where massive lay-offs (due to the liquidation, restructuring, or privatisation) affect more than a quarter of the active population permanently domiciled in the respective areas; and where the area lacks an adequate infrastructure. New enterprises (whether trading companies with majority private capital, family associations or individuals) may be exempted from profits tax and all taxes on imported equipment, buildings, transport and land. West Region is much involved in this programme. In Hunedoara County, where problems have arisen in connection with Brad (mining of non-ferrous ores) and the Jiu Valley (coal production), two LFAs were created in 1998. And in Caras-Severin County, mine closures have affected Anina (coal), Bocşa (iron ore), Moldova Nouă (non-ferrous ores), Ruschiţa (lead-zinc ores) and Sasca Montană (copper ores). Accordingly, the relevant areas were grouped into LFAs, covering Bocşa, Moldova Nouă and Rusca Montană in 1999. Up to the end of March 2000 84 certificates were issued to allow fiscal concessions in respect of 5,687.2 m.lei (Table 6.6).

The Jiu Valley in the southern part of Hunedoara County is a major problem area. It is Romania's most important pitcoal producer, but demand is falling and imported coal is cheaper. So the government has been making heavy cut-backs, combined with relatively generous redundancy terms and attempts to

Table 6.6 Less Favoured Areas in Romania's West Region

Area	A	B	C	D	E	F	G	H	I	J	K	L	M	N	O
Bocşa	4.99	286.2	25.0	9	–	8	1	7	2	–	104.4	52.4	11.4	5.9	55
Brad	12.98	1318.4	52.9	18	–	13	5	16	2	–	96.0	10.0	15.1	12.1	93
Moldova Nouă	4.99	1926.2	61.9	14	–	9	5	13	–	1	280.7	10.0	11.5	6.3	55
Rusca Montană	4.99	154.4	2.2	1	–	1	–	1	–	–	2.0	0.0	7.6	7.6	2
Valea Jiului	12.98	996.0	162.6	42	1	24	17	36	5	1	5204.1	0.0	17.2	19.7	151
Total	–	4681.2	304.6	84	1	55	28	73	9	2	5687.2	72.4	na	na	356

Source: West Region Regional Development Agency

Notes
A Zone declared (month and year)
B Area (ha)
C Population (,000)
D Certificates issued
E Investments in the primary sector
F Ditto manufacturing, energy and construction
G Ditto services
H Investments under 10 m. lei
I Ditto 10–100 m. lei
J Ditto over 100 m. lei
K Total investment (m. lei)
L Ditto foreign investment
M Unemployment January 1999
N Ditto April 2000
O New jobs created

introduce alternative employment. The run-down of the industry is deeply resented by a workforce privileged under communism. The miners' attempted march on Bucharest in January 1999 demonstrated the scale of the opposition. However, the five-week general strike left the mines in an extremely bad condition and a viable industry is still a long way off. Meanwhile, diversification is under way. Before the revolution some attention was given to the food-processing and textile industries, although well-paid jobs for men reduced the demand for female employment. But now a professional training and reconversion centre is operating and a business consulting service is also being provided in Petrosani by a local foundation established in 1997 for the promotion of SMEs and supported by the UN Development Programme and the local business community. Recent research has looked into the coping strategies for redundant workers and progress made in retraining and small-business creation. This has highlighted the need for confidence building in the creation of partnerships and the need for small incubators and subsidised credits.

Investment

LFA status has attracted considerable investment. Very prominent is the Lupeni cigarette factory of 1998: a DM7.0 m. project by Bulgar Tabac/Romned International, using Romanian capital and Dutch/Italian technology. Here, 200 locals (plus 70 key employees from the Targu Jiu cigarette factory) are employed in a modernised two-storey building packing cigarettes imported from Bulgaria, though the firm intends to invest in a tobacco plantation in Gorj which lies immediately to the south. Meanwhile, the storage-battery enterprise Acumulatorul is linked with the mining industry, while the industrial joinery complex ('Fabrica de Tâmplarie') uses PVC imported from Italy. Tourism has a good future in view of the Retezat National Park nearby and the planned road link with Băile Herculane which will in itself create many male jobs. Investment is also going into the local water and power systems.

The Brad zone has attracted a number of small manufacturing ventures. A lingerie factory operates with Italian equipment in a former vegetable store in Brad to supply the Italian market, while meat processing takes place in a new building in the town. A low-technology woodprocessing unit has been installed in an empty building at Lunciou de Jos, producing furniture from laminated panels imported from Italy, while another empty building in Băiţa produces cosmetics again using imported materials and low technology. Meanwhile, in Caras-Severin there has been little progress at Rusca Montană although there is considerable potential for working marble at Ruschiţa quarry dating back to 1886: the stone is attractive and similar to that of Carrara in Italy. In Bocşa the situation is also very difficult because in addition to the closure of the local iron ore mines, the town has witnessed the near total demise of the engineering enterprise, the country's largest producer of metal structures: bridges on transport routes and swing bridges in ports, as well as other welded structures for the mining industry and for thermal and nuclear power stations, and farm machinery. This is despite quality certification in welding and collaboration with a range of foreign companies such as Ansaldo, Chyoda-Mara, Krupp, MAN Babcock and Mitsubishi. The situation has not been helped by a large debt of $3.35 m. arising from goods supplied to an iron ore plant in Krivoi Rog (Ukraine) in 1989–90.

Surplus space is now being used for new investments including a French firm making car parts, an Italian clothing firm, the printing works by Mazzolin of Timişoara and the workshop of individual entrepreneur Florentin Carpanu making aluminium and wood panels. The German Kurt Bluml wood-processing firm is linked with an existing enterprise established in 1994, while another unit is concerned with printing ink cartridges (based on Italian technology) and a jewellery workshop has been set up in a converted apartment. Finally Avicola, the old state poultry farm, survives as a new chicken-rearing enterprise using 40 former workers, while another food enterprise is concerned with meat processing, though it has difficulty in getting raw material. Mineral water is being bottled at Calina near Dognecea. Ornamental rocks could support some local economic diversification and there are some 250 m m^3 reserves of marble at Dognecea and Ocna de Fier. There are also further opportunities in wood processing.

The Moldova Nouă area was originally delimited to include Oraviţa and all the communes between Oraviţa and the Danube, extending eastwards through the Almaj Depression and the Danube Defile to Iablaniţa and Mehadia, and the commune of Ciudanoviţa to the north of Oraviţa. However, in order to concentrate the resources in the areas of greatest need, the area was reduced to two separate districts within this zone: Anina and Oraviţa with Bozovici, Ciudanoviţa, Mehadia and Prigor; and Moldova Nouă, with Berzasca, Cărbunari, Coronini, Sasca Montană and Sicheviţa communes. Oraviţa, which offers additional fiscal concessions and free professional help, has already diversified through Normarom, a Franco-Romanian enterprise producing garden furniture. A German enterprise producing furniture based on the local beechwood is a possibility. In Anina, which retains a furniture factory and a sawmill in addition to its long-established screw factory, an Italian capitalist is interested in opening a sawmill, producing for export, an Austrian firm is thinking about a furniture factory, and a Maltese entrepreneur is considering a clothing enterprise.

There is less interest in Moldova Nouă despite a commitment to the area by the EU and Nordrhein-Westfalen. The aim is to develop the port on land previously part of the local copper mine and then establish a 'free port' regime. The infrastructure is a problem since the road from Orşova through the Danube Defile is still unsurfaced, though it has the potential to provide a new route from Bucharest to Belgrade, with a frontier post at Socol. Moldova also lacks a rail link although the reopening of the route from Iam (near Oraviţa) to Baziaş and its extension along the Danube could solve this problem and, at the same time (given modernisation of the line to Berzovia in the north), make it easier for the Reşiţa engineering and metallurgical works to despatch rails and heavy equipment. The rural areas have also gained little, although Ciudanoviţa has been badly affected by the closure of a uranium mine and is dependent on EU assistance for the national uranium company with regard to a unique conservation project involving some 50 ha around the mine which are being prepared for agricultural use. One significant local advantage here is the spare apartment accommodation made available after many of the workers in the former uranium mine left the village.

Growth potential

Backed by strong political pressure the LFAs have made a modest impact. Regional Development Agency staff have built up good relations with investors but there have been problems over training, involving the country labour-training agency, and the local authorities are not always closely involved with projects and may even be ambivalent over making a case for their areas to benefit from the legislation in the first place. Publicity material for Caras-Severin's LFAs has been produced through Soros Foundation support for a Romanian organisation promoting democracy ('Fundaţia pentru o Societate Deschisă') which cooperates with the County Hall ('Prefectura') and the CCI in Reşiţa. Opportunities are seen in industry, particularly in food processing, timber and furniture and textiles, while agriculture features with regard to livestock rearing, milk production with cereals, fruit growing and viticulture in appropriate areas. The organisation responsible for promoting the mineral exploitation ('Inspectoratul Zonal pentru Resurse Minerale Caransebeş') is trying to develop interest in this domain. There are reserves of brown coal at Bozovici and Mehadia, in addition to the hard coal at Anina, Baia Nouă and Cozla where some production continues. Reference should also be made to the gold-silver ores at Oraviţa, the alluvial gold in the Bistra, Nera and Timiş valleys, ornamental granite at Topleţ, refractory sand at Anina and mineral waters in various locations. These resources are assessed by Formin, the geological research company based in Caransebeş. The company concerned with mineral water at Dognecea is considering a further venture at Prednicova in the Miniş Valley and it may also be feasible to rework slag heaps in several industrial locations. Such efforts form part of a scenario for maximising investment in LFAs in general which should hinge on local institutional capacity while requiring a 'strategic approach to inward investment as well as integration with a broader longer-term regional development strategy that focuses on existing industrial strengths' (Amin and Tomaney 1995: 218).

TOURISM

Agrotourism can be based on the scenery of the mountains and the Danube Defile, with particular interest attached to the national parks and protected areas including many natural monuments: the Anina Mountains (reaching 1,160 m at Vf.Leordiş) display karstic forms and an interesting vegetation which includes Mediterranean species. There is also scope for hunting and cultural/religious interests (through churches, monasteries and festivals) especially in the context of ethnic diversity. The industrial history is greatly undervalued. Oraviţa railway station of 1849 is the oldest in Romania, as is the Anina railway dating back to 1853: originally extending to the Danube but now only to the Yugoslav frontier at Iam. The Anina railway could be developed as a major tourist asset with interpretation to cover the phases of construction, the use of horse and locomotive traction, the historic station at Anina (1864) and locomotive depot at Oraviţa (1898) and some vintage equipment retained at wayside stations. Industrial archaeology also offers substantial opportunity through the remains of the furnace for the non-ferrous metals near Oraviţa dating to 1857 and other mining legacies in Anina and Dognecea. There is a tourist project in the Dognecea–Ocna de Fier area (inspired by experience at the Wieliecka salt mine near Kraków) to make use of six lakes (including two at Dognecea refurbished in the 1980s), created in connection with the washing and sorting of ore, and an underground transport passage between Dognecea and the former narrow-gauge railhead at Ocna de Fier. In Dognecea the local furnace site is available for reclamation because, although most of the old buildings are covered by slag, it would be possible to present the ruins of one of the furnaces, closed in the late nineteenth century in order to concentrate production in Reşiţa. The private Gruescu museum ('Muzeul de Mineralogie Estetică Constantin Gruescu') lies adjacent to Paulus Mine and comprises a remarkable collection of beautiful rocks collected from the local mines.

Oraviţa also offers the oldest theatre in the SEECs: designed by Viennese architect Johann Neumann and built between 1789 and 1817 in a town that was once the centre of Caras County. There are eighteenth-century churches: Catholic 1713, and Orthodox 1755 (the latter painted in 1867). The Calugara Monastery was built in 1860–1 and the Town Hall ('Primărie') dates to 1880 (though modified in 1911). Meanwhile, Anina has Roman Catholic churches dating to 1772 (Steierdorf) and 1828 (Anina), while the historic 'Farmacia Steierdorf' now houses a history museum. Finally, Bocşa, which lies at a lower altitude of 170 m in a depression in the Dognecea Mountains, offers the ruined Cetatea Cuieşti: a fourteenth-century fortress destroyed by Turks in 1658 – hence the name Buza Turcului. Bocşa also has one of the oldest Roman Catholic churches in Romania, while Orthodox churches in the area were built in 1755, 1808 and 1815. The most notable is that of Sf. Nicolae, built 1795, painted in 1810 and restored in 1938. Vasiova monastery opened in 1905 and textile workshops were subsequently built during the period 1933–8. When these were taken over by the state in 1959 the monastery closed. It reopened in 1990 and its important collection of religious books can now be appreciated. Indeed, Bocşa also has a particular diversity in terms of religion with 14 different cults represented. There is also a rich industrial history based on iron smelting which began in 1719 and the hydropower station (still operating) which took over the water works on the Barzava when the last blast furnace was closed.

Border regions

The West Region lies adjacent to areas of Hungary and Yugoslavia which used to form part of a unified Banat province under pre-First World War Habsburg administration. Now, in the context of an enlarged interdependent Europe, there are benefits in developing cohesion. Links between local government in Csongrád

(Hungary) and Timiş County in 1992 grew by stages into the first version of the protocol for the Euroregion Danube–Criş–Mureş–Tisza (DCMT) in 1994 and its final realisation in 1997 with an area of 77,600 km^2 and a population of 5.2 m. people (Figure 6.11). It is intended that there should be a coordinated strategy of regional economic development, with a regional commercial centre in Timişoara and a greatly improved supply of information coming from a single business centre ('Directorul de Afaceri'), although there will also be publications ('Zilele') representing the nine CCIs. There will have to be training for young businesspeople and efforts to achieve uniform technological standards. There are already forwarding companies in southeastern Hungary, exporting into neighbouring countries as a result of exhibitions organised by the CCIs which points to the benefits of the Euroregion in terms of cross-border institutions. Others are needed to organise cultural programmes, including twinning arrangements, the production of tourist maps and media coverage for the Euroregion as a whole.

Border areas in general have been blighted by depopulation and the loss of young people, but Hungarian research suggests that new border crossing points generate a stimulus for growth as commercial activity encourages provision of accommodation and better cultural, medical and leisure facilities as well as a general cleaning up in 'no-man's land'. There is a desire to improve connections between the largest cities in the three national components – Novi Sad (Yugoslavia), Szeged (Hungary) and Timişoara (Romania) – which are also the second cities in their respective countries. Szeged's rail connections with the other two cities could be greatly improved by restoring the bridge over the Tisza and reopening communication with Kikinda and Subotica along railways dating back to 1857 but closed in 1940. The Kikinda–Jimbolia section, disused over recent years due to breakdown of services between Timişoara and Zagreb during the Yugoslav war, would also have to be refurbished. Other ways of linking Szeged and Timişoara would require restoration of the bridge at Cenad on the Mureş or a short overland section at Valcani. However, whichever route is preferred it is interesting to note that mending the breach at Szeged and restoring the link eastwards from Subotica towards Baja and Kaposvár in Hungary would provide a new east–west rail route between the Black Sea at Constanţa and the Adriatic at Rijeka. Of more local importance is the possibility of restoring the railway between Oraviţa and Baziaş – closed in 1950 – with a possible extension to Moldova Nouă (already referred to). A rail link between Timişoara and Zrenjanin is also mooted to aid the trickling down of growth to the smaller towns.

Meanwhile, the European road corridors will bring big improvements to the Euroregion. Some link roads have already been improved in Hungary and similar works in Romania could greatly improve access in the south where the main highways are presently unsurfaced. Cross-border roads are increasing all the time and many more are proposed. Of great symbolic importance is the possibility of a new facility at Beba Veché–Kűbekháza–Rabé where the three frontiers merge. Meanwhile the Danube has enormous potential despite the present closure due to NATO bombing of the bridges at Novi Sad during the Kosovo

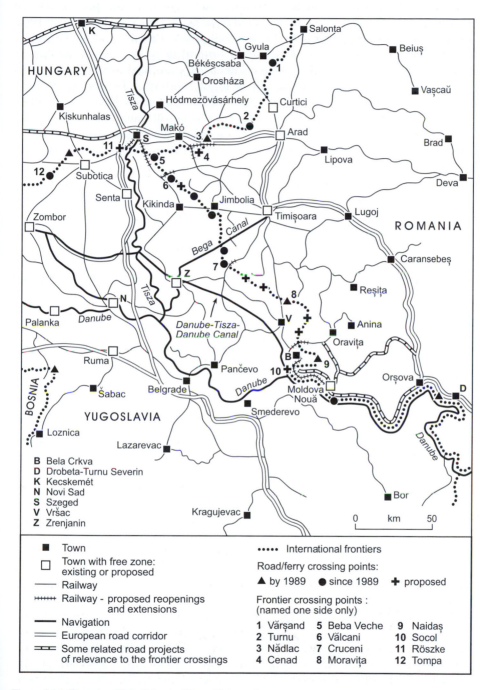

Figure 6.11 Danube–Criş–Mureş–Tisza Euroregion

Source: West University of Timişoara

War in 1999. The Tisza is also navigable with a modernised port at Szeged (also a logistical centre with railway 'piggy back' services to Wels in Austria) and there are plans to reopen the Bega Canal (disused since 1958) to Timişoara and establish a customs-free zone: one of several proposed across the region. The problem lies with the lock at Sânmihaiu Român and the degradation of Pragul Costei, as well as silting and the state of the banks. Finally, the network of regional airports is one of the strengths of the region. In Romania, in addition to Timişoara's established role, Arad is seeking international status airports and there are domestic facilities at Caransebes (Caras-Severin) and Deva (Hunedoara) which should boost the prospects of the metallurgical centres of Hunedoara and Reşiţa.

7   Conclusion

An overview of the region

The communist era has left an indelible impression on the region. In terms of human effort the achievements were in many respects heroic, and contemporary Western literature reflects an element of the triumphalism that marked some of the early development projects. Social scientists appreciated that the West was not typical of the world as a whole and that the Soviet Union was inevitably generating a distinctive economic and social system. But with decision-making impossible to scrutinise and travel complicated by bureaucratic restrictions it was difficult to separate reality from communist party propaganda claims which were the nearest approximation to 'official publications'. It was only in the 1980s in the context of the Western offensive over human rights, fortified by the structuralist movement seeking greater political awareness in social science, that the 'unbalanced' nature of the system became widely understood. Even so, there continued to be a mismatch between some Western scholarship engaging with elegant models of socialist construction and the thrust of underground literature from the region pointing to human rights abuses, environmental crises and, above all, a priority for personal leadership goals in policy implementation. Material from the region may have been high on reports of insensitive bureaucracy (not necessarily restricted to communist systems!) but Western assessments remained all too circumspect regarding the barriers to economic efficiency.

Such was the obsession with validation of the 'totalitarian model' that it seemed a veil could be drawn across the Stalin period (with its acknowledged excesses), leaving authoritarian rule through monopoly parties and central planning to be legitimated as a corrective to a wayward capitalism and a precondition for prosperity over the longer term. Yet nobody in the West could see 'the accounts' (indeed we now know that not even Gorbachev himself was fully aware of the parlous financial situation in the Soviet Union) and appreciate how much economic activity was unprofitable, as politically-motivated management strategies continued to operate in a semi-feudal environment where progress depended primarily on personal connections within the 'nomenklatura' hierarchy. The Soviet socialist 'war economy', with its political imperatives, became economically counterproductive, but although ideology was widely trivialised by young technocrats in the party, vested interests remained a stumbling block for change. Thus when the Polish communist party reformists sat down with radical intellectuals early in 1989, both sides could not believe that (despite revoking the 'Brezhnev Doctrine') the Soviets really would allow the Communist Party to relinquish ultimate control. It was Gorbachev's crucial concession on this point that opened the way towards the present post-socialist era.

After the fall of communism and the removal of political constraints on reform, there was a clear consensus for change. But adjustment was gravely complicated by the trade 'shocks' associated with the reunification of Germany (which made East Germany a hard-currency trading area), the collapse of the Soviet Union, and the wars and sanctions affecting former Yugoslavia. Moreover, assuming that the broad objective might be a Western-style mixed economy appropriate for membership of the EU, there was initially no blueprint for getting there. It was a great surprise to see the region (and soon after the former Soviet states) embroiled in Europe's greatest non-wartime depression leading to the loss of a third of GNP in seven years. The shock was less severe in ECE than elsewhere (with 1997 GDP averaging 85 per cent of 1989 in 10 states) with the northern countries (including Slovenia) almost back to parity while five SEECs (excluding B&H and Yugoslavia) were down to 70 per cent (Adamowicz 2000). Nevertheless, the setbacks were traumatic. Reform has produced goods and services and has provided opportunity without oppression, but there has also been unemployment and insecurity with crime and corruption (notably through mafia activities). Reform has also been uneven across the region and despite progress with privatisation SOEs continue to absorb disproportionate amounts of state funding in some countries. Enormous variations in family fortunes breed some nostalgia for the old system in which virtually everyone had a place in society, a role in production driven by the arms race and an income that could easily cope with transport costs, rents and taxes.

Compromises are now being found to balance the interventionism necessary for regulation and welfare with the 'de-invention of the state' that has been deemed necessary to encourage the reconstruction of civil society and to participate in the global competition for investment. As Berend (1996: 380) observed, 'by the mid-1990s the bulk of the countries of the region will certainly end their economic decay and a new sustained growth will improve the crisis ridden economies'. But will there be a lasting prosperity based on radical technological change and competitive exports, or a struggle for survival in the EU's backyard with a widening income gap between east and west? And hence the historical question of whether the region 'will be an equal part of the new Europe or only a peripheral part of it, as it has been in its entire early and modern history' (ibid.). It will certainly be many years before the transition states climb the 'wage precipice' on the frontiers of Austria, Germany and Italy. Hence, even if cultural differences can be overcome, there is scepticism in ECE as to the cohesion of an enlarged Europe, despite the lack of obvious alternatives, for already the SEECs are finding themselves as peripheral to CEFTA as they were before within Comecon. The region might be placed 30–60 years (or one Kondratieff cycle) behind a Western Europe that is currently in the fifth cycle or post-industrial age. But enough progress has been made since 1989 to suggest that ECE in the 1990s may not be altogether dissimilar to the Mediterranean states that were seeking EU membership two decades ago.

As for the politics, the reform agenda in the context of 'New Right' influence in the West encouraged neo-liberal economic thinking which accelerated the decline in the state's coordinating role, although there was always going to be a conflict in the region between the political imperative of dismantling communist dictatorship to prevent any kind of restoration scenario and the economic need for continuing state intervention. On this basis the political parties tended to be too confrontational, understandably lacking a willingness to compromise, which bred popular scepticism of the whole political system. As new governments struggled for legitimacy and communist hierarchies re-emerged under the cachet of social democracy, Lithuania, Poland and Hungary lurched back to the left in the early 1990s (hence the 'LPH syndrome') with more interventionist policies and ambivalence towards foreign penetration. Meanwhile, Romania had yet to shake off a neo-communist image. Thus the politicians who brokered the revolutions were themselves displaced, given the dynamism and voluntarism of the transition. While Solidarity in Poland was an important spiritual and emotional phenomenon in the 1980s, it has failed to transform itself into a lasting institution.

Yet effective choices are being offered and election turnout, though variable, has often been encouraging: viable Centre-Left and Centre-Right coalitions have been formed in virtually all states. Indeed, elections inevitably introduce new governing parties but with a broad pro-European consensus, since a nagging sense of peripherality has not produced convincing alternatives. Natural resources – which encourage delayed reform in some CIS states – are rarely abundant enough to keep trade in balance, let alone generate the wealth to restructure and modernise the SOEs. Yet progress in reaching a social consensus on what constitutes a modern European state is slow. With high poverty levels, a policy of low corporate taxation is controversial given the pressure of debt repayments, the burden of education, health and welfare spending and the costs of supporting large armies as well as ailing banks and non-viable heavy industries. Despite the heavy legacy of the past, most people are looking forward. Those who were persecuted under the old regime will have few reservations about democracy, while politically-marginalised elements of the former elite used their social capital invested in privileged networks to gain a disproportionate stake in the new economic order.

A management style has emerged whereby market institutions are being built and much of the state sector is privatised, while FDI constitutes a major source of wealth to set alongside modest taxation revenues and the flow of funds from international financial institutions which help to modernise the infrastructure and stimulate enterprise through SMEs. After some highly idiosyncratic policies such as precipitated the collapse in Bulgaria in 1997, not to mention the war years in former Yugoslavia, there is a readiness to accept the scrutiny of the EU and IMF in return for a measure of stability. But the pressure on the population has been severe and distortions have arisen. External influences are assuming that the region is advancing towards Western democratic and free market structures, yet there are extensive areas of extreme weakness through poverty and a lack of institutions. Political parties and NGOs have succeeded in bringing many traditional tensions into the arena of political debate and the change in the way that ethnic conflict is now handled indicates something of a seismic shift (Stroschem 2001). Given the absence of effective social integration policies to overcome or narrow the gap between rich and poor, it is still too early to assume that the 'shatter zone' of ECE will now be fully assimilated into a wider Europe.

Throughout the region there remain urgent tasks because spending on welfare and education has been limited by financial rectitude. The region is no longer maintaining its population, even with migration discounted. J. Witkowski (1993) refers to the poorer demographic performance of Poland's 'endangered towns' which include many of the older regional centres, but the urban sector as a whole is now suffering natural decrease which reflects pollution, but also inadequate access to health care and a falling birthrate. Rural fertility is higher than in the towns but is declining towards the national average, while rural death rates also tend to be higher now – the reverse of the past – due to the ageing of the population and the ruralisation of degenerative diseases (cancers and circulatory disease). Given the bad situation in the towns, natural increase is higher in rural areas in Poland but lower in Romania. Either way a great improvement in medical services is needed and the problems of both town and country need to be addressed in order to maintain the present level of population (although the consequences of EU membership in terms of migration to Western Europe – after a transition period of possibly seven years – are potentially destabilising). It is also necessary to boost birthrate by creating attractive conditions so that women can have children as well as a career. Meanwhile, despite good education standards, only half the population has attained a cultural level adequate for an active role in the transformation process. Rapid growth of the education system is necessary following a severe decline in spending on science and R&D potentials. The decline is attributed partly to actions by foreign capital in closing laboratories and R&D centres not up to world standards despite significant potential. At the same time national science and technology policies have been inadequate.

North–south contrasts

These differences are perpetuated through the relative ease with which the northern states have moved towards the EU and now enjoy good prospects of early entry, whereas in the SEECs only Bulgaria and Romania are candidates – and even these two states face a relatively lengthy waiting period. The creation of a 'Visegrád Group' (the Czech Republic, Hungary, Poland and Slovakia) in the 'fast lane' for EU membership has encouraged the other states to project their reformist credentials and seek membership of what is now a much wider free trade association (CEFTA), extending southwards to Slovenia in 1995, Romania in 1997 and Bulgaria in 1999. The northern states have also done relatively well and have attracted a disproportionate share of the FDI sunk in the region. The jury is still deliberating on an overall verdict, but it is clear that autonomous regional development may be constrained when FDI projects establish strong external links without becoming significantly 'embedded' in the local area. The case can be exaggerated because while some ventures geared to export may impose factory regimes which minimise the scope for the workforce to contribute to policy making and survive only until wage differentials are eroded, others have involved large investments which demonstrate commitment.

Even if foreign-owned enterprises remain largely disembedded 'palaces in the desert', there is a distinct possibility – as in the case of Daewoo at Lublin – that new work practices will be established to improve competition in areas that previously may have had only limited exposure to Western business. It is also claimed that American capital-intensive developments may tend to support strong regions because of the availability of quality producer services as well as the skilled labour and infrastructure. Further benefit could accrue by recruiting ambitious local people who have bypassed the local employment market in favour of commuting to larger cities. Meanwhile, there seems little doubt that companies with foreign capital participation have a growing influence in foreign trade balances, accounting for 24 per cent of Polish exports in 1994 but 47 per cent in 1998. Exports of manufactured goods are now prominent whereas raw materials predominated in the communist period. Furthermore, the companies voted 'national champions' are frequently locally-managed companies that have done well with foreign help – like the VSŽ steelworks in Slovakia (sold to US Steel), the Hungarian mobile phone company 'Westel' and the Bulgarian distributor of lubricants 'Prista Oil'.

By contrast with the north, SEE is an area of great poverty and also a major zone of instability, usually governed by fragile coalitions that cannot support reform wholeheartedly and score indifferently for their transparency and accountability. The Yugoslav wars opened the way for illegal trade and trafficking and a criminalisation of politics. The SEECs have done relatively badly over investment, reflecting acute sensitivity by the business community over political and economic risk including the presence of 'investor unfriendly' elements which may influence a bureaucracy to protect domestic interest groups. Indeed, as already noted Bulgaria and Romania are the only EU candidates in a region which has served as part of Western Europe's periphery in modern times. Instability continues to emanate from the material gulf that separates the SEECs from the West – given the failure of communist industrialisation to bridge a historic divide. With 'the summit of economic difficulties shifting eastwards and southwards' (Nefedova and Trejvis 1994: 38) very substantial Western economic intervention is needed to prevent a possible 'Peronist' internalisation of economic and demographic problems.

Economically, there is a choice between open borders, with a growth of trade and progressive enlargement of the EU, against a high level of protectionism and limited effectiveness for international organisations. This would have implications for economic planning. On the one hand there could be foreign investment on the main transit routes and the approaches to major ports and the emergence of some airports with hub status or a more limited autarkic approach – and at the same time, a growth of foreign capital in the development of each country, with joint ventures, foreign-owned banks and regional offices for multinational companies. Or, on the other hand, development could be based on domestic capital and related very much to home market demand, without the economies of scale that would arise from regional specialisation. Integration might bring low unemployment with rising welfare provision and limited international migration, against a situation where economic stagnation might produce high unemployment with low welfare provision and higher levels of out-migration, although the short-term costs of restructuring could be high, as is apparent in parts of the region today. Politically there is a choice between national chauvinism with the orchestration of territorial disputes and intransigence towards ethnic minorities, and a pragmatic approach in which the drawbacks arising from the existing boundaries are minimised by cross-border cooperation and acceptance of internationally-sponsored codes of practice in treating minority groups. All this has implications for the level of defence spending and support for unprofitable strategic industries (Carter *et al.* 1996).

The highly uneven distribution of investment, both between countries and regions, is unfortunate because stabilisation and IMF support can be compromised if FDI does not follow hard on the heels of reform. Experts suggest that success depends largely on a strong national economy combined with social cohesion which would seem to offer hope for regions which have achieved good inter-ethnic relations like Romania and Slovenia. The former has acquired foreign-owned companies that are taking useful initiatives in the social sphere as evidenced by entrepreneurial training offered to redundant employees by Lafarge-Romcim at Medgidia. Here, a business incubator was placed under the management of 'The Foundation for the Assistance of the Underprivileged Social Categories' for a five-year period. Progress is less likely in regions experiencing stress, for restructuring is tentative where new job creation fails to make an impact on rising unemployment, because unions are loath to accept investment that generates redundancies and will prefer low technology and low productivity to retain jobs. Even if it were to be realised, EU membership may not boost investment, for while risk will be reduced, it may be hypothesised that the dividend will be modest (especially where the 'risk bonus' quickly wears off) because it will be easier to supply home markets from outside and more investment might therefore relate to export potential (especially newly-privatised companies). The scope for trade with the Mediterranean and Middle East as well as the CIS is already reflected in the growing scale of Greek and Turkish business. SEE is also a small part of wider 'east' that is given coherence by the Black Sea – an area of relatively weak multi-ethnic states on the periphery of Europe – and the approaches to the Caucasus and Central Asia.

The southeast 'combines the advantage of being geographically close and historically linked to the rest of the European continent and the disadvantage of being remote enough, at the periphery, not to affect direct interests' (Clement 2000: 75). However, there are signs of change sufficient to 'inspire attempts to reconceptualise the ways in which the different parts of that European mosaic fit together' (Sadler 1999: 275), as both the EU and USA now agree that more must be done to stabilise the Balkans. A start was made with the 'Process of Stability and Good Neighbourliness in South East Europe' linked with a wider 'Pact on Stability in Europe' signed in 1995 under the OSCE umbrella. And a Southeast European Cooperative Initiative was launched in 1996 to provide 'economic and financial assistance through private-funded projects in order to support the modernisation of the local private sector and reintegrate the region into European and broader Western structures', particularly with regard to border crossings, energy and environmental protection (Clement 2000: 82). In the aftermath of the Kosovo War, support has been provided through the NATO-backed Stability Pact for South Eastern Europe. Meanwhile, some bottom-up initiatives are now emerging. Since 1996 there have been ministerial 'Conferences on Stability, Security and Cooperation': in 1998 Albania, Bulgaria, Macedonia, Romania and Turkey – joined by Greece later – agreed on a Multinational Balkan Rapid Reaction Force, and cooperation with Caucasus and Central Asia is developing through TRACECA as already noted. Romania is committed to the process but clearly wants regional cooperation within the mainstream of Euro-Atlantic evolution. But all individual countries tend to retain greater faith in wider European institutions than in organisations confined to the region (e.g. Bosnia & Hercegovina with regard to the Dayton peace agreement).

Integration is slowly moving forward. With road reconstruction seen as a 'must' if investment is to be attracted, the Stability Pact is also playing a key role with regard to the modernisation of transport. Thus the Calafat–Vidin bridge (already mentioned) is part of a coordinated scheme of projects – supported by the EIB – which also include the Bucharest–Constanţa motorway, the Bucharest–Constanţa and Bucharest–Giurgiu railway upgrades, the Constanţa LPG facility, and work concerned with Danube ecology and navigation. Telecoms are improving and there is better banking and bank regulation. More inclusive social policies are beginning to benefit the Roma as the politics of equal respect in multicultural societies move closer to centre stage. The politics of moderation in Serbia which emerged after 2000 are a positive factor, but important territorial questions await resolution and the present uncertainty was highlighted when the stronger position of the Albanians in Kosovo threatened to destabilise Macedonia during 2001. There is a chance of a breakup of the 'new' Yugoslavia into Kosovo, Montenegro and Serbia, reflecting a majority view in Montenegro and even more unequivocally in Kosovo, where a peaceful resolution of the problem is now sought. Unable to defend itself and lacking economic viability, an independent Kosovo could hardly be a factor for stability (and partition along the Ibar line, enabling the northern part to merge with Serbia, could trigger further radicalisation). Hence a regional approach has been suggested to square the circle: an independent Kosovo (likewise Montenegro) tied into 'a web of economic, political and security relationships' with other former Yugoslav territories including Bosnia & Hercegovina, Macedonia and Serbia (Sell 2001: 12).

Regional policy

This theme is of particular geographical interest and is highlighted as one of the outstanding deficiencies of the transition to date. Growth has been concentrated in a few regions and in the larger cities with insufficient attention to achieving greater mobility of people and capital between regions and settlements. Despite the efforts of some inter-ministerial councils, regional policy was constrained after 1989 by the prevailing neo-liberal outlook and the priority for sectoral policies. There was support for a number of uncoordinated programmes (like Poland's special economic zones, offering land and fiscal concessions) as well as concern for greater cohesion in border regions. And local authorities were free to promote themselves, albeit with limited resources. So typically there was decentralisation, yet also fragmentation and weak institutional capacity with local budgets dependent on central decisions. There has also been limited ability to absorb funds and PHARE projects have sometimes overloaded administrative capacities (Bachtler *et al.* 2000: 372). As a result, strong contrasts emerged between the western growth zone – 'the Central European boomerang' – and the eastern periphery that no longer gains any stimulus from neighbouring countries of the FSU. The gap between rich and poor widened in the absence of an adequate social integration policy.

However, during this period CBC programmes provided an element of regional planning that was generally absent elsewhere. This is particularly the case with programmes that included regional planning as a specific objective. On the German–Polish border a Spatial Planning Commission was established in 1992 and three years later it produced a policy for improved infrastructure and environmental rehabilitation to promote decentralised urbanisation. Planning among the Baltic countries produced a concept of spatial development and CBC through an Oder region, including the Berlin and Dresden areas of Germany and extending to Szczecin, Zielona Góra and Legnica in Poland (Van de Boel 1994). This drew particular attention to the Pomeranian coastlands (especially Szczecin/Świnoujście) attractive to German and Scandinavian investors, especially in the light of a possible Oder waterway linked with the European system. Through the Baltic arena Poland also became involved in cooperation with neighbouring countries as part of the 'Green Lungs of Europe' project – growing out of the initial 'Green Lungs of Poland' project – to coordinate activities in networks of national parks and other protected areas.

Further south, the logic of an integrated Bratislava–Vienna city region is underpinned by the Danube as well as by cooperation over airports and the restoration of railway links. Further scope exists for integration to extend to the Brno area in the Czech Republic and Győr in Hungary to form a wider working community. Both these cities are already benefiting from Austrian investment and have plans for motorway links with Vienna. A special economic zone is emerging in Bratislava's Petrzalka suburb to attract more Austrian investment. The significance of this complex could increase with possible canalisation along the Morava and Váh valleys to link the Danube with the Elbe and the

Oder–Vistula. Reference should also be made to an Austro-Hungarian Cross-Border Regional Council founded in 1992 which has addressed Austria's concerns over prostitution, crime and drugs arising from more open borders. Also public utility maps have been harmonised to facilitate joint planning of infrastructure including sites for solid-waste disposal. In the run-up to EU enlargement CBC issues are appearing in all levels of strategic planning and there is a good basis here for a joint vision of an integrated regional economy, social cohesion and good neighbourly relations.

From 2000–2, however, regional plans from the candidate countries are required by Brussels as a basis for PHARE funding and payments to national and regional development councils, which are now functioning. With stronger regional development programmes there will be greater pressures for more even development through the settlement structure and equalisation of social and economical differences between regions. Regional self-government is gaining momentum and regional plans are to some extent negotiable. Poland's national strategy 2001–6 is based on a government support programme to which the regions respond with their individual strategies in order to produce the final 'contract'. However, there is still uncertainty over the competencies of state and local administration: in Hungary, 'the county level [below the provincial NUTS II level on which regional planning is based] is still fragmented and state-dominated and cannot fulfil its role as an "intermediary" between the central government and the lowest tier administrative units' (Szigetvari 2001: 299). However, even with a 're-strengthening' of the counties it is difficult to see how the balance between Budapest and the 'antipodean' provincial cities can be significantly changed.

Thus in the future, regional variations could increase and the most under-developed regions could find themselves in a situation of irreversible depression, despite more EU finance to improve cohesion. Although there is a system in place, it will take time to build capacity and develop a sense of partnership between central government and the regions. According to Bachtler *et al.* (2000: xv), 'it is still an open question, however, whether regional policies are regarded merely as a presentational device to fulfil EU requirements or whether they will be able to make a difference to those regions and groups disadvantaged by economic liberalisation' in spite of the promotional efforts of local government in general and entrepreneurial cities in particular. There is no clear answer, but it would appear that regional policy is making a difference, given the special incentives available for development in LFAs which are encouraging investors to take full advantage of relatively low labour costs and high unemployment. Moreover, EU funding is likely to be very substantial after accession because the vast majority of ECE regions are well below the EU Objective 1 threshold of 75 per cent of the EU15 average. Even the metropolitan areas struggle to approach EU average incomes even on a 'purchasing power parity' basis.

Meanwhile, CoE has also become involved through a 1997 'European Conference of Ministers Responsible for Spatial Policy' to establish guidelines for future spatial development. Seven large overlapping regions (and four smaller ones) have been identified including 'Central Adriatic Danubian and South

Eastern European Space' (CADSES). These broader dimensions can be extended in the context of EU financial support for border regions now extend to the frontiers of candidate countries, in line with a 'European Spatial Development Perspective' (ESDP) published in 1997 as 'Europe 2000+', seeking economic and social cohesion as well as parity of access to infrastructure and knowledge. In addition to E180 m. of PHARE funding for 'INTERREG III' (2000–6), there is support for the borders of candidate countries and FSU states (Belarus, Moldova, Russia and Ukraine) through the Tacis CBC programme which started in 1996. The region can now gain further from the experience of cross-border cooperation (CBC) in Western Europe dating back to the 1950s coordinated through the Association of European Border Regions (1971) and its observatory on networking and good practice – Linkage Assistance and Cooperation for the European Border Regions (LACE) – which produces assessment reports, strategies and operational programmes for entire frontiers between pairs of countries, including maritime frontiers in the case of the Baltic. Programmes pay attention to job creation, since border regions tend to have above-average unemployment: exemplified by Jesenice and Maribor in Slovenia's borderland with Austria. Furthermore, Euroregions are now very numerous in the northern half of the region and the cumulative impact in terms of cohesion is considerable.

The rural

Policy also needs to address rural problems, for despite a difficult housing situation migration to the main towns continues while the settlement crisis in the smaller villages deepens. The present village revitalisation programme will have to achieve a major improvement in technical infrastructure if rural industrial estates are to become a possibility, for example, if 'important information related mainly to social, economic, cultural development, projects, funds or foundations of the third sector is not disseminated here at all or is considerably delayed' (Pasiak *et al.* 2001: 335). Both schools and potential exporters need modern telecoms for better Internet access, while the younger generations should build up their network of social capital (weakened by emigration of cultural elites to larger towns) and create beneficial partnerships which transcend the 'clientelism and protectionism' that remain endemic. Closer relations between groups of municipalities is suggested by the greater dynamism of the larger administrative units where a 'concentration of teachers creates a cultural centre around which interest groups are concentrated' (ibid.: 336). The political process must take note of these concerns for although global financial institutions are very insistent in their demands for 'monetarist' policies and private capital seeks low taxation and repatriation of profits as well as cheap labour, governments need to control the rate of 'de-invention' in the face of public opinion that seeks more interventionist policies.

Country people have experience of non-agricultural employment and are well educated, yet there is now relatively little non-agricultural employment available which means that there are many small pluriactive (often welfare-dependent) subsistence farmers alongside a minority of commercially-minded family farmers, cooperatives and estates. Few non-agricultural businesses have been started by local people and progress relates more to knowledge, skills and contacts gained in the socialist sector under communism than to recent support policies for SMEs (Swain 2000). And where there is a migration balance in favour of rural areas it usually indicates suburbanisation on the edge of the capitals and large provincial cities. Many more jobs outside farming are needed to allow for a measure of consolidation given the removal of remaining controls on land sales (like Hungary's 1994 Land Act which prohibits land purchase by corporate farms) and the availability of land mortgage schemes. Yet there is also a school of thought that sees pluriactivity as 'a consciously constructed nexus that allows on the one hand the continuation of farming; on the other hand it makes for the reproduction of other economic activities that would be impossible if they had to be grounded on stable and full-time employment relations' (Van Der Ploeg and De Rooij 2000: 48). While the region faces stiff competition from EU and US surpluses, is it desirable to invest in commercial farming and create 'new hinterlands for West European agribusiness' (ibid.: 52) or strengthen the peasantry through low-cost farms (of the kind being researched in parts of Western Europe) that would harmonise with the sustainability agenda discussed below? It is no surprise that the poorer rural regions show the greatest concern over EU membership. The scale of the 'small farm' problem can hardly be reconciled with the continued adoption of the Common Agricultural Policy (CAP). Polish farmers are particularly worried about profitability in the EU if current levels of protection are eroded, as well as the possibility of losing competitiveness in their home market through a flood of imports – not to mention the large share of foreign capital participating in the privatisation process of agrifood-processing industries.

Sustainable development

Looking beyond the status of the region as a whole and its salient regional problems, it is appropriate in this conclusion to highlight the way in which traditional environmental concerns, which dominated the last communist years and the early phases of transition (Carter and Turnock 2002), are being overtaken by wider debates about sustainability to integrate policy and ensure that Europe as a whole moderates the pressure placed on global resources. Lafferty and Meadowcroft (2000: 451–2) refer to a 'broader intellectual frame' linked with an international effort to resolve dilemmas of environment and development. Environmental policy therefore merges with general societal decision processes in the quest for sustainable development, defined by the UN Conference on Environment and Development (held at Rio de Janeiro in 1993) as meeting the needs of the present without compromising future generations. This calls for new regimes of production and consumption which cannot be simply delivered by politicians, but require acceptance and commitment from society as a whole. 'Paradigmatic change' is needed to break the long-term process of 'material intensive growth' so that economic development can continue without higher rates of resource use.

This is plainly challenging in the light of the European Environmental Agency statement in its 'Dobřiš Assessment' that 'the decline of Europe's biodiversity in many regions derives mainly from highly intensive partially industrial forms of agricultural and silvicultural landuse, from an increased fragmentation of remaining natural habitats by infrastructure and urbanisation and the exposure to mass tourism as well as pollution of water and air' (Gyulai 1998: 45). The hotter, drier summers experienced by the SEECs during the last two decades have revealed the vulnerability of the lowlands to climatic extremes in the light of modern management. Forest fires plagued Croatia's Dalmatian coastal localities during the dry summers of 2000–1, while severe drought in Macedonia threatens to leave some localities without drinking water. Albania's dependence on hydropower is problematic under these conditions, with low water precipitating power cuts in summer, in addition to those in the winter, arising from frozen rivers. While historical perspectives reveal environmental problems as a necessary and inevitable part of resource management, the great pressures exerted under communism through 'tight' planning meant that capital was insufficient for a sustainable approach through the filtering of emissions and the purification of waste water, while the failure to attach value to resources needed by industry made it difficult to account for environmental damage.

With sustainability seen as environmentally-respectful economic development, the Amsterdam Treaty of 1997 requires the integration of environmental concerns into all EU policy areas. This now impinges on ECE where 'membership of the Union has become a determining factor in shaping current environmental policy' (Baker 2000: 319), ensuring that the aim of developing functioning market systems is not achieved at the expense of sustainable development. Environmental clauses were built into cohesion funding during 1994–9, but this is being done more explicitly during the present period 2000–6. Thus the national and regional development plans already referred to (initially for 2000–2) must take environmental issues into account and will form the basis for accession funding in future. ENGOs are taking a keen interest to ensure that SAPARD funding for rural development supports 'win–win' projects which are good in terms of both socio-economic development and biodiversity conservation. However, the challenge goes much deeper in view of the threat of chronic road congestion and inefficient use of energy, for the ESDP anticipated a substantial increase in traffic after enlargement that would have a negative impact on some natural areas.

So the question arises whether ECE can move quickly towards a model of dematerialised sustainable development as opposed to following the material-intensive growth trajectory of Western Europe. There are ENGOs which are highly committed to educating the public, but while there is tacit support for EU membership, there is inadequate discussion of sustainability issues and the wider political and economic framework that forms the basis for the present association arrangements and ultimately for membership: a situation which 'hinders the development of civil society and democratic participation in the process of change' (Baker 2000: 319). Capacity building is a priority which is being addressed in certain areas. For example, in the Šumava the planning system is being revamped in Česke Krumlov after an NGO-sponsored facilitator was able to maintain a volunteer Strategic Planning Team (SPT) in 1995–6. The SPT's task was to formulate a sustainable development strategy based on enhancing the town's independence and administrative capacity, with the town as a single harmoniously-integrated unit in symbiotic relationship with the region's natural, social and cultural attributes. There is now a strategic planning committee to influence the bioregion, manage amenity migration and tourism, and improve environmental awareness and hosting skills (Moss 1996).

As part of the EU's readiness to provide a stronger lead following Rio, 'Birds and Habitats Directives' strengthen conservation measures for natural habitats and their species, while 'Natura 2000' seeks a regional network of special protected areas where member states will have to prevent deterioration. So ECECs are now drawing up lists of habitats to harmonise with the EU classification system, with much support from a Regional Environment Centre – established at Szentendre in 1990 – which handles much of the money going into PHARE multi-country environmental projects. However, a sustainable future is still a long way off. As in other candidate countries, a new generation of civil servants (e.g. in Hungary) should be fully geared up to appropriate environmental standards. Yet in this country regional plans are still being prepared separately from environmental strategies and, despite the Hungarian Commission on Sustainable Development created in 1993, 'there are no national programmes incorporating principles of sustainable development into the routines and regular decisions of the government or the implementers of environmental efforts' (O'Toole and Hanf 1998: 101).

As a strong message to accession countries, it is necessary to reinforce the 'second pillar' of the CAP through 'multi-functional agriculture', which includes rewards to farmers for countryside management (and especially direct support payments in LFAs). Organic and other low-input sustainable agriculture (LISA) systems are important for sustainable agriculture and with pre-accession funding now linked increasingly with environmental issues (ISPA and SAPARD), national policy choices for agriculture and rural development are crucial at the present time. The Czech Republic, Hungary, Poland, Slovakia and Slovenia already provide support for marginal areas. Such action could be complemented by zoning systems to relate cropping to the characteristics of natural regions and secure the required production in the most sustainable fashion. Similar concerns underpin the conflict between traditional (nineteenth-century) 'rational' forest management of regular rotational felling and the modern ecological view which accepts the diverse functions of forests apart from wood production. The political 'tug of war' is well illustrated by the celebrated case of Slovak ENGO 'Vlk' being fined for not cutting wood in its reserve in the Čergov Mountains, established with private donations to preserve natural forest growth in accordance with current conservation practice. Vlk wants to set aside a fifth of all forests for non-timber-producing functions, including buffers for streams and springs, while selective logging elsewhere would leave the undergrowth undisturbed. Less timber would mean higher prices, but yield a net benefit through savings in costs arising from erosion, loss of biodiversity, disturbed water regimes and unstable landscapes. Finally, 'industrialised architecture' in large village centres, with high-density buildings containing much underused space, raises 'the urgent need of revitalisation in our whole countryside regarding the human, natural and architectural environment' (Ruda 1998: 93).

The Sudetes

Such crucial issues for the future of the mountain regions may be illustrated with reference to the Sudetes. They comprise an old industrial region where the search for alternative income sources has extended to rural tourism and the fuller use of textile plants and medicinal herbs. A portrait for the 1980s would highlight the destabilisation of woodland ecosystems – with 'forest death' occurring over an area exceeding 12 km^2 where reconstruction areas now provide an indication of the extent and location of the damage (Figure 7.1). In addition to transboundary pollution, there was damage from local factories, but local heating plants and domestic fireplaces were the main source of harmful emissions. Pollution hazards were increased by temperature inversions in the Jelenia Góra Valley, with a population surpassing 100,000, and in the Swieradow Depression. Meanwhile, in agriculture there was quite heavy grazing pressure as a result of activities by highly subsidised state farms. The rivers were heavily polluted and only a limited area in the headwaters maintained first-class quality for pollution began at quite high levels through discharge of sewage in holiday centres. Sewerage was available for less than half the population: mechanised waste-water-treatment installations in the fast-growing towns were getting worn out by 1989, while some places had no facilities at all. Less than a fifth of industrial sewage was treated, with severe problems in the case of some industrial enterprises, like the paper mill at Karpacz.

The bulk of the sewage was generated in the towns but over 70 per cent of rural households discharged their waste into the natural river system. The situation was particularly unsatisfactory where houses had piped water (relatively cheap to provide) but not mains sewerage because consumption – and therefore waste levels – rose without the means for efficient disposal. Yet a positive feature arises from the construction of additional reservoirs along the mountain foreland, partly for defence against floods triggered by deforestation. Some agricultural land was lost during the 1980s as a result of this programme which went some way to establish a new equilibrium. The capacity is still not sufficient, given the rapid run-off which precipitated the Oder valley floods in 1997 which engulfed 180,000 ha in Poland, made 45,000 people homeless and seriously damaged the cities of Opole and Wrocław (based on islands in the river). The costs of protection are huge and while Poland's ten-year 'Odra 2006' programme has secured dyke reconstruction there is a lack of money for additional water storages and the modernisation of the locks. More storages will be needed in the Sudetes, but while they offer flood relief and provide for a growth of tourism and recreation facilities, which are part of the region's sustainable future in the aftermath of pressure from mining and heavy industry, there is sensitivity over further interference with the natural river system. Czetwertynski-

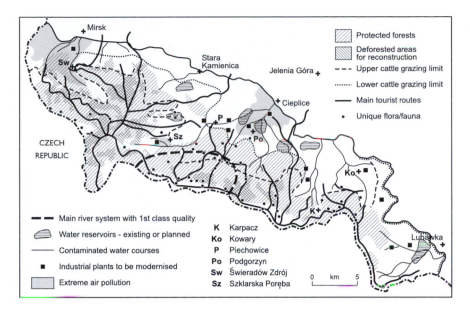

Figure 7.1 The environment of the Sudetes in the 1980s
Source: Czetwertynsk-Sytnik *et al.* (2000: 277)

Sytnik *et al.* (2000) document the work of ENGOs with respect to environmental education and conservation in a region that will offer useful experience for other mountain areas.

The Danube

There is still much controversy over the Danube which attracted international action when pollution concerns gave rise to the Danube River Protection Convention (DRPC) adopted in 1994 and activated in 1998 (Figure 7.2). But the pollution issue must be seen in the context of a deepening tension between development and conservation. The principal valleys of the region are obvious transport axes and in this connection the European Commission's Regional Policy Directorate has commissioned a 'Danube Space Study' from the Austrian Institute for Space Planning. However, the rivers themselves have come under scrutiny in connection with the EU's Transport Infrastructure Needs Assessment (TINA), under review at a base in Vienna since 1996. The Danube itself is recognised as a transport corridor and this implies a further phase of development following the regulation of the last two centuries which has transformed what was once a wide branching river into a navigation canal with a high degree of flood control linked with hydropower generation. However, TINA has raised the possibility of a Danube–Oder–Elbe Canal, intersecting with an east–west Oder–Warta–Vistula–Bug Canal (extendable through Ukraine to the Black Sea over the longer term). These waterways will have to be quite massive and while they will serve the needs of 'combined transport' they pose a severe threat to biodiversity.

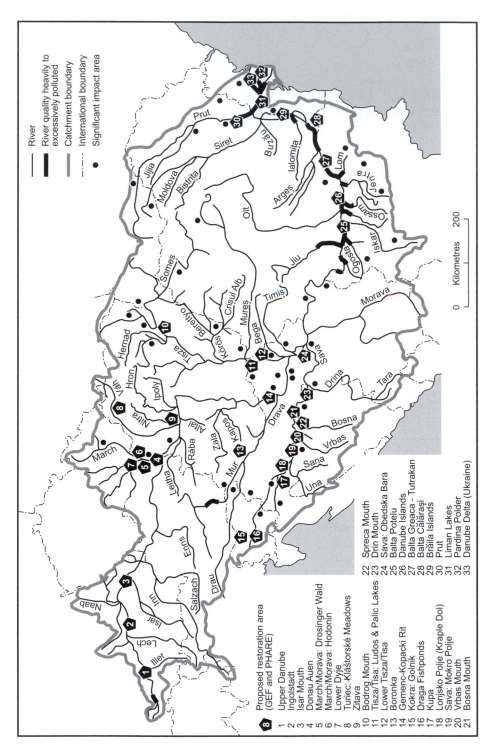

Figure 7.2 Danube Basin: pollution and wetland conservation

Source: World Wide Fund for Nature and Blue Danube

The conservationists have an alternative vision based on floodplain protection and restoration, after four-fifths of such lands were destroyed during the twentieth century and many species, such as the Dalmatian pelican and White-tailed eagle, have become endangered as a result. This approach is now being adopted because of its relevance to biodiversity but also because it has been demonstrated that wetlands absorb nutrients very effectively. A handful of early projects covering the Lower Morava, the Danube–Drava confluence, Bulgaria's Danube Islands and the delta are now being integrated into a 'green corridor' through a chain of other conservation projects. This work is now being expanded into a Danube–Carpathian Programme through a Carpathian initiative undertaken by the World Wide Fund for Nature (WWF). And there is likely to be a spin-off to other valleys where wetlands are threatened, such as the Lower Vistula where Niesawa is one of a series of controversial dams which will transform one of the few natural rivers still remaining in the region. Since the additional dams are necessary for the proposed navigation schemes the choices are particularly stark.

The local

Despite the potency of national and global forces, the local shows a capacity to innovate to meet new challenges. While a worldwide web of interdependencies might signal the end of geography, the outcome seems to be 'the reinforcement or even re-invention of traditions and local identities as an answer to the fear of loss of identity through homogenisation' (De Haan 2001: 366). There is a historical component in this because while the notion of 'path dependency' – grounded in communist networks and mind-sets – will lose much of its relevance as the transition develops momentum, we may follow Maurel's (2000: 157) remarks about countryside processes which have important points in common but nevertheless show regional variations while patterns at the regional level 'appear to owe a great deal to historical processes whose roots go far back in time in the areas and societies in which they emerged'. Yet it is difficult to see the local as a structured 'nested hierarchy', building up from locality, through drainage basin to region, since dynamic and unpredictable 'horizontal' links – comprising global connections – may operate at all levels. Hence, there are 'constantly shifting regions, each with its own specificity of nature, pace and direction of economic, social and cultural change, spatial arrangements and land use' (ibid.). These regions are substantially influenced by global investment decisions – linked with the eminently dispassionate evaluation of local resources – which are anathema to those who require all legitimate forces of change to be cosily embedded in stable regional structures.

Local actions will inevitably constitute engagement with government policies at each moment, including the fiscal regime. The scope for innovative action was high during the 'de-invention' phase, while centralising moves, e.g. by the Mečiar governments in Slovakia, restricted local attempts to develop cross-border cooperation, and 'spontaneous activities in the newly established independent republic started to collide with substantial political obstructions, fears and administrative barriers from the perspective of the ruling power' (Pasiak *et al.* 2001:

358). There will also be 'mindsets' conditioned by the potentials of centre and periphery, so that while the West Pannonia Euroregion is stimulating business successfully, the Carpathian counterpart finds 'the true integrating market forces are still underdeveloped [and] it is only the underground (black) economy where intensive "cooperation" can be observed' (Szigetvari 2001: 305). This situation has not been eased by the Russian crisis in 1998 and measures to protect 'Schengen space', which have complicated the movements of small traders into eastern Poland where Białystok once supported the greatest trade fair in the region. The periphery shows some signs of dynamism in the provincial cities, but elsewhere enterprise that does not have Internet access to potential export markets in the West has few illusions about local spending power.

The potency of the local was quite evident under communism. In Romania the large industrial 'centrals' enjoyed sufficient autonomy for the Reşiţa steelworks to implement a small hydropower project for greater energy self-sufficiency (crucial for the survival of the enterprise today!) within 30 km of a massive state investment to install 900 MW of generating capacity in a thermal power station burning bituminous schist and natural gas. At the same time, the local authorities in Maramureş were confident enough over their 'distance' from presidential scrutiny to evade the worst excesses of the draconian rural planning system programme of 'sistematizare', while the community at Jina near Sibiu gained considerable wealth through a wool trade lying outside formal regulation systems which normally included official registration of all animals. However, the situation is now very different because there is positive encouragement of private enterprise which must forge a productive partnership with the state, whereas communism consigned the 'private' to a marginal, largely domestic domain beyond the reach of activists and authorities. Although talk of a myriad of separate transformations across the region can become obsessive, it is nevertheless true that the 'local' is alive and well; not only through poverty and unemployment, but also through systems of governance, identity creation, development strategies and business environments.

This private domain is now becoming a positive force within civil society, whereas in the early years of transition it generated a more spontaneous and unpredictable development processes. Arguably, 'bottom up' is taking precedence over communism's 'top down'. Local situations vary greatly regarding the production profile – arising from the resources and the nature of post-communist restructuring – as well as infrastructure and local services. Inclusiveness requires that all stakeholders should be involved, but this will involve a range of regulation systems in which stakeholder coalitions – representing local and external interests (including foreign capitalists in many cases) – are combined in a myriad of different ways, while also engaging, with more or less cohesion, with neighbouring communities, with regional centres of administration, cross-border cooperation networks, and the wider global economy. In the first instance reference may be made to local resource management, for each area has its specific profile in terms of natural and human resources. Included in this is a range of environmental problems which may relate to deforestation and heightened flood risks.

Engagement with tasks of environmental management

The sensitive Subcarpathian environment in southeastern Romania shows a propensity for landslides and mudflows given the succession of clays, sands and shales in the context of geological uplift, downcutting by rivers and consequent steepening of slopes (Figure 7.3). Settlement is attracted to the old landslides because of the fertility of the immature soils, but periodic adjustments are inevitable and logging in the late nineteenth century in the context of heavy population pressure to extend agricultural land exposed unstable slopes to erosion and caused much damage to housing and infrastructure after heavy rain. Each family and community has made its own response in terms of housebuilding and agriculture so that, despite some reorganisation under communism, settlement is still highly dispersed across the hillsides whereas under different economic and demographic conditions the land might be much more heavily forested. Constant adjustments are also being made on the Hel Peninsula (in Poland's Gdańsk Bay) which evolves naturally in relation to the balance between erosion and the longshore drift of material. Forest helps to protect the spit and maintain communications from Władysławowo at the base of the peninsula to the fishing villages of Jastarnia and Kuźnica, the small tourist promenade at Jurata and Hel at the far end. There is an ongoing management problem and the trapping of material by a new fishing harbour at Kuźnica has been countered by efforts to maintain the spit by the construction of breakwaters and the bulldozing of sand dunes as defences. On the other hand, the erosion of the spit and its division into islands (which has happened at times in the past) would allow more water into the bay and reduce pollution – and might even improve the local fishing.

Responses to the region's pollution 'hot spots' may also be considered. Overall ECE pollution levels are not excessive, but Poland has 27 'ecologically threatened areas', including a section of the tri-frontier 'Black Triangle' – suffering extensive forest damage through insect attacks and parasitic fungi in the context of acid rain to which the spruce forests were particularly susceptible – and copper smelting at Głogów where several villages have been evacuated due to contaminated soil. Reference may also be made to Bulgarian and Macedonian non-ferrous metal smelters impacting on urban and agricultural areas in rich farming areas, the legacy of Czech uranium mines and toxic waste spillages in the Tisza basin (especially from the Baia Mare area of Romania) (Figure 7.4). The UN Environment Programme has identified one of the Balkans' most polluted spots: an abandoned chemical plant in Durrës where 3.0 km^2 is heavily polluted with 20,000 t of dangerous and health-threatening chemicals. Under such circumstances the local may have to make decisions balancing conservation and employment. Thus the people of Zlatna in the Apuseni Mountains accept heavy pollution from their copper smelter, with no immediate prospect of modernisation, rather than campaign for closure and a consequent loss of jobs, even though pollution discourages any growth of light industry that the town's LFA status might otherwise encourage. NIMBY attitudes to garbage incinerators make it difficult to expand waste management systems, but fortunately in Warsaw the Góra Kalwaria community will accept financial

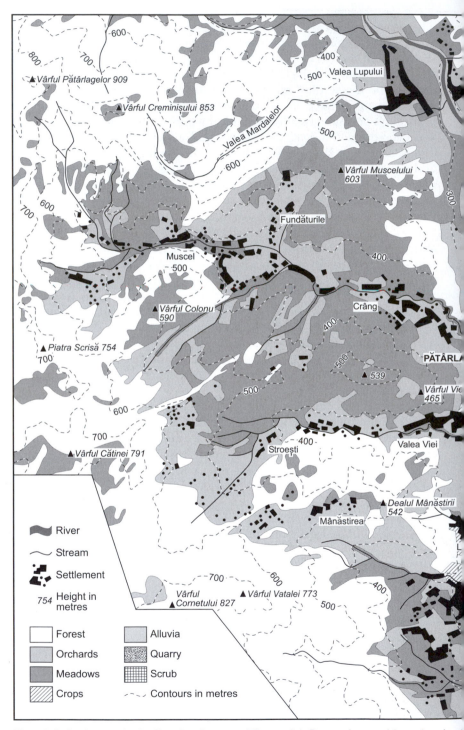

Figure 7.3 Settlement in the Pătârlagele area of Romania's Buza subcarpathions showing
use and dispersed settlement on unstable terrain prone to landslides and mudfl

Source: Topographical maps and field work

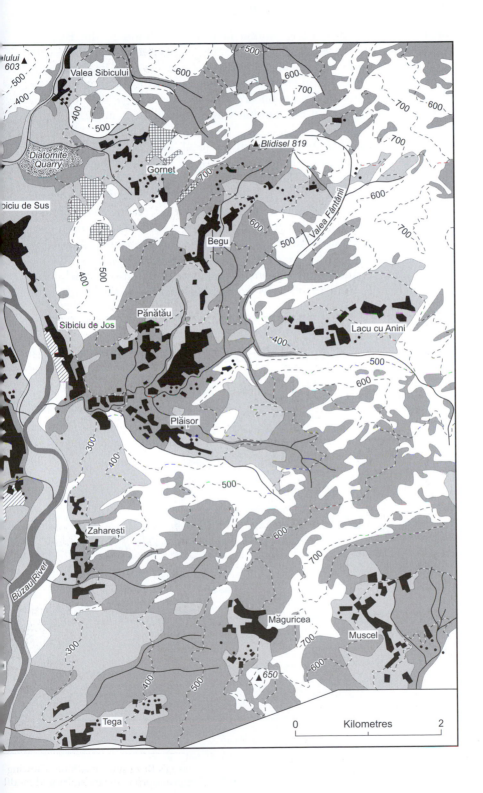

inducements to extend an existing landfill facility. In relation to the EU, the national level becomes the local in winning concessions over compliance with the environmental 'acquis': acceptable standards of waste water treatment can be deferred until after EU membership is secured: 2010 in the Czech Republic and 2015 in Hungary and Poland (while Poland will not achieve the norms for integrated pollution prevention and landfill waste management until 2010 and 2012 respectively).

Some powerful manifestations arise when the local 'bites back' in anger over new pollution threats from Trans-European Transport Corridors that offer little benefit to rural communities, e.g. the road development proposals for the České Středohoří Protected Landscape Area where the Prague–Dresden motorway, running parallel to an upgraded railway, will pose a threat to biodiversity. Similar controversy, based on inadequate environmental impact analysis, has arisen over the Bulgarian General Roads Authority's desire to route the Struma motorway through the Kresna Gorge which is rich in biodiversity. The local is also active in routine planning matters and regional development initiatives which now embrace an increasing number of trans-frontier regions, especially in the light

Figure 7.4 Hazardous waste storage in the Tisza Basin

Source: International Commission for the protection of the Danube River

of projects seeking to develop civil society and reduce national government's involvement in 'firefighting' to ameliorate local problems. The development of institutions is particularly important where official government does not provide an adequate hierarchy, as in Slovenia where there is no layer of administration interposed between the central state and 192 municipalities that have limited technical, financial and leadership resources. Disadvantaged rural areas are being assisted by VITRA – the Centre for Sustainable Development – which works with the Croatian ENGO 'Eko Liburnia' to share knowledge and experience with rural areas and generate 'a vision for revitalisation': preserving landscape and architectural heritage while developing family businesses in organic farming and green tourism. In Croatia organisations are sending strong signals in favour of tourism and organic farming which have prompted the government to bar genetically-modified organisms (GMOs) on the grounds that they constitute a threat to biodiversity.

Furthermore, EU structures lay great emphasis on local organisations which are positively encouraged to generate project proposals, after criticisms of PHARE in the early years, for preferring large projects instead of clusters of locally-administered small projects. Hence the widespread belief that future pre-accession funding for ECE should reflect the status of agri-environmental schemes as a key policy instrument throughout the EU (Baldock *et al.* 2001: 6), following the 1996 Cork Declaration – seeking integrated programmes of sustainable rural development for each region – and the LEADER approach of targeted/tailored programmes for specific problems and areas (most recently spelt out under Article 33 of Regulation 1257/1999). SAPARD funding should be contingent on grass-roots' participation and on connections with protected areas networks and Natura 2000. According to Lowe (2000: 1), commercial agriculture must be promoted but it will not increase rural employment significantly: 'what is needed in addition is an approach that promotes a whole range of grassroots orientated rural development programmes', as provided for under the LEADER programme and the new Regulation just referred to. Given the high habitat value of farmland in the region, where traditional forms of land use have persisted in contrast to EU, there is much appreciation of the environmental and food quality benefits of organic farming, supported by enhanced 'capacity' through cultural change in institutions and appropriate training/advice for farmers. Steps are being taken in agriculture ministries in most accession countries and a number of projects are now under way.

Measures have been taken since 1998 by the Hungarian Ministry of Agriculture and Rural Development to introduce a Hungarian Agri-Environmental Programme (AEP): initially a landuse zonation study identifying target areas for projects. There are 'horizontal' schemes for environmentally friendly production (through reduced fertiliser use – which does not necessarily mean a commensurate reduction in yield – and nature-oriented landuse systems aiming at quality food production) and also zonal/regional schemes for low-input/ extensive farming and specific nature conservation objectives. Hungary received PHARE money in 1999 for pilot testing and preparations for pre-accession funding of projects under SAPARD. This country also has the benefit of supportive NGOs like MME (the

Hungarian Ornithological Society), E-Miszio, Duna Kor, Holocen and Somogy which all have a strong influence on nature conservation and are trusted by the public to a greater extent than the statutory authorities. Meanwhile, the Czech Republic is progressing with agri-environmental schemes linked with SAPARD, including projects linked with the conservation of species-rich meadows of the White Carpathians.

In the latter area of Moravia, Hostětín stands as a model community for rural sustainable development based on semi-natural beech woodland and organically-farmed species-rich grassland. Boasting a rural development centre built by Veronica Ecological Institute, Brno, the community is sustaining a remarkably cohesive coalition under an innovative mayor, including community groups, environmental organisations, farmers, business people, government agencies and local and international funders. New orchards are being planted with reference to the history of fruit growing in an area which has 250 naturally-occurring fruit varieties. Pure apple juice is now sold under the 'Traditions of the White Carpathians' label which covers a range of other food products. Solar collectors have been installed on homes and public buildings under a project developed by the Veronica Institute, while a Czech–Dutch government project has secured a 700 Kw biomass district heating system which burns sawmill waste in a high-tech boiler. This supplies four-fifths of the village's 80 houses with hot water while the ash from the boiler can be used as fertiliser.

Situated in the Carpathians on the Czech Republic–Slovakia border, this project has much significance for the WWF's Carpathian Ecoregion programme with its long-term 'vision', looking forward to a 10–15-year conservation plan covering specific priority areas, backed by sustainable development across the whole region in the hands of local stakeholder coalitions supported by a network of highly-active ENGOs associated with the Environmental Partnership for Central Europe (Figure 7.5). Increased adoption of good practice in land stewardship and the use of timber certification to gain premiums in the marketplace for sustainable forestry are also contributing. Sustainable tourism will also have an important role including the use of the education infrastructure during holiday periods. Cross-border networks have also often managed to rise above historic tensions, e.g. around Babia Góra on the Polish–Slovak border where a coalition of stakeholders, local governments and schools are developing an international biosphere reserve. On this basis the local is poised to deliver a rich harvest of empirical detail from which a digest of recommended practices can be drawn.

Economic and social issues

Of course the ecological perspectives by no means exhaust the expressions of the local. Yoder (2001) has pointed to a rejuvenation of regional diversity in East Germany and a rediscovering of identities strong enough to block the federal government's wish to amalgamate the Länder of Berlin and Brandenburg, where the revamped communist party continues to attract a significant level of support through advocating eastern interests. In the broader context of Westernisation

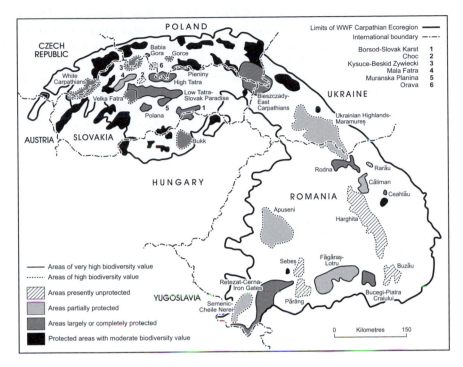

Figure 7.5 Carpathian Ecoregion: protected areas
Source: World Wide Fund for Nature

and globalisation, there is a contest in Poland between the modernisers and the conservative/national groups that fear the loss of Polish identity with further integration. Cross-border cooperation has undoubtedly achieved greater cohesion, yet historic tensions remain quite close to the surface on the Czech–German frontier. Many people in East Germany still feel estranged from the West and regret the loss of communist social structures despite their reservations over the political system. Economic management is needed to resolve contests between stakeholder groups who have their various interests in local resources (e.g. logging, grazing, fruit picking, recreation and conservation in the case of woodlands). Hence the myriad of local situations each reflecting unique mix of natural resources and types of business – arising from the workings of the privatisation process – which determines the level of investment and employment (Staddon 2001). Foreign investors are exhorted to be sensitive to the need for improved local regulation and aim at 'institutional gradualism' through discussion and negotiation of change so that local elites can regain recognition and legitimation and thereby become more active in regional development (Uhlir 1998: 683).

Still further scope for the local arises from the relations between social and ethnic groups and the varying degrees of inclusion and exclusion and the preferences emerging in local housing markets. While conflict between Roma and mainstream ethnic groups has produced fatalities and attempts at exclusion best represented by the wall in Ústí nad Labem, local self-government in Hungary (under the Minorities Law of 1993) is demonstrating how ethnic relations can be handled peacefully through local politics, although this particular institution, seeking to reconnect the poor, appears to work best where there is already a degree of cohesion (Schaft 2000). Elsewhere, despite NGOs actively advocating minority rights, 'local decision-makers are often imprisoned in their feelings of isolation when coping with historically-rooted community conflicts' (Kovacs 2000: 151) and there is resistance among local leaders to applying Western approaches to conflict resolution. Finally, communities have their own ideas concerning such communist legacies as street names and monuments associated with the former ideology: both can be removed quite easily but there is considerable toleration and most communist memorials, where they were not destroyed at the height of the revolution, have simply been marginalised by neglect or removed to museums or other special repositories like Budapest's suburban 'Statuepark'. While contest should not override consensus, it is important to end with a reminder that the EU project highlighted throughout this book is producing both positive responses and a range of backward glances, nourished not only by the interface of the 1980s and 1990s, but more deep-seated historical experiences.

References

For a set of additional references, a list of books published since 1989 (arranged by country) and a list of websites, related to the material covered in this book, please go to the following website: www.reference.routledge.com/research. Then click on the subject link for 'Environment & Society' series, then choose the link to the 'Routledge Studies in Human Geography series' and then click on the links to the extra material, which are beside this book's title.

M. Adamowicz 2000, 'Institutional restructuring and rural development' in B. Brown and A. Bandlerova (eds), *Rural development in Central and Eastern Europe* (Nitra: Slovak Agricultural University) 158–63.

J.B. Allcock and K. Przeclawski 1990, 'Tourism in the centrally-planned economies: introduction', *Annals of Tourism Research* 17, 1–6.

F-L. Altmann 2001, 'Economic reconstruction in Southeast Europe: a Western view', *Southeast European and Black Sea Studies* 1, 114–18.

A. Amin and J. Tomaney 1995, 'The regional development potential of inward investment in the less favoured regions of the European Community' in A. Amin and J. Tomaney (eds), *Behind the myth of European union: prospects for cohesion* (London: Routledge) 201–20.

R. Andorka 1997, 'The development of poverty during the transformation in Hungary' in I.T. Berend (ed.), *Long-term structural changes in transforming Central and Eastern Europe: the 1990s* (Munich: Sudosteuropa-Gesellschaft in cooperation with the Center for European and Russian Studies, University of California) 75–100.

G. Andrusz, M. Harloe and I. Szelenyi (eds) 1996, *Cities after socialism: urban and regional change and conflict in post-socialist societies* (Oxford: Blackwell).

J. Bachtler 1992a, 'Regional problems and policies in Central and Eastern Europe', *Regional Studies* 26, 665–71.

J. Bachtler (ed.) 1992b, *Socio-economic situation and development of the regions in the neighbouring countries of the Community in Central and Eastern Europe* (Brussels/Luxembourg: Commission of the European Communities Directorate-General for Regional Policies, Regional Development Studies 2).

J. Bachtler, R. Downes and G. Gorzelak (eds) 2000, *Transition, cohesion and regional policy in Central and Eastern Europe* (Aldershot: Ashgate).

S. Baker 2000, 'The European Union: integration, competition, growth – and sustainability' in W.M. Lafferty and J. Meadowcroft (eds), *Implementing sustainable development: strategies and initiatives in high consumption societies* (Oxford: Oxford University Press) 302–36.

V. Balaz 1994, 'Tourism and regional development in the Slovak Republic', *European Urban and Regional Studies* 1, 171–6.

D. Baldock, H. Bennett and G.W. Verschuur 2001, *Agri-environmental policy development in CEE: synthesis report of a multi-partner project* (Wommels, The Netherlands: Avalon Foundation).

V. Baran (ed.) 1995, *Boundaries and their impact on the territorial structure of region and state* (Banska Bystrica: University M. Bel, Faculty of Natural Sciences).

G. Barta 1992, 'The changing role of industry in regional development and regional development policy in Hungary', *Tijdschrift voor Economische en Sociale Geografie* 58, 372–9.

J. Batt 1991, *East Central Europe from reform to transformation* (London: RIIA/Pinter).

A Bazydko 1999, 'Special Economic Zones of Poland', *Acta Facultatis Rerum Naturalium Universitatis Comenianae Geographica Supplementum* 2(2), 299–304.

V. Beckmann and K. Hagedorn 1997, 'Decollectivisation and privatisation policies and resultant structural change of agriculture in Eastern Germany' in J.F.M. Swinnen, A. Buckwell and E. Mathijs (eds) 1997, *Agricultural privatisation land reform and farm restructuring in Central and Eastern Europe* (Aldershot: Ashgate) 105–60.

P. Beluszky 1999, *The Hungarian urban network at the end of the second millennium* (Pecs: Centre for Regional Studies of the Hungarian Academy of Sciences Discussion Papers 27).

I.T. Berend 1996, *Central and Eastern Europe 1949–1993: detour from the periphery to the periphery* (Cambridge: Cambridge University Press).

I.T. Berend 1997, 'Transformation and structural changes: Central and Eastern Europe's adjustment in a historical perspective' in I.T. Berend (ed.), *Long-term structural changes in transforming Central and Eastern Europe: the 1990s* (Munich: Sudosteuropa-Gesellschaft in cooperation with the Center for European and Russian Studies University of California) 9–28.

I. Berenyi and Z. Dovenyi 1996, 'Historische und ackuelle Entwicklung des ungarischen Siedlundsnetzes' in A. Mayr and F.-G. Grimm (eds), *Stadte und Stadt systeme in Mittel- und Sudosteuropa* (Leipzig: Institut für Länderkunde) 106–71.

T. Bergmann 1992, 'The reprivatisation of farming in Eastern Germany', *Sociologia Ruralis* 32, 305–16.

J. Berry and S. McGreal 1995, 'Berlin' in J. Berry and S. McGreal (eds), *European cities, planning systems and property markets* (London: Spon) 371–94.

I. Bicanic and M. Skreb 1991, 'The service sector in East European economies: what role can it play in future development?', *Communist Economies and Economic Transformation* 3, 221–33.

I. Bicik and A. Gotz 1998, 'Czech Republic' in D. Turnock (ed.), *Privatisation in rural Eastern Europe: the process of restitution and restructuring* (Cheltenham: Edward Elgar) 93–119.

D. Binder 2000, 'Why the Balkans?', *Woodrow Wilson International Center for Scholars: East European Studies Newsletter*, May–June, 3–10.

O. Blanchard 1994, 'Unemployment in Eastern Europe', *Finance and Development: A Quarterly Publication of the International Monetary Fund and the World Bank* 31(4), 6–9.

J. Blazek 1997, 'The Czech Republic on its way towards West European structures', *European Spatial Research and Policy* 4(1), 37–62.

I. Blumi 1997, 'The politics of culture and power: the roots of Hoxha's postwar state', *East European Quarterly* 31, 379–98.

J. Blumi 1998, 'The commodification of otherness and the ethnic unit in the Balkans: how to think about Albanians', *East European Politics and Societies* 12, 527–69.

A. Bochmann 1995, 'The city of Chemnitz in Saxony: building its new economic profile', *GeoJournal* 37, 539–56.

J. Bodnar 1996, '"He that hath to him shall be given": housing privatisation in Budapest after state socialism', *International Journal of Urban and Rural Research* 20, 616–36.

S. Böhmer-Christiansen 1998, 'Environment-friendly deindustrialization: impacts of unification on East Germany' in A. Tickle and I. Welsh (eds), *Environment and Society in Eastern Europe* (London: Addison-Wesley-Longman) 67–96.

J. Borocz 1990, 'Hungary as a destination', *Annals of Tourism Research* 17, 19–35.

M. Bucek 1992, 'Regional policy of the Slovak Republic in the period of transition' in T. Vasko (ed.), *Problems of economic transition: regional development in Central and Eastern Europe* (Aldershot: Avebury) 1–17.

P.J. Buckley and S.F. Witt 1990, 'Tourism in the centrally-planned economies of Europe', *Annals of Tourism Research* 17, 7–18.

R. Cappellin 1992, 'Theories of local endogenous development and international cooperation' in M. Tykkylainen (ed.), *Development issues and strategies in the new Europe* (Aldershot: Avebury) 1–20.

F.W. Carter 1996, 'Central Europe: fact or geographical fiction?' in F.W. Carter, P. Jordan and V. Rey (eds), *Central Europe after the fall of the Iron Curtain: geopolitical perspectives, spatial patterns and trends* (Frankfurt am Main: Peter Lang Europaischer Verlag der Wissenschaften) 9–44.

F.W. Carter and D. Turnock (eds) 2002, *Environmental problems of East Central Europe* (London: Routledge).

F.W. Carter, D.R. Hall, D. Turnock and A.M. Williams 1996, *Interpreting the Balkans* (London: Royal Geographical Society, Geographical Intelligence Paper 2).

P. Chalupa 1993, 'Synergetic conceptions of regional population and social-democratic processes taking place in the Czech Republic', *GeoJournal* 31, 435–8.

D. Chirot 2000. 'How important is the past?: interpreting Eastern Europe's transitional failures and successes', *Woodrow Wilson International Center for Scholars: East European Studies Newsletter*, Nov–Dec, 3–9.

D. Christopulos 1994, 'Minorities' protection: towards a new European approach', *Balkan Forum : International Journal of Politics, Economics and Culture* 2, 155–74.

S. Clement 2000, 'Subregionalism in Southeastern Europe' in S.C. Calleya (ed.), *Regionalism in the post-cold war world* (Aldershot: Ashgate) 71–98.

N.J. Cochrane 1994, 'Farm restructuring in Central and Eastern Europe', *Soviet and Post-Soviet Review* 21, 319–35.

E. Commisso 1997, '"Is the glass half full or half empty?": reflections on five years of competitive policies in Eastern Europe', *Communist and Post-Communist Studies* 30(1), 1–22.

P. Cooke 1997, 'Inward investment and economic development in Poland', *European Planning Studies* 5, 691–7.

O. Crisan 1999, 'The concept of the interdisciplinary applied in village development' in V. Surd (ed.), *Rural space and regional development* (Cluj-Napoca: Editura Studia) 200–4.

B. Csatari 1993, 'Crisis signs of the Hungarian small towns' in A. Duro (ed.), *Spatial research and the social-political changes* (Pecs: Centre for Regional Research) 97–102.

B. Csatari 2001, 'European rurality and the Great Plain', *Alfoldi Tanulmanyok* 18, 44–63 (in Hungarian with an English summary).

Z. Csefalvay 1994, 'The regional differentiation of the Hungarian economy in transition', *GeoJournal* 32, 351–61.

L. Czetwertynski-Sytnik, E. Koziol, K. Mazurski and D. Turnock 2000, 'Settlement and sustainability in the Polish Sudetes', *GeoJournal* 50, 273–84.

T. Czyz 1993, 'The regional structure of unemployment in Poland', *Geographia Polonica* 61, 479–96.

A. Dawson 1997, 'Two maps of Poland' in A. Dingsdale (ed.), *Transport in transition: issues in the new Central and Eastern Europe* (Nottingham: Nottingham Trent University, Trent Geographical Papers 1) 30–41.

J. Dawson 1996, *Eco-nationalism: anti-nuclear activism and national identity in Russia, Lithuania and Ukraine* (Durham, N.C.: Duke University Press).

L.J. De Haan 2001, 'The question of development and environment in geography in the era of globalisation', *GeoJournal* 50, 359–67.

L. De Silva 1993, 'Women's emancipation under communism: a re-evaluation', *East European Quarterly* 27, 301–15.

L. Dobraca 1999, 'Embryons urbains en Roumanie: les localités rurales à fonction de marché' in V. Surd (ed.), *Rural space and regional development* (Cluj-Napoca: Editura Studia) 183–8.

B. Domanski 2000, 'The impact of spatial and social qualities on the reproduction of local economic success: the case of path dependent development of Gliwice' in B. Domanski (ed.), *Studies in local and regional development* (Krakow: Instytut Geografii Uniwersytetu Jagiellonskiego Prace Geograficzne 106) 35–53.

P. Dostal and M. Hampl 1994, 'Development of an urban system' in M. Barlow, P. Dostal and M. Hampl (eds), *Territory, society and administration: the Czech Republic and the industrial region of Liberec* (Amsterdam: University of Amsterdam/Charles University Prague/Czech Academy of Sciences Prague) 191–224.

Z. Dovenyi 1994, 'Transition and unemployment: the case of Hungary', *GeoJournal* 32, 393–8.

V. Drgona, A. Dubcova and H. Kramarekova 1998, 'Slovakia' in D. Turnock (ed.), *Privatisation in rural Eastern Europe: the process of restitution and restructuring* (Cheltenham: Edward Elgar) 251–73.

M. Dunford 1998, 'Differential development, institutions, modes of regulation and comparative transitions to capitalism: Russia, the Commonwealth of Independent States and the former German Democratic Republic' in J. Pickles and A. Smith (eds), *Theorising transition: the political economy of post-communist transformations* (London: Routledge) 76–111.

W.G. East 1961, 'The concept and political status of the shatter zone' in N.J.G. Pounds (ed.), *Geographical essays on Eastern Europe* (Bloomington, Ind.: Indiana University Press) 1–27.

P. Eberhardt 1994, 'Distribution and dynamics of rural population in Central Eastern Europe', *Geographia Polonica* 63, 75–94.

N. Eberstadt 1993, 'Mortality and the fate of communist states', *Communist Economies and Economic Transformation* 5, 499–518.

E. Ehrlich and G. Revesz 2001, 'Hungarian economy: prediction and outcome' in G. Gorzelak, E. Ehrlich, L. Faltan and M. Illner (eds), *Central Europe in transition: towards EU membership* (Warsaw: Scholar Publishing House) 39–74.

G. Ekiert 1998, 'Why some succeed and others fail: eight years of transition in Eastern Europe', *Woodrow Wilson International Center for Scholars: East European Studies Newsletter*, March–April, 5.

C. Ellger 1992, 'Berlin: legacies of division and problems of unification', *Geographical Journal* 158, 40–6.

G. Enyedi 1990, *New basis for regional and urban policies in East-Central Europe* (Pecs: Centre for Regional Studies).

G. Enyedi 1996, 'Urbanization under socialism' in G. Andrusz, M. Harloe and I. Szelenyi (eds), *Cities after socialism: urban and regional conflict in post-socialist societies* (Oxford: Blackwell) 100–18.

European Bank for Reconstruction and Development 1999, *Transition report 1999* (London: EBRD).

European Conference of Ministers of Transport (ECMT) 1991, *Prospects for East–West European transport* (Paris: ECMT Publications).

Eurostat 1999, *Regional GDP in the Central European Countries 1999 edition* (data 1993–6) (Luxembourg: Statistical Office of the European Communities).

Eurostat 2000, *Statistical yearbook on candidate and South East European countries* (data 1994–8) (Luxembourg: Statistical Office of the European Communities).

Eurostat 2001, *Statistical guide to Europe* (data 1994–9) (Luxembourg: Statistical Office of the European Communities).

G. Evans and S. Whitefield 1998, 'The structuring of political cleavages in post-communist societies: the case of the Czech Republic and Slovakia', *Political Studies* 26, 115–39.

Z. Ferge 1997, 'Is the world falling apart?: a view from the east of Europe' in I.T. Berend (ed.), *Long-term structural changes in transforming Central and Eastern Europe: the 1990s* (Munich: Sudosteuropa-Gesellschaft in cooperation with the Center for European and Russian Studies University of California) 101–20.

M. Foucher 1994, *Minorities in Central and Eastern Europe* (Strasbourg: Council of Europe).

R.A. French and F.E.I. Hamilton (eds) 1979, *The socialist city: spatial structure and urban policy* (Chichester: Wiley).

F. Fukuyama 1992, *The end of history and the last man* (New York: Free Press).

I.R. Gabor 1989, 'Second economy and socialism: the Hungarian experience' in E.L. Feige (ed.), *The underground economies* (Cambridge: Cambridge University Press) 339–60.

J. Galombos 1993, 'An international environmental conflict on the Danube: the Gabčikovo–Nagymaros dams' in A. Vari and P. Tamas (eds), *Environment and democratic transition: policy and politics in Central and Eastern Europe* (Boston: Kluwer) 203–25.

B. Gorz and W. Kurek 1998, 'Poland' in D. Turnock (ed.), *Privatisation in rural Eastern Europe: the process of restitution and restructuring* (Cheltenham: Edward Elgar) 169–99.

G. Grabher 1995, 'The elegance of incoherence: economic transformation in East Germany and Hungary' in E.J. Dittrich (ed.), *Industrial transformation in Europe* (London: Sage) 33–53.

G. Grabher and D. Stark 1998, 'Organising diversity: evolutionary theory network analysis and post-socialism' in J. Pickles and A. Smith (eds), *Theorising transition: the political economy of post-communist transformations* (London: Routledge) 54–75.

J. Green 1998, *Energy and the environment in the EU: training document for NGOs in accession countries* (Brussels: Friends of the Earth Europe).

K. Grime and G. Weclawowicz 1981, 'Warsaw' in M. Pacione (ed.), *Urban problems and planning in the developed world* (London: Croom Helm) 258–91.

I. Gyulai 1998, *The European Union and biodiversity* (Brussels: Friends of the Earth Europe and European Environmental Bureau).

A. Halasiewicz 1995, 'A Polish village in the process of transformation towards a market economy' in D.A. Kideckel (ed.), *East European communities: the struggle for balance in turbulent times* (Boulder, Col.: Westview) 85–97.

D.R. Hall 1990, 'Stalinism and tourism: a study of Albania and North Korea', *Annals of Tourism Research* 17, 36–54.

D.R. Hall and J. Kowalski 1993, 'Introduction and overview' in D.R. Hall (ed.), *Transport and economic development in the new Central and Eastern Europe* (London: Belhaven) 1–33.

M. Hampl 1999a, 'The development of regional system and societal transformation in the Czech Republic' in M. Hampl, *Geography of societal transformation in the Czech Republic* (Prague: Charles University Department of Social Geography and Regional Development) 27–130.

M. Hampl 1999b, 'Current tendencies in the hierarchy of regional systems: example of transformation in the Czech Republic', *Acta Facultatis Rerum Naturalium Universitatis Comenianae Geographica Supplementum* 2/1, 31–42.

M. Hampl and J. Muller 1999, 'Geographical aspects of societal transformation in the Czech Republic' in F.W. Carter and W. Maik (eds), *Shock-shift in an enlarged Europe: the geography of socio-economic change in East-Central Europe after 1989* (Aldershot: Ashgate) 27–37.

J. Hardy and A. Rainnie 1996, *Restructuring Krakow: desperately seeking capitalism* (London: Mansell).

M. Harloe 1996, 'Cities in transition' in G. Andrusz, M. Harloe and I. Szelenyi (eds), *Cities after socialism: urban and regional conflict in post-socialist societies* (Oxford: Blackwell) 1–29.

H. Haussermann 1996, 'From the socialist to the capitalist city: experiences from Germany' in G. Andrusz, M. Harloe and I. Szelenyi (eds), *Cities after socialism: urban and regional conflict in post-socialist societies* (Oxford: Blackwell) 214–31.

M. Heffernan 1998, *The meaning of Europe: geography and geopolitics* (London: Arnold).

W. Heller 1998, 'Theoretial perspectives' in W. Heller (ed.), *Romania: migration, socio-economic transformation and perspectives of regional development* (Munich: Sudosteuropa-Gesellschaft, Sudosteuropa Studien 62) 14–22.

A. Herbst 1998, 'Stumbling blocks to ecumenism in the Balkans', *Religion, State and Society* 26, 173–80.

G. Horvath 1998, *Regional and cohesion policy in Hungary* (Pecs: Centre for Regional Studies Discussion Papers 23).

R. Hudson, M. Dunford, D. Hamilton and R. Kotter 1997, 'Developing regional strategies for economic success: lessons from Europe's economically successful regions?', *European Urban and Regional Studies* 4, 365–73.

A. Iancu 2000, 'Some problems of the strategy of development and integraton into the EU', *Romanian Business Journal* 7(9), 6.

C. Ingrao 1999, 'Understanding ethnic conflict in Central Europe: an historical perspective', *Nationalities Papers* 27, 291–318.

C. Ionete and V. Dinculescu (eds) 2000, *National human development report: Romania 1999* (Bucharest: Romanian Academy for the United Nations Development Project).

K. Ivanicka 1996, *Slovakia: genius loci* (Bratislava: Korene Press).

K. Jasiewicz 1998, 'The 1997 parliamentary elections in Poland: how much déjà-vu?', *Woodrow Wilson International Center for Scholars: East European Studies Newsletter*, May–June 3.

K. Jowitt 1996, 'Dizzy with democracy', *Problems of Post-Communism* 43(1), 3–7.

L. Kabat and K. Hagedorn 1997, 'Privatisation and decollectivisation policies and resulting structural changes of agriculture in Slovakia' in J.F.M. Swinnen, A. Buckwell and E. Mathijs (eds), *Agricultural privatisation, land reform and farm restructuring in Central and Eastern Europe* (Aldershot: Ashgate) 229–79.

G. Karasimeonov 1998, 'Conceptions and misconceptions of political change in East-Central Europe' in M.J. Bull and M. Ingham (eds), *Reform of the socialist system in Central and Eastern Europe* (Basingstoke: Macmillan Press) 60–76.

Z. Karpaty 1986, 'Peripheral settlement in Hungary: the example of Baranya County' in G. Enyedi and J. Veldman (eds), *Rural development issues in industrialized countries* (Pecs: Hungarian Academy of Sciences, Centre for Regional Studies) 128–33.

G. H. Kats 1991, 'Energy options for Hungary', *Energy Policy* 19, 855–68.

J. Keep 1997, 'The Gorbachev era in historical context', *Studies in East European Thought* 49, 271–86.

B. Keresztesi 1993, 'Hungary' in A. Mather (ed.), *Afforestation: policies planning and progress* (London: Belhaven) 59–71.

M.R. Khan 1995, 'Bosnia-Hercegovina and the crisis of the post-war international system', *East European Politics and Societies* 9, 459–98.

T. Kingsley 1993, 'Housing reform in Czechoslovakia: promise not yet fulfilled', *Cities* 10, 224–36.

E. Kiss 2000, 'Rural restructuring in Hungary in the period of socio-economic transition', *GeoJournal* 51, 221–33.

Y. Kiss 1993, 'Lost illusions?: defense industry conversion in Czechoslovakia 1989–1992', *Europe–Asia Studies* 45, 1045–69.

H. Kitschelt 1992, 'The formation of party systems in East Central Europe', *Political Studies* 20, 7–50.

M. Klemencic and C. Schofield 1996, 'Croatia's territorial consolidation and prospects for the future', *GeoJournal* 38, 393–8.

K. Kobayashi 1996, 'Structural changes and renovation programs in rural areas of the Former East Germany centring on Mecklenburg-Western Pomerania state' in H. Sasaki (ed.), *Geographical perspectives on sustainable rural systems* (Tokyo: Kaisei Publications) 389–97.

J. Kochanowicz 1997, 'New solidarities?: market change and social cohesion in a historical perspective' in I.T. Berend (ed.), *Long-term structural changes in transforming Central and Eastern Europe: the 1990s* (Munich: Sudosteuropa-Gesellschaft in cooperation with the Center for European and Russian Studies University of California) 55–74.

G. Kolankiewicz 1993, 'Poland' in S. Whitefield (ed.), *The new institutional architecture of Eastern Europe* (London: Macmillan) 99–120.

B. Koulov 1992, 'Tendencies in the planning of territorial development in Bulgaria', *Tijdschrift voor Economische en Sociale Geografie* 58, 390–401.

B. Koulov 1999, 'Minorities states and conflict' in P. Heenan and M. Lamontaigne (eds), *The Central and Eastern Europe handbook* (London: Fitzroy Dearborn) 195–205.

P. Kovacs 2000, 'Innovative solutions to promote multi-ethnic coexistence in Central and Eastern Europe' in B. Brown and A. Bandlerova (eds), *Rural development in Central and Eastern Europe* (Nitra: Slovak Agricultural University) 147–51.

M. Kovats 1998, 'Gypsy self-governments in Hungary' in S. Bridger and F. Pine (eds), *Surviving post-socialism: local strategies and regional responses in Eastern Europe and FSU* (London: Routledge) 124–47.

T. Kowalik 1994, 'The "big bang" as a political and historical phenomenon: a case study of Poland' in I.T. Berend (ed.), *Transition to a market economy at the end of the twentieth century* (Munich: Sudosteuropa-Gesellschaft) 115–23.

J.M. Kramer 1999, 'Energy and environment in Eastern Europe', *Problems of Post-Communism* 46(6), 47–56.

S. Kratke 1992, 'Berlin: the rise of a new metropolis in a post-Fordist landscape' in M. Dunford and G. Kafkalas (eds), *Cities and regions in the new Europe* (London: Belhaven) 213–38.

S. Kratke 1996, *Where east meets west: prospects for the German–Polish border region* (Frankfurt/Oder: Europa Universität Viadrina Faculty of Cultural Sciences).

J. Kraus 1994, 'Agricultural reform and transformation in the Czech Republic' in J.F.M. Swinnen (ed.), *Policy and institutional reform in Central European agriculture* (Aldershot : Avebury) 107–34.

J. Kruczala 1990, 'Tourism planning in Poland', *Annals of Tourism Research* 17, 69–78.

W. Kurek 1996, 'Agriculture versus tourism in rural areas of the Polish Carpathians', *GeoJournal* 38, 191–6.

W.M. Lafferty and J. Meadowcroft (eds) 2000, *Implementing sustainable development: strategies and initiatives in high consumption societies* (Oxford: Oxford University Press).

M. Lagerspetz 2001, 'From "Parallel Polis" to "The Time of the Tribes": post-socialism self-organization and post-modernity', *Communist Studies and Transition Politics* 17(2), 1–18.

M.A. Landesmann and I.P. Szekely 1996 'Introduction' in M.A. Landesmann and I.P. Szekely (eds), *Industrial restructuring and trade reorientation in Eastern Europe* (Cambridge: Cambridge University Press) 1–22.

A. Langer 1990, *Ecological bricks for our common house in Europe: global challenges network* (Munich: Verlag für Politische Okologie).

M. Lavigne 2000, 'The economics of the transition process: what have we learned?', *Problems of Post-Communism* 47(4), 16–23.

D. Light and D. Phinnemore (eds) 2001, *Post-communist Romania: coming to terms with transition* (Basingstoke: Palgrave).

N. Lindstrom 2001, 'The Southeastern enlargement of the EU: what is at stake for Croatia and Slovenia?', *Woodrow Wilson International Center for Scholars: East European Studies Newsletter*, March–April 3–9.

A.E. Liolin 1997, 'The nature of faith in Albania: towards the 21st century', *East European Quarterly* 31, 181–94.

L. Lorber 1997, 'The role of the integration of science and technology (Technological Parks) in restructuring the economy of Slovenia', *Acta Geographica Croatica* 32, 111–24 (in Croatian with an English summary).

P. Lowe 2000, 'Challenges for rural development in CEE', *Rural Areas Newslink: Newsletter for Agriculture Environment and Rural Development in CEE* 9, 1.

J. Luxmoore 1997, 'A review of religious life in Albania, Bulgaria, Romania, Hungary, Slovakia, Czech Republic and Poland', *Religion, States and Society* 25(1), 89–101.

L.W. Lyde 1926, *The continent of Europe* (London: Macmillan).

R. Maggi and P. Nijkamp 1992, 'Missing networks and regional development in Europe' in T. Vasko (ed.), *Problems of economic transition: regional development in Central and Eastern Europe* (Aldershot: Avebury) 29–49.

G. Mangott 1998, *The elections of 1994–1997 in Central and Southeastern Europe* (Vienna: Osterreichisches Ost- und Sudosteuropa-Institut, Atlas Ost- und Sudoseuropa).

A. Marga 1993, 'Cultural and political trends in Romania before and after 1989', *East European Politics and Societies* 7, 14–32.

M.-C. Maurel 2000, 'Patterns of post-socialist transformation in the rural areas of Central Europe' in G. Horvath (ed.), *Regions and cities in the global world* (Pecs: Centre for Regional Studies) 141–58.

I. Merdjanova 2000, 'In search of identity: nationalism and religion in Eastern Europe', *Religion, State and Society* 28, 233–62.

R. Meszaros 1996, 'The future of villages in the process of economic change in Hungary' in H. Sasaki (ed.), *Geographical perspectives on sustainable rural systems* (Tokyo: Kaisei Publications) 406–11.

B. Meyer and P. Geschiere 1999, *Globalization and identity: dialectics of flow and closure* (Oxford: Blackwell).

J.P. Moran 1994, 'The communist torturers of Eastern Europe: prosecute and punish or forgive and forget?', *Communist and Post-Communist Societies* 27, 95–109.

W.B. Morgan 1992, 'Economic reform: the free market and agriculture in Poland', *Geographical Journal* 158, 145–56.

L.A.G. Moss 1996, *Cesky Krumlov: sustainable development indicators project proposal* (New York: Foundation for a Civil Society/Ceske Krumlov: Mayor's Office).

R.D. Mueller 1993, *Market changes and their impact on the structure of food distribution systems in the impaired economies of Central Europe* (Leicester: De Montfort University Leicester Business School Occasional Paper 10).

H. Muent 2000, 'Localised cooperation between small firms in the socio-economic transition process in Poland: two cases from the Poznan region' in J.J. Parysek and T. Stryjakiewicz (eds), *Polish economy: spatial perspectives* (Poznan: Bogucki Wydawnictwo Naukowe) 73–88.

C.K. Nagy 1993, 'Local development strategies' in A. Duro (ed.), *Spatial research and the social political changes* (Pecs: Centre for Regional Studies) 103–7.

G. Nagy 2001, 'Information society and the Great Plain: chances for adaptation', *Alfoldi Tanulmanyok* 18, 90–108 (in Hungarian with an English summary).

G. Nagy and D. Turnock 1998, 'The future of East European small towns', *Regions: Newsletter of the Regional Studies Association* 214, 16–22.

T. Nefedova and Trejvis 1994, *First socio-economic effects of transformation in Central and Eastern Europe* (Vienna: Osterreichisches Ost- und Sudosturopa-Institut).

J.D. Nelson 1997, 'Hungary at the cross-roads: challenges for transport policy in the 1990s' in A. Dingsdale (ed.), *Transport in transition: issues in the new Central and Eastern Europe* (Nottingham: Trent Geographical Papers 1) 42–63.

J. Nemes Nagy 1994, 'Regional disparities in Hungary during the period of transition to a market economy', *GeoJournal* 32, 363–8.

T. Nicolaescu 1993, 'Privatisation in Romania: the case for financial institutions' in D. Fair and R.J. Raymond (eds), *The new Europe: evolving financial and economic systems in Eastern Europe* (Dordrecht: Kluwer) 101–9.

J. Nipper and M. Nutz 1995, 'Break or redevelopment: changes in retail trade in medium-sized towns in the Harz foothills six years after the opening of the wall', *Europa Regional* 3, 15–24 (in German with an English summary).

D. Njegac and A. Toskic 1999, 'Rural diversification and socio-economic transformation in Croatia', *GeoJournal* 46, 263–9.

T. Nowotny 1997, 'Transition from communism and the spectre of Latin Americanisation', *East European Quarterly* 31, 69–91.

D. Ostry 1988, 'The Gabčikovo–Nagymaros dam system as a case study in conflict of interest in Czechoslovakia and Hungary', *Slovo: Journal of Contemporary Soviet–East European Affairs* 1(1), 11–24.

L. O'Toole and K. Hanf 1998, 'Hungary: political transformation and environmental change' in S. Baker and P. Jehlicka (eds), *Dilemmas of transition: the environment, democracy and economic reform in East Central Europe* (London: Frank Cass) 93–112.

A. Pailhe 2000, 'Gender discrimination in Central Europe during systemic transition', *Economics of Transition* 8, 505–35.

M. Pak and V. Brecko 1999, 'Problems of agriculture in Slovenia with special reference to Cirkovce', *GeoJournal* 46, 257–61.

M. Palairet 1995, '"Lenin" and "Brezhnev": steelmaking and the Bulgarian economy 1956–1990', *Europe Asia Studies* 47, 493–505.

J. Pasiak, P. Gajdos and L. Faltan 2001, 'Regional patterns in Slovak development' in G. Govzelak, E. Ehrlich and L. Faltan (eds), *Central Europe in transition: towards EU membership* (Warsaw: Scholar Publishing House) 330–63.

B. Pavlin 1991, 'Contemporary changes in the agricultural use of land in the border landscape units of the Slovene littoral', *Geographica Slovenica* 22, 1–119.

P. Pavlínek 1992, 'Regional transformation in Czechoslovakia: towards a market economy', *Tijdschrift voor Economische en Sociale Geografie* 83, 361–71.

P. Pavlínek 1995, 'Regional development and the disintegration of Czechoslovakia', *Geoforum* 26, 351–72.

P. Pavlínek 1998a, 'Privatisation and the regional restructuring of coal mining in the Czech Republic after the collapse of state socialism' in J. Pickles and A. Smith (eds), *Theorising transition: the political economy of post-communist transformations* (London: Routledge) 218–39.

P. Pavlínek 1998b, 'The role of foreign direct investment in the Czech Republic's transition to capitalism', *Professional Geographer* 50, 71–85.

P. Pavlínek and J. Pickles 2000, *Environmental transitions: transformation and ecological defence in Central and Eastern Europe* (London: Routledge).

P. Pavlínek and A. Swain 1998, 'Internationalization and embeddedness in the East European transition: the contrasting geographies of inward investment in the Czech and Slovak Republics', *Regional Studies* 32, 619–38.

E. Perger 1989, 'An overview of East European developments' in R. Bennett (ed.), *Territory and administration in Europe* (London: Pinter) 93–110.

G.C. Petrakos 1996, *The new geography of the Balkans: cross-border cooperation between Albania, Bulgaria and Greece* (Chania: University of Thessali Department of Planning and Regional Development).

P. Petrov 1996a, 'Prospects for developing rural and agri-tourism in the central Pre-Balkan region' in G. Niculescu and C. Muica (eds), *Southern Carpathians and Stara Planina (Balkan) Mountains: geographical studies* (Bucharest: Academia Romana Institutul de Geografie, Geographical International Seminars 3) 156–60.

P. Petrov 1996b, 'A concept of construction of the tourist zone between the towns of Levetch and Veliko Tarnovo (Central Prebalkan karst zone)' in G. Niculescu and C. Muica (eds), *Southern Carpathians and Stara Planina (Balkan) Mountains: geographical studies* (Bucharest: Academia Romana Institutul de Geografie, Geographical International Seminars 3) 174–80.

J. Pickles 1998, 'Restructuring state enterprises: industrial geography and Eastern European transitions' in J. Pickles and A. Smith (eds), *Theorising transition: the political economy of post-communist transformations* (London: Routledge) 172–96.

F. Pine 1998, 'Dealing with fragmentation: the consequences of privatisation for rural women in central and southern Poland' in S. Bridger and F. Pine (eds), *Surviving post-socialism: local strategies and regional responses in Eastern Europe and the Former Soviet Union* (London: Routledge) 106–23.

F.T. Pine and P.T. Bogdanowicz 1982, 'Policy response and alternative strategy: the process of change in a Polish highland village', *Dialectical Anthropology* 7(2), 67–80.

D. Plut 1997, 'Slovenia and its adapting to European processes of sustainable development', *Hrvatski Geografski Glasnik/Croatian Geographical Bulletin* 59, 35–47.

H. Puwak 1992, *Poverty in Romania: territorial distribution and the intensity of poverty level* (Bucharest: Romanian Academy, Institute for Quality of Life).

Ramboll Group 1997, *Profiles of the Romanian development regions* (Bucharest: RCG for PHARE Programme, Regional Development Policy).

T. Ratinger and E. Rabinowicz 1997, 'Changes in farming structure in the Czech Rep. as a result of land reform and privatisation' in J.F.M. Swinnen, A. Buckwell and E. Mathijs (eds), *Agricultural privatisation, land reform and farm restructuring in Central and Eastern Europe* (Aldershot: Ashgate) 63–104.

B. Redding and A.R. Ghambari Parsa 1995, 'Budapest' in J. Berry and S. McGreal (eds), *European cities: planning systems and property markets* (London: Spon) 291–320.

V. Rey (ed.) 1994, *Czechoslovakia: transition, fragmentation, recomposition* (Fontenay-St Cloud: Ens Editions) (in French with an English summary).

V. Rey and M. Bachvarov 1998, 'Rural settlements in transition: agricultural and countryside crisis in Central-Eastern Europe', *GeoJournal* 44, 345–53.

G. Ruda 1998, 'Rural buildings and environment', *Landscape and Urban Planning* 41, 93–7.

D. Sadler 1999, 'A divided European future?' in R. Hudson and A.M. Williams (eds), *Divided Europe: society and territory* (London: Sage) 269–76.

D. Sadler, A. Swain and R. Hudson 1993, 'The automobile industry and Eastern Europe: new production strategies or old solutions?', *Area* 25, 339–49.

K. Schaft 2000, 'Old walls/new walls: the construction and mediation of post-socialist rural inequality in the 1990s' in B. Brown and A. Bandlerova (eds), *Rural development in Central and Eastern Europe* (Nitra: Slovak Agricultural University) 141–6.

K. Schmeidler 1998, 'Housing estates in the Czech Republic', *European Spatial Research and Policy* 5(1), 71–4.

K-D. Schmidt and P. Naujoks 1996, 'Deindustrialisation or reindustrialisation?: on the future of the eastern German economy' in M.A. Landesmann and I.P. Szekely (eds), *Industrial restructuring and trade reorientation in Eastern Europe* (Cambridge: Cambridge University Press) 127–48.

G. Schopflin 1988, 'The Stalinist experience in Eastern Europe', *Survey, Journal of East–West Relations* 30(4), 124–47.

L. Sell 2001, 'Kosovo: getting out with peace and honor intact', *Problems of Post-Communism* 48(2), 3–14.

S. Shepley and J. Wilmot 1995, 'Core versus periphery' in A. Amin and J. Tomaney (eds), *Behind the myth of European union: prospects for cohesion* (London: Routledge) 51–82.

K. Skubiszewski 1993, 'Nationalism in Europe today', *Polish Quarterly of International Affairs* 2(3), 11–26.

B. Slee 1999, *A tourism development plan for the Stuzhytsa-Ushanski Park, Ukraine* (Aberdeen: University of Aberdeen Department of Agriculture).

A. Smith 2000, 'Employment restructuring and household survival in "postcommunist transition": rethinking economic practices in Eastern Europe', *Environment and Planning A* 32, 1759–80.

A. Smith and J. Pickles 1998, 'Introduction: theorising transition and the political economy of transformation' in J. Pickles and A. Smith (eds), *Theorising transition: the political economy of post-communist transformations* (London: Routledge) 1–24.

A. Smith and A. Swain 1997, 'Geographies of transformation: approaching regional economic restructuring in Central and Eastern Europe' in D. Turnock (ed.), *Frameworks for understanding post-socialist processes* (Leicester: Leicester University Geography Department Occasional Paper 36) 32–7.

T. Snyder and M. Vachudova 1997, 'Are transitions transitory?: the types of political change in Eastern Europe since 1989', *East European Politics and Societies* 10, 1–35.

C. Staddon 2001, 'Restructuring the Bulgarian wood-processing sector: linkages between resource exploitation capital accumulation and redevelopment in a post-socialist locality', *Environment and Planning* 33A, 607–28.

D. Stark 1995, 'Not by design: the myth of designer capitalism in Eastern Europe' in J. Hausner, B. Jessop and K. Nielsen (eds), *Strategic choice and path dependency in post-structuralism* (Aldershot: Edward Elgar) 67–83.

R. Sternberg 1995, 'Assessment of innovation centres: methodological aspects and empirical evidence from Western and Eastern Germany', *European Planning Studies* 3, 85–97.

M. Stewart 1997, '"We should build a statue to Ceauşescu here": the trauma of decollectivisation in two Romanian villages' in S. Bridger and F. Pine (eds), *Surviving post-socialism: local strategies and regional responses in Eastern Europe and FSU* (London: Routledge) 66–79.

J.E. Stiglitz 1992, 'The design of financial systems for the newly emerging democracies of Eastern Europe' in C. Clague and G.C. Rausser (eds), *The emergence of market economies in Eastern Europe* (Oxford: Blackwell) 161–86.

S. Stroschem 2001, 'Measuring ethnic party success in Romania, Slovakia and Ukraine', *Problems of Post-Communism* 48(4), 59–69.

T. Stryjakiewicz and T. Kaczmarek 2000, 'Transborder cooperation and development in the conditons of great socio-economic disparities: the case of the Polish–German border region' in J.J. Parysek and T. Stryjakiewicz (eds), *Polish economy in transition: spatial perspectives* (Poznan: Bogucki Wydawnictwo Naukowe) 49–71.

I. Suli-Zakar 1996, 'The role of small agrarian entrepreneurs in the sustainability of Hungarian rural systems' in H. Sasaki (ed.), *Geographical perspectives on sustainable rural systems* (Tokyo: Kaisei Publications) 411–16.

W. Surazska 1996, 'Transition to democracy and the fragmentation of a city: four cases of Central European capitals', *Political Geography* 15, 365–81.

W. Surazska, J. Bucek, L. Malikova and P. Danek 1997, 'Towards regional government in Central Europe: territorial restructuring of postcommunist regimes', *Environment and Planning C: Government and Policy* 15, 437–62.

V. Surd and V. Zotic 1999, 'Similarities and differences in Europe's rural space' in V. Surd (ed.), *Rural space and regional development* (Cluj-Napoca: Editura Studia) 223–7.

A. Swain 1998, 'Governing the workplace: the workplace and regional development implications of automotive foreign direct investment in Hungary', *Regional Studies* 32, 653–71.

N. Swain 1994, *Agricultural development policy in the Czech Republic: is one really necessary?* (Liverpool: University of Liverpool Centre for Central and East European Studies, Working Papers Rural Transition Series 12).

N. Swain 2000, 'Rurality in modern societies with a particular focus on the countries of Central and Eastern Europe' in B. Brown and A. Bandlerova (eds), *Rural development in Central and Eastern Europe* (Nitra: Slovak Agricultural University) 21–34.

J.F.M. Swinnen and E. Mathijs 1997, 'Agricultural privatisation land reform and farm restructuring in Central and Eastern Europe: a comparative analysis' in J.F.M. Swinnen, A. Buckwell and E. Mathijs (eds), *Agricultural privatisation land reform and farm restructuring in Central and Eastern Europe* (Aldershot: Ashgate) 333–73.

L. Sykora 1993, 'City in transition: the role of rent gaps in Prague's revitalisation', *Tijdschrift voor Economische en Sociale Geografie* 84, 281–93.

A. Szakolczai 1996, *In a permanent state of transition: theorising the East European condition* (Florence: European University Institute Working Paper SPS 96/9).

I. Szelenyi 1996, 'Cities under socialism – and after' in G. Andrusz, M. Harloe and I. Szelenyi (eds), *Cities after socialism: urban and regional conflict in post-socialist societies* (Oxford: Blackwell) 286–317.

T. Szigetvari 2001, 'Regional development in Hungary' in G. Gorzelak, E. Ehrlich, L. Faltan and M. Illner (eds), *Central Europe in transition: towards EU membership* (Warsaw: Scholar Publishing House) 287–309.

A. Tickle and I. Welsh 1998, 'Environmental politics civil, society and postcommunism' in A. Tickle and I. Welsh (eds), *Environment and Ssociety in Eastern Europe* (London: Addison-Wesley-Longman) 156–85.

V. Tismaneanu and D. Pavel 1994, 'Romania's mystical revolutionaries: the generation of Anst and adventure revisited', *East European Politics and Societies* 8, 402–38.

J. Tomes and M. Hampl 1999, 'The development of regional differentiation in Eastern Central European countries during the transformation era' in M. Hampl, *Geography of societal transformation in the Czech Republic* (Prague: Charles University Department of Social Geography and Regional Development) 131–52.

D. Topalovic and M. Krleza 1996, 'Croatia and Central Europe: the geopolitical relationship', *GeoJournal* 38, 399–405.

J. Toth 1992, 'Villages, cities, counties, regions and peculiarities of urbanization in Hungary' in W. Zsilincsar (ed.), *Zur okonomischen und okologischen Problematik der Stadte Ostmitteleuropas nach der politischen Wende* (Graz: Karl Franzens-Universität Institut für Geographie) 10–46.

J. Toth 1993, 'Historical and today's socio-economic conditions of regionalism in Hungary' in A. Duro (ed.), *Spatial research and the social-political changes* (Pecs: Centre for Regional Studies) 15–28.

V. Trnkova 1994, *Social policy in the Czech Republic with special reference to the rural situation* (Liverpool: University of Liverpool Centre for Central and East European Studies, Working Paper – Rural Transition Series 12).

D. Turnock 1997, *The East European economy in context: communism and transition* (London: Routledge).

D. Uhlir 1998, 'Internationalization and institutional and regional change: restructuring post-communist networks in the region of Lanskroun, Czech Republic', *Regional Studies* 32, 673–85.

A. Vaishar 1992, 'Ethnic structure of the Czech Republic in the census of 1991 and its connections', *Geographica Slovenica* 23, 385–401.

A. Vaishar 1993, 'Ethnic, religious and social problems of frontier districts in the Czech Republic', *Geographica Slovenica* 24, 167–77.

S. Van Der Boel 1994, 'The challenge to develop a border region: German–Polish cooperation', *European Spatial Research and Policy* 1(1), 57–72.

J.D. Van Der Ploeg and S. De Rooij 2000, 'Agriculture in Central and Eastern Europe: industrialization or repeasantisation' in B. Brown and A. Bandlerova (eds), *Rural development in Central and Eastern Europe* (Nitra: Slovak Agricultural University) 45–53.

I. Vasary 1990, 'Competing paradigms: peasant farming and collectivization in a Balaton community' in C.M. Hann (ed.), *Market economy and civil society in Hungary* (London: Frank Cass) 163–82.

K. Verdery 1993, 'Ethnic relations, economies of shortage and the transition in Eastern Europe' in C.M. Hann (ed.), *Socialism: ideals, ideologies and local practice* (London: Routledge) 172–86.

K. Verdery 1994, 'The elasticity of land: problems of property restitution in Transylvania', *Slavic Review* 53, 1071–109.

C. Von Hirschhausen 1995, 'From privatization to capitalization: industrial restructuring in post-socialist Central and Eastern Europe' in E.J. Dittrich (ed.), *Industrial transformation in Europe* (London: Sage) 54–78.

H. Von Zon 1996, *The future of industry in Central and Eastern Europe* (Aldershot: Avebury).

M. Vrabete and C. Popse 1999, 'Strategical approach for the valuation of potential at the county level' in V. Surd (ed.), *Rural space and regional development* (Cluj-Napoca: Editura Studia) 258–65.

T. Vuics 1992, 'The new poverty in the environment of Pecs' in W. Zsilincsar (ed.), *Zur okonomischen und okologischen Problematik der Stadt ostmitteleuropas nach der politischen Wende* (Graz: Karl Franzens Universität Institut für Geographie) 53–66.

H.G. Wanklyn 1941, *The eastern marchlands of Europe* (London: Philip).

P. Watson 1993, 'Eastern Europe's silent revolution: gender', *Sociology* 27, 471–87.

G. Weclawowicz 1996, *Contemporary Poland: space and society* (London: UCL Press).

R. Whitley 1995, 'Transformation and change in Europe: critical themes' in E.J. Dittrich, G. Schmidt and R. Whitley (eds), *Industrial transformation in Europe* (London: Sage) 11–30.

C-C. Wiegandt 2000, 'Urban development in Germany: perspectives for the future', *GeoJournal* 50, 5–15.

A.M. Williams and V. Balaz 1999, 'Privatisation in Central Europe: different legacies, methods and outcomes', *Environment and Planning C: Government and Policy* 17, 731–51.

O.J. Wilson 1998, 'East Germany' in D. Turnock (ed.), *Privatisation in rural Eastern Europe: the process of restitution and restructuring* (Cheltenham: Edward Elgar) 120–44.

J. Witkowski 1993, 'The quality of the natural environment and demographic processes in the large towns in Poland', *Geographia Polonica* 61, 367–77.

H. Wollmann 1997, 'Institution building and decentralization in formerly socialist countries: the cases of Poland, Hungary and East Germany', *Environment and Planning C: Government and Policy* 15, 463–80.

J.A. Yoder 2001, 'West–East integration: lessons from East Germany's accelerated transition', *East European Politics and Societies* 15(1), 14–38.

Index

ABB 136, 141, 173
accessibility xxii, 66, 72, 114, 140, 145, 185, 207, 211–12, 219, 234, 247, 253, 260, 265, 280, 294, 300, 331–2, 349, 353
acquis communautaire *see* European Union
administration/administrator *see also* local government and regulation: political economy and social issues: xxii–xxiii, 1, 6, 12, 22, 31, 35–6, 58–9, 87, 97–8, 103; production and tertiary sector 112, 152, 178, 183–4, 187, 189, 191, 193–4, 202; settlement and regional issues 252, 256, 266, 273, 290, 298–9, 307, 311, 313–8, 327, 330, 338–9, 341, 361, 364, 369, 372, 377, 381, 387
Adriatic 23–4, 26, 34, 40, 192, 204–6, 210, 212, 214, 243, 339, 362, 373
advice *see* information
agricultural machinery 102, 148, 152–6, 165, 176, 184, 207, 294, 299, 304
agriculture: production and tertiary sector 117, 137, 148–70, 174–6, 183–8, 229, 231, 249; regional issues 309–11, 313, 320–1, 324–5, 330–1, 333–5, 339, 346, 348, 350, 352, 354–5, 360; socio-political issues xxii, 2, 24, 61, 66, 69, 72, 84, 86–8, 101–2, 104–5; settlement 256, 290, 293, 297, 299–305; sustainability 376–7, 383, 387–9
agri-environment programmes 387–8
agrotourism 153, 176, 185, 249, 295, 301, 335, 360
Ahold 207, 274
aid 10, 12, 31–2, 34, 37, 39, 54, 60, 65, 148–70, 174–6, 184–6, 318–29
airport/air transport *see* aviation
Albania/Albanians: political economy xx–xxi, 1, 3, 5–7, 10–11, 21–2, 34–8, 43–5, 47–9, 53; production 117–18,

121–2, 126, 139, 151, 154–5, 157, 160, 163–70; regional issues 312, 315, 318, 333, 335; settlement 252, 254–5, 258, 261–3, 286–9, 292, 294–5, 302; social issues 58–60, 65, 68, 71–2, 74, 76–9, 82–3, 86–9, 93, 97, 100, 101–3, 107; sustainability 371, 376; tertiary sector 187–8, 191, 195–6, 198–200, 205–6, 210–1, 213, 218–20, 224, 227–8, 231, 235, 243, 246–8, 251
Alcatel 136, 173, 219, 354
alcohol 28, 58, 66–7, 72–3, 84, 96, 209, 264, 272, 294, 354
Amoco 173–4
Ansoldo 239–40
Antall, J. 105–6
apartment *see* housing
arable *see* crops
architecture *see* buildings
armaments/arms 17, 28, 102, 124, 366, 370
armed forces 2, 8, 12, 14, 16, 20–39, 46, 74, 97, 102, 110, 124, 140, 156, 170, 203, 210, 216–7, 241, 275, 286, 319, 334, 349, 367, 370
arts *see* culture
assimilation 17, 91, 96
assistance *see* aid
Auchan 274, 277
Audi 108, 137
Austria/Austrians: political economy and social issues xx, 2, 8, 16, 22, 26, 32, 43, 88; production issues 122, 127, 134, 139–40, 150, 156, 181, 183–4; settlement and regional issues 273–4, 278, 321, 330, 333–5, 337, 339, 366, 372–4, 379; tertiary sector 194, 204–5, 207–8, 210, 213, 223, 234–5, 237–9, 242–3, 248–9; *see also* Habsburg Empire
autarky *see* self-sufficiency

automobile industry xxi, 13, 31, 105, 115, 121–2, 126–31, 135, 140, 143, 181, 193–5, 197, 227, 279, 319, 329; *see also* cars and trucks

autonomous region/autonomy 21–4, 26, 39, 61–2, 90–2, 95, 112–3, 133, 193, 293, 299, 307, 318, 329, 354, 369–70, 382

aviation 73–4, 90, 124, 135–6, 139, 141, 143, 181, 191–4, 214–8, 244, 257, 277–81, 318–9, 347, 349, 354, 364, 370, 372

backwardness/backward region 16, 40, 111, 147, 186, 192, 248, 300, 303, 308, 318–29, 332

Balkan Reconstruction Agency 9–10

Balkans xvii, 1, 4–5, 20, 39–40, 59, 74, 89, 92, 101, 205–6, 210, 232, 242, 250, 292, 302, 321, 332–3, 339, 371; *see also* named Balkan states

Baltic 2, 4, 192, 208, 210, 212–3, 227–8, 237, 245, 321, 337, 339, 372, 374

Baltic States 8–10, 139, 147, 203, 209

Banat 82, 143, 338, 350, 352–61

Banja Luka 25, 28, 31

banking/bank loan: production 113–4, 116, 136, 140, 143, 152, 165, 171, 175–7, 180–1, 184, 186; settlement and regional issues 267, 269, 272, 274–5, 277, 286, 298, 301, 311, 313, 321, 325, 354, 367, 370–1; socio-economic issues 24, 32, 34, 36–7, 46, 50, 52, 67, 84, 102–3, 106; tertiary sector 194, 215, 229, 231, 238, 247, 249

bankruptcy 69, 130, 152, 155, 175, 181, 347

Banská Bystrica 129, 138, 141, 286, 312, 317, 321

barter 41, 240

Belarus/Belorussians xx, 9, 92–3, 139, 196, 203, 207–8, 294, 335, 337, 374

Belgrade xx, 10, 16, 20–1, 24–5, 33, 36–9, 135, 187, 191, 196, 209, 215, 278, 312, 318, 359, 363

Berisha, S. 5, 101–3

Berlin/Berlin Wall: political economy and social issues xx, 2–3, 12, 15–6, 86, 103; production and tertiary sector 134, 136–7, 191–2, 196–7, 201–3, 208–9, 212, 214, 216, 241; regional issues 312, 318, 339, 341–9, 372, 388; settlement 259, 261, 264–5, 269–70, 277, 278–81, 287, 296

biodiversity 231, 234, 248, 250, 335, 340, 360, 371, 376–81, 377–81, 386–8

Bitola 206, 230, 312

Black Sea 2, 34, 174, 178, 192, 204, 206, 208, 210, 213, 216–24, 228, 232, 244, 339, 362, 370, 379

Black Sea Economic Cooperation Group 7, 9, 178

Black Triangle 339–40, 383

blockade *see* sanctions

Boeing 215–6

Bohemia 63, 122, 124, 158, 210, 214, 224, 228–9, 245, 292–3, 296, 308, 320, 349

Bombardier 134, 216

border *see* frontier

Bosnia & Hercegovina/Bosniaks: political economy xviii, xx–xxi, 4–7, 10, 20–32, 36, 44–5, 47–9; production 118, 139–40, 155, 157, 163, 166–70; regional issues 312, 315, 335, 337; settlement 254–5, 261–3, 288–9, 292; social issues 59, 67, 74, 78, 82, 87, 89, 92, 97, 107; sustainability 366, 371; tertiary sector 196, 205, 215, 223; *see also* Federation of Bosnia & Hercegovina, Hercegovina and Republike Srpska boundary *see* frontier

Brandenburg 12, 15–6, 86, 216, 265, 280, 346–8, 388

Bratislava xx, 16, 18, 116–7, 128–9, 140–1, 196–7, 208, 213–4, 217, 275, 278, 278, 284–5, 312, 317–8, 321, 339, 372

brewing *see also* alcohol 32, 103, 117, 126, 174, 176, 184, 272, 293

Brezhnev, L./Brezhnev Doctrine 3, 366

Brno 133, 141, 149, 196–7, 208, 275, 279, 312, 339, 388

brown coal *see* coal

Bucharest: political economy and social issues xx, 58, 82; production and tertiary sector 164, 170–1, 173, 179, 181–2, 184–5, 196, 201, 204–5, 209, 213, 217, 221, 226; settlement and regional issues 258, 264, 274, 278, 284, 286, 312, 350, 354, 358–9, 371

Budapest: regional issues 312, 321–2, 324–5, 334, 336, 373, 390; settlement 252–3, 258, 261, 269–70, 273–8, 278–80, 284–5; socio-political issues and production xx, 66–7, 134–5, 138, 141, 147, 150 – tertiary sector 201, 204, 207–9, 213–4, 243, 245

budget xxi, 10, 36, 46, 50, 69, 73, 85, 135, 152, 171, 177, 238, 290, 294, 299, 314, 372 *see also* finance

building industry/materials *see also* housing
and offices: socio-political issues and
production 32, 84, 120, 124–5, 131–3,
140, 149, 151–2, 175, 181, 185; tertiary
sector 187–9, 191, 224, 229, 238, 245,
249, 251; settlement 258, 261, 264–71,
277, 280, 282, 284–6, 295–7, 299;
regional issues 309, 314, 321, 331, 341,
347, 349, 354, 356–8, 377, 388
Bulgaria/Bulgarians: political economy
xx–xxi 1, 3, 5–11, 22, 32–4, 43–5, 47–9;
production 117–20, 149–51, 153–4, 157,
162–70; regional issues 309–10, 312–5,
333, 335, 337, 358; settlement 254–5,
261–3, 273, 287–9, 292–3, 302; social
issues 63, 65, 67–8, 71, 74, 76–9, 82,
83–4, 87, 90–3, 96–7, 103–5, 107;
sustainability 367–9, 371, 381, 383, 386;
tertiary sector 188, 191, 194–200,
205–6, 209–10, 213, 215, 217–9, 221–2,
224–5, 228–30, 232, 235–8, 240–4,
246–8, 250
bureaucracy *see* administration
Burgas 191, 196, 210, 213, 228–9, 312
bus 128–9, 130–1, 192, 195–6, 200, 206,
212, 215, 217, 266, 278–9, 296, 300
business/business community/services *see
also* private economy: socio-political
issues 11, 46, 51–4, 64, 66, 69–70, 72–3,
84, 90, 96, 98, 103, 105–6; regional
issues 307, 314, 319, 321, 324, 327–31,
334, 339, 344, 346, 348, 354, 356,
358–9, 362, 369–70, 382, 387–9;
production 112–5, 117, 137–8, 140–1,
144, 151–2, 171, 174, 177, 180–1;
tertiary sector and settlement 187, 210,
222–4, 226, 252, 257, 264–9, 267–9,
271–2, 293, 295–6, 300–1
business centre *see* offices

canals *see* waterways
capital/capitalist *see also* foreign investment
and investment: production 114, 123,
126, 132, 134, 155, 171, 173, 184;
regional issues 307–8, 319, 322, 329,
349, 356, 358, 365, 369–70, 372, 376;
socio-political issues 4, 6, 11, 14, 41–2,
50, 52, 54, 59, 105, 108; tertiary sector
and settlement 218–9, 223, 243, 247–9,
256, 257–8, 275, 285, 296, 299, 301,
303
capital cities *see also* named cities 61, 86,
99, 132, 134–6, 146, 192–3, 216, 252–3,
260, 262–4, 267, 270, 278, 285, 308,
311, 375

cargo *see* freight
Carpathians 82, 87, 158, 176, 184–5, 187,
200, 204, 231, 233, 241, 247, 252–3,
280, 297, 301, 304, 306, 333–5, 337–8,
340–1, 381–5, 388–9
Carrefour 274, 281
cars/car parking: production and tertiary
sector 117, 119–20, 125, 126–31, 137–9,
141, 145, 174, 192, 194–5, 206, 209,
211, 227; settlement and regional issues
258, 265, 267, 269, 271, 273–5, 277–80,
286, 293, 296–7, 300, 351; socio-
political issues 13, 18, 73, 78, 102,
Ceauşescu, N. 58, 100, 131–2, 170, 247,
258, 292
cement *see* building materials
central business district (CBD) 259, 265–9,
271, 277, 279–80, 282, 327, 343–6
Central European Free Trade Association
7, 9, 104, 165, 183, 208, 222, 340, 354,
366, 368
central planning 1–2, 6, 42, 51, 62, 111–3,
126, 131–2, 134, 150, 156, 171, 191–2,
242–9, 257, 260, 292, 352, 365
cereals 39, 78, 149, 153, 156, 158–9, 162,
164–6, 170, 176, 184–6, 212–3, 285,
330, 347, 360
České Budějovice 238, 296, 311–2,
333
chain *see* network
Chamber of Commerce & Industry 22,
137, 139, 318, 328, 360–2
chemical industry 13, 23, 131, 139, 156,
172, 181, 183, 212–3, 223, 256, 304,
320, 330, 383
Chemnitz 13, 88, 256–7, 278–9, 312
children 65, 72–3, 78, 84–5, 114, 150, 185,
277, 310, 353, 356, 368
Christian Democrats 18, 20, 23
churches *see also* religion 12, 29, 35, 38,
59–61, 97, 138, 148, 170, 265, 285–6,
293, 298, 303, 330, 333, 335, 342,
360–1
cigarettes *see* tobacco
cities *see* urbanisation and named cities
citizenship *see* civil society
city centre *see* central business district
civic liberties *see* civil society
civil defence *see* defence
civil service *see* administration and local
government
civil society 3, xix, xxii, 10, 35–6, 41–2,
52, 54, 56–110, 115, 144, 147, 153,
178, 232, 264, 300, 328, 376, 382,
387

civil war 4–6, 10, 20–39, 42, 59, 78, 89, 104, 129, 139, 171, 194, 206, 247, 335, 362, 367, 369, 371

climate 37, 39, 148, 155, 157, 165, 167, 173, 221, 231, 242–3, 245, 285, 292, 302, 376, 378

clothing xxi, 24, 125–6, 183, 272, 277, 359

Cluj-Napoca 95, 145, 173, 179, 182, 190, 217, 233, 253, 312, 386

coal 31, 67, 104, 119, 123, 125, 138, 141, 143. 160, 191, 196–7, 203, 221–2, 224, 226–7, 229–31, 238–9, 253, 256, 292, 311, 320, 339–40, 347, 352, 356, 360

coalition 9, 15, 18–20, 23, 25, 29–30, 34–5, 37, 57, 95, 98, 100–1, 103, 106, 109, 176–7, 367, 369, 382, 388

coast 207, 210–3, 242–7, 376, 383 *see also* Adriatic/Baltic/Black Seas and ports

Coca-Cola 137, 173, 181

cohesion *see* integration

Cold War xix, 333

Colgate-Palmolive 114, 135, 173, 181

collective farm *see* cooperative farm

Comecon 3, 13–4, 40–1, 111, 192–4, 218, 221, 232, 264, 366

command economy *see* central planning

commerce *see* shopping, trade etc.

communism/Communist Party: political economy xviii–xix, xxii, 1–3, 15–7, 19–22, 35, 40, 51–5; production 111, 116, 125, 128, 131–5, 137–8, 146, 148, 151–2, 155–6, 158, 160, 162, 164, 170–2, 175–6; regional issues 307–8, 311, 314–5, 321, 324, 332–3, 339, 352, 358; settlement 252–61, 264–6, 269–72, 275, 278, 280, 284–7, 293, 299, 302; social issues 56–7, 59–60, 60–2, 64–5, 67, 69, 71–2, 74, 84–5, 88–9, 91–2, 96–7, 99–101, 103–5, 108–10; sustainability 365–7, 369, 375–6, 381–2, 388–90; tertiary sector 191–3, 197, 206, 208, 213–5, 218, 221, 226, 235, 242–5, 248–9

commuting 15, 86, 144, 155, 191, 252–3, 258, 265, 287, 290, 293, 296, 347, 369

competition: production 113–4, 116–7, 121–2, 125–6, 128, 130, 143, 153, 156, 165, 171–2, 175, 177–8, 181, 183–4; settlement and regional issues 257, 260, 264, 271–2, 278, 284–5, 311, 318, 331, 339, 347, 354, 366, 369, 375; socio-political issues 2, 41, 50, 54, 63–4, 67; tertiary sector 193–4, 201, 211–4, 216–9, 223–4, 227–9, 235, 242

components 127–31, 133, 147, 181, 216, 324

computers *see also* information technology 135, 143, 219, 227

confederation 21, 26, 28, 30–2, 92

Confederation of Independent States (CIS) 89, 135, 183, 194, 209, 211–2, 222, 367, 370

conflict *see* war

congestion 135, 137, 308, 376 do urban

Conoco 229, 274

conservation 138, 164, 221–2, 235, 238, 244, 248–9, 259, 261, 284–6, 301–2, 335, 342, 360, 377, 379, 381, 383, 387–90

Constanţa 72, 80, 159, 165, 179, 182, 184, 192, 196, 204, 209, 213, 217, 228–9, 312, 362, 371

construction *see* building

consumer/consumer goods/industries: production 111, 114, 126, 128, 137–9, 156, 165, 172–3, 175, 177–8, 180, 183, 185; regional issues 313, 322, 330–2, 351, 356, 368, 370, 373, 375, 378, 382; settlement 257, 260, 270, 272, 275, 282, 287, 293–4, 306; socio-political issues 40, 52–3, 65, 78, 84, 86, 90, 101; tertiary sector 191, 211, 215, 218–9, 221, 223–4, 226, 228, 231–2, 235, 245, 248

Continental 143, 354

cooperation/cooperative *see* aid, cross-border cooperation, foreign relations etc: regional issues 321, 327–30, 332, 334–5, 338, 340–1, 349, 358, 360, 371–2, 382; socio-political issues and production 9–10, 60, 94, 137, 139, 150, 178, 181, 184–5; tertiary sector and settlement 197, 209, 217, 222, 229, 248, 285, 293, 298, 300, 304

cooperative 46, 69, 109, 111, 124, 259, 264, 271, 287, 293

cooperative farm 148–55, 174–5, 287, 290–1, 295, 301, 375

Cora 274, 277

corruption xxi, 19, 25, 51, 58, 60, 73, 101–4, 106, 366 *see also* crime

costs *see also* rent, taxation etc: production 111, 114–6, 119, 123, 125–7, 129–31, 133–5, 139, 147, 149, 152, 173–4, 180, 183–4; settlement 257, 259, 265, 267, 279, 287, 296, 300; socio-political issues: 31, 40–1, 71, 86; regional issues 321, 325, 330–1, 333, 354, 356, 366–7, 370, 373, 375, 377; tertiary sector 191–2, 194, 201, 210, 213, 222, 227, 229, 231, 235, 238, 240

Council of Baltic Sea States 7–9
Council of Europe 7, 9, 25, 72, 94–5,
    332–3, 338, 373
counterurbanisation 86, 146, 294–6, 299,
    302
crime 8–9, 29, 34–7, 39–40, 46, 53, 58,
    65–6, 73–4, 88, 90–1, 96, 101–4, 106,
    136, 139, 176, 209–10, 248, 251, 264,
    273, 277, 281, 285, 332, 334, 351, 353,
    366, 369, 373
Croatia/Croatians: political economy
    xx–xxi, 4–7, 10–1, 16, 20–32, 36, 38,
    43–5, 47–9; production 116, 118, 139,
    157, 163–4, 166–70; regional issues 312,
    315, 318, 333, 335, 337; settlement
    254–5, 261–3, 264, 274, 288–9, 295;
    social issues 60, 65, 67–8, 76–9, 82–3,
    89, 92–3, 95, 97, 102, 107–8;
    sustainability 376, 387; tertiary sector
    188, 195–6, 198–200, 205, 208, 210,
    213–5, 219, 223, 228–31, 237, 240–1,
    246–7
crops *see also* cereals and industrial crops
    154, 160, 163, 168, 185, 295, 298, 377
cross-border cooperation 8, 94, 108, 139,
    144, 302–3, 320, 322, 332–41, 348–9,
    361–4, 370–4, 381, 388–9
Csurka, I. 100, 108
culture: production and tertiary sector 135,
    139–41, 143, 172–3, 175–6, 180, 243–5,
    248–50; settlement 257–8, 260, 265,
    267, 269–70, 277, 285–6, 293–4, 300–2;
    socio-political issues: xxiii, 1, 8, 20, 30,
    34–5, 37–8, 50, 52, 56, 58–61, 70–1,
    90–6, 98–9; regional issues 315, 322,
    329, 331–5, 338, 341, 344, 348, 350–1,
    353, 356, 360–2, 366–8, 374, 377, 381,
    387
currency *see* finance
customs *see* tariff
Czechoslovakia/Czecho-Slovakia/
    Czechoslovaks: political economy and
    social issues xviii, 2–4, 6, 9, 11, 16–21,
    74, 89; production issues 119–20, 126,
    155–7, 163, 166–7, 169; settlement and
    regional issues 256, 262–3, 273, 275,
    287, 292, 308–11, 319–20; tertiary
    sector 188, 198, 204, 209, 211, 213–5,
    221–2, 232, 234–5, 238–9, 243
Czech Lands/Czech Republic/Czechs:
    political economy xix, xx–xxi, 4–8, 11,
    16, 18–20, 43–9; production 117, 120–1,
    123–4, 126–7, 129–31, 134–5, 139–40,
    143–4, 153–4, 157, 162–70; regional
    issues 308, 311–5, 319–20, 331–3, 335,

337, 339, 349; settlement 254–5, 261–3,
    271, 273, 288–9, 293, 295–6; social
    issues 62, 65, 67–8, 69, 71, 74, 76–7, 79,
    82–3, 88–9, 92–3, 96–7, 106–7;
    sustainability 368, 372, 377, 383, 386,
    388; tertiary sector 188, 195–7,
    198–201, 206–8, 211, 215, 218–9, 222,
    225, 226, 228–31, 236, 238–40, 246–9

Daewoo 128–31, 143, 174, 180–1, 183,
    267, 369
Daimler–Benz 136, 181, 201
Dalmatia 29, 230, 235, 242, 247, 318, 376
Danube 7–8, 10, 104, 106, 119, 158, 165,
    192, 194, 204, 209, 221, 232–5, 248,
    277, 284, 295, 332, 334–5, 337, 339,
    359–60, 362–3, 371–3, 379–81
data *see* information
Dayton Agreement 25, 27–31, 36, 98, 371
Debrecen 137, 147, 197, 204, 208, 252,
    261, 274, 312, 322, 336, 386
debt 24, 46, 50, 106–7, 111, 116, 123–4,
    136, 148, 152, 171, 176, 180, 183, 215,
    227, 359, 367
decentralisation 13, 20, 51, 56, 61–2, 100,
    113, 132–4, 146, 256, 260, 290, 299,
    301, 308, 311, 314, 318, 320, 322, 324,
    327–9, 331, 339, 341, 346, 349, 372
defence *see* armed forces and security
deindustrialisation 137–8, 187, 264, 266,
    344
demand *see* consumption and market
democracy 17–8, 34, 38, 42, 52, 56–9, 62,
    64, 69, 99, 101, 151, 291, 307, 311, 318,
    330, 360, 366–8, 376; *see also* election
demography *see* population
deportation *see* migration
depression *see* recession
devolution *see* decentralisation
diaspora 108, 144, 146, 180, 245, 275,
    314
disembedded *see* embedded
dispersal *see* decentralisation
distilling *see* alcohol
distribution *see* transport
diversification/diversity 67, 105, 113, 136,
    149–50, 160, 176, 185, 215, 227, 251,
    295, 303–4, 311, 318, 320, 325, 329,
    339, 353–6, 358–9
Djindjić, Z. 5, 37
Djukanović, M. 5, 38–9
Doboj 29–30, 312
downsizing *see* deindustrialisation and
    restructuring
Drava 231, 334, 337, 380–1

Dresden 15–6, 86, 196–7, 202–3, 207, 214, 217, 221, 257, 261, 279, 287, 296, 312, 348–9, 372, 386
Drnovšek, J. 5, 23
drought *see* climate
drug trafficking *see* crime
Durrës 86, 131, 205, 210–1, 213, 223, 286, 312, 383
Dzurinda, M. 5, 18

East Germany/East Germans: political economy xx–xix, 3–4, 11–6, 54; production 116, 118, 120, 125–7, 136, 152–3, 156–7, 162, 171; regional issues 309, 312, 319, 337, 341–9; settlement 252, 257–61, 264–5, 270–2, 275, 278, 287, 292–4, 296; social issues 58, 66, 78, 85–6, 89–90, 222–3; sustainability 388–9; tertiary sector 191–4, 196–8, 200–2, 206–7, 211, 214, 216, 218, 221–4, 230, 235, 237, 247
ecofarming *see* organic farming
ecology *see* environment
economic development/policy/reform/ economy: production and tertiary sector 111–251; regional issues 319–21, 328–32, 334, 338–9, 344, 348, 350, 354, 359, 365, 367, 369–71, 373–4, 376; settlement 257, 260, 299–300; socio-political issues 6, 12, 20–1, 24, 37, 53, 66, 69, 87, 94, 96, 102–6
economic zones 125, 128, 140–3, 194, 217, 227, 269, 319, 333, 372 *see also* industrial estate
Economist Intelligence Unit xxii, 44–5, 47–9, 68, 107, 118, 157
ecotourism *see* tourism
education: production 114, 134, 138, 140, 143–4, 148, 162, 174, 176, 185, 187; socio-political issues 3, 14, 17, 24, 29–30, 35, 40, 46, 52, 59–60, 63–4, 67, 71–2, 78, 84–8, 92, 95–6, 106; regional issues 319, 324, 351, 354, 356, 367–8, 374, 376, 378, 388; tertiary sector and settlement 210, 257, 270, 272, 280, 290, 297–9, 303 efficiency: production 111, 113, 117, 123, 131–4, 151, 156, 162, 164, 172, 175, 180, 184; settlement and regional issues 260, 279, 282, 287, 294, 299, 311, 318, 330, 340, 344, 339–40, 352, 365, 376, 378; socio-political issues 50, 52–3, 56, 58, 62, 69; tertiary sector 193–4, 201, 212, 215–6, 224, 229
effluent *see* sewage and waste

egalitarianism *see* equity
Elbasan 103, 117, 121–3, 131, 232, 286, 312
Elbe 202, 211, 214, 372, 379
electricity/electrification *see also* nuclear power and water power xxi 31, 37, 99, 104, 120–1, 123, 131–2, 145, 160, 172, 180, 183, 186, 181, 197, 226, 228–41, 256–7, 282, 290, 298, 330, 340, 352, 382
election 3–4, 12–3, 17–20, 23–5, 30–2, 37, 39, 51–3, 62, 96–110, 101, 103–4, 112, 116, 170–1, 175, 176–8, 180, 184, 189, 201–5, 216, 220–40, 294, 311, 314–5, 319, 367
electronics 135, 183, 218, 319, 324–5, 354
elites xxii, 3, 8, 31, 40, 50–1, 54, 56–7, 60, 65, 70, 97–8, 115, 148, 170, 257, 270, 277, 294, 327–9, 344, 365, 367, 374, 389
embeddedness 42, 52–3, 66, 113–6, 129, 209, 329, 369, 381
emigration *see* migration and refugees 86–90, 95
emissions *see* pollution
Employee–Management Buy-Out (EMBO) 13, 15, 112, 123, 136, 181, 344
employment: production 119, 122, 124–5, 126, 129, 133, 137, 141, 146, 152, 154, 162, 172, 180, 185; regional issues 308–9, 313, 329, 332, 352, 358, 369, 375, 383, 387, 389; socio-political issues 13, 40, 43, 51, 53, 56, 64, 66, 69–71, 73, 78, 82–6, 89, 91, 96, 99, 105, 110–11; tertiary sector and settlement 187, 190, 226, 242, 248, 252, 287, 290, 294–6, 300–2
energy *see also* coal, electricity, oil and water power xxii, 19, 23, 26, 65, 104–5, 111, 123, 126, 145, 162, 172–3, 180, 185, 187, 193, 220–41, 247, 287, 339–40, 357–8, 371, 376
engineering xxi, 17, 41, 111, 113, 119, 123–4, 130, 133, 139–40, 155, 183, 210, 222–4, 235, 242, 257, 271, 290, 319, 321, 354, 358–9
enterprise *see* business, innovation, market economy, private enterprise and SMEs
entertainment *see* leisure
entrepreneurialism *see also* business and private enterprise 42, 98, 136–9, 144, 146, 152, 155, 174, 183–6, 257, 265, 286, 295–6, 301, 304, 321–4, 331, 344, 373

environmentalism/environmental issues/
environmentalists *see also* biodiversity,
pollution, sewage etc: production 116,
136–8, 141, 147, 162, 175; regional
issues 271, 277, 280–2, 284, 294–5, 303,
308, 311, 319–21, 327–31, 333–5, 349,
354; socio-political issues xxii, 3, 8–10,
14, 29, 40, 57–8, 106, 110; sustainability
365, 371–2, 377, 382, 387–8; tertiary
sector 208–10, 214, 221–2, 224, 226–7,
230–2, 234–5, 237, 242, 244, 248
environmental impact assessment 10,
237–9, 277
environmental non-governmental
organisations (ENGOs) 58, 84, 231–2,
238, 240–1, 248, 377, 379, 388
equilisation/equity 131–3, 260, 287, 308,
318–29
erosion 102, 158, 234, 248, 301, 303, 377,
383
ethnic cleansing/ethnicity *see also*
minorities: production and tertiary sector
115–16, 172, 177–8, 180, 217, 234, 249;
settlement and regional issues 259, 273,
299, 304, 314–5, 329, 331–4, 340, 350,
356, 360, 368, 370–1, 389; socio-
political issues 1, 4, 8–9, 17–8, 20–39,
41, 52, 54, 56, 58–61, 66, 87, 89, 90–6,
99–100, 106
Eurocorridors 194, 196, 209–11, 332, 379,
386
European Bank for Reconstruction &
Development xxi, 7, 45, 109, 122–3,
147, 184, 195, 211, 218, 264, 271,
286
European Investment Bank 217, 371
European Union: *see also* Euroregion,
Eurostat and Poland-Hungary . . .
(PHARE): enlargement xix, 6–8, 10, 16,
18–9. 25–6, 31, 34, 43, 60, 94, 104,
108–9, 136, 178, 181, 185–6, 194,
237–8, 313, 322, 332, 334, 366–8, 370,
374–6; political economy xviii–xix, 1, 6,
10, 15, 18, 22, 30, 33–4, 38, 42, 52–5;
production 121–2, 126, 128–9, 136–7,
140, 143, 151, 156, 158, 165, 172, 183;
regional issues 308, 311, 313–5, 320,
327–8, 332–3, 338–9, 348–54, 359;
social issues 58–9, 61–2, 71, 74, 88–90,
99, 103, 105; sustainability 366–7, 369,
373–5, 386–7, 390; tertiary sector and
settlement 210, 229–31, 238, 249, 264,
275, 281, 304
Euroregion 94, 139, 144, 295, 334,
336–41, 348–9, 362–4, 374

Eurostat xxii, 43, 65, 71, 76–7, 79, 83,
163, 188, 195, 199–200, 219, 246
exports: production 114, 117, 119, 126–31,
137, 139, 144, 148, 165, 167, 171–3,
175, 177–8, 180–4, 186; settlement and
regional issues 273, 308, 311, 320, 348,
362, 366, 369–70, 374, 382; socio-
political issues xxi, 7, 14, 18, 26, 36,
38–9, 41, 46, 53, 70, 72, 99, 105–6, 108,
111; tertiary sector 201, 212, 220–2,
228, 230, 238, 240, 243
expulsion *see* migration

factory *see* industry
family/family farming xxi, 65–7, 69–70,
74, 84–5, 89, 91, 101, 146, 150, 152,
154–5, 162, 174–6, 184–6, 260, 265–6,
270–1, 277, 279, 290, 295, 297,
299–302, 304, 353, 356, 366, 374,
387
farm/farming *see* agriculture, crops and
livestock
federal government/federation xviii, 4, 8,
16–40, 89, 216, 278, 318, 342, 388
Federation of Bosnia & Hercegovina 28–32
ferry *see* shipping
fertilizer 102, 106, 152, 155–6, 158, 172,
176, 333, 387–8
Fiat 126–7, 130, 133
finance/financial policy/services: political
economy xxi, xxiii, 4, 8, 13–14, 22, 24,
28, 30, 36, 38, 46, 49–50, 52–3;
production 111, 114, 119, 122, 132,
140–1, 146–7, 151–2, 155–6, 171, 175,
177–8, 180–1, 184, 186; regional issues
308, 311, 313–14, 319–23, 325, 327–30,
333, 335, 338, 340–1, 347, 349–50,
353–4, 356, 359, 366–8, 371–4, 376,
378, 381, 387–8; settlement 270–2, 278,
278–9, 281, 286, 295, 304; social issues
61–3, 69–70, 72, 84, 86, 98, 100, 104,
109; tertiary sector 187, 189, 193–4,
208–10, 212–5, 217, 221–3
First World War 1, 4, 16, 140, 151, 175,
300, 325, 333
fiscal policy *see* finance and taxation
fishing 301, 309, 333, 383
food/food processing: production 117, 121,
125–6, 131, 134, 137, 139, 144, 148,
151–2, 155–6, 162, 165–7, 175–6,
184–6; settlement and regional issues
260, 270, 272, 277, 290, 293–5, 299,
301, 303–4, 319, 330, 339, 347, 352,
356, 358–60, 375, 387; socio-political
issues 40, 65, 72, 78, 104

Food and Agriculture Organisation (FAO)
xxii, 78, 157, 162, 166–9, 288–9
footwear xxi, 13, 125–6, 131, 134, 183
Ford 127, 130
foreign investment: political economy xxi,
10, 14, 17–19, 22, 26, 31, 34, 38, 41,
43, 45–6, 51–3; production 112, 114–19,
121–4, 126–8, 130–47, 151, 165, 173,
177–8, 180–1, 183–4; regional issues
311, 313, 320, 324–5, 329–31, 333–4,
348–9, 351–3, 357, 367–72, 375; social
issues 56, 59, 66, 70, 98–100, 102,
104–6, 109–10; tertiary sector and
settlement 193, 212, 218, 222–3, 273,
275, 299
forestry *see* woodland
Former German Democratic Republic *see*
East Germany
Former Soviet Union *see* Soviet Union
Former Yugoslav Republic of Macedonia
see Macedonia
free zone *see* economic zone
freight 139–40, 183, 192–9, 203–4, 206–7,
211–3, 215–7
frontiers *see also* cross-border cooperation:
production and tertiary sector 136, 139,
147, 193–5, 197, 201–4, 206–10, 217,
221; settlement and regional issues
274–5, 293–5, 297, 300, 311, 313, 320,
322, 324–8, 332–6, 348–9, 359–64,
369–74, 378, 386, 388; socio-political
issues 2, 4–5, 8, 10, 12–13, 25–6, 29, 32,
34–5, 67, 74, 88, 90, 94–5, 102, 106–8
fruit 78, 126, 150, 153, 156, 162, 164–5,
175–6, 184, 277, 286, 300–1, 306,
346–7, 356, 360, 388–9
fuel *see* coal, energy, oil etc.
furniture 124–5, 172, 277, 298, 330,
359–60

Galaţi 80, 119, 121, 158–9, 179, 182–3,
312
garden city *see* new town
gas *see also* air pollution and sulphur dioxide 33, 123, 146, 172, 174, 192, 212–3,
221–2, 224–31, 238, 293, 299, 322–3,
325, 327, 330, 340, 356, 371, 382
Gastarbeiter 88, 270
Gdańsk 123, 137–8, 144, 196–7, 203, 206,
211–2, 219, 224, 227–8, 237, 245, 285,
312, 316, 319, 383
gender 55, 65, 70–1, 73, 82–5, 141, 155,
186, 295, 358
General Electric Company (GEC) 117,
124, 201, 222

General Motors 117, 127–8, 130, 141
gentrification 259, 270, 343, 346
Georgievsky, L. 5, 34
German Democratic Republic (GDR) *see*
East Germany
German Empire/Germany/Germans: *see
also* East Germany and Prussia: political
economy xviii, xx, 1–2, 4, 8, 15–6, 19,
24, 26, 42–3, 54; production 112, 120,
122, 124, 127, 130, 134, 139–40, 153,
156, 162, 174, 181–2, 184; regional
issues 320, 324, 330–3, 335, 337–40,
342, 348, 352–3; settlement 273–4, 278,
286–7, 293, 295; social issues 66, 73–4,
85, 88, 90, 92–3, 95–6, 106;
sustainability 369, 366, 372; tertiary
sector 192, 196, 198, 201–3, 206–9, 211,
213, 217, 219, 221–3, 239, 241, 243,
245, 249; unification xviii, 12–6, 42, 71,
87, 138, 171, 200, 202, 218, 235, 257,
261, 271–2, 319, 341–9, 366
Gierek, E. 259, 307
global corporations *see* multinational
companies
global economy/globalisation xix, 6, 52,
54–5, 58, 69, 98–9, 105, 113, 115–7,
119, 127, 130, 136, 140, 146–7, 214,
223, 226, 229, 231, 264, 284, 299, 331,
338–9, 342, 352, 366, 374–5, 381–2, 389
Global Environment Facility (GEF) 229,
335
global institutions *see* individual institutions
by name
Gorbachev, M. 2–3, 365–6
government intervention/policy *see* policy
and regulation: production and tertiary
sector 112, 152, 156, 158, 165, 170–86,
192, 215, 229, 231, 237, 240, 243–4;
settlement and regional issues 252, 260,
294, 307, 313–4, 319, 322, 327, 338,
343, 356, 367, 369, 373–4, 387–8; socio-
political issues 16–40, 43, 46, 51, 53–5,
57, 62, 72, 88, 95–110
gradualism xxii, 19, 53–4, 115, 171, 389
Greece/Greeks xx, 2, 9–10, 22, 32–4, 43,
87–9, 93, 102, 180, 194, 196, 205–6,
210, 218, 223, 230, 242–3, 247, 302,
332–3, 335, 337–8, 370–1
green *see* environment
greenfield sites 127, 134, 136–7, 140, 144,
258–9
Gross Domestic/National Product xxi *see
also* production 24, 44, 103, 105, 110,
177–8, 187, 230, 270, 299, 308, 312,
323, 351, 353

growth centre 331, 354, 356
Gyor 105, 117, 127–8, 130, 135, 138, 141,
   143–4, 147, 149, 197, 208, 220, 252,
   274, 312, 322, 336, 339, 372
Gypsies *see* Roma

Habsburg Empire xix, 1, 8, 16, 21, 39, 81,
   157, 214, 350, 361
Halle 197, 201–2, 208, 278, 285, 312
handicraft 51, 176, 249–50, 266, 295, 301,
   319, 321, 361
Hapag-Lloyd 211, 245
Harz 200, 248–9
Havel, V. 5, 17, 285
health 18, 32, 46, 52, 61, 72–4, 78, 84–5,
   99, 173, 180, 189, 210, 257, 286, 290,
   293–4, 298, 351, 353, 356, 362, 367–8
heating 229, 231, 278, 282–3, 297, 378,
   388
heavy industry 46, 86, 172, 229, 339, 367,
   378 *see also* chemicals, engineering and
   metallurgy
Herceg-Bosna/Hercegovina 24, 26, 28–9
   see also B&H and FB&H
heritage *see* conservation
hightech *see* technology
holidays *see* tourism
Horn, G. 100, 105, 235
hotels and catering 90, 137–8, 146, 186,
   193–4, 217, 241–8, 266–7, 277, 279,
   281, 285, 290, 295, 343
household *see* family
housing: production and tertiary sector
   126, 133, 186, 242; regional issues 308,
   319, 325, 339, 341–4, 346–7, 360, 374,
   383, 388–9; settlement 252–3, 257–60,
   264–7, 270–1, 277, 279–81, 284–7, 290,
   293, 295–6, 299, 301–2; socio-political
   issues xxii, 24, 29, 40, 65, 67, 70, 78, 86,
   91, 96, 99
housing estate *see* suburb
Hoxha, E. 100–1
human capital/resources 56, 90, 140, 173
human rights 3, 9, 35, 54, 64, 91–2, 94,
   178, 365
Hungary/Hungarians: political economy
   xix, xx–xxi, xxiii, 3, 5–9, 11, 16, 18,
   22–3, 41, 43–5, 47–9, 53; production
   113, 115, 117–20, 125–7, 129–31, 135,
   139–40, 144, 148–9, 151–2, 154, 156–8,
   163–71, 177–8, 180; regional issues
   307–15, 318–9, 321–3, 327–9, 331,
   333–7, 339–40, 350, 354, 361–4;
   settlement 252, 254–5, 257, 261–3, 266,
   269, 271, 273, 278, 280, 285, 288–90,
   292, 295–6, 300, 302; social issues 62–4,
   65–8, 69, 73–4, 76–9, 82–91, 93, 95–7,
   100, 105–8; sustainability 367–8, 372–3,
   375, 377, 386–90; tertiary sector 188,
   191–2, 194–201, 204, 206–9, 211–6,
   218–20, 222–3, 225, 228, 230–6, 238,
   240, 242–6
hydrocarbons *see* gas and oil
hydropower *see* water power
hypermarket *see* supermarket

Iaşi 72, 80, 159, 179, 182, 190, 217, 274,
   312
IBM 108, 135, 280
identity xix, xxiii 6, 59, 62, 91, 96, 99, 116,
   135–6, 173, 247, 299–300, 311, 328,
   332, 340, 348, 350, 381–2, 388–9
ideology *see* communism and neo-liberalism
Ikea 125, 274, 277, 330
illegality *see* crime
immigration 73–4
image/image building *see* identity
imperialism *see* geopolitics
imports 14, 24, 31, 41, 88, 102, 119, 126,
   128, 130, 145, 148, 156, 165, 167,
   171–3, 175–6, 181, 183, 206, 211,
   221–2, 224, 226–30, 232, 238–9, 272,
   320, 347, 349, 356, 358, 375
income/income tax 33, 41, 64, 78, 84, 99,
   104, 145, 149, 153, 162, 165, 178, 185,
   187, 220, 243, 257, 264, 270–1, 277,
   286, 296–7, 301–2, 311, 343, 354, 366
   see also profit and wages
incubator 51, 71, 114, 134, 175, 322, 331,
   356, 358, 370
industrial archaeology 360–1
industrial crops 153, 156, 162, 165–75
industrial estates 136–7, 140–5, 147, 265,
   277–8, 314, 320, 322, 325–7, 330–1,
   346, 349, 354–5, 374
industrial location 131–47, 266, 269
industrial policy/production/region/
   industry *see also* deindustrialisation and
   named sectors: socio-political issues
   13–6, 27, 34, 36, 38–40, 46, 51, 53, 60,
   65, 67, 71–2, 85–8, 101; production
   111–48, 158, 162, 172–4, 180–4, 176,
   180–5; tertiary sector 187–8, 191, 203,
   207, 213, 217, 220–1, 223, 231–2, 237,
   240–1, 249–50; settlement 252–3,
   257–60, 266–7, 269, 273, 278, 281–2,
   284, 286–7, 290, 292, 295–301, 303–4;
   regional issues 307, 309–11, 313–4,
   319–21, 324–5, 333–4, 339, 344, 348–9,
   352–4, 356, 358, 360, 369, 378, 382

inequality *see* regional development

inflation xxi, 18–9, 24, 46, 48, 65, 98, 104, 106, 110, 123, 171–2, 176–7

informal sector *see* second economy

information/information technology *see also* Internet 52, 63–4, 88, 116, 137, 140–1, 154–6, 174, 185, 193, 195, 207, 220, 227, 250, 278, 280, 300, 302–4, 320–2, 328, 331, 349, 362, 374–5, 387

infrastructure *see also* energy, transport etc: production 114–5, 132, 134, 136–7, 140–1, 146–7, 158, 164, 173, 175, 181, 183, 186; regional issues 308, 311, 313, 319–22, 327, 329–34, 338, 348–9, 354, 356, 367, 369, 372, 374, 376, 382–3; socio-political issues xxi, 22, 26, 29, 41, 61, 70, 100, 102; tertiary sector and settlement 192, 218, 244–5, 248–50, 256, 257–8, 265, 281–4, 287, 293–6, 303–4

innovation 51, 53–4, 70–1, 111, 134–5, 139–41, 155, 202, 210, 213, 295, 319, 330–1, 333–5, 338, 343, 371, 381, 388

instability xviii, 1–2, 23, 53, 94, 112, 114, 172, 177, 192, 247, 332, 354, 368–9, 378

institutions 8, 50, 52–4, 58–60, 62–3, 69–70, 94–7, 100, 102, 106, 114–6, 154, 171, 174, 271–2, 302–3, 308, 314, 321, 324, 327–9, 331, 338, 348, 354, 360, 362, 367–8, 372, 387, 390

Instrument for Structural Policies for Pre-Accession Aid (ISPA) 185, 377

integration *see also* EU and restructuring: production 117, 119, 132, 155–6, 164, 173, 177; regional issues 308, 314, 318–20, 322, 325, 328–9, 331–2, 334–5, 338–9, 343, 349–50, 354–6, 361–4; settlement 257, 264, 279, 284, 293, 295, 299–304; socio-political issues 10, 54, 59, 61–2, 66, 69, 85, 87, 90, 92; sustainability 366, 368, 370–4, 376, 382, 387–90; tertiary sector 191, 194, 207–8, 213, 217, 222, 232

interest 51, 53, 103, 106, 114, 177, 319

international *see* frontier, trade and named organisations

International Criminal Tribunal for the Former Yugoslavia 25–6, 30–1, 37

International Monetary Fund (IMF) 6, 20, 32, 34, 37, 60, 69, 98, 104, 109, 171, 177–8, 245

Internet 218–20, 303, 374, 382

investment/investment fund *see also* capital and foreign investment: production 117, 119, 122, 128, 130, 134–6, 139, 143, 145, 147, 150–1, 154, 156, 158, 165, 171–5, 181, 183–4, 186; regional issues 311, 314, 319–29, 331–2, 352, 357–60, 366, 370, 373, 381–2, 389; settlement 253, 256, 264–5, 279, 282, 290, 299–300, 304; socio-political issues 43–4, 69–71, 87–8, 114; tertiary sector 191, 194–5, 201, 218, 224, 227, 231, 244–9

inward investment *see* foreign investment

irrigation 39, 102, 155–6, 158, 164–5, 184, 347

Islam 5, 26–32, 38, 59, 88, 91–3, 292

Istria 23–4, 205, 247, 318

Italy/Italians xx, 2, 8, 10, 22, 26, 31, 43, 53, 63, 87, 93, 103, 117, 122–4, 130–1, 139, 183, 196, 206, 208–11, 213, 217, 232, 319, 331–2, 337, 358–9, 366

Jews 91, 95–6, 264, 286

job/job creation *see* employment 69–72, 78, 87, 96, 98, 141, 180, 256, 293, 295, 319–20, 322, 347, 354, 357–8, 370, 374, 383

joint venture 46, 89, 113, 116–7, 127–30, 134, 136, 144, 146, 154–5, 174, 181, 193, 201, 213, 215, 218–9, 222–3, 232, 239, 243, 245, 269, 275, 278, 313, 319, 328–30, 348–9, 370

Kadar, I. 105, 307

Kaliningrad 203, 208, 319, 332

Katowice 122, 136, 141–2, 196, 204, 207–8, 212, 219, 226–7, 256, 316, 348

Klaus, V. 18–9, 97

knowledge *see* information

Košice 18, 116–7, 119, 124, 129, 138–9, 141, 196, 204, 308, 312, 317, 321, 330, 337, 386

Kosovo/Kosovars xviii, 5. 10, 26, 32, 34–7, 39, 53, 59, 89, 101–3, 121, 133, 178, 206, 217–9, 335, 362, 371

Koštunica, V. 5, 25, 37

Kragujevac 129, 133, 312

Krajina 23–5, 36

Kraków 1, 122–3, 136, 143, 196–7, 203, 217, 219, 256, 274, 279, 286, 312, 316, 348, 361

Kwaśniewski, A. 5, 108

labour/labour market *see also* employment and unemployment: production 111–12, 114–16, 119, 124–5, 126–7, 129, 134–5, 137, 139, 143–4, 147–8, 151, 154, 172, 181, 183; socio-political issues 2, 32, 34, 38, 40–1, 56, 58, 64–7, 71–2, 78, 84–5, 88, 106; regional issues 308, 311, 313, 322, 324, 330–1, 333, 340–1, 344, 349, 353–4, 360, 369, 373–4; tertiary sector and settlement 191, 201, 243, 248, 257, 290, 294, 300, 302

land/land reform/land restitution/ landownership: production 137–8, 141, 144, 146, 148, 151–5, 162–3, 172, 174–5, 178, 180, 184–6; regional issues 321, 324, 330, 333, 339, 347, 349, 353, 372, 375, 387; socio-political issues 39–40, 42, 61, 87, 91–2, 101–2, 104; tertiary sector and settlement 231, 244, 252, 257–60, 265–9, 277, 281, 287, 294–6, 302–3

language *see* culture

Latvia/Latvians xix, 9, 74 *see also* Baltic States

law/law enforcement xxi, 37, 50, 52, 54, 56–7, 61, 67, 73–4, 97, 101–2, 104, 173–5, 178, 185, 209, 231, 239, 249, 231, 239, 249, 257, 264, 279, 298, 322

leasing 152–3, 165, 215, 247, 249, 266–7, 269, 272, 294–5, 303

legal system/legality/legislation *see* law

legislature *see* parliament

Leipzig 16, 136, 138, 197, 202, 208, 214, 217, 222, 265–6, 270, 278, 296, 312

leisure 15, 126, 138, 144, 153, 164, 241–3, 267, 269, 275, 277, 281, 284, 293, 296, 334, 346–7, 354, 356, 362, 378, 389

Less-Favoured Areas (LFAs) 143, 181–2, 186, 305, 356–60, 373, 377, 383, 387

liberalism *see* pluralism

light industry *see also* named sectors 113, 125–6, 134, 137, 145, 150, 181, 226, 290, 335, 339, 348–9, 354, 383

lignite *see* coal

linkages *see* joint ventures and networks

liquidation *see* bankruptcy

literature *see* publicity

Lithunia/Lithuanians xix, 4, 74, 82–3, 95, 100, 208, 228, 294, 337, 367 *see also* Baltic States

livestock/livestock products 148–51, 153, 155–7, 164–5, 167–8, 176, 184, 193, 231, 293, 295, 301–2, 330, 360, 378

living standard 41, 64–6, 70, 72, 88, 96, 104, 106, 115, 133, 148, 172, 185, 221, 229, 252, 271, 290, 296–7, 307, 329, 353, 366 *see also* consumption and welfare

Ljubljana xx, 22, 196, 209, 217, 278, 312

loan *see* banking

local government: socio-political issues 15, 18, 30, 33, 51, 55, 57, 61–2, 73, 94–5, 97, 103, 109; production 111, 113–16, 126, 131–2, 136–8, 140, 143–4, 146–8, 174, 180; regional issues 315, 320–3, 325, 327–32, 334, 340–3, 341–50, 354, 360–1, 372–4, 383, 390; settlement 259, 264–6, 270–1, 277–8, 281–2, 285–6, 293, 299, 304; tertiary sector 195, 212, 222, 231, 240, 244, 248

localism xix, xxii, 6, 15, 42, 52, 55, 58–9, 61–2, 64, 115–6, 147, 178, 304–5, 354, 369

location/location policy 131–47

Łodz 74, 136–8, 208–9, 267, 278, 284, 286, 312–3, 316, 330

logistics *see* transport

Lucchini 122, 279

Macedonia/Macedonians political economy xx–xxi, 5–7, 10, 20, 22, 32–5, 43; production 118, 121, 123, 157, 163–4, 166–70; regional issues 312, 315, 335; settlement 254–5, 261–3, 288–9, 292; social issues 60, 65, 67–8, 74, 76–9, 82–3, 84, 86, 92–3, 96–7, 102–4, 107 – sustainability 371, 383; tertiary sector 195–6, 198–200, 205–6, 210–1, 217, 219, 229–30, 232, 246, 248

machinery *see* agricultural machinery and engineering

Mafia 3–4, 53, 60, 73, 112, 136, 176, 209, 366

Magdeburg 198, 202, 214, 296, 312

Magyars *see* Hungarians

Majko, P. 101–2

Makro 207, 274

mall *see* supermarket

management *see also* business: production 111–13, 115–16, 121–2, 124, 133, 147–8, 152–3, 155–6, 164, 171; regional issues 320, 325, 327, 330–2, 334–5, 342, 344, 365, 367, 369–70, 376, 389; socio-political issues 13–14, 41, 50–2, 58, 67, 70, 91, 103; tertiary sector and settlement 193–4, 216, 218, 239, 271, 273, 275, 281, 301

manpower *see* labour

manufacturing *see* industry
Maramureş 143, 200, 302, 353, 382
Maribor 129–30, 312, 374
market economy/marketing: political
 economy xviii–xix, xxi, 6, 14, 18–19, 42,
 46, 50–2; production 111, 113–15,
 117–19, 121–2, 124–5, 126, 129–30,
 135, 137, 144, 146–50, 152, 154–6, 158,
 162, 164, 171, 175, 176, 178, 181,.
 183–5; regional issues 311, 313, 324,
 328–31, 333–4, 339–40, 344, 349, 352,
 356, 368, 370, 375, 382, 388; settlement
 260–1, 265–6, 269–70, 272–3, 275, 277,
 279, 294, 297–300, 302–4; social issues
 56–8, 62–3, 78, 88, 94, 98, 100–1, 103,
 105, 108; tertiary sector 195, 201,
 226–7, 235, 242–4, 247, 249
McDonalds 137, 181, 271, 285
McDonnell-Douglas 181, 215
Mečiar, V. 17–8, 53, 116, 314, 321, 381
Mecklenburg–Vorpommern 12, 15, 348
media 22, 24, 35, 57, 60, 63–4, 84, 96,
 101, 113–14, 136, 272, 293, 300, 311,
 328, 341, 351, 353, 362
medical services *see* health
Mejdani, R. 5, 101
Mercedes–Benz 128, 280
merchants *see* traders
Meta, I. 5, 102
metallurgical industry xxi. 23, 31, 38, 103,
 111–12, 116, 119–23, 139, 141, 143,
 172, 183, 204, 223, 253, 256, 320,
 349–50, 352, 359, 361, 364, 369, 382
metro 125, 193, 201, 275, 278–9
Metro 273–4
migrants/migration: production and
 tertiary sector 150, 171, 178, 180, 210,
 243; regional issues 308, 318, 333–4,
 341, 344, 349, 351, 353–4, 368, 370,
 372, 374–5, 377; settlement 256–7,
 259–61, 265, 287, 290, 292–3, 295–7,
 299; socio-political issues 1, 4, 15–16,
 19, 32, 38, 56, 67, 78–82, 86–90, 92,
 102
military *see* armed forces
military-industrial complex 40, 64, 105,
 111, 137, 365–6
Milošević, S. 16, 21, 25, 27, 32, 35–9, 59,
 95
minerals *see* mining
mineral water 241, 247, 301, 359–60
mining 59, 66–7, 82, 101, 103, 111, 122,
 138, 141–4, 171, 177, 181, 185, 187,
 224, 226, 238, 256–7, 290, 292, 306,
 339, 350, 352–3, 356, 358–60, 378

minorities *see* ethnicity
Miskolc 141, 146–7, 197, 208, 252, 312,
 322, 336–7
mobility *see also* migration and transport
modernisation: production 121–3, 139,
 156, 164–5, 172, 175, 178, 180–3;
 regional issues 322, 359, 364, 367, 371,
 378, 383; settlement 252, 256, 264–5,
 270–2, 275, 278, 282, 287, 290, 293–5,
 302–4; socio-political issues 40–2, 60,
 99; tertiary sector 197, 207–12, 214,
 216–8, 225, 227, 229–30, 232, 235,
 237–8, 244–5, 247, 249
Moldavia 80, 82, 170–1, 175, 184, 306,
 327, 331, 339, 341–3, 347, 349–50
Moldova/Moldovans xx, 4, 9, 74, 90, 139,
 156, 165, 196, 240, 337–8, 374
monetary policy/money *see* banking,
 finance and taxation 50
monitoring 9–10, 27, 319, 321
monopoly 2, 20, 41, 63, 113, 155, 176,
 193, 219, 222, 365–6
monopoly party *see* communism
Montenegro/Montenegrins: political
 economy 1, 5, 10, 20–1, 23, 25, 35,
 38–9, 53; settlement and regional issues
 335, 371; social issues and production
 97, 102, 107, 139–40; tertiary sector
 205–6, 215, 228, 232, 248
Morava 315, 337, 339, 372, 380–1
Moravia 16, 62, 121, 124, 210, 229, 297,
 301, 320, 324, 337, 388
mortgage 152, 375
Mostar 24, 28–30, 191, 205, 312
motorway 15, 23, 29, 116, 125, 135, 144,
 147, 192–5, 205, 207–10, 212, 244, 256,
 265–6, 268, 275, 278, 280, 320, 347–8,
 354, 371–2, 386
mountains 158, 160, 164, 176, 185–6, 200,
 231, 241–2, 247, 250, 293, 301–2,
 304–6, 311, 335, 350, 352–3, 356,
 360–1, 378 *see also* Carpathians and
 other named mountains
multinational companies/enterprises *see also*
 named companies 98, 106, 114–15, 173,
 223, 264, 370
municipalities *see* local government
Muslims *see* Islam

Nano, F. 58, 101, 103
national income *see* income
nationalism/nationalist parties 4, 10,
 16–39, 53, 56, 59, 61, 90–6, 99, 102,
 105–6, 109, 116–7, 132, 171, 177–8,
 234, 350, 370

national park 147, 200, 241, 247–8, 302, 334–5, 340, 348, 358, 360, 372
nations *see* ethnicity, minorities and named nations/states
Natura 2000 377, 387
natural gas *see* gas
natural increase *see* population
navigation *see* shipping
navy *see* armed forces
neo-liberalism 6, 17, 19, 43, 46, 54, 58, 61, 98, 108, 114, 270, 319, 367, 372
network/working: production 112–13, 115, 125, 127–9, 134, 139, 147, 152, 154–6, 185; regional issues 321, 324–5, 328–9, 331–2, 334–5, 338, 340, 354, 364, 367, 372, 375, 377, 381, 387–8; socio-political issues 28, 31, 42, 55, 97; tertiary sector and settlement 191–3, 210, 257, 261, 280, 297, 300
New Economic Mechanism (NEM) 41, 113, 134, 252
new town 128, 143, 257, 261, 292
Nitra 141, 226, 265–6, 275, 291, 312, 317, 321
Nokia 135, 138
nomenklatura *see* elite
non-ferrous metals xxi, 30, 34, 36, 38, 103, 111, 122–3, 127, 144, 205–6, 256, 352, 356, 359–61, 383 *see also* metallurgy
non-governmental organisations 18, 84, 98, 185, 244, 311, 322, 328–30, 340, 368, 377, 377, 387–8, 390
North Atlantic Treaty Organisation (NATO) 1–2, 5–8, 10, 12, 16, 19, 27, 34–5, 37, 39, 94, 102, 104, 106, 108, 110, 124, 129, 136, 178, 180, 213, 217–8, 362, 371
Novi Sad 20, 135, 213–4, 312, 362–3
nuclear power/nuclear weapons 2, 8, 26, 104, 221–2, 224–6, 230, 235–41, 319, 332–3, 358

occupations *see* employment
Oder 4, 12, 192, 214, 245, 281, 284, 339, 348–9, 372–3, 378–9
offices 135–7, 265–9, 275, 279, 281, 295, 343–4
Ohrid, Lake 10, 34, 232
oil 18, 28, 32, 34, 40, 102, 135, 183, 192–3, 209, 211–13, 215, 221–9, 238, 339, 349
Oltenia 80, 129–31, 143
one party rule *see* communism
Opel 127–8, 227
organic farming 301, 377, 387

Organisation for Security & Cooperation in Europe 7, 9, 37, 94, 371
Osijek 24, 115, 140, 209, 312, 334
Ostrava 89, 119, 122, 124, 308, 312, 321, 333
Ottoman Empire 1, 10, 21, 59, 61, 91, 114, 191
output *see* production

park 138, 258, 266–7, 269, 281–2, 284, 286, 343 *see also* national park
parliament 5, 12, 18–9, 24, 37–8, 50, 57, 59, 64, 95–6, 101–2, 104, 108, 177, 180, 231, 239, 258, 264, 284, 308, 315, 322, 342 *see also* law and regulation
participation *see* civil society
partnerships *see* joint ventures
passengers 191, 193–4, 196–7, 200, 201–3, 206, 212, 216–7, 278–9
path dependency xxii, 60, 112, 311, 381
peasantry 24, 39–41, 91, 148–50, 151, 154–5, 174–5, 176, 273, 287, 295, 311, 313, 375
Pécs 67, 138, 197, 220, 252, 280, 312, 322, 336
pensions 18, 26, 65, 67, 69
periphery 6, 52–3, 69, 86–7, 90, 99, 121, 129, 139, 144, 146–7, 253, 256, 292, 296, 308, 331, 346, 350, 366–7, 369–70, 372, 382
Peugeot 127–9
Philips 108, 135
pipeline 33–4, 103, 113, 173, 183, 191–2, 195, 199, 209, 213, 223–4, 228–9
planning 46, 54, 138, 185, 211, 222, 247, 265, 267, 277–8, 280–5, 293, 299, 304, 325, 327–8, 340–64, 368, 372–3, 376–7, 382 *see also* cross-border cooperation and central planning
Plovdiv 123, 196, 312
pluralism 3, 16, 22, 56–7, 66, 94, 96–110, 293, 300, 366–8
pluriactivity 87, 150, 154, 176, 301, 314, 374
Plzeň 113, 140, 196, 202, 208, 222, 241, 275, 282, 311–2
Podgorica 23, 38–9, 191, 205, 209, 215, 217, 232, 312
Poland/Poles: political economy xix, xx–xxi, 1–9, 42–9; production 117–20, 122–6, 129–31, 134, 137, 139, 141–5, 149, 151, 153, 156–8, 160–70; regional issues 307–16, 319, 329–35, 337–40, 348–9, 364; settlement 252–5, 257, 259, 261–3, 266–9, 273, 275, 281–9, 292–7,

299, 301 – social issues 60, 62–3, 65–6,
68–74, 76–9, 82–5, 88–93, 95, 97,
99–100, 106–10; sustainability 365–9,
372–3, 377–8, 382–3, 386, 389; tertiary
sector 188, 191–201, 203–4, 206–7,
209–11, 214–16, 218–19, 221–31, 233,
235–6, 239–40, 243, 246–7

Poland and Hungary: Assistance for
Restructuring the Economy (PHARE)
33, 71, 137, 154, 164, 174, 185, 207,
211, 264, 271, 293, 301, 304, 320–1,
334, 372–4, 377, 387

police *see* law enforcement

policy *see* economic/foreign/regional/social
policy

political pluralism *see* pluralism

political culture/politics *see* civil society

political issues *see* foreign relations,
regulation etc

polluter pays principle/pollution 13, 72,
132, 135, 138, 158, 172–3, 181, 193,
226, 229, 231, 238, 266, 282, 308, 320,
332–4, 340–1, 347, 353–4, 368, 376,
378–9, 383, 386 *see also* water

Pomerania 200, 337, 372

population *see also* labour, migration etc:
production 132, 145, 148, 167, 152–3,
160–1, 167, 169–70, 177–8, 185–6;
regional issues 308, 310–11, 314, 318,
321–3, 325, 331–3, 341, 343, 347,
350–4, 356–7, 362, 368–9, 385; socio-
political issues xxi, 3, 15–16, 36, 40–1,
56, 62, 67, 72, 74–82, 101–2, 104–5;
tertiary sector and settlement 208, 220,
252, 257, 259, 261–2, 271, 279, 282,
287, 292–4, 300–4

ports 25, 29, 33, 38, 134–6, 192–4, 205,
211–14, 227–8, 234, 281, 319, 339, 359,
364, 369

poultry 155, 176, 359

poverty xxi, 53–4, 61–2, 65–70, 73, 103,
114, 123, 144, 153–4, 160, 178, 264,
270, 281, 294, 302–5, 350, 367–8, 369,
375, 382, 390

power station *see* electricity and energy

Poznań 114, 127, 136, 153, 165, 196–7,
203, 208, 219, 282, 308, 312, 316, 348

Prague xx, 16, 18, 63, 113, 133–4, 180,
196–7, 202, 207–8, 215, 218, 224, 232,
238, 241, 245, 269–70, 272, 275, 277,
278–9, 285, 296, 308, 312, 348, 386

presidency 5, 17–18, 24–6, 33, 37, 59, 97,
101, 103, 108–9, 116, 264, 382

Prešov 129, 141, 312, 317

press *see* media

prices 1, 19, 36, 41, 46, 48, 131, 137, 140,
148–9, 156, 164, 171, 175–6, 183–4,
221–4, 226–8, 231, 238, 259–60, 266–9,
272–3, 275, 277, 279, 281, 295–6, 377
see also costs

private enterprise/property/sector/
privatisation: political economy xxi–xxii,
18–19, 22, 24, 26, 42–6, 50–1, 53;
production 112–13, 119, 121–2, 134,
137, 143, 146, 156, 171–4, 177–8,
180–1. 183–4; regional issues 259–60,
264, 266, 269–70, 272, 282, 294, 296,
356, 366–7, 370–1, 374–5, 382, 389;
social issues 57, 64–6, 70–2, 87, 101,
103–6, 109; tertiary sector 193–4, 200,
207, 209, 211, 215–16, 218–19, 222–3,
231–2, 243–5, 247–9

private farm/plot 149–50, 151–5, 160,
162, 176, 184, 248, 257, 259–60, 301–2,
353, 374

Procter & Gamble 135, 354

producer services *see* business and financial
services

production/productivity *see also* by named
branches: industry and agriculture
111–70, 172, 175, 178, 180–1, 183–4;
settlement and regional issues 252, 257,
260, 295, 299, 321–2, 331, 352, 366,
370, 375, 377; socio-political issues
21–2, 31, 40–1, 50, 56, 58, 66, 101–2,
104, 106; tertiary sector 191, 195–6,
206, 221, 224, 226–7, 235

professionals 32, 154–5, 214, 219–20, 257,
265, 271, 304, 308, 325, 354, 356, 359

profitability *see* efficiency, finance and
restructuring 53, 69, 73, 113, 123, 128,
148, 153–4, 156, 158, 162, 174, 193,
212, 215, 222, 226, 252, 259, 270, 285,
299, 331, 347, 356, 365, 370, 374–5

proletariat *see* labour

propaganda 14, 70–1, 172, 175, 193, 293,
341

property *see* building and land

prosperity *see* wellbeing

protected areas *see also* national parks

protection *see* tariff

protectorate xviii, 5, 10, 16, 36, 102–3

Prussia/Prussian Empire 1, 39, 261

public opinion/support 3, 8, 22, 36, 57,
66, 171, 185, 222, 230, 234, 240, 280,
330, 341, 348 *see also* coalition and
election

public transport 86, 192, 243, 249, 258,
279–80, 284, 290, 293, 300, 320 *see also*
aviation, buses and rail

purchasing power *see* consumption
pyramid investment 53, 101, 103, 108–9,
    116, 173–4, 295

quality 125–7, 130, 133, 136, 143, 165,
    172, 185, 193, 218, 226, 243, 247, 250,
    260, 271–3, 294, 300–1, 304, 329, 334,
    348–9, 369
quality of life *see* living standard

Racan, I. 5, 25
radioactivity *see* nuclear power
railway/railway equipment xxi, 24, 29, 34,
    119, 126, 132, 135–9, 142–4, 172,
    278–9, 281, 284–5, 330, 341, 349, 352,
    355, 359–64, 371, 386
raw materials 2, 39–40, 46, 111, 119, 132,
    134, 148, 150, 165, 172–3, 181, 183,
    191–200, 201–7, 209, 213–4, 257, 259,
    265–6, 269, 275, 313, 369
real estate *see* building and land
recession 2, 64, 104–5, 117, 139, 177, 216,
    247, 279, 353, 366, 370, 373
reconstruction *see* restructuring
recreation *see* leisure and sport
redundancy 66, 96, 172, 181, 185, 197,
    319, 353, 356, 358, 370
referendum *see* public opinion
reform: *see also* economic reform, land
    reform, restructuring and social change:
    production and tertiary sector 126, 149,
    155, 170–2, 175–8, 180, 183; settlement
    and regional issues 259, 294, 302, 314,
    331, 366–7, 369–70; socio-political issues
    3, 22, 39, 41–3, 46, 50, 52–4, 56, 58,
    62, 66, 101–2, 104–5, 108–9
refugees 10, 28, 30, 32–4, 90, 95, 247, 295,
    332, 334, 341
refurbishing *see* restoration
regional development/policy/regions *see
    also* local government and named
    regions: production 111, 115–16, 124,
    132, 135–6, 140, 143, 147, 173, 178,
    181; settlement and regional
    development 57, 260, 297, 307–64, 366,
    368–76, 381, 387–9; socio-political issues
    15–16, 18, 52–3, 57, 59, 61–3, 66, 72,
    74, 92, 94, 98–9, 105–6, 109; tertiary
    sector 187, 207, 220, 232, 248–51
regulation *see also* administration and
    government xxii, 8, 42, 46, 54, 70, 98,
    100, 114, 147, 149, 260, 264, 318, 366,
    371, 379, 382, 389
rehabilitation *see* modernisation
religion 20–39, 59–60, 91, 175, 249, 361
    see also Islam and Jews

remoteness *see* accessibility
Renault 126, 129–30, 143, 180, 330
renewable energy *see also* water power
    230–5, 388
rent 19, 67, 86, 102, 129, 134–5, 152, 154,
    259, 264, 267, 269–72, 274, 277, 279,
    281, 294, 366,
reorganisation *see* restructuring
Republika Srpska 5, 27–32, 36, 215, 315
research and development 9, 13, 112, 115,
    140, 147, 174, 185, 222, 279, 319,
    322–3, 331, 339, 354, 358, 360, 362,
    368
resources 2, 40, 61–2, 94, 111, 121, 307,
    319, 322–5, 338, 360, 367, 372, 375–6,
    381–2, 389 *see also* agriculture, land, raw
    materials etc.
restaurants *see* hotels and catering
restitution xxii, 64–5, 134, 150, 151,
    153–5, 174–5, 185, 264–5, 270, 272,
    286, 295, 300 *see also* land restitution
restructuring 13–4, 18–9, 31, 37, 46, 57,
    66, 100–10, 112–3, 115, 117, 119,
    121–3, 134–5, 138, 141, 155, 172, 177,
    180–1, 184–5, 201, 207, 214–5, 226,
    278, 294, 302, 314, 318, 320, 352–4,
    356, 367, 370, 382 *see also* shock therapy
retailing *see* shopping
retraining *see* training
Rewe 274, 324
Rhodope 91, 337
Rijeka 23, 196, 204–5, 209–10, 213, 223,
    228, 312, 362
risk *see* security
rivers *see* waterways and named rivers
road/road haulage xxi, 102, 126, 135, 139,
    143–6, 186, 191–4, 199, 206–14, 217,
    280, 282–4, 290, 296, 298–9, 318, 327,
    330, 334, 344–6, 349, 355–6, 358, 362,
    376, 386 *see also* motorways
Roma 10, 67, 73, 86, 89, 91–3, 95–6, 178,
    259, 371, 389–90
Romania/Romanians: political economy
    xx–xxi, xxiii, 1, 3–11, 39, 43–5, 47–9,
    51; production 117–20, 123, 129–30,
    133, 139, 143–4, 153–7, 163–4, 166–86,
    176–86; regional issues 304–6, 309–13,
    315, 334, 337–9, 350–64; settlement
    252–5, 258, 261, 274, 287–9, 292, 295,
    297, 299–301; social issues 58, 60, 62–3,
    65, 68, 71–4, 76–9, 80–1, 82–4, 87–93,
    95–7, 99–100, 106–7; sustainability
    367–8, 370–1, 382–6, 389; tertiary
    sector 187–90, 192, 194–200, 204,
    206–9, 213–17, 219, 221, 223–31, 233,
    235–6, 239–40, 243, 246–7

rural areas *see also* agriculture: production and tertiary sector 132–4, 144–70, 185–7, 191, 218, 220, 226; regional issues 319, 321, 324, 330–1, 354, 359, 368, 374–5, 382, 387–8; settlement patterns 257, 259, 265, 286–306; socio-political issues xxii, 13, 15–6, 40, 51, 63, 66, 69, 74, 82, 84–6, 89; tourism 244, 249, 295, 301–2, 304–6, 308, 313, 377–8, 387–8

Russia/Russians xviii–xx, 2, 6, 8, 16, 20–1, 33–4, 37, 43, 60, 63, 73–4, 93, 105, 112–14, 124–5, 135, 139, 180–1, 201, 203–4, 208, 211–6, 228–9, 240, 247, 286, 332, 337, 374, 382

Ruthenians 333–5

safety *see also* security 197, 209, 218, 235, 237–9, 329

Saint Gobain 125, 141

sales *see* market

sanctions 3, 10, 32–3, 111, 171, 203, 206, 213, 333, 366

Sarajevo xx, 129, 191, 196, 205, 209, 285, 312, 334

satellite states *see* named states

Sava 26, 29, 205, 209, 214, 334, 337, 380

savings 103, 293, 296 *see also* capital

Saxe-Coburg, S. 5, 105

Saxon/Saxony 12, 15, 226, 250, 261, 332, 339, 348

Scandinavia 204, 212–13, 340

school *see* education

Schuster, R. 5, 18, 116

science *see* research and technology

secondary sector *see* industry

second economy 41–2, 51, 57, 103, 112–14, 257, 329, 382

Second World War xxii, 1–2 , 16, 21, 59, 134, 191, 221, 241, 256, 260, 279, 285, 295, 342

security xviii, xxii, 1–2, 8–9, 16, 32, 34, 39–40, 53, 60, 66, 73–4, 90, 94, 105–6, 151, 210, 214, 232, 237, 240, 243, 295, 343, 366, 369–71, 387, 389 *see also* NATO

security police 14, 58, 64, 101, 148, 171, 180, 247, 265, 307

self-determination/government *see* autonomy and local government

self-sufficiency 40, 121, 142, 148, 153–4, 164, 176, 184, 221–2, 290, 294–6, 300, 382

Serbia/Serbs: political economy 1, 4–5, 20–39, 53; settlement and regional issues 261–3, 274, 295, 319, 335, 339, 371; social issues and production 60, 73, 89, 92, 95, 97, 99, 104, 107, 140; tertiary sector 197, 215, 229, 247

services/service sector *see also* education, health, infrastructure, shopping, transport etc: production and tertiary sector 115, 134–5, 138, 140, 155, 185; regional issues 310, 314, 319, 321, 324–5, 329–30, 332–3, 339, 343–4, 346, 348–9, 353–4, 366, 369, 382; settlement 256–7, 259–60, 266, 269, 281, 290, 293, 295–9, 303–4; socio-political issues xxii, 13, 27, 51–2, 58, 62, 70–1, 84–5

settlement *see* rural, urban and named cities

sewage 138, 281–3, 299, 320, 325, 330, 343, 356, 378

shareholding 51, 173, 193, 218, 223–4, 282, 354

shatter belt/zone 1–2, 368

Shell 173–4, 207, 274

Shkodër 53, 210, 248, 286, 312, 335, 337

shock therapy xxii, 14, 46, 54, 100, 104–5, 108

shipbuilding/shipping 10, 26, 106, 119, 123–4, 136, 139, 143, 165, 172, 181–2, 191–4, 204, 210–4, 234, 247, 318, 339, 354

shopping 15, 23, 32, 50, 90, 114, 137–8, 186–7, 194, 217, 248, 250, 257–9, 264–7, 269, 271–81, 286, 290, 293, 296–7, 299–300, 324, 330, 334, 340, 342, 346, 356

shortage 40, 51, 259–60, 302

Siemens 113, 127, 173, 180, 216–7, 219, 237, 278, 354

Silesia 62, 114, 122, 124. 133, 136, 139, 141–3, 160, 191, 197, 203, 207, 226, 250, 253, 256, 311, 321, 337

skill *see* training 66, 70–2, 86–7, 96, 102, 114–5, 128, 134, 137, 147, 154–6, 165, 172, 181, 186, 201, 222, 240, 243, 250, 296, 311, 329, 349, 354, 369, 375, 377

Škoda 113, 127, 129–30, 140, 222

Skopje xx, 33, 196, 205, 210, 214–5, 217, 312

Slavonia 23–4, 29, 125

Slovakia/Slovaks: political economy xx–xxi, 4–8, 11, 16–8, 42–3, 53; production 116–8, 120, 123–4, 127–9, 135, 139–41, 152, 154, 157–8, 160, 162–4, 166–70, 180; regional issues 312–3, 315, 317–8, 320–1, 333–5, 337–9, 349; settlement 261–3, 265, 274, 278, 285, 288–9, 291, 295, 300, 303; social issues 60, 65, 67–8, 73–4, 76–7, 79, 82–3, 85, 88–9, 91–3, 95–7, 106–8; sustainability 368–9, 377, 381, 386, 388–9; tertiary sector 188, 192, 194–6, 198–200, 226, 228–40, 246, 248–9, 254–5

Slovenia/Slovenes political economy xix–xxi, 5–9, 11, 16, 20–3, 26. 32, 42–5, 47–9; production 116–18, 120, 123–4, 129, 139–40, 155, 157, 163–4, 166–70; regional issues 308, 311–15, 318, 333, 337, 339; settlement 254–5, 261–3, 278, 288–9, 293; social issues 65, 67–8, 71, 76–9, 82–4, 89, 92–3, 97, 106–8; sustainability 366, 368, 370, 374, 377, 387; tertiary sector 188, 195–6, 198–200, 204, 208, 211–12, 215–16, 219, 225, 228, 230, 236, 246

Small- and Medium-sized Enterprises (SMEs) 19, 22, 31, 50, 51, 66, 69, 71, 98, 106, 108, 111, 113–4, 123, 126, 132, 134, 145, 147, 173, 181, 186, 220, 226, 243, 300, 302, 319–22, 329–30, 354–6, 358, 367, 375

smuggling *see* crime

social change/social differentiation/social issues: production 111–5, 123, 137, 140, 144, 146, 152, 167–72, 178, 185–6; regional issues 319, 321, 327–33, 338, 344, 356, 365–76, 377, 381, 389; socio-political issues 3, 24, 40–1, 56–110, 101, 104, 110; tertiary sector and settlement 221, 226, 230–1, 248–50, 258–60, 264, 270–1, 287, 293–4, 296, 299, 301–2

social democracy 19, 23, 25–6, 30, 33–4, 38, 63, 85, 99, 101, 108–9, 176–7, 180, 183–4, 237–9, 367

socialism *see* communism

socialist city 257–60, 266, 282

socialist farm *see* cooperative farm and state farm

social services *see* welfare

Sofia xx, 23, 87, 123, 191, 196–7, 205, 210, 214, 217–8, 286, 302, 312

Solidarity (Poland) 3, 70, 108–10, 123, 192, 256, 367

Sony 134, 342–3

Soros, G./Soros Foundation 10, 330, 360

sourcing 127–7, 130, 207

South Eastern European Countries (*SEECs*): political economy xviii, xxii, 9–10, 52; production 114–5, 129, 132, 134, 139, 146, 153, 165; regional issues 308, 313, 324, 332, 335, 338, 351, 354, settlement 260, 265, 272, 274, 286, 294, 297, 300, 302–3; social issues 59–61, 65–6, 70, 74, 78, 89, 99, 111; sustainability 366, 368–71, 374, 376; tertiary sector 191, 197, 205, 207, 209, 212, 216–18, 221, 228, 231, 240, 242, 245, 247–8

Soviet Union (USSR): political economy xviii–xix, xx, xxii, 1–4, 6, 12, 20–1, 40, 42, 53–4; production 111, 123–4, 126, 133, 139, 147–8, 151, 165, 171, 177; regional issues 329, 338, 349, 365–6, 372, 374; settlement 256, 257–8, 261, 273, 275, 279, 292; social issues 57, 64, 73, 88, 90, 100; tertiary sector 191–3, 201, 203, 212, 215–16, 221–2, 228, 232, 235, 238, 240, 243

spa 138, 241, 245, 249

Spar 274, 278

spatial development/planning *see* planning and regions

Special Accession Programme for Agriculture & Rural Development (SAPARD) 160, 185, 304, 376–7, 387–8

specialisation 14, 117, 137, 149, 155, 191, 253, 277, 300, 328, 339, 346, 370

spending *see* consumption

Split 29, 116, 139, 196, 213, 244, 312, 318

sport 126, 138, 242, 244, 281, 330, 341 *see also* winter sports

stabilisation/stability *see also* security: production and tertiary sector 140, 146, 221, 227, 247–8; settlement and regional issues 272, 293, 302, 304, 311, 313; socio-political issues 9–11, 17, 22–34, 52, 54, 87, 94, 98–9, 102–4, 106

stabilisation force 30, 32, 34–5, 37, 171, 338, 340, 348, 367–8, 371

Stability Pact for South Eastern Europe 7, 9–10, 37, 94, 178, 232, 335, 371

Stalin, J. xxii, 2, 40, 56, 149, 365

state capitalism *see* communism

state farm 86, 148, 150, 151, 154, 160, 162, 174–6, 184, 245, 290, 294, 298, 301, 330, 378

state intervention *see* regulation

state-owned industries/state sector 18–9,
    30, 34, 38, 41, 46, 50–1, 53, 55, 66, 72,
    99, 105, 108, 111–16, 134, 137, 146,
    171–3, 177, 180–1, 184, 193, 226, 242,
    244–5, 272, 274, 293, 313, 319, 366–7
state socialism *see* communism
states *see also* named states
steel *see* metallurgy
subsidies 10, 46, 69, 71, 85–6, 89, 113,
    123, 135, 158, 165, 171–2, 175, 193,
    195, 221, 223, 245, 247, 260, 270, 279,
    281, 313, 378
subsistence *see* self-sufficiency
suburbs 259–60, 265–6, 269–71, 274, 277,
    296, 327, 342–3, 346–8, 354, 375
subventions *see* subsidies
Sudetenland/Sudetes 4, 20, 160–2, 247,
    250, 287, 340, 378–9
Šumava 241, 249, 337, 376–7
supermarket 90, 137–9, 266, 271, 273–8,
    280, 324, 330, 348
superpower xix, 8, 40, 54
sustainability 164, 167, 175, 186, 229, 232,
    248–51, 302, 322, 330, 333, 335,
    339–40, 366, 375–81, 387–8
Szeged 73, 139, 197, 208, 220, 252, 274,
    312, 322, 327, 334, 336, 362–4

tariff 15, 21, 62, 65, 121, 126, 129, 156,
    184, 209, 211, 217, 279, 301, 320, 364
Tatra 216, 221–22, 242, 248, 301, 335,
    337
tax/taxation 26, 46, 53, 62, 69, 88, 91–2,
    102–5, 109, 112, 114, 128–9, 135–6,
    141, 174, 231, 249, 264, 272, 275, 277,
    299, 301, 306, 313, 318–20, 325, 327,
    330–3, 342, 356, 366–7, 374
technocrats 18, 25, 46, 98, 104, 307
technology, technology transfer: production
    111, 113–5, 117, 123–5, 128–31, 136,
    139–41, 143–4, 146–7, 151, 155, 173–5,
    180–1, 183; settlement and regional
    issues 257, 260, 269, 282, 311, 313,
    318–22, 328, 330–1, 333–4, 339, 344,
    348–9, 358–9, 362, 366, 368, 370, 374,
    387–8; socio-political issues 3, 8–9, 14,
    17, 38, 46, 51, 63, 71, 87, 94; tertiary
    sector 193, 197, 201, 212–3, 215, 218–9,
    221–2, 224, 227, 229, 231–2, 235,
    237–40, 242
telecommunications xxi, 19, 34, 52, 103,
    117, 135, 140, 146, 173, 180, 189, 193,
    218–20, 245, 267, 287, 297, 299, 303,
    319, 321–3, 330, 351, 356, 369, 371
tension *see* security

territory *see* land
tertiary sector *see* services
Tesco 207, 273–5, 277
textiles 13, 24, 32, 117, 125, 131, 137, 139,
    181, 183, 266, 286, 290, 295, 298, 306,
    320, 329, 358, 361, 378
Thuringia 12, 15, 197, 230, 266
Thyssen 121–2, 223
timber industry *see* wood processing
Timişoara 170, 196, 179, 181–2, 209, 213,
    215, 217, 266, 312, 354–5, 359, 362–4
Tirana xx, 53, 86, 102–3, 117, 131, 196,
    210–1, 214, 217, 267, 278, 292, 312
Tisza 214, 295, 337, 339, 362–4, 380, 383,
    386
Titograd *see* Podgorica
tobacco industry 28, 34, 38, 53, 72–3, 105,
    137–8, 149, 156, 176, 209, 250, 294,
    330, 358
totalitarianism *see* communism
tourism *see also* agrotourism: production
    and tertiary sector 138–9, 150, 160, 162,
    173, 181, 185, 187, 197, 200, 212,
    214–5, 241–51; regional issues 311, 318,
    321, 330, 333, 335, 339–40, 343, 348,
    351, 354–6, 358, 360; settlement 259,
    273, 281, 290, 295; socio-political issues
    xxii, 19, 26, 38, 51, 64, 101, 103;
    sustainability 362, 376, 377, 383, 388
towns *see* urban areas
Toyota 127–8
trade/trader *see also* exports, imports and
    markets: production 111, 119, 123, 126,
    136–7, 139–40, 148–50, 164–7, 171–2,
    176; regional issues 332–4, 339–43, 346,
    348–9, 356, 362, 366–7, 369, 382; socio-
    political issues 1, 3, 8, 10, 14, 17, 33, 42,
    46–7, 60, 64, 70, 89–90, 102, 105;
    tertiary sector and settlement 189,
    193–4, 267, 272, 293, 295
trade centre *see* offices
trade unions 46, 51, 70, 108–9, 127, 173,
    177, 180–1, 226, 245, 256, 264–5, 370
traffic *see* road/rail etc
training 14, 51, 69–72, 86, 89, 114, 138,
    140, 143, 147, 154, 165, 174, 217, 243,
    248–9, 302, 319, 321, 348–9, 354, 358,
    360, 362, 387
tramways 269, 278–9
transformation/transition *see* civil society,
    economic issues etc.
transit 90, 102, 194, 197, 202, 204–5, 213,
    217, 228, 240, 278, 280, 341, 344, 369
trans-national companies *see* multinational
    companies

transport *see also* Eurocorridors, transit and
    individual transport modes: production
    111, 116, 119, 126, 130–1, 133, 137,
    139–40, 143–6, 156, 173; regional issues
    319, 321, 328, 332–3, 344–6, 348, 356,
    366, 371, 379; socio-political issues xxii,
    22, 26, 34, 40–1, 50–2, 101; tertiary
    sector and settlement 187–220, 245,
    247, 265–6, 271, 273, 278–80, 298, 300,
    303, 305
Transylvania xxiii, 80, 82, 89, 95, 173–4,
    241, 273
travel agencies *see* tourism
trucks 13, 19, 128–30, 137, 194–5, 206–7,
    211, 273, 280–2
Tujman, F. 24–7, 116
Turkey/Turks xx, 8–9, 22, 30, 32, 38, 74,
    87–8, 91–3, 103, 105, 156, 176, 183,
    196–7, 204, 206, 210, 213, 232, 274,
    287, 303, 337, 341, 370–1 *see also*
    Ottoman Empire

Ukraine/Ukrainians xx, 73–4, 88, 90, 93,
    95, 105, 119, 128, 139, 147, 156, 165,
    177, 196, 203–4, 207–8, 210–2, 214,
    216, 240, 274, 287, 302, 334–5, 337,
    340, 348, 359, 374, 379, 386, 389
underdevelopment *see* poverty
unemployment: production and tertiary
    sector 112, 127, 129, 135, 141, 143, 155,
    171, 177, 180, 248; regional issues 311,
    313, 319–22, 327, 331–2, 347–8, 350–1,
    354, 357, 366, 370, 373–4, 382; settle-
    ment 264, 294–6, 299, 302, 306; socio-
    political issues 14–5, 18–9, 22, 24, 33–4,
    37, 39, 46, 65–71, 73, 84–7, 90, 105, 108
unification/union *see also* Germany 4, 9,
    21, 29
United Nations xviii, 5, 16, 23–5, 27, 30,
    32–3, 89, 111, 120, 157, 171, 186, 198,
    214, 358
United States of America (USA) 9, 27, 33,
    36–7, 54, 60, 63, 71, 73, 117, 121–2,
    140, 150, 183, 210, 214–7, 219, 238–9,
    243, 275, 278, 285, 340, 369, 371, 375
universities 30, 50, 51, 71–2, 140, 143,
    174, 185, 265, 280–1, 291, 321, 331,
    333, 340–1, 348, 354
Upper Silesia *see* Silesia
uranium 239, 241, 257, 339, 359–60, 383
urban areas/urbanisation *see also* named
    cities: production 111, 133, 137–40, 144,
    146, 150, 154, 156, 162, 164, 172, 179,
    186; tertiary sector 187, 191–3, 218,
    229, 244–5; redevelopment 253, 257,
    259–61, 265–9, 272, 285; regional issues
    307–8, 311, 318, 324–5, 327–9, 331–3,
    336, 351–6, 362, 368–9, 372, 374, 376,
    377–8, 382–3; settlement patterns
    252–80, 287, 290, 292, 294, 296–302;
    socio-political issues 3, 15–6, 26, 40, 51,
    56, 61–4, 66, 69–70, 72, 82, 85–6,
    101–2, 104, 109
USA
USAID 229
US Steel 117, 121, 369
utilities *see* services

value *see* price
value-added tax *see* taxation 53, 156, 184
Varna 115, 119, 129, 139, 191–2, 196,
    210, 213, 228, 312
vegetables 156, 162, 165, 286, 300–1, 306,
    347, 371–2
vehicles *see* automobiles
village *see* rural settlement
Visegrad *see* CEFTA
Vistula 192, 207, 212, 214, 256, 279, 285,
    339, 373, 379, 381
Vojvodina 21, 33, 35, 95, 108, 228
Volkswagen 13, 117, 127–9, 319
Volvo 130–1, 278
voucher privatisation *see* privatisation

wages *see also* incomes: production and
    tertiary 115–6, 127, 139, 149–50, 162,
    172, 174, 176, 180, 183–4, 187, 220,
    227; regional issues 311, 333, 341, 349,
    351, 354, 356, 366, 369; socio-political
    issues 14, 36, 39, 41, 54, 56, 58, 65–7,
    69–70, 72, 78, 85, 88, 108
war *see* civil war and First/Second World
    War
Warsaw xx, 67, 73, 122, 127–8, 130, 135,
    138, 141, 143–4, 147, 149, 191, 196–7,
    200, 203, 207–8, 217, 219, 231, 244,
    256, 258, 261, 265–9, 271–3, 277–81,
    284–6, 308, 312, 319, 348, 383
Warsaw Pact 2–3, 124, 133
waste/waste management 71, 138, 231,
    240–1, 282, 298, 307, 330, 334, 346,
    354, 360, 373, 376, 378, 383, 386
water/water power: production 121, 124,
    131, 138, 156, 165, 185–6; regional
    issues 330, 334, 343, 351, 356, 358, 361,
    376–7, 379, 381–3; settlement 259,
    281–3, 285, 290, 293–4, 295, 298–9,
    302–3; socio-political issues 23, 37, 103,
    106; tertiary sector 192, 209, 213, 221,
    223, 225, 227–8, 230–5

waterway 139, 192, 194–5, 199, 211, 212–4, 232, 253, 354, 363–4, 378–9, 381

welfare/well-being 4, 18, 36, 46, 53–4, 56, 59, 61, 64–7, 69–71, 78, 90, 98, 100, 103, 108–9, 148–9, 152, 171, 271, 294, 302–3, 318–9, 332, 366–8, 370, 374

West/Western Europe political economy xviii–xix, 1–3, 6, 10, 20, 24, 30, 33, 35, 37–40, 52–5; production 111, 113–14, 119, 124–8, 130–1, 134–5, 138–40, 144, 146, 170, 172, 175–7, 181, 183; regional issues 311, 314, 330, 332, 338–9, 365–76, 389–90; settlement 257–8, 260, 267, 270, 272–3, 278, 284–5, 295, 301, 304; social issues 56, 58–9, 65, 69, 71–2, 74, 84, 86, 90, 94, 101; tertiary sector 187, 191–4, 201, 203, 206–8, 212–7, 220–3, 226–9, 235, 237–8, 240, 242–5, 247–8

see also European Union

Westinghouse 237–8

wind 230–1

wine 301–2, 360

winter sports 243, 248, 302

woodland/wood processing: production 124–5, 131, 134, 138, 148, 160, 162–4, 173, 183; regional issues 309, 335, 339–40, 346, 351, 358–60, 377–9, 383, 388–9; settlement 252, 256, 266, 284, 290, 295, 299, 301–3, 305; socio-political issues 40, 82, 102; tertiary sector 195, 212, 231, 247–9

workers *see* labour

workshops *see* handicraft

World Bank 60, 71, 73, 98, 173, 183, 186, 210–1, 218, 248, 253, 282, 334–5

world economy/system *see* global economy

World Trade Organisation 7, 34

World Wide Fund for Nature 334, 381, 388

Wroclaw 99, 131, 136, 144, 196–7, 201, 207–8, 245, 261, 265, 280–4, 312, 316, 330–1, 348, 378

yield *see* production

youth 67, 155, 258, 270, 272, 294, 296, 299, 313, 325, 330, 344, 353, 362, 374

Yugoslavia/Yugoslavs: political economy xx–xxi, 5–7, 10–1, 35–9, 43–5, 47–9; production 117–18, 120, 123, 129, 155, 157, 163, 165–70; regional issues 312, 315, 333, 337, 354, 361–4; settlement 261–3, 274, 288–9; social issues 65, 67–8, 74, 76–7, 79, 82, 92–3, 97, 104, 106–8; sustainability 366, 386; tertiary sector 197–200, 208–9, 241, 246

Yugoslavia (Former): political economy xviii, 3–4, 6, 9–10, 16, 20–39, 42; production 129, 139–40, 149, 155, 157, 163, 166–7, 169, 171; settlement and regional issues 252, 261–3, 278, 292, 309–10, 354, 362, 367, 369; social issues 59, 73–4, 82, 88, 92, 94, 99; tertiary sector 188, 191–2, 194, 198, 197, 205, 210, 215, 217, 223, 228, 235, 237, 242–3, 247

Zagreb xx, 10, 20, 22, 24–5, 29–30, 116, 140, 144, 196, 205, 209, 217, 230, 247, 303, 312, 318, 362

Žilina 141, 196, 312, 317, 321